THE COASTAL ENVIRONMENT

THE COASTAL ENVIRONMENT

Toward Integrated Coastal and Marine Sanctuary Management

Gary A. Klee
San José State University

Prentice Hall
Upper Saddle River, New Jersey 07458

Library of Congress Cataloging-in-Publication Data

Klee, Gary A.
The Coastal Environment: Toward Integrated Coastal and Marine Sanctuary Management / by Gary A. Klee
p. cm.
Includes bibliographical references and index.
ISBN 0-13-080034-1
1. Coastal zone management—United States. 2. Marine parks and reserves—United States—Management. I. Title.
HT392.K54 1999 98-26263
333.91'7'0973—dc21 CIP

Executive Editor: Robert A. McConnin
Editorial Assistant: Grace Anspake
Executive Managing Editor: Kathleen Schiaparelli
Assistant Managing Editor: Lisa Kinne
Art Director: Jayne Conte
Cover Designer: Bruce Kenselaar
Cover Photo: Gary A. Klee
Manufacturing Manager: Trudy Pisciotti
Production Supervision/Composition: WestWords, Inc.

© 1999 by Prentice-Hall, Inc.
Simon & Schuster/A Viacom Company
Upper Saddle River, New Jersey 07458

All rights reserved. No part of this book may be reproduced, in any form or by any means, without permission in writing from the publisher.

Printed in the United States of America

10 9 8 7 6 5 4 3 2 1

ISBN 0-13-080034-1

Prentice-Hall International (UK) Limited, *London*
Prentice-Hall of Australia Pty. Limited, *Sydney*
Prentice-Hall Canada Inc., *Toronto*
Prentice-Hall Hispanoamericana, S.A., *Mexico*
Prentice-Hall of India Private Limited, *New Delhi*
Prentice-Hall of Japan, Inc., *Tokyo*
Simon & Schuster Asia Pte. Ltd., *Singapore*
Editora Prentice-Hall do Brasil, Ltda., *Rio de Janeiro*

I would like to dedicate this book to our first grandchild from my lovely daughter Laura and her wonderful husband, Jeff. May their children, and the children of other cultures and generations from around the world, have the opportunity to play on a clean, safe, white sandy beach; frolic in ocean waves; and generally marvel at the wonders of a healthy *coastal environment.*

Contents

Preface

The Coastal Environment was written specifically for an upper division course in coastal resource management taught in a Department of Environmental Studies. Unlike more traditional disciplines (such as Geology, Biology, Chemistry), the emphasis in Environmental Studies is on multidisciplinary and interdisciplinary education, taking into consideration science and policy, as well as management. To put it another way, the discipline takes a rigorous, systematic, and *integrative* approach to looking at environmental problems and issues. Therefore, the field (and this book) are very *broad in scope.*

Consequently, this book is for the professor, student, or individual who wants to go beyond the confines of a traditional discipline or perspective on the world. It is not for geology professors who may be looking for a good traditional textbook for their course in coastal geology. There are many excellent books already on the market for that purpose. Nor is this book for marine scientists who may be looking for a new textbook for their introductory course in marine science. Rather, this book is designed primarily for Environmental Studies, Environmental Science, Natural Resource Management, and Geography Departments that offer courses with titles like *Coastal Resource Management, Coastal Environments, Coastal Field Studies,* and *Coastal Geography and Planning.* It is also hoped that an occasional professor or two from one department (e.g., Geography, Urban Planning, Marine Studies) may stumble upon this book and say, "We don't offer a course like this, but maybe we should. We might even try team teaching it with some of our colleagues from different departments, schools, or colleges!" Over the 26 years that I have been a professor (University of Wisconsin-Superior, San Diego State University, The Pennsylvania State University, and, since 1977, San José State University), I have developed many new courses based on a book that gave me an interesting new perspective. Although the notion of multidisciplinary and interdisciplinary collaboration among colleagues is difficult to say the least, its acceptance as a twenty-first century mode of doing business is gaining popularity. Scientists, social scientists, and engineers are slowly recognizing that they need each other's expertise to arrive at viable solutions to many of the world's most daunting environmental problems, including those affecting our coasts and marine sanctuaries.

Furthermore, the book is designed to appeal to anyone interested in the link between coasts and marine sanctuaries, since each major chapter is approximately one-third science, one-third policy, and one-third management. Within the professional community, the book should prove of special interest (and possible use) to coastal and marine sanctuary managers (such as the offices of NOAA), environmental planners (such as the offices of Coastal Commissions), and coastal/marine educators (such as aquariums, and nonprofit coastal protection advocacy groups, such as the Center for Marine Conservation, Save Our Shores, and Coastal Advocates).

Why the need for such a coastal book, and at this time in American history? America's coasts are now under siege. Approximately 75 percent of Americans already live within 80 km (50 mi.) of a coast, and population pressure on these shorelines is mounting. Every imaginable environmental problem, from El Niño-intensified storm damage, to offshore oil tanker spills, to floating plastic bags with used hypodermic needles, are affecting our beaches and coastal waters. In fact, the National Oceanic and Atmospheric Administration

(NOAA) is so troubled by these trends that, in Febru ary 1998, it launched a major new effort to identify the problems that affect America's shoreline and marine ecosystems. The "State of the Coast" study, scheduled to be published in early 1999, will include essays by university professors, government scientists, and private researchers on the condition of coastal areas, as well as successes in prevention or restoration efforts. As evident by this recent NOAA project, there is an increasing national consensus that America's shorelines and adjacent marine sanctuaries are national treasures that must be protected.

The Coastal Environment is an introduction to **coastal and marine sanctuary management**—*the art and science of seeing the interrelationship between coastal land issues and adjacent marine sanctuary concerns.* Specifically, it is about understanding, monitoring, and managing these two related areas for their mutual benefit. Traditionally, coastal areas and marine sanctuaries have been studied by two different groups of people: In one camp have been those *concerned with the land*—the "dry-side" scientists (such as coastal geologists), planners (such as urban planners), and policy implementers (such as park, forest, and wildlife managers). In the other camp have been those *concerned with the ocean*—the "wet side" scientists (e.g., marine scientists, oceanographers) and marine educators (e.g., educators working for marine sanctuaries, marine labs, and aquariums). Until recently, both groups usually stayed in their separate territories—remaining in their separate departments, attending their own professional meetings, and publishing in their own highly specialized journals. It was generally believed that "wet side" and "dry side" specialists cannot (and should not) mix.

This has certainly been true in academia. For example, the coastal geologist (Geology Department) has traditionally clung to discussions of plate tectonics, sea level change, and coastal processes; the urban planner (Urban Planning Department) has kept to land use issues and policy formation; and the marine scientist (Marine Science Department) has emphasized the science of ocean dynamics, marine habitats, and marine organisms. Authors and publishers of university textbooks have generally written to match those specific needs. For years, I, too, taught my coastal resource management class in the tradition more typical of the way that geographers (my professional Ph.D. training) normally handle the class, with an emphasis almost exclusively toward the land—the "dry side."

In 1992, however, my perspective began to change on what should be taught in a class on coastal resource management, particularly for those university students that live near a marine sanctuary. This was the year the 13,798 sq km (5,328 mi.) Monterey Bay National Marine Sanctuary was designated along California's coast. The official designation awakened me to the fact that this was just the newest in a series of existing marine sanctuaries along America's shores. Since that date, additional marine sanctuaries have been established along the coasts of Massachusetts, Texas-Louisiana, Hawaii, and the state of Washington, and more marine sanctuaries are being proposed.

I slowly began to learn that the sanctuary managers had a number of environmental concerns—that many of their sanctuary problems are directly attributed to what took place on the land. In other words, marine sanctuary managers were finding that simply drawing a "protective" boundary or zone on a map did not protect a marine habitat and its organisms; they had to have an input into the policy and management decisions affecting land, not just the sea. For example, marine sanctuary officials had to get involved with *watershed management* (traditionally the domain of "dry-side" specialists), for what flows into the sea affects the marine habitat.

I know of no other book that uses the U. S. National Marine Sanctuary Program as a vehicle to discuss the physical nature of America's coasts, their environmental problems, and various other conservation strategies associated with the field of coastal resource management. In particular, this book looks at the environmental impacts and corrective management techniques dealing with coastal hazards, coastal pollution, ocean dumping, fossil fuel development and transport, and the disappearance and degradation of open space. While the book focuses on these general coastal issues, primary examples come from within (and adjacent to) those sanctuaries within the U. S. National Marine Sanctuary Program. *The Monterey Bay National Marine Sanctuary is highlighted more than any other, since it is one of the newest and more extensive units in the Program, and the third largest sanctuary in the world.* Covering an area approximately the size of the state of Connecticut, this jewel of the sanctuary system is already being called *the model* that scientists around the world will follow in their efforts at coastal and marine sanctuary management.

Consequently, this is the first book to discuss and analyze: (a) marine sanctuaries as a conservation strategy within the larger context of the field of coastal resource management, (b) the environmental problems and conservation lessons from sanctuaries within the U. S. National Marine Sanctuary Program, and (c) the Monterey Bay as a sanctuary—its environmental concerns and associated management strategies. Since marine sanctuaries now exist along America's Atlantic, Pacific, and Gulf Coast shores (as well as Hawaii and

American Samoa), this book is *national in scope,* and even provides conservation lessons for coastal and marine sanctuary managers in other countries.

I should also make it very clear that this book only begins to touch on this integrative approach, thus the book's subtitle—*Toward Integrated Coastal and Marine Sanctuary Management.* Coastal scientists, planners, and sanctuary managers have only begun these discussions about "integration." Consequently, there are only a few examples to draw upon. In this book I have merely provided "some hints" of what has been done, and what can be done. The reader should not look for 250 plus pages of examples of integrated coastal and marine sanctuary management. A plethora of examples simply does not exist—not yet, anyway! Perhaps in time, with enough of us pushing in that direction, a national trend toward integrated coastal and marine sanctuary management will emerge.

Gary A. Klee

Acknowledgments

This book would never have been completed without the encouragement and advice of others. I would like to first thank Dr. Lester Rowntree, Chair, Department of Environmental Studies, San José State University, and especially Karin Sandberg, Field Editor, Science and Mathematics, Prentice Hall, for encouraging me to approach Bob McConnin, former Executive Editor of Geology at Prentice Hall. It was only after a very pleasant and informative meeting with Bob McConnin in Monterey, California that this project got seriously launched. I am, of course, indebted to Bob for he saw value in the book that I was proposing to Prentice Hall. He then carefully guided me up to the book production stage right before his retirement. Others at Prentice Hall also deserve recognition for their contributions, such as Lisa Kinne, Grace Anspake, Leslie M. Cavaliere, and Cheryl Adam. In one form or another, they have been regularly involved with this project.

Once in production, the majority of the manuscript was turned over to WestWords, Inc. in Utah, under the project management of Jennifer Maughan. Wow, what an efficient and impressive lady! She had me on a production schedule that made my head spin... and my students think I am a rigorous taskmaster! After copyeditor Charles R. Batten helped clarify my awkward prose, Jennifer pulled all the loose ends together and magically made what I consider to be an attractive book (considering the budget restrictions we were under) from the weathered pages of a lengthy manuscript. I would love to work with Jennifer and Charlie again; they are true professionals.

Between the book prospectus stage with Prentice Hall, and the production stage with WestWords, Inc., there have been many years of professional reviews and hundreds of re-drafted manuscript pages. I would like to thank Joann Mossa, University of Florida; Randall W. Parkinson, Florida Institute of Technology; Gary B. Griggs, University of California, Santa Cruz. Dr. Griggs' comments were especially useful because of his tremendous knowledge of the Central California coast, a key focus of this book. I would also like to thank all of the many individuals, institutions, and government agencies that provided comments, data, maps, and diagrams for use in this text. All the following people (and others that I have shamefully neglected to record) have made this book possible:

Graduate assistants: Without the help of my excellent graduate assistants (whom I also consider friends), this project definitely would never have been completed. I cannot thank enough the energy, enthusiasm, hard work, and dedication of graduate researchers Laureen Ferguson, Chris Flynn, Maureen Nelson, Karl Schwing, and my two creative graduate illustrators, Susan Giles and Char Holforty. In addition to their data collection abilities and illustrative talents, they simply made this project fun. Thank you, thank you, thank you.

Marine sanctuary managers and staff: This book focuses on marine sanctuaries, and it was consequently necessary to constantly hound sanctuary managers or staff for information, comments, or suggestive criticisms. In no particular order, please recognize the following individuals for their contributions: William Douros, Patrick Cotter, Dr. Holly Price, and Liz Love *(Monterey Bay NMS)*; Ed Ueber and Karina Racz *(Gulf of the Farallones [and Cordell Bank] NMS)*; Ann Walton, Edward Cassano, and Colleen Angeles *(Channel Islands NMS)*; John Broadwater *(Monitor NMS)*; Punipuao Nagalapadi *(Fagatele Bay NMS)*; June Cradick *(Florida Keys NMS)*, Sheeley Du-Puy *(Flower Garden Banks NMS)*; Alex Score and Cherl L. Callahan *(Gray's Reef NMS)*; Allen Tom *(Hawaiian Islands Humpback Whale NMS)*, Nancy Beras *(Olympic Coast NMS)*, and Brad Barr *(Stellwagen Bank NMS)*.

Federal, state, regional, local agencies and environmental organizations: Again, in no particular order of importance, please recognize contributions in the form of reviews, information, or advice from the following individuals: Sharene Azimi and Ann Northoff *(Natural Resources Defense Council)*; Mark Silberstein and Becky Christensen *(Elkhorn Slough National Estuarine Research Reserve)*; Tom Moss *(State Park Resource Ecologist, Monterey, Ca.)*; Christine Boltz *(Public Affairs Assistant, National Solid Wastes Management Association)*; Dr. Thomas H. Suchanek *(Research Ecologist, Division of Environmental Studies, University of California, Davis)*; Thomas Culliton *(Geographer, Strategic Assessment Branch, National Oceanic and Atmospheric Administration)*; David Boyd *(Senior State Park Resource Ecologist, Santa Rosa, Ca.)*; Donna Bradford *(Resource Planner, Planning Department, County of Santa Cruz)*; Emily Kishi *(Planner, Farmland Mapping and Monitoring program, Department of Conservation, Resources Agency, State of California)*; Barbara Harris *(Technical Information Advisor, National Park Service)*; Roger K. Rector *(Superintendent, Assateague Island National Seashore)*; Daniel Basta *(Chief, Strategic Environmental Assessments Division, National Oceanic and Atmospheric Administration)*; Monty Knudsen *(Branch of Coastal and Wetlands Resources, U.S. Fish and Wildlife Service)*; Julia Bott *(Central Coast OCS Regional Studies Program)*; Justin Kenney *(Writer/Editor, National Oceanic and Atmospheric Administration)*; Mitchell J. Katz *(Editor, SEA/NOAA, National Oceanic and Atmospheric Administration)*; Dan Farrow *(Chief, Pollution Sources Characterization Branch, Strategic Environmental Assessment Division, NOAA)*; and Lee Otter *(Chief Planner, California Coastal Commission)*.

Borrowed photographs: If you have ever written a textbook, you know how difficult it is to get the right photograph for a certain section of the book. I must thank many individuals for giving me permission to use their wonderful photographs, such as Dr. Gary Griggs *(Director, Long Marine Lab, Santa Cruz, Ca.)*; Christine Kook *(former graduate student and now Environmental Planner, Laughing Creek Environmental Planning, Los Gatos, Ca.)*; Keith Angell *(a sailor who has his boat next to mine in Santa Cruz Harbor, Ca.)*; Michael Blackford *(Acting Director, International Tsunami Information Center, Honolulu, Hawaii)*; Jim Cunningham *(Department of Transportation, Oakland, Ca.)*; Mara Tongue *(ESIC, Chief, U.S. Geological Survey)*; and Kenton Parker *(Elkhorn Slough National Estuarine Research Reserve)*.

Finally, I would like to thank my wife Helen—*my inhouse computer whiz*. Without her assistance when I periodically reached my expertise (and patience) with "the machine," I would have probably reverted back to my old "IBM Selectric" and never have finished this book. When it comes to writing acknowledgments, there is obviously a tendency to want to thank everyone and everything, from my closest friend, greatest critic, and veteran professional educator—Ray Schultz; to my "puppy" (and pal) named *Asia* that is sleeping on my lap as I write these very words. But, obviously, it is now time for you to tackle the actual text of *The Coastal Environment*—a serious subject to be read, mulled over, and acted upon.

Gary A. Klee

THE COASTAL ENVIRONMENT

1 The Coastal Environment

Most Americans have visited the coast, and some are even fortunate enough to live in close proximity to the sea. In fact, nearly one in every two Americans is living within 80 km (50 mi) of a coastline (Owen and Chiras 1995). The word "coast" may bring up positive images of white sandy beaches, rocky shores, and headlands, or wetlands teaming with wildlife. The same word may also cause one to visualize inappropriate development (e.g., oversized hotels built down to the water's edge); overcrowding (e.g., wall to wall sunbathers on a too-small beach; motorhomes jammed together on a coastal parking strip), or higher risk development (e.g., nuclear power plants built on a fault zone). The coast, of course, is all of these things. For whatever reason, whether it be its sheer beauty, the unique aesthetic experience, or the abundance of natural resources, Americans are fascinated with the coastal environment. Perhaps better than any other, Rachel Carson (1983, p. 11) has captured the confrontation between land and sea:

> *The edge of the sea is a strange and beautiful place. All through the long history of Earth it has been an area of unrest where waves have broken heavily against the land, where the tides have pressed forward over the continents, receded, and then returned. For no two successive days is the shoreline precisely the same. Not only do the tides advance and retreat in their eternal rhythms, but the level of the sea itself is never at rest. It rises or falls as the glaciers melt or grow, as the floors of the deep ocean basins shift under its increasing load of sediments, or as the Earth's crust along the continental margins warps up or down in adjustment to strain and tension. Today a little more land may belong to the sea, tomorrow a little less. Always the edge of the sea remains an elusive and indefinable boundary.*

Americans are not alone in being drawn toward the coast. Two-thirds of the world's population lives near the coast. As humans increasingly inhabit coastal areas, they bring their houses, schools, agricultural systems, businesses, factories, and power plants, which in turn elevate the levels of coastal pollution and coastal areas used as dumping grounds for waste.

The very fact that so many people in the world live in coastal areas and use (and often abuse) their natural resources is reason enough to study coastal resource management. However, the coastal zone is important for other reasons as well, such as:

- *a habitat for wildlife* (e.g., the myriad of plants and animals that depend on coastal estuaries);
- *a natural filter* (e.g., wetlands that filter impurities from waters that pass through them);
- *a safety barrier* (e.g., barrier islands, beaches, dunes, and cliffs that buffer residents along the coast from high winds and seas);
- *a food source* (e.g., coastal fisheries that provide a food source for millions of people worldwide);
- *a recreation area* (e.g., open space for beach-combing, sunbathing, swimming, boating, fishing, and relaxing of whatever sort);

- *aesthetics* (e.g., inspiration for the painter, nature photographer); and
- *a source for psychological and spiritual renewal* (e.g., a mental "time-out" from the daily grind to regenerate one's soul).

Of course, all of these demands upon the coast take a heavy toll. Beaches, sand dunes, and elements within the coastal zone have a **carrying capacity** (population and impact limits) that must be respected. Just as you can only put so many cattle or sheep on a pasture before the grassland ecosystem deteriorates, there are limits to which coastal ecosystems can withstand natural impacts (e.g., winds, coastal storms, hurricanes, wave surge), and especially human impacts, such as population pressures (e.g., urban development, tourism, loss of open space), coastal pollution, offshore oil development, and ocean dumping. As will be illustrated throughout this book, determining the carrying capacity of a site is not easy, and, consequently, is open to much scientific and political debate.

There is no debate, however, that understanding the coastal environment—its major elements and how they interact—is the first step toward alleviating the pressures on the coastal zone and marine environment. Consequently, this opening chapter covers the following background information: *The Coastal Zone* provides a working definition of the coast, and identifies the five major issues facing America's coastal areas. *The Coastal System* introduces the reader to various ways to classify coastal types and processes, as well as major subdivisions of the coast. Natural processes affecting these coastal elements are also discussed, such as wind, coastal storms, and sediment transport. *Coastal Formations* identifies and briefly discusses the primary physical formations that can be found in the coastal zone, ranging from inland coastal mountains to open ocean habitats. Managing these phenomena, including their associated economic (e.g., fisheries, tourism) and cultural (e.g., shell middens & other archaeological treasures) resources, requires a new way of thinking, which is the theme of this book.

THE COASTAL ZONE

Defining the Coast

The **coast** refers to the area where land, water, and air meet. Most definitions of the coast merely have it beginning with the **shoreline** or **coastline** (the line of intersection of a water body with land) and extending inland to the limit of tidal or sea-spray influence, or to where the terrain shows signs of major change; but in this book, the term coast or **coastal zone** will also include the nearshore waters that extend from the shoreline to the outer limit of the continental shelf. The coast is of indefinite width (from several hundred feet to several miles) and varies with season and time. The coastal zone has the characteristics of mixing or adjustment, and change (Carter 1988). It is, in other words, where the terrestrial environment influences the marine or lacustrine (lake) environment, and vice versa. Hence, it is an **ecotone**—a transition area or border where two ecological communities meet.

In addition to defining the coast in physical or ecological terms, the coast or coastal zone can also be defined in terms of management or planning boundaries. For example, **coastal areas** (or coastal counties) in the United States are defined by political and cultural elements, as well as by physical features. According to definitions established by the Federal Coastal Zone Management Program, which is managed by the National Oceanic and Atmospheric Administration (NOAA), "coastal areas" of the United States include 30 coastal states (which includes those bordering the Great Lakes) and their 451 coastal counties.

The Federal Coastal Zone Management Act (CZMA), which created the Federal Coastal Zone Management Program, uses ecological, as well as political boundaries, to legally define the coastal zone. According to the act, the coastal zone encompasses the state's coastal waters and shorelands including "islands, transitional and intertidal areas, salt marshes, wetlands, and beaches" (CZMA 1994). The coastal zone also "extends inland from the shorelines only to the extent necessary to control shorelands, the uses of which have a direct and significant impact on the coastal waters, and to control those geographical areas which are likely to be affected by or vulnerable to sea level rise" (CZMA 1994). The coastal zone is further defined as the "connecting waters, harbors, roadsteads, and estuary-type areas such as bays, shallows, and marshes" of the Great Lakes, and in other coastal regions as "those waters, adjacent to the shorelines, which contain a measurable quantity or percentage of sea water, including, but not limited to, sounds, bays, lagoons, bayous, ponds, and estuaries" (CZMA 1994).

Meanwhile, political coastal zone boundaries extend "seaward to the outer limit of state title and ownership" as established by legislative action, such as the Submerged Lands Act enacted by Congress in 1953. This act allowed every coastal state to extend its boundaries to 4.8 km (3 mi) from its coasts. In the Great Lakes region, the states' coastal zone extends to the "international boundary between the United States and Canada" (Clark 1977).

States and territories subject to these definitions are those which are "bordering on, the Atlantic, Pacific, or Arctic Ocean, the Gulf of Mexico, Long Island Sound, or one or more of the Great Lakes," as well as "Puerto

Rico, the Virgin Islands, Guam, the Commonwealth of the Northern Mariana Islands, and the Trust Territories of the Pacific Islands, and American Samoa" (CZMA 1994). As indicated earlier, there are 30 states and 451 counties which are *directly defined* as "coastal areas." However, expanding upon these regions to include geographic areas that are "influenced" by the coast, 1,569 non-coastal counties, the District of Columbia, 23 boroughs or census areas in Alaska, and 42 independent cities in Virginia and Maryland, become "coastal areas" as well, as defined by the Coastal Zone Management Act.

To further complicate matters, individual state coastal management programs have sometimes modified the federal definition to suit their local political, economic, or cultural situation. However, local and state governments must have the same definition as the federal government, if they wish to participate in the Federal Coastal Zone Management program. For example, the California Coastal Act, the enabling legislation for California's coastal program, expands upon the coastal zone definition found in the Federal Coastal Zone Management Act. This definition specifies that the coastal zone includes inland areas "generally 1,000 yards from the mean high tide line of the sea." In addition, where "significant coastal estuarine, habitat, and recreational areas" exist, the coastal zone "extends inland to the first major ridgeline paralleling the sea or five miles from the mean high tide line of the sea, whichever is less." Furthermore, "in developed urban areas the zone generally extends inland less than 1,000 yards." Interestingly, California's Coastal Act also excludes an area, the San Francisco Bay, normally considered "coastal" by the Federal Coastal Zone Management Act, covering management of this region under a separate governmental authority known as the San Francisco Bay Conservation and Development Commission (California Coastal Act, Chapter 2, Section 30103).

Coastal Issues and Management Concerns

Coastal resource managers are faced with a number of pressing issues. Individual chapters in this book address the five major environmental concerns facing America's coastal areas: coastal hazards (Chapter 4), coastal pollution (Chapter 5), ocean dumping (Chapter 6), offshore oil development (Chapter 7), and disappearance of open space (Chapter 8). Recent (non-traditional) issues are integrated into the above chapters when appropriate. For example, the topic of "social equity" (i.e., the displacement of minority and/or poor residents by new coastal developments) will be discussed in the chapter on open space management. The "recent" coastal management concern—the need for integrated coastal and marine sanctuary management—is addressed throughout each of the major chapters.

- *Coastal hazards.* The prevention and mitigation of coastal hazards is a subfield of coastal resource management. It includes such problems, issues, and concerns as *storm hazard mitigation* (e.g., through warning systems and evacuation strategies; integrated coastal planning); *controlling shoreline erosion* (e.g., through traditional techniques and innovative strategies for erosion control protection; limits on permissible development); and *projecting and preparing for sea-level rise* caused by global warming (e.g., planning for inundation of coastal communities; coastal reinforcement or strategic retreat). The last issue may be the most crucial one facing coastal scientists today (Carter 1988).
- *Coastal pollution.* This category of concern includes all the types and sources of coastal pollution. For example, it requires understanding, minimizing, or mitigating concerns regarding *point source pollutants* (e.g., sewage outfalls, pollutants from marinas, industrial waste water), *nonpoint sources* (e.g., agricultural lands, urban areas, marine debris), and *physical and hydrological modifications* (e.g., harbor dredging, groundwater withdrawal, and dam construction and irrigation—all human activities that exacerbate saltwater intrusion).
- *Ocean dumping.* Though also a form of coastal pollution, "ocean dumping" is generally thought of as a separate subfield of coastal resource management. For over 100 years, America's coastal waters have been used as a dumping ground for the nation's waste materials, including *dredged materials, sludge, solid waste (garbage), industrial waste, military waste, nonmilitary radioactive wastes, and ocean incinerated wastes.* This subject area also includes the *deep ocean disposal debate* over whether or not to intensify the disposal of waste materials in deep ocean areas.
- *Offshore oil development and transport.* Developing coastal fossil fuel resources can also be a source of coastal pollution. In fact, many of the marine sanctuaries within the U.S. National Marine Sanctuary Program were established to prevent offshore oil development from occurring within the region. Coastal resource managers, local governments, and the average "coastal citizen" need to understand the components and associated impacts of offshore oil development, as they relate to exploration activities, offshore development and production schedules, subsea and onshore pipelines, vessel traffic, and oil spill response programs.
- *Open Space Preservation and Management.* This final major subfield of coastal resource management requires an understanding of human population demographics and how the urbanization process has impacted coastal areas, such as with the disappearance of coastal agricultural land, the loss of coastal wildlife habitats, declining

coastal recreational resources, loss of coastal village or small town character, and declining respect for coastal property. Coastal resource managers who are interested in this subfield may work on projects to sustain the productivity and diversity of coastal ecosystems (e.g., aquaculture, fisheries management), to reclaim wetland and estuary ecosystems, or to restore recreational opportunities (e.g., beach access) for coastal dwellers and visitors, and even cultural historic projects to help an area maintain its sense of place.

In order to move toward resolving the above issues, coastal resource managers will need to have a degree of scientific knowledge, an understanding of legal terms, an ability to help resolve conflicts, and skills within the arena of public education.

THE COASTAL SYSTEM

Coastal Types and Processes

Different coastal types are the result of various processes (e.g., tectonics, exposure to wind and waves, sediment supply and transport) working at varying geographic scales. These processes have been categorized in a hierarchy as first, second, and third-order processes. (See Table 1-1). Each of these classification scales will be discussed briefly in this chapter to give the reader a better understanding of how our coasts are formed and how susceptible they are to environmental changes such as global warming and sea level rise.

First-order processes. First-order processes are those which act at a *global* level, affecting coastal lengths of 1000 km (621 mi) or more (Inman and Nordstrom 1971). These processes include climate changes, sea level variation, and plate tectonics.

(a) Climate Changes

Global climate and temperature changes resulting from natural variations in atmospheric conditions (e.g., ice ages), or as a result of human interference (e.g., greenhouse warming), affect global sea level and influence second-order processes such as increased sediment deposition or erosion.

(b) Sea Level Variation

King (1972) and others have suggested that coasts can be classified according to sea-level change. During the Pleistocene, variations in sea level greatly influenced coastal forms. For example, a coast could be described in terms of *submergence* (e.g., a subaerial valley flooded by sea-level rise such as a **fjord**) or in terms of *emergence* (e.g., an uplifted, wave-cut platform such as a **marine terrace**). Many coasts, however, do not fit these generalizations (Hansom 1988). It is still important to understand the effect sea level changes have on processes shaping present day coasts.

Sea level oscillations can result from many environmental changes. These environmental changes range from short-term processes, such as atmospheric pressure changes which can last several months, to long term processes, such as advancing and retreating glaciation lasting millennia. (See Figure 1-1).

For example, the massive quantities of water locked in the world's ice sheets responds to environmental changes, releasing or decreasing the volume of water in the world's oceans. Today, sea level is also being affected through human extraction of fluids (e.g., oil, water, gas) and mass building on "soft" ground (e.g., coastal development in Venice, Italy; and Hawaii, USA). These

TABLE 1-1
The three major coastal scales

Order	*Dimensions*		*Controls*	*Results*
1st order	Length	c. 1000 km	Plate tectonics	Coastal plain and continental shelf
	Width	c. 100 km		
	Height range	c. 10 km		
2nd order	Length	c. 100 km	Erosion and deposition modifying 1st order features	Deltas, coastal dunefields estuaries
	Width	c. 10 km		
	Height range	c. 1 km		
3rd order	Length	1-100 km	Wave action and sediment size	Beaches longshore bars, mudflats
	Width	10 m-1 km		
	Height range	?		

Note: 1st and 2nd order factors define the coastal zone, 3rd order factors define the shore zone
Source: Viles & Spencer 1995, 20, adapted from Inman and Nordstrom 1971.

human impacts alone are causing sections of cities to sink below sea level (e.g., New Orleans) and thus require further intervention with artificial levees and pumps.

Sea level change also results from a change in volume of the ocean basins through oceanic plate spreading and continental uplift or lowering. One method by which the ocean basin volume is changing today is due to lifting or lowering of the earth's continents in response to the weight released from the last glacial melting. Areas once covered by glaciers are still rising in response to the removed weight, while surrounding areas previously bulging up around the weight of the glacier are now lowering (Davis 1994).

(c) Plate Tectonics

Given that the earth's crust is made up of continental and oceanic plates which are in constant motion with respect to each other, Inman and Nordstrom (1971) devised a method for the classification of coastal types dependent upon their position on a crustal plate. Using this method, they have categorized coastal areas as either leading-edge coasts, trailing-edge coasts, or marginal coasts.

i. Leading-Edge Coasts (active or collision coasts)

Definition: A leading-edge coast develops at the margin of a landmass, near the collision of active crustal plate margins (Davis 1994). (See Figure 1-2).

Characteristics: Leading-edge coasts are characterized by narrow continental shelves, deep ocean basins and trenches, rugged shores and sea cliffs, nearby coastal mountain ranges, and volcanic and earthquake activity. Because of the mountain slopes, the streams and small rivers flow relatively straight and fast, eroding their beds and dumping sediment directly into the ocean (Davis 1994). Of course, not all these characteristics occur in all leading-edge coastal areas.

15,000 Years Ago
(End of last Ice Age—Sea level approximately 400 feet below present level; rivers not shown)

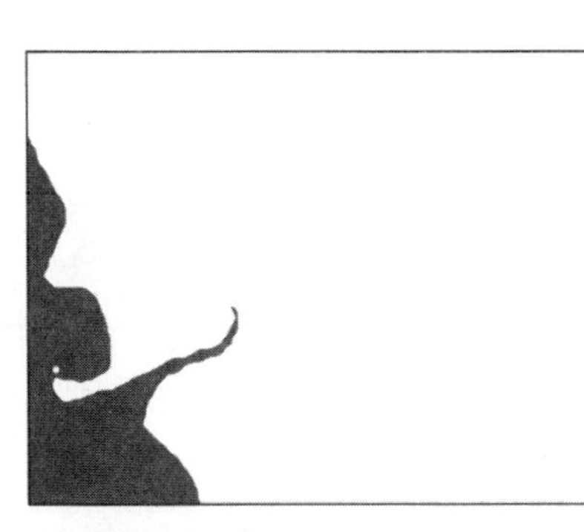

10,000 Years Ago
(Formation of Farallon Islands and intrusion into "Golden Gate")

5,000 Years Ago
(Formation of Bay and Delta basins)

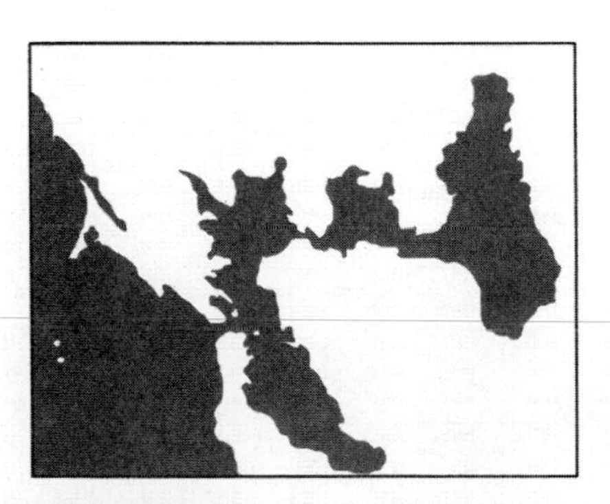

125 Years Ago
(Landward edge of undiked tidal marsh)

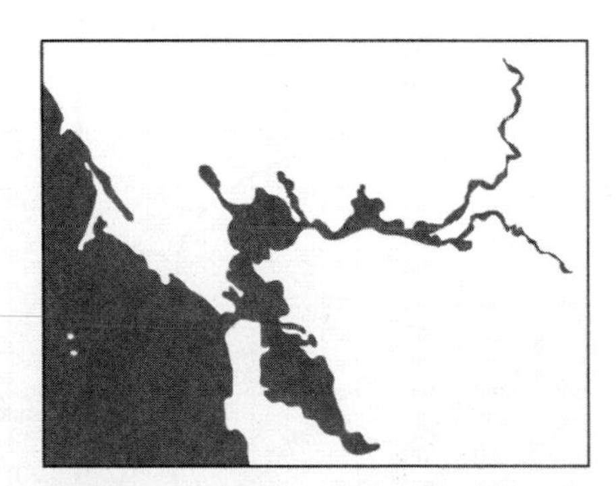

Today
(Includes changes due to hydraulic mining sediment deposition, land reclamation, and filling of wetland areas)

FIGURE 1-1 Sequential sea level rise in the San Francisco Bay/delta estuary. (*Source:* Association of Bay Area Governments 1992, 12–13)

Examples: West coast of U.S.; Andean coast of South America; East and West Indies (Davis 1994). (See Figure 1-3).

ii. Trailing-Edge Coasts (passive or plate-imbedded coasts)

Definition: Trailing-edge coasts are formed on the edge of a land mass, moving with a crustal plate away from the divergent plate margin, and situated very far from a plate margin (i.e., in a very tectonically stable portion of the plate).

Characteristics: Trailing-edge coasts typically have broad continental shelves without ocean trenches. Their coastal plains are wide, have shallow waters, and often contain low-lying lagoons and barrier islands. Erosional forces are slow during normal wave activity due to the shallow water and resulting weak wave action (Davies 1980). (Of course during the hurricane season, erosional forces are much greater). Sub-categories of trailing-edge coasts exist, depending on the time since the plate separated from it's prehistoric neighbor; or stated differently, depending on the maturity of the coastline and shelf width (Davis 1994).

Examples: East coast of U.S., particularly the coasts of North and South Carolina; the Gulf of California; Greenland; the Bay of Bengal.

iii. Marginal-Sea Coasts (also plate-imbedded)

Definition: Marginal-sea coasts develop along shores of a sea, enclosed by continents and island arcs which further protect it's shores and continental shelves from wave action (Davies 1980).

Characteristics: Marginal coasts typically have wide continental shelves with shallow seas to break incoming waves. The coastal plains vary in width but are often bordered by hills and small mountains. The rivers are

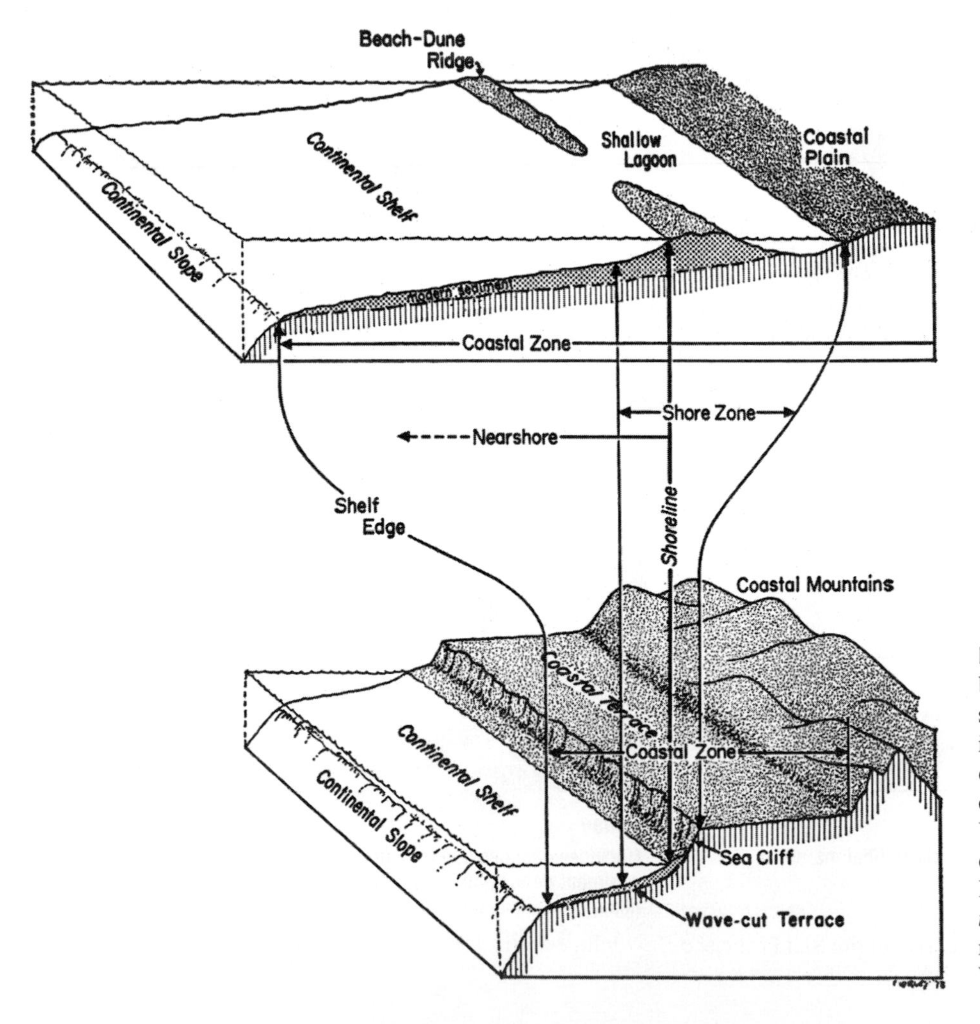

FIGURE 1-2 Coastal classification based on tectonic situation. Wide-shelf plains coast (upper) characteristic of the U.S. east coast (trailing edge), and narrow-shelf mountainous coast (lower), characteristic of the U.S. west coast (collision or leading edge), based on Douglas l. Inman and Brush 1974. (*Source: Coastal Ecosystems Management* by Clark © 1977, reprinted by permission of John Wiley & Sons, Inc.)

often large and carry huge sediment loads, resulting in large deltas at the river mouths.

Examples: the South China Sea protected by the Philippine Islands; the Gulf of Mexico protected by the volcanic area of Central America and the adjacent Caribbean island arcs.

Second-order processes. Second-order processes are those which act at a more local level, effecting coastal lengths of 100–1000 km (62–621 mi) (Refer back to Table 1-1). These processes include erosion, **biogenic** activity (e.g., coral reef building), **cryogenic** activity (e.g., seasonal carving of ice-laden coasts), and deposition forces which result in large local features like deltas, dunefields, coral reefs, and ice fields of arctic coasts. Taking these second-order processes into consideration with the tectonic first-order classifications, Inman and Nordstrom (1971) further defined subclassifications for coastal areas as shown in Table 1-2.

Third-order processes. Third-order processes are those that give classifying features to a local length of shore. These processes (e.g., wind, waves) react with land formed by the previous first- and second-order processes (e.g., sea level, sediment deposition, tectonics) (Viles and Spencer 1995). Wave action is considered the most significant third-order process shaping our shore zones today.

Because of the importance of wave action, Davies (1972) even suggested a separate classification scheme which grouped coasts according to three types: (a) protected sea environments; (b) storm wave environments; and (c) swell wave environments. (See Figure 1-4).

His classification is actually a grouping of the world's coastal areas by their wave environment, or degree of exposure to wave energy. Given that wave heights are controlled by wind speed, direction, duration, and the distance the wind blows (i.e., fetch), this diagram of wave patterns and intensity can help in predicting coastal features at local levels (Viles and Spencer 1995). Note that these factors will cause local waves to have predominant directions, with deviations occurring during events such as storms and typhoons.

It is important to note that *all scales of classification must be considered when studying a local coastal area.* Steep cliffs and narrow beaches, though appearing to be a first-order leading-edge coast, may in fact be a trailing-edge coast which has been greatly impacted by second- and third-order processes. This is the case regarding the southern coast of Australia (Davis 1994). In addition, one

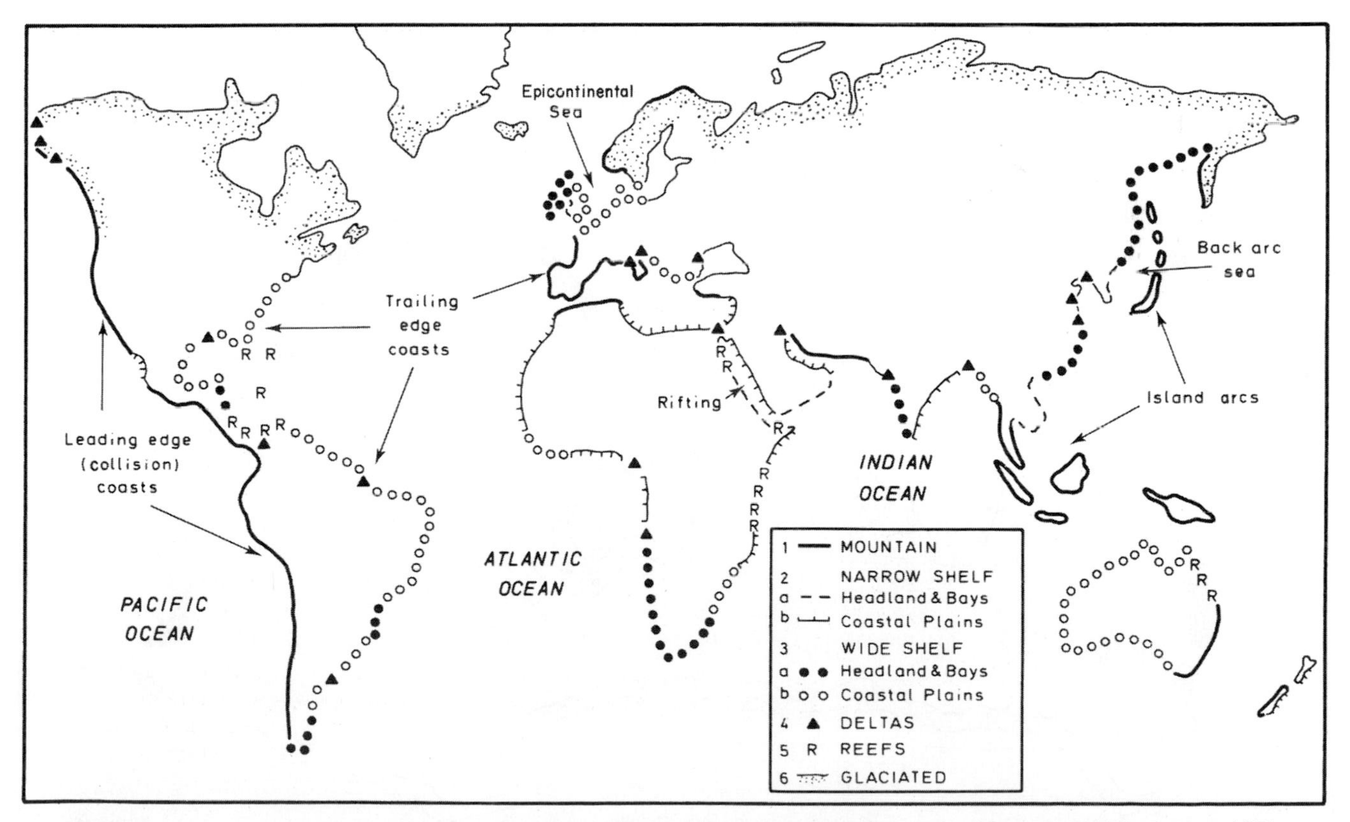

FIGURE 1-3 The tectonic classification of the world's coastlines, based on the work of Inman and Nordstrom 1971. (*Source:* Reprinted from *Coastal Environments,* 1988 by Carter, by permission of Acadmic Press Limited, London)

must not assume that if you find a trailing-edge coast on one side of a continent, you will find a leading-edge coast on the other. Coasts have features of different size and scale, and all three orders of coast-forming processes must be considered before any conclusions of origin are drawn.

TABLE 1-2
Morphological classification of coasts

Coast type	*Characteristics*
Mountainous coast	Shelf < 50 km wide, coastal mountains > 300 m high, rocky shore zone with pocket beaches. Mainly on collision coasts
Narrow-shelf hilly coast	Shelf < 50 km wide, coastal hills c. 300 m high or less, occasional headlands and beaches, some barriers
Narrow-shelf plains coast	Shelf < 50 km wide, low-lying coastal plains, barrier beaches, occasional low cliffs
Wide-shelf plains coast	Shelf > 50 km, low-lying coastal plains and wide shore zone, often with barrier beaches
Wide-shelf hilly coast	Shelf > 50 km wide, coastal hills c. 300 m or less, barrier beaches and occasional headlands and cliffs
Deltaic coast	Sediment deposited where river enters sea; low-lying coastal bulge
Reef coast	Organic origin, resistant; fringing or barrier type
Glaciated coast	Coastal features dominated by erosional effects of glaciers, precipitous cliffs and fjords common

Source: Viles and Spencer 1995, 22, adapted from Inman and Nordstrom 1971.

Subdivisions of the Coast

Hansom (1988) has identified common nomenclature for the subdivisions of the coast based on wave process zones. There are three major zones—backshore, nearshore, and

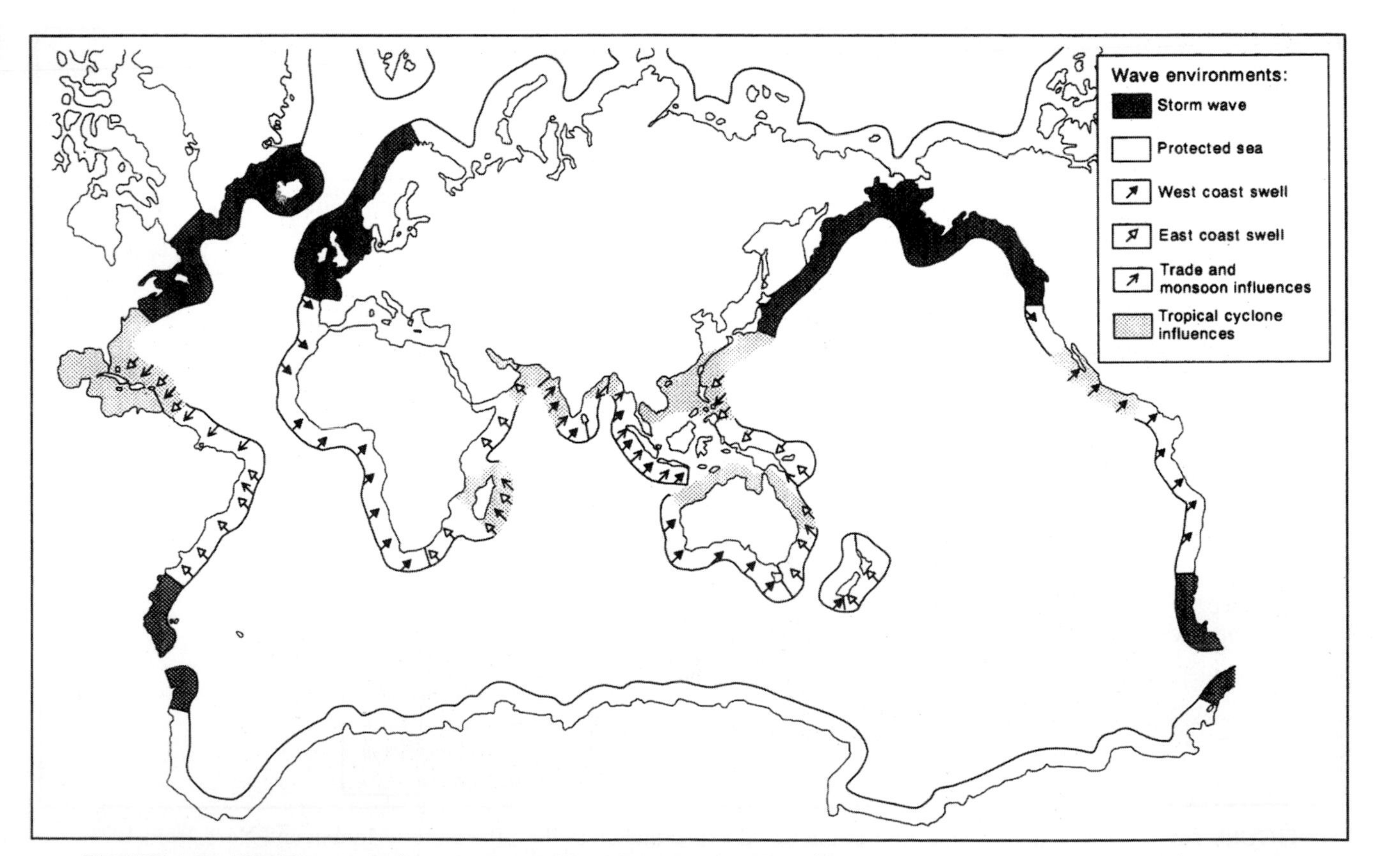

FIGURE 1-4 World wave environments. Modified from Davies 1972. (*Source:* Viles and Spencer 1995, 27)

offshore, each of which can be further subdivided. (See Figure 1-5).

The **backshore zone** extends from sand dunes or beach ridges down to the high water mark (HWM). This is an area that generally lies above the level of wave reach, except during heavy storms.

The **nearshore zone** extends from the HWM to the offshore zone. This is the area that is affected most by waves and where sediment movement occurs. The nearshore zone has three subdivisions: the swash zone; the surf zone; and the breaker zone. The **swash zone** is the area of the shore that is alternately covered by swash (wave uprush) and backwash (wave recession). Of course, the location of the swash zone changes with the level of the tide. The **surf zone** is where over-steepened waves topple; here, broken waves travel towards the shore, depositing (and removing) sediment in the process. The third subdivision, the **breaker zone,** is the most seaward area where the tallest breaking waves occur.

The **offshore zone** extends beyond the breaker zone out to sea, where water depths are greater than half the wave length of the incoming waves. In this zone, wave-induced sediment movement is limited.

Coastal Dynamics

The coastal zone is under constant change due to three major groups of natural factors: (1) terrestrial factors; (2) marine factors; and (3) biological factors (Hansom 1988), in addition to anthropogenic factors, discussed later.

Terrestrial factors. The *geologic structure* of the site is a primary terrestrial factor. As indicated earlier, tectonic forces that determine whether a coast is a collision coast, trailing edge coast, or marginal sea coast have an impact on the nature of the structural base upon which the coastal features develop. For example, a coastal outline may have the appearance of a drowned valley landscape because of large-scale folding in its geological past, or it may have a very straight coastline due to geological faults in the area. It could also have a fjord coastline caused by the erosional forces of glacial activity and sea level changes.

According to Hansom (1988), the *climatic regime* is also another terrestrial factor that impacts coastal areas in three primary ways: (1) The climate of an area affects temperature and rainfall amount and intensity, vegetation type, and consequent discharge of *fluvial sediment* from the coastal **hinterland**—the "uplands" away from the coast, such as watersheds, water courses, plains, and hills (Clark 1996). (2) Climate also affects such *geomorphic processes* as weathering, mass movement, erosion, as well as biological processes in the zone between the shore and the coastal hinterland. (3) Finally, the present atmospheric climate directly affects *shore processes* themselves, such

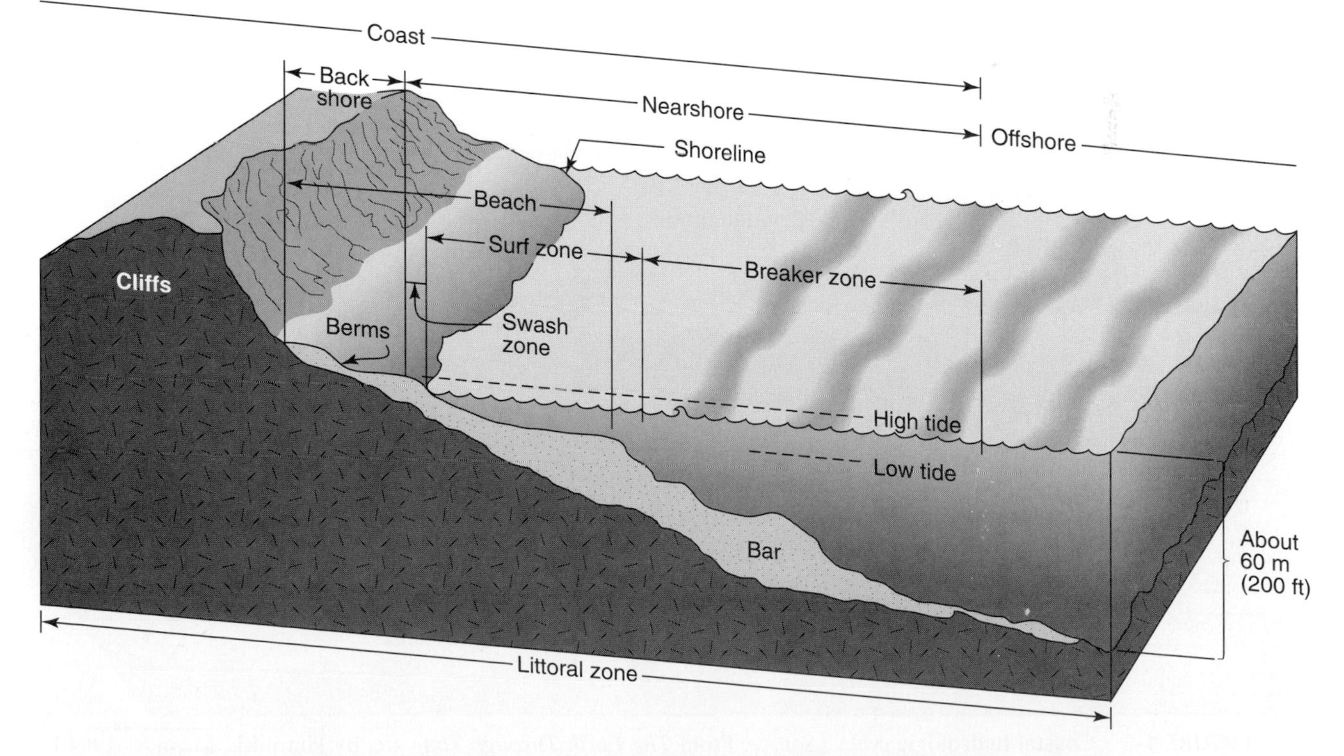

FIGURE 1-5 Coastal wave zones and beach features. (Diagram by Heather Theurer)

as platform weathering and dune building (Davies 1980). One can easily see why a coastal resource manager must have a good understanding of the *coastal hydrologic cycle,* particularly rates of precipitation and evapotranspiration, surface water flow, percolation and infiltration, groundwater flow, saltwater intrusion, and so on. (See Figure 1-6).

Marine factors. The coastal zone is also affected by marine factors, such as tides, storm surges, waves, tsunamis, currents, upwelling, and sea level rise.

Tides (the daily oscillations in sea level resulting from the physical relationships of sun and moon) are a major marine factor affecting coastal dynamics, as well as such human economic activity as traditional food gathering, commercial fishing, recreational tide-pooling, and, more recently, generating electricity by harnessing tidal power. Every day, many coastlines experience two high tides (rising tides, known as **flood tides**) and two low tides (falling tides, known as **ebb tides**). All sites have a **tidal range**—the difference between high and low tides. The greatest tidal range on Earth exists at the Bay of Fundy in Nova Scotia, where the maximum tidal range reaches 15.4 m (51 ft) (Viles and Spencer 1995). The lowest tidal range exists in Lake Superior in the United States, with a tidal variation of only approximately 5 cm (2 in) (Christopherson 1997).

The tidal range for the Santa Cruz harbor on the shores of the Monterey Bay National Marine Sanctuary reaches only a maximum of 2.4 m (8 ft). Tidal prediction is especially important to deep-hulled commercial ships. In some cases, passage in and out of a commercial port is possible only during high tide. Extreme high tides, when linked with storm surges, can lead to flooding, property damage, and loss of life.

Waves are another marine factor affecting coastlines. **Waves** can be defined as undulations of water that generally result from wind friction on the surface of an ocean or bay. As mentioned earlier, coastlines are sometimes classified as either *storm wave coastlines* (short high waves from various directions), *swell wave coastlines* (long and low waves from a consistent direction), or *protected coastlines* (no significant wave action) (Davies 1980). One type of wave that greatly influences coastlines is the tsunami. Rather than being generated by wind friction on the surface of the ocean, a **tsunami** is the result of an undersea disturbance due to seismic activity (e.g., a submarine fault movement). *Tsunami* is a Japanese word meaning "harbor wave," so named because of its devastating effect on harbors. Further details about this highly destructive type of wave are given in Chapter 4, *Coastal Hazards.*

When normal (non tsunami) waves enter shallow waters close to the beach, the wave height increases rapidly, reaches a point when its height exceeds its vertical stability, and the wave then falls into a characteristic **breaker** crashing on the beach. There are four types of breaking waves: spilling, plunging, collapsing, and surging. *Spilling breakers* occur when steep waves meet flat beaches; *plunging breakers* are associated with steep beach-

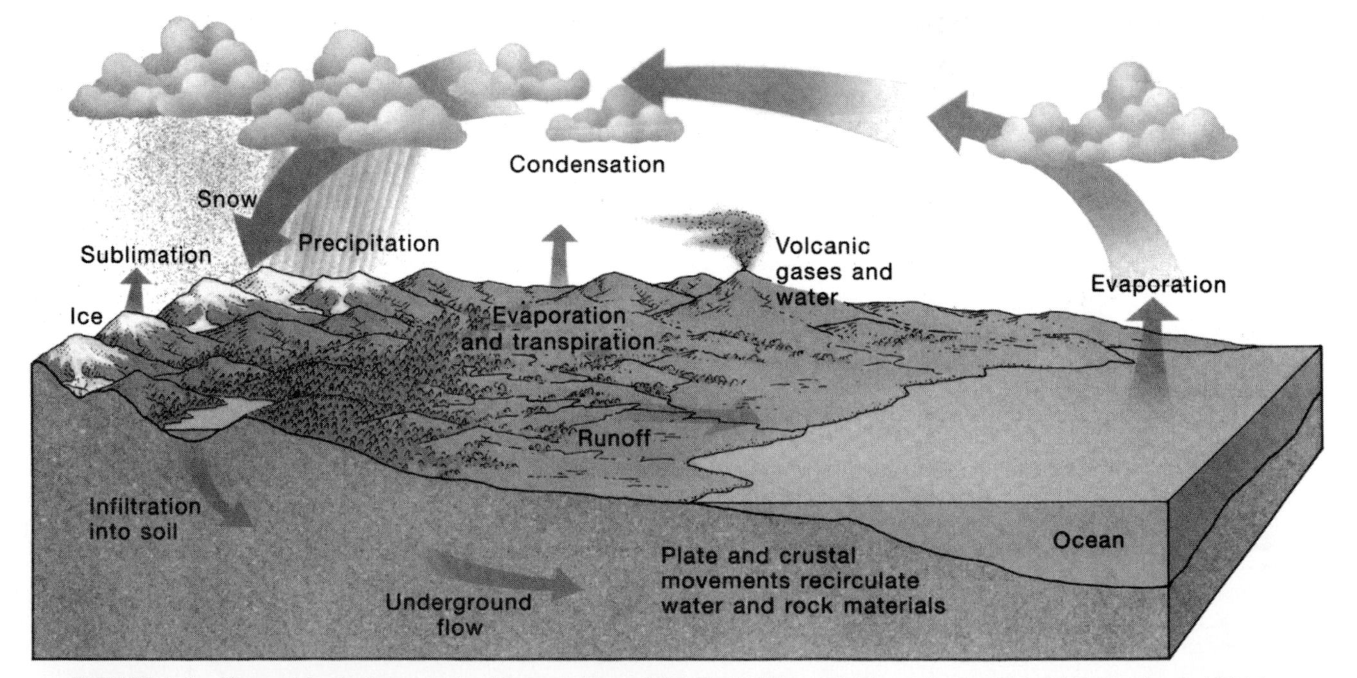

FIGURE 1-6 Coastal hydrologic cycle. (*Source:* From *The Earth Through Time* 4/e, by Harold L. Levin, © 1994 by Saunders College Publishing, reproduced by permission of the publisher)

es that cause the leading edge of a wave to become almost vertical prior to the top curling over and plunging forwards; *collapsing breakers* occur when swell waves break on steep beaches, causing the leading edge of the wave to collapse before it can form a curl; and *surging breakers* occur when waves of low steepness approach very steep beaches (Davis 1994). (See Figure 1-7).

When waves approach the surf zone and shallower water at an oblique angle, a longshore current can result. (See Figure 1-8). This current works in combination with

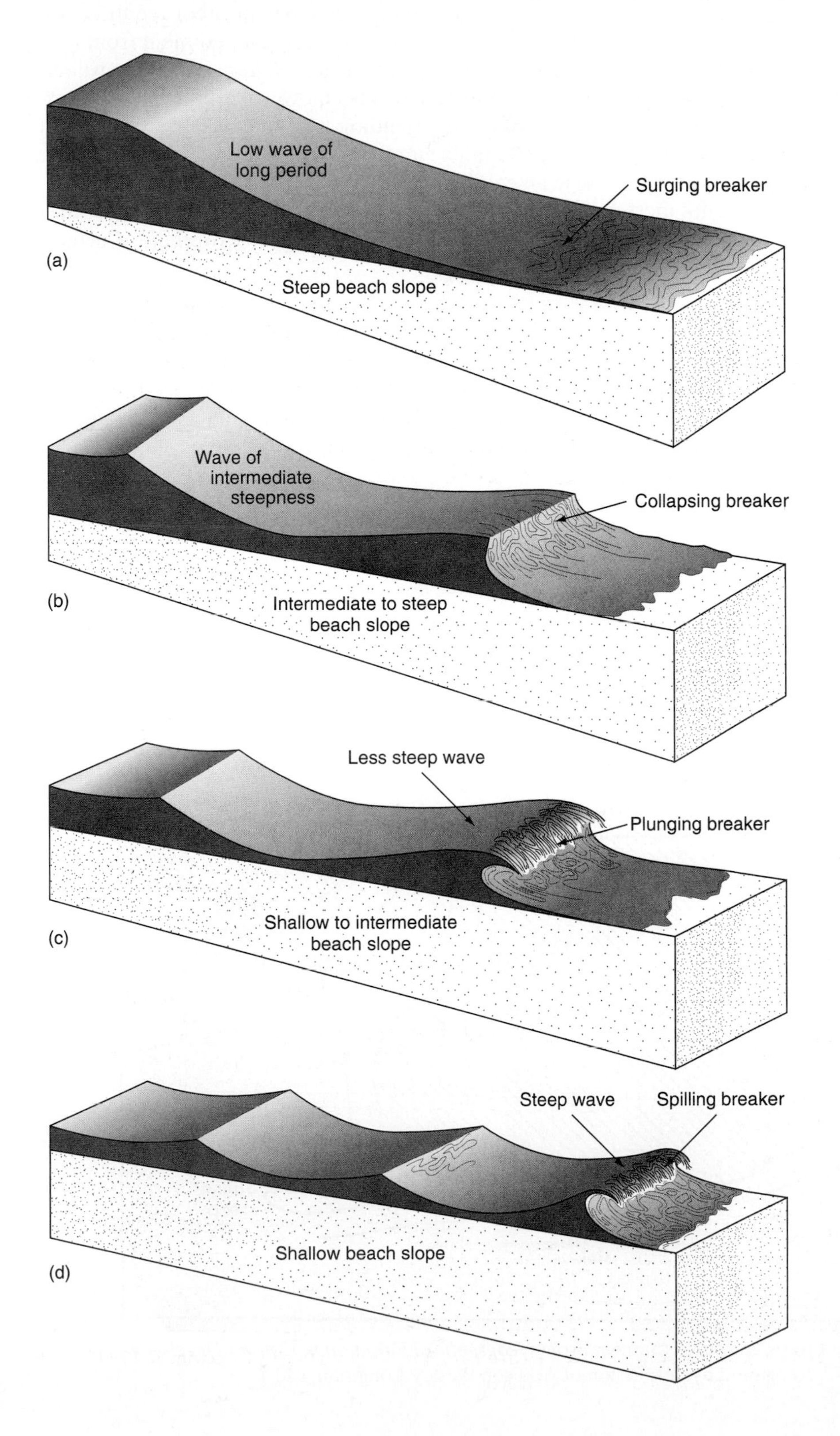

FIGURE 1-7 Types of breaking waves and the beaches on which they commonly occur. (a) surging breaker, (b) collapsing breaker, (c) plunging breaker, (d) spilling breaker. (*Source:* Diagram by Heather Theurer. Adapted from Davis 1994.)

wave action to create **littoral drift**—the transport of large amounts of sand, gravel, sediment, and debris along the shore. With each *swash* and *backwash* of surf, **beach drift** may also occur—the movement of sand particles on the beach in the direction of the littoral drift. In other words, waves can move sediment on the beach as well as within a longshore current. These transported sediments can represent a significant volume, filling inlets and harbors.

Sediments come to rest in four main sorts of situations—reentrant trap; salient trap, equilibrium trap, or deep sink (Davies 1977). (See Figure 1-9). *Trap* refers to the temporary escape of sediments from the transport system (e.g., littoral drift). In other words, some if not most of the sediment may eventually get freed and remobilized back into the transport system. Sediment lost into a "sink" (e.g., the submarine canyon of Monterey Bay, California; the submarine canyon of southern California) is removed permanently from the transport system, at least until the present day sea level changes. In a *reentrant trap* situation, sediments are trapped in a bay or other reentrant where land jutting seaward prevents the sediment from moving further along the coast. *Salient traps* can occur where the shore suddenly trends landward, producing a spit or tail like sand bar extending into deep water. Where approximate equilibrium of forces occur, sediment accumulates in an *equilibrium trap.* (See Figure 1-10).

Coastal storms and hurricanes are also major marine factors that alter the coastal environment. Certain ge-

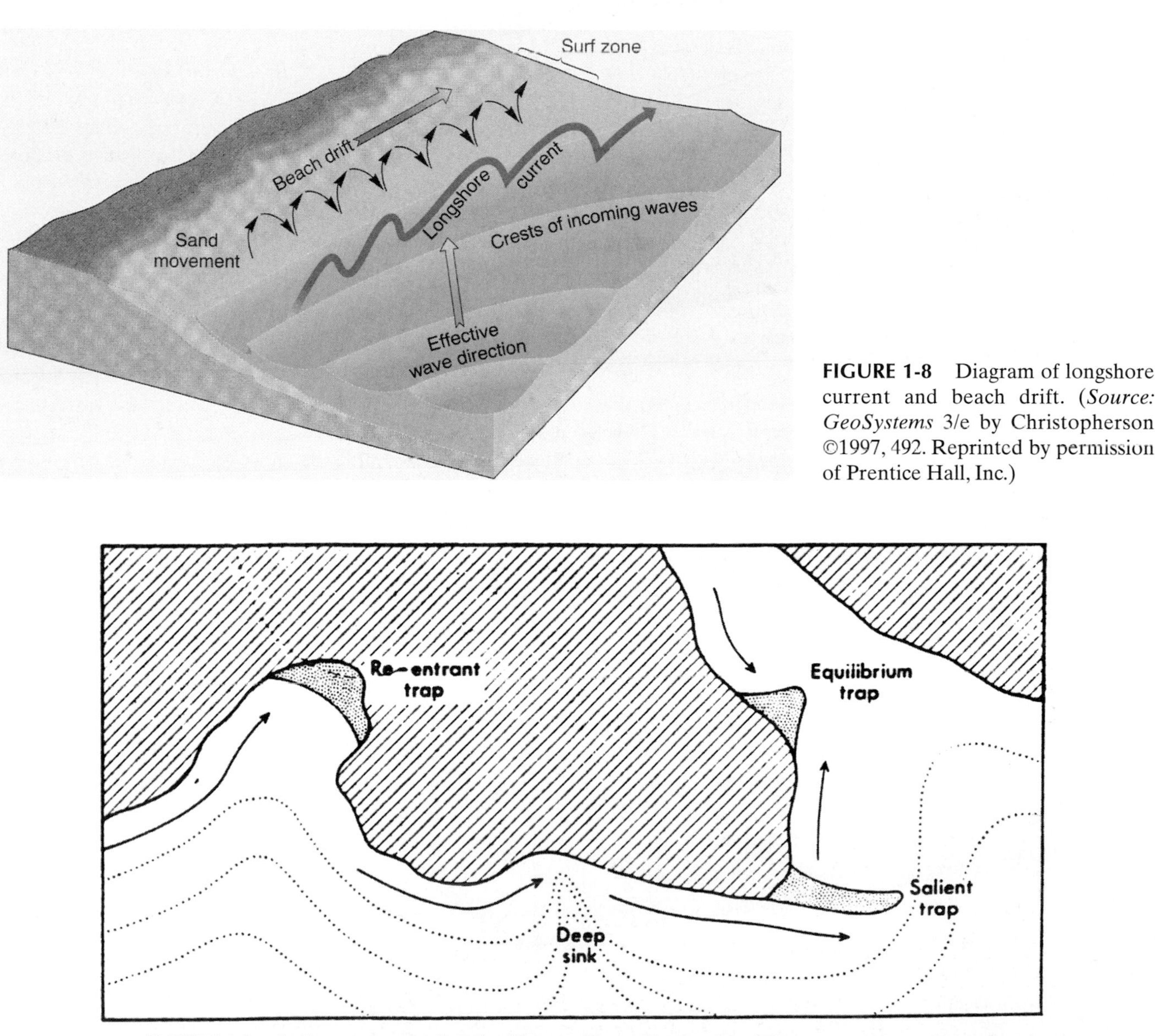

FIGURE 1-8 Diagram of longshore current and beach drift. (*Source: GeoSystems* 3/e by Christopherson ©1997, 492. Reprinted by permission of Prentice Hall, Inc.)

FIGURE 1-9 Sediment traps and sinks. (*Source:* From *Geographical Variation in Coastal Development* by Davies, 1977. Reprinted by permission of Addison Wesley Longman, Ltd.)

ographic regions are more prone to *landfalls* (the system reaching the land) than others. Along America's shores, Florida has received the greatest number of hits by hurricanes. In 1992, for example, Hurricane Andrew struck Southern Florida, altering shorelines, local ecosystems, and devastating homes. Some 75,000 homes were destroyed, leaving 250,000 people homeless. Texas, Louisiana, and North Carolina are also heavily hit by hurricanes, **storm surges** (heightened wave action due to hurricanes and coastal storms) and **storm flooding** (heightened flooding due to hurricanes and coastal storms). It is easy to visualize how coastal storms or hurricanes (which have a minimum sustained wind speed of 75 mph) can erode shorelines (or cause new beaches), create or close inlets, damage coastal vegetation, and affect already endangered wildlife species (See Chapter 4, *Coastal Hazards,* for further details).

Finally, *sea level change* is a major marine factor that must not be overlooked. Over geologic time, fluctuations in sea level have exposed a great range of coastal landforms to wave and tidal processes. Relative sea level change results from either *tectonic forces* that lower or raise coastlines, or from *glacio-eustatic processes* that alter the amount of water locked up in ice. Throughout most of the recent geologic epoch (the last 10,000 years, known as the Holocene), the sea level has been rising. Today, most scientists believe that global temperatures are increasing, which will further melt glacial ice, thereby further raising sea levels worldwide. Although it is generally believed that the rise in sea level will be less than 2 m (6.5 ft), a rise of only 0.3 m (1 ft) would cause shorelines worldwide to move inland an average of 30 m (100 ft) (Christopherson 1997). Planning for such a change to coastlines will be a political and economic nightmare, especially when scientists continue to debate about the degree of sea level change that might occur.

Biological factors. In addition to terrestrial and marine factors, there are biological factors that affect coasts. They are commonly divided into two types: plant and animal. The impact of humans on the coast will be summarized at a later point in this chapter.

Vegetation affects coasts by either enhancing the depositional process or as one of many factors in the erosional process of a coastal landform. Marram grass (*Ammophilia arenaria*) and other salt-tolerant species that are the first colonizers of a sandy beach or sand dune help in the build-up of these landforms; their horizontal and vertical shoots or rhizomes act as a net that helps hold the landform in place. Marsh samphire (*Salicornia spp.*) and other salt-marsh plants perform a similar function in promoting deposition and stabilization in mudflat environments in the middle latitudes, as do mangrove trees in the hot and wet tropical regions. Vegetation can also perform the opposite function by enhancing erosion (e.g., blue-green algae that erode rock by secreting oxalic acid which dissolves calcium carbonate) or by "pulling down" coastal cliffs (e.g., as does the heavy hottentot fig on a bluff).

Animals are also an important factor affecting coasts. Like vegetation, animals can either be a constructive or erosional force. Coral polyps are one of the most constructive marine organisms on the planet. By taking up

FIGURE 1-10 Equilibrium trap at west jetty of the Small Craft Harbor Santa Cruz, CA. In the right corner, two police officers are dislodging the body of a lost fishermen that also drifted into the equilibrium trap. In the center of the inlet is the harbor district's main dredge. (*Source:* Photo by author)

calcium carbonate from the sea water and secreting it into various hard skeletal structures, coral polyps have built impressive coral reefs on coasts around the world. For example, the Great Barrier Reef on Australia's eastern coast extends for approximately 1,750 km (1087 mi). Coral reefs are generally classified as either a **fringing reef** (coral rock platforms that adjoin the coast; e.g., Hawaii Island, Hawaii); a **barrier reef** (coral rock platforms that lie offshore, generally surrounding a lagoon; e.g., Chuuk [formerly Truk] or Belau [formerly Palau], Micronesia); or an **atoll** (circular or ring-shaped coral rock platforms that encircle a lagoon, e.g., Bikini atoll, Micronesia). These reefs not only represent organic deposition but their presence also helps to protect the shore against storm waves and strong currents.

Marine organisms can also be an erosional force affecting coasts. For example, several types of mollusks, such as the piddock (*Pholas dactylus*) and the wrinkled rock borer (*Hiatella arctica*), bore into coastal rock for shelter. Riddled with bore holes, the coastal rocks become more susceptible to erosion. Other intertidal and subtidal mollusks browse on algae that inhabit rock surfaces. In the process of browsing, the marine animals accidentally remove rock in the process (Hansom 1988).

COASTAL FORMATIONS

The U.S. coast includes a varied terrain that includes mountain ranges, rivers and streams, marine terraces, estuaries and bays, bluffs and headlands, rocky intertidal shores, coastal sand dunes, sandy beaches, nearshore waters, islands, and offshore rocks. What follows is a brief introduction to these terrain types. Entire books have been written about each formation type, and where appropriate, references for more detailed information are cited. (See Figure 1-11).

Coastal Mountains and Watersheds

Coastal mountains. Coastal ranges have many effects: They can serve to separate one geographic region from another. For example, California's Coast Ranges separate the state's coastal areas from the Great Central Valley and the deserts of the interior. Second, coastal ranges can separate one climatic region from another. These same coastal barriers dramatically affect California's climate: Pacific storms bring rain to the western or windward slopes of California's Coast Ranges, while the leeward

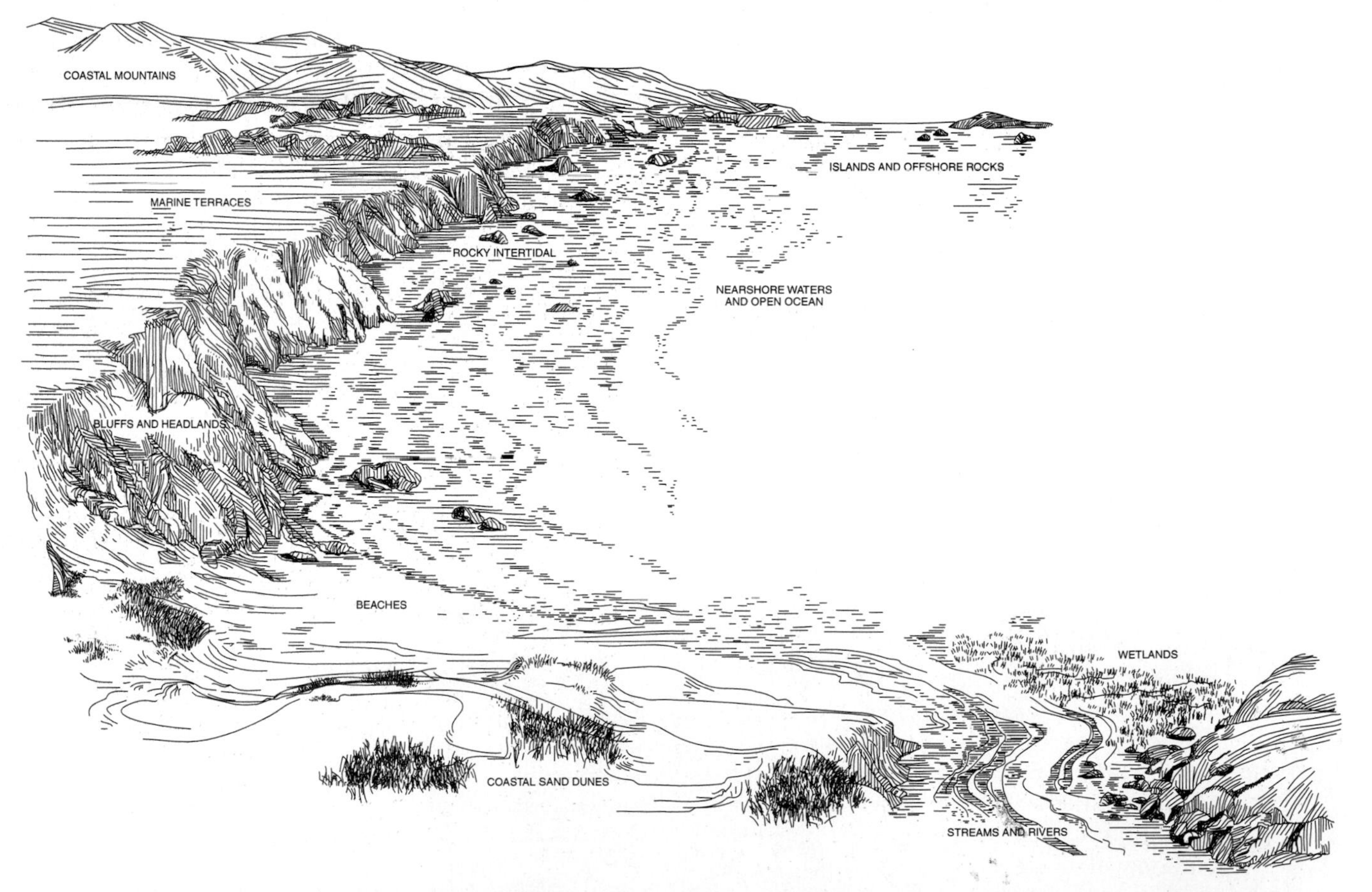

FIGURE 1-11 Diagram of major coastal habitats. (*Source:* Adapted from California Coastal Commission 1987, 13–14)

slopes remain relatively dry. Third, mountains mean dramatic variations in elevation and climatic zones, which in turn, result in a diversity of plant life. This explains why the windward slopes of California's Northern Coast Ranges are cloaked with redwood and Douglas fir trees. The heavy winter rains, summer fog, and moderate temperatures in this area have produced redwood groves where 2,000-year-old trees tower more than 91 m (300 ft). By contrast, the drier regions of the Southern Coast Ranges are vegetated with oaks, pines, and drought-resistant chaparral species (e.g., manzanita, chamise, and sage). Fourth, coastal mountains provide an aesthetic wonderland. It would be hard to imagine California's famous Big Sur Coast on the shores of the Monterey Bay National Marine Sanctuary without the back drop of the Santa Lucia Mountains, nor Highway 1 in San Mateo County, California, without the problems of Devil's Slide (See Chapter 4, *Coastal Hazards,* for further details).

Watersheds, rivers, and streams. Coastal rivers, streams, and creeks flow through the canyons and valleys of coastal mountains until they reach the sea. These drainage ways transport oxygen, nutrients, and sediments through the watershed, giving life to the area. Riparian woodlands and associated wildlife develop along stream banks and floodplains, while wetlands and estuaries form where these drainage ways meet the sea. Without coastal waterways, anadromous fish such as steelhead and salmon could not migrate to well-oxygenated fresh water streams and gravel streambeds that they need for spawning.

Coastal rivers play other crucial roles, such as depositing the sediments that form broad floodplains. These floodplains contain rich, deep soils for agriculture (e.g., the Salinas River floodplain in CA), and have often been the sites of urban development (e.g., the Los Angeles Basin which was formed by the deposition of sediment from the Los Angeles, San Gabriel, and Santa Ana rivers). Coastal rivers also replenish sand lost from beaches, unless of course, these rivers have undergone dam construction, channelization, or some other form of water diversion for agricultural irrigation or the increased water demands of a growing urban population.

Marine Terraces, Coastal Bluffs, and Rocky Headlands

One of the most distinctive differences between the Atlantic and Gulf coasts versus the Pacific coast of North America, is that the former have gently sloping seashores—the result of gradual *submergence* of the continent's edge, whereas the Pacific coast has elevated marine terraces, coastal bluffs, and rocky headlands—the result of *emergence,* specifically abrupt faulting and uplift in combination with the erosive power of waves, winter rainstorms, and wind.

Between the coastal mountains and the sea cliffs on America's western shores, one may find marine terraces. Marine terraces are seaward-sloping wave-cut benches or platforms that have been uplifted above sea level. (See Figure 1-12).

It is possible to have several terrace levels at one coastal site where geologic processes have been at work for millions of years. Along the California coast, for example, the oldest and highest terraces were created by the same processes that forced up the Coast Ranges some 1–2 million years ago. A terrace 396 m (1,300 ft) above sea level can be found along the sides of the Palos Verdes Hills in Los Angeles County, and twenty stepped terraces have developed along the coast of San Clemente Island, California. Along the shores of the Monterey Bay National Marine Sanctuary, marine terraces are visible just north of the city of Santa Cruz. Terrace soils are composed of rock debris, shells, and other marine fossil fragments that were deposited on the once-submerged terrace. Once emerged from the sea, coastal streams and rivers cross the terraces depositing a thick layer of nutrients, sand, and gravel, making them excellent areas for coastal agriculture.

The seaward edge of a marine terrace is known as a **coastal bluff.** Coastal bluffs are steep coastal slopes shaped by the erosive power of waves at its base, and uplifted from the ocean floor. They are often prone to erosion, since many are composed of highly erodable material (e.g., sandstones, shale, siltstones, and mudstones). Moving up the California coast, examples of sedimentary coastal bluffs are the shale cliffs of Point Loma in San Diego County, the alluvial cliffs of La Jolla, and the sandstone bluffs of Santa Cruz. Some coastal bluffs are eroded to the point where arches have been created, and have later fallen due to further erosion (e.g., Natural Bridges State Beach, Santa Cruz, California). However, not all coastal bluffs are made up of highly erodable material, such as the highly resistant granite bluffs of the Monterey Peninsula, California.

Development on coastal bluffs has increased the rate of coastal erosion. Everything from drain pipes and septic tanks that saturate soils with runoff, to lawns and gardens that are over-irrigated, to road construction too close to the water's edge have added to nature's own rate of eroding coastal bluffs.

Rocky headlands occur on high-energy coasts where mountains meet the sea. They are composed of igneous rocks (e.g., granites and basalts) that are resistant to wave erosion. On the highly active tectonic California coast, an example of a *granitic headland* can be found at Point Reyes headland in Marin County, California. Along California's shores, one can also find examples of *basaltic lava headlands,* such as at Morro Rock in San Luis Obispo County, and Point Sur in Monterey County.

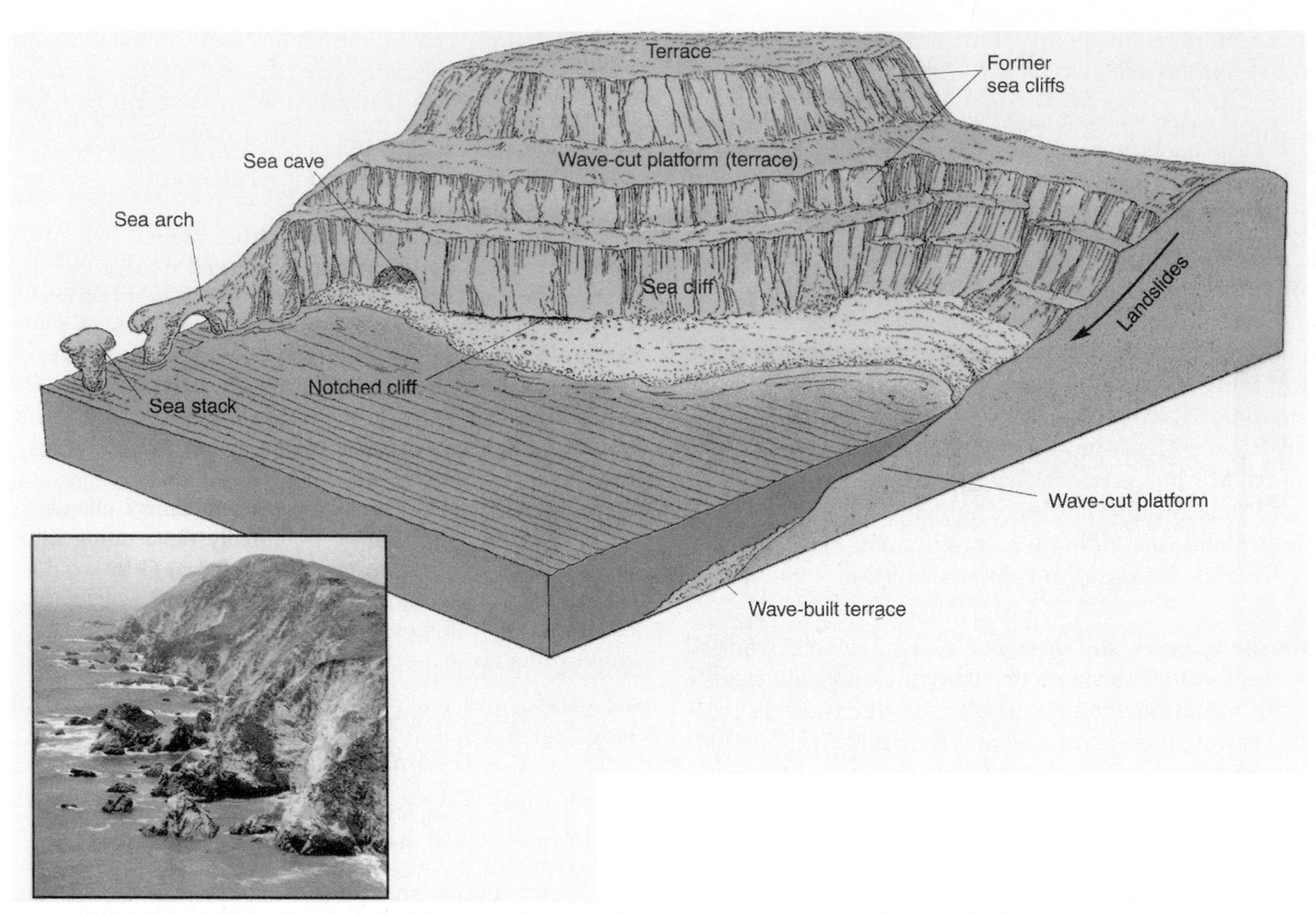

FIGURE 1-12 Erosional coastal features. Characteristic coastal erosional landforms: platforms, terraces, caves, arches, and stacks. (*Source: Geosystems* 3/e, by Christopherson 1997, 494) The photo shows the California coast, a typical erosional coastline. (*Source:* photo by Robert Christopherson, in *Geosystems* 3/e, by Christopherson 1997, 494. Reprinted by permission of Prentice Hall, Inc.)

Coastal Sand Dunes

Sand dunes are among the more interesting, dynamic, and fragile coastal ecosystems. They are classified as aeolian contour bedforms which can be defined as mounds, ridges, or hills of loose sand that have been heaped up by the wind (Griggs and Savoy 1985). Dunes can be formed from both beach sand blowing inland and land sand blowing seaward. Dunes shift over time until a veil of pioneer plants stabilizes the drifting sand. Even after that occurs, however, the dunes still change form under the stress of wind, storm waves, and the traffic of human activity.

Distribution, formation, patterns. Coastal sand dunes occur worldwide, though they are most common in temperate climates, especially along those coasts that have strong onshore winds and a plentiful supply of sand-sized sediment (Carter 1988). Sediment deposited at the mouths of rivers and offshore sandbars provides an important source of material for dune building. The littoral current carries the sediment along the shore until the particles are deposited on a beach by wave action. The dry sand particles on the beach are then blown landward from the beach until driftwood and plants interrupt the wind flow. The drift gradually builds up to a sizable mound. Coastal dune fields may take several shapes, such as *transverse ridges* (parallel ridges perpendicular to the prevailing winds; e.g., Northern California coast), or *parabolic or conical ridges* (U-shaped dunes with the concave side facing the prevailing wind; e.g., Pismo Beach in Central California). The most recently formed dune near the beach is known as the *primary* or *foredune,* whereas the older, more vegetated and thus stabilized dune is called the *backdune or stable dune.* Some backdunes may be as much as 18,000 years old.

Function of sand dunes. Coastal sand dunes have several functions. (See Figure 1-13). First and foremost, they buffer the shore against extreme winds and waves. Second, they replenish beaches and nearshore areas that have had their sand supply depleted during and after heavy storms. The sands eroded from dunes and beaches during winter

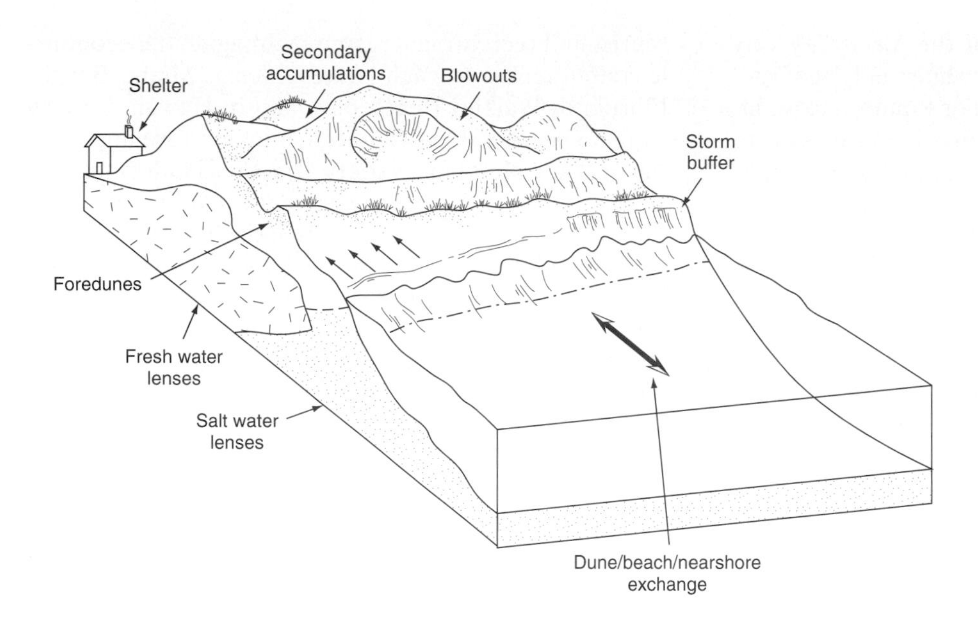

FIGURE 1-13 The coastal sand dune system. (*Source:* Diagram by Heather Theurer.)

storms often form a sandbar offshore. During the calm summer season, the sand is gradually returned to the beach. Third, coastal dunes shelter inland residences and settlements. Fourth, they help in keeping salt-water intrusion from contaminating fresh water lenses (underground freshwater supplies). Coastal dunes are habitats for wildlife (e.g., gray foxes, striped skunks, and mule deer). Finally, they are recreation areas for humans.

Human impact. According to Carter (1988), sand dunes have suffered the greatest degree of human pressure of all the coastal ecosystems. Humans impact dunes by building dams on rivers that trap river sediment, thereby depleting the sand supply to the dune field. Humans also construct groins, jetties, and seawalls that disrupt the natural sediment flow in the littoral drift as well as the recycling of sand from sandbar to beach. Foot traffic, horses, and off-road vehicles damage the dune plants that act as a protective veil; the result often becomes wind erosion and **blowouts**—wind hollows or basins within dunes. Development on our nation's coastal dunes has brought severe damage, particularly to the vulnerable foredune where development should never occur.

Beaches

Of all the coastal formations associated with a depositional coastline, the most familiar is the beach. Most people think of the beach as merely that sandy area on the shore where they sunbathe, picnic, or play in the surf. Technically, however, a **beach** is that area of the shore, on average, from 5 m (16 ft) above high tide to 10 m (33 ft) below low tide (Christopherson 1997), and is composed of sand or pebbles (Refer back to Fig. 1-5). [It should be noted, however, that *coastal scientists often disagree as to the exact definition of the beach area, as well as other subdivisions of the coast.* Consequently, Schwartz (1982, p. 140) maintains, "the best working definition of a beach is the accumulation of unconsolidated sediment that is limited by low tide on the seaward margin and by the limit of storm wave action on the landward side."]

Regardless, more of "the beach" can be underwater than visible to the person strolling along its shore. Seasonal cycles determine the appearance of the visible beach. In summer, beaches are generally wide and gently sloping; they become steep-fronted and narrow from winter storms. With violent storm waves, beaches may be stripped of all their sand, exposing larger rock particles or rock pavement. The sand that is removed from beaches in the winter is deposited in offshore sandbars. During mild summer months, gentle swells push the offshore sediment back to the exposed shore.

Source and type of sediment. The beach is an area where sediment is in motion. Constructive action is from the **swash** of the wave, while destructive action occurs from its **backwash.** Sediment is deposited by breaking waves, often with the aid of littoral drift. The sand, pebbles, and fragmented shell originate from several sources, including sediment from rivers entering the sea, erosion of cliffs, and debris recycled from other beaches. In some California sites, river sediments are the source of 80 to 90 percent of beach sand (California Coastal Commission 1987).

The color of the sand gives a clue to its mineral content, as well as to the origin of the eroded sediments that make up the sand supply. Some of the world's beaches are stark white, dominated by sands of quartz and feldspar minerals, such as at Carmel Beach, California, which is located on the shores of the Monterey Bay National Marine Sanctuary. Other "white sandy beaches" are actually more amber-colored. For example, close inspection of the beach at Sand City just a few miles north of Carmel Beach reveals the presence of iron minerals; pale quartz grains; flecks of black mica; and green, pink, or white feldspar. In volcanic areas, such as Hawaii and Iceland, wave-processed lava creates black-sand beaches. On some shores, beaches are almost nonexistent, such as on the coast of Maine, portions of the Atlantic provinces of Canada, and the southern coast of Monterey Bay, California. Here, the scenically rugged shores are dominated by resistant granite rock, rather than long and wide sandy beaches.

Function. A beach has several functions, but the primary one is to stabilize a shoreline by absorbing wave energy. It can do this because of two principal features: First, the *sloping surface* of a beach gradually dissipates the energy of a wave as the wave makes its way up shore. Second, the fact that a beach is *made of sand* allows it to be flexible, changing its slope and contour as the waves change.

Types of sandy beaches. There are three major types of sandy beaches: mainland, pocket, and barrier. **Mainland beaches** stretch unbroken for several miles along the edges of a "mainland," or major land mass. They are generally low standing, prone to flooding, and are often backed by steep headlands. Sediment arrives from nearby rivers and eroding bluffs.

Examples of mainland beaches can be found in southern California, northern New Jersey, and the coasts of the Great Lakes. **Pocket beaches** are found in "pockets"—small bays or alcoves protected by surrounding rocky cliffs or headlands. The coasts of New England and the Pacific Northwest have numerous pocket beaches. By contrast, one would look along the east and southeast coast of the U.S., the Gulf of Mexico, and much of Alaska to find examples of barrier beaches. **Barrier beaches** are part of a complex integrated system of beaches, dunes, bays, marshes, tidal flats, and inlets. **Barrier islands** and associated beaches will be discussed in greater detail below.

Ecological habitat and human impact. One often forgets that a beach is an ecological habitat for numerous organisms, as well as a place for humans to sunbathe, swim, surf, picnic, and fish. For example, beaches are inhabited by numerous invertebrates and insects: Crustaceans, bivalve mollusks, and tube-building worms exist in the surf zone; sand crabs scavenge in the sun-dried kelp inland from the surf zone; beetles and pesty kelp flies roam the beach foreshore; and air-breathing pill bugs and beach hoppers inhabit the dry upper beach.

Beaches, especially the sandy ones, are of considerable importance to humans. They are obviously a major tourist and recreational resource. Imagine the economic consequences to such beach cities as Miami Beach, Florida or Waikiki Beach on Oahu in Hawaii, if there were no beaches. Unfortunately, humans have impacted this coastal habitat in a variety of ways, including the discharge of raw domestic sewage; oil pollution from offshore drilling rigs or tanker spills; industrial effluent; and the disruption of sediment flow by sand mining, or the construction of such items as jetties, groins, and breakwaters. All of these negative impacts can result in lost or closed beaches, sediment filled (and thus closed) harbors, falling coastal highways, and condemned beach houses.

Coastal Barriers

Another coastal formation is the coastal barrier. A **coastal barrier** is a natural geomorphic feature that forms offshore roughly parallel to the coast. It acts as a barrier, taking the brunt of storm energy, thereby protecting the mainland. These protective coastal features can be found along 4,345 km (2,700 mi) of our nation's shoreline, from the rocky headlands of Maine to the arid salt flats of south Texas. (See Figure 1-14). Coastal barriers are composed of sand and other sediments that are supplied by longshore currents, tides, and waves.

Types of coastal barriers. There are three major types of coastal barriers: barrier spits, bay barriers, and barrier islands. A **barrier spit** is an elongated depositional sand barrier attached to a headland that partially crosses the mouth of a bay, creating a **lagoon.** An example of this is Siletz Spit on the Oregon coast. (See Figure 1-15).

When a bank of sand or shingle extends all the way across a bay from headland to headland, a **bay barrier** has developed. In this case, the bay is cut off completely from the ocean. The barrier at Edgartown Great Pond on the Martha's Vineyard coast is an excellent example of a barrier. Some bay barriers may have been formed by the convergence of spits, growing in opposite directions from each end of the bay. Bay barriers may also be the result of onshore migration of offshore bars, with the aid of a rising sea level. Regardless, what is characteristic of bay barriers is that they connect headlands and occur in shallow waters. Other terms are also used for bay barriers, such as *bay bars* or *baymouth bars.* If the sandy bar is in the form of an island—not attached to a headland, nor spanning headlands—it is called a **barrier island.** The often used example of a barrier island chain is that which is found along the North Carolina coast, from Virginia Beach in the north to Cape Lookout in the south. (See Figure 1-16). This area includes Kitty Hawk where the Wright Brothers gave flight to their first aircraft, as well as Cape Hatteras—one of three local na-

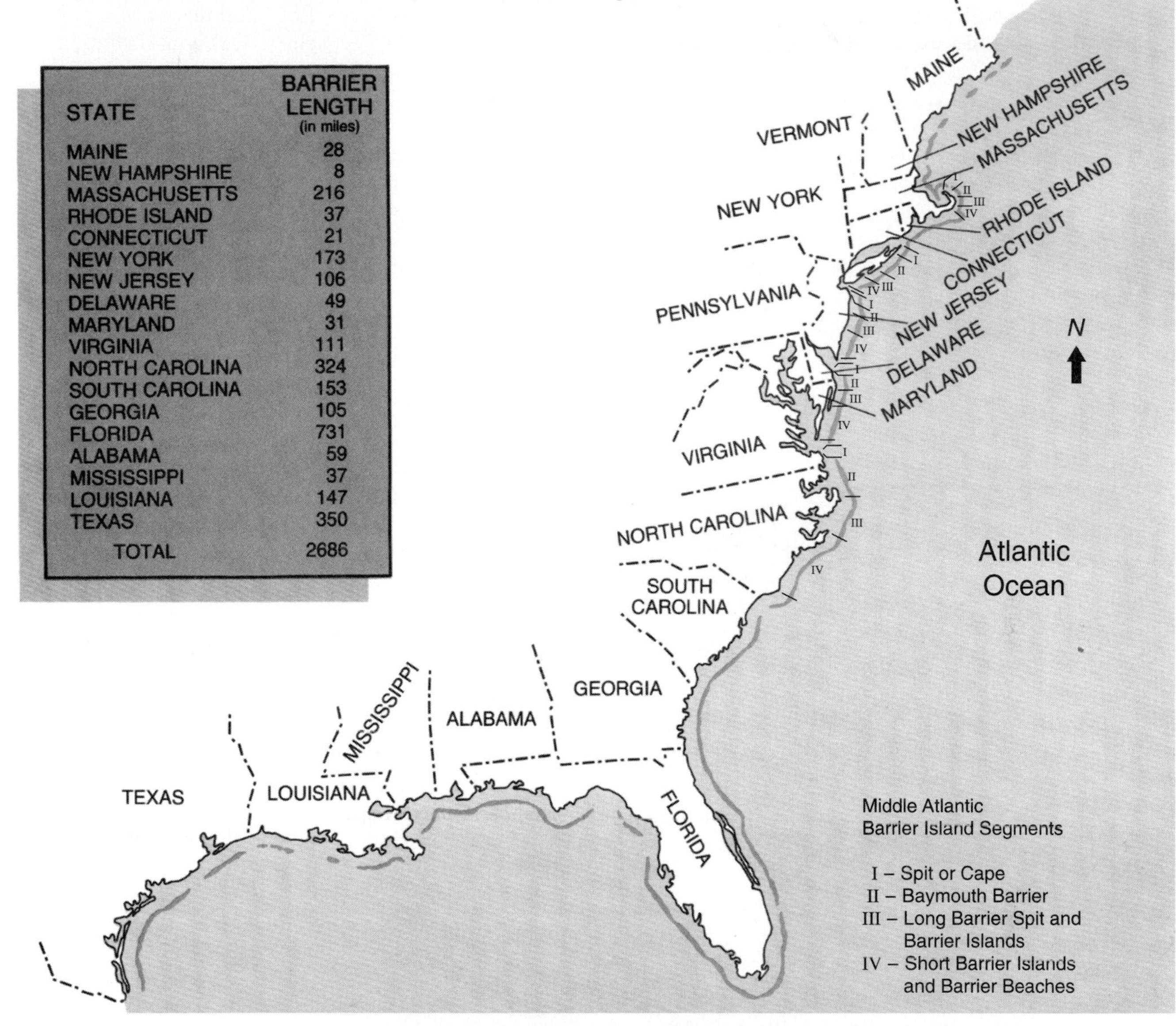

STATE	BARRIER LENGTH (in miles)
MAINE	28
NEW HAMPSHIRE	8
MASSACHUSETTS	216
RHODE ISLAND	37
CONNECTICUT	21
NEW YORK	173
NEW JERSEY	106
DELAWARE	49
MARYLAND	31
VIRGINIA	111
NORTH CAROLINA	324
SOUTH CAROLINA	153
GEORGIA	105
FLORIDA	731
ALABAMA	59
MISSISSIPPI	37
LOUISIANA	147
TEXAS	350
TOTAL	2686

FIGURE 1-14 Coastal barriers along the Atlantic and gulf coasts, illustrating middle Atlantic barrier island segments. (*Source:* Adapted from Wells and Peterson, undated; 4)

tional seashore reserves managed by the National Park Service.

Barrier islands. Barrier islands are the most complex of the various coastal barriers, since they include all of the environments also found in other coastal barriers, and they have been the most used (and abused) by humans. Approximately 12–15 percent of the Earth's outer coastline has barrier islands, ranging from the north slope of Alaska to the tropics of South America and Australia (Davis 1994). In order to form, barrier islands need three key ingredients—*sediment* (an abundant supply of sand); *a transport agent* (wind, waves, and currents to carry the sediment), and *an accumulation site* (a gently sloping continental shelf where the sediment can accumulate). All three ingredients are found on stable, trailing edge coasts. Consequently, the Atlantic coast of the United States—a classic example of a trailing edge coast, is basically one continuous barrier island system. For a discussion of various theories on barrier island formation, see Davis 1994.

Value of coastal barriers. As mentioned earlier, coastal barriers protect the mainland from storm waves and surge. However, this *buffering service* against wave energy provides another value—*habitat formation.* In other words, coastal barriers allow marshes, estuaries, and other ecological habitats to develop. For example, a well

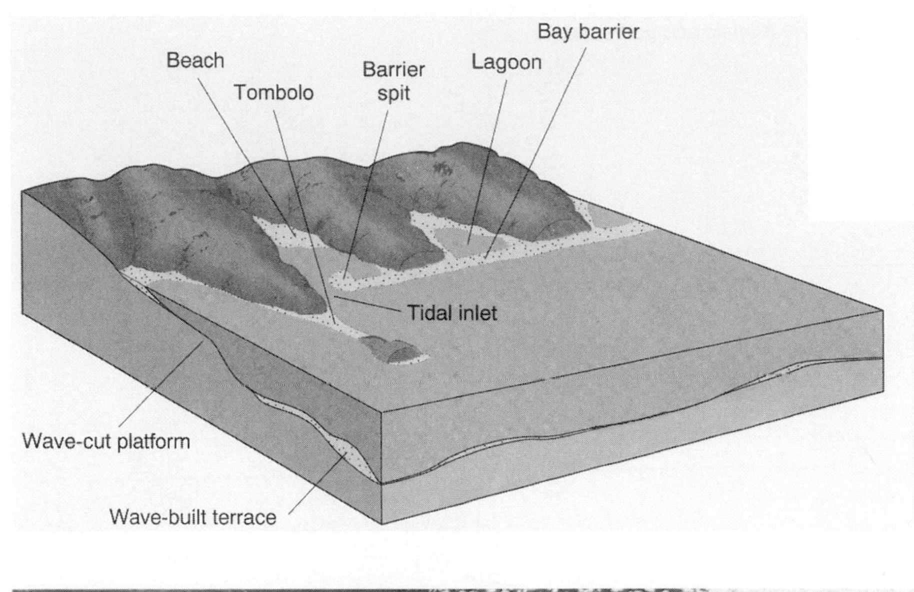

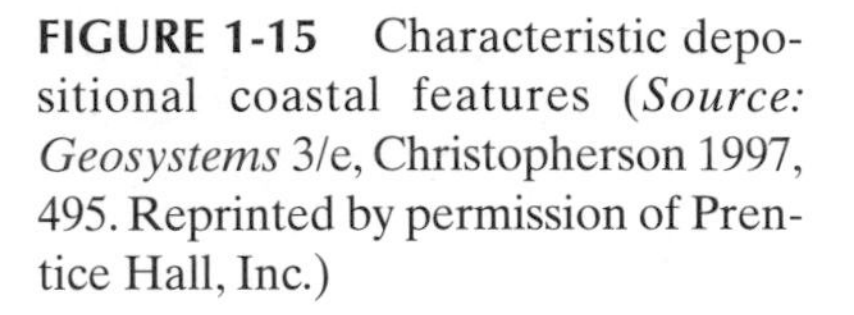

FIGURE 1-15 Characteristic depositional coastal features (*Source: Geosystems* 3/e, Christopherson 1997, 495. Reprinted by permission of Prentice Hall, Inc.)

FIGURE 1-16 *Landsat* image of barrier island chain along the North Carolina coast. Hurricane Emily swept past Cape Hatteras in August 1993, causing damage and beach erosion. (*Source:* Image from NASA.)

developed barrier island and nearby mainland has six major coastal ecosystems: (1) a *coastal marine ecosystem,* which includes the intertidal and subtidal zone, as well as the beach; (2) a *seaward maritime ecosystem,* which includes the dune field, shrub thicket and maritime forest on the ocean side of the island; (3) a *freshwater ecosystem,* which contains rivers, lakes, and associated marsh or swamps; (4) *a landward maritime ecosystem,* containing a maritime forest; (5) an *estuarine ecosystem,* which contains marsh or mangroves on its seaward side, and seagrass beds, sand and mud flats, and oyster reefs in the center and landward side; and (6) a *mainland ecosystem,* which is the upland habitat adjacent to barrier islands. (See Figure 1-17).

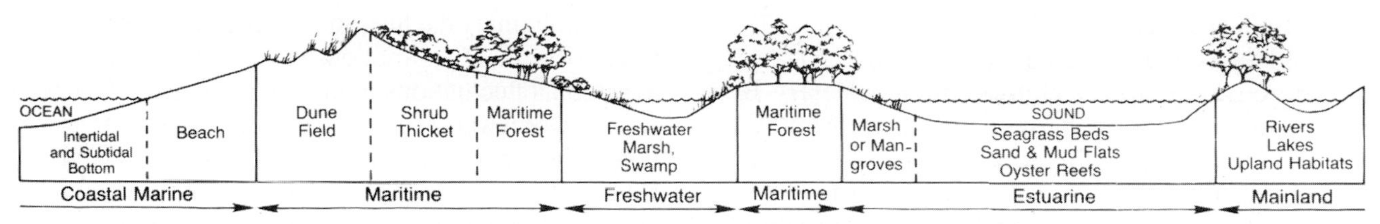

FIGURE 1-17 Cross section of typical barrier island and nearby mainland. (*Source:* Wells and Peterson [undated])

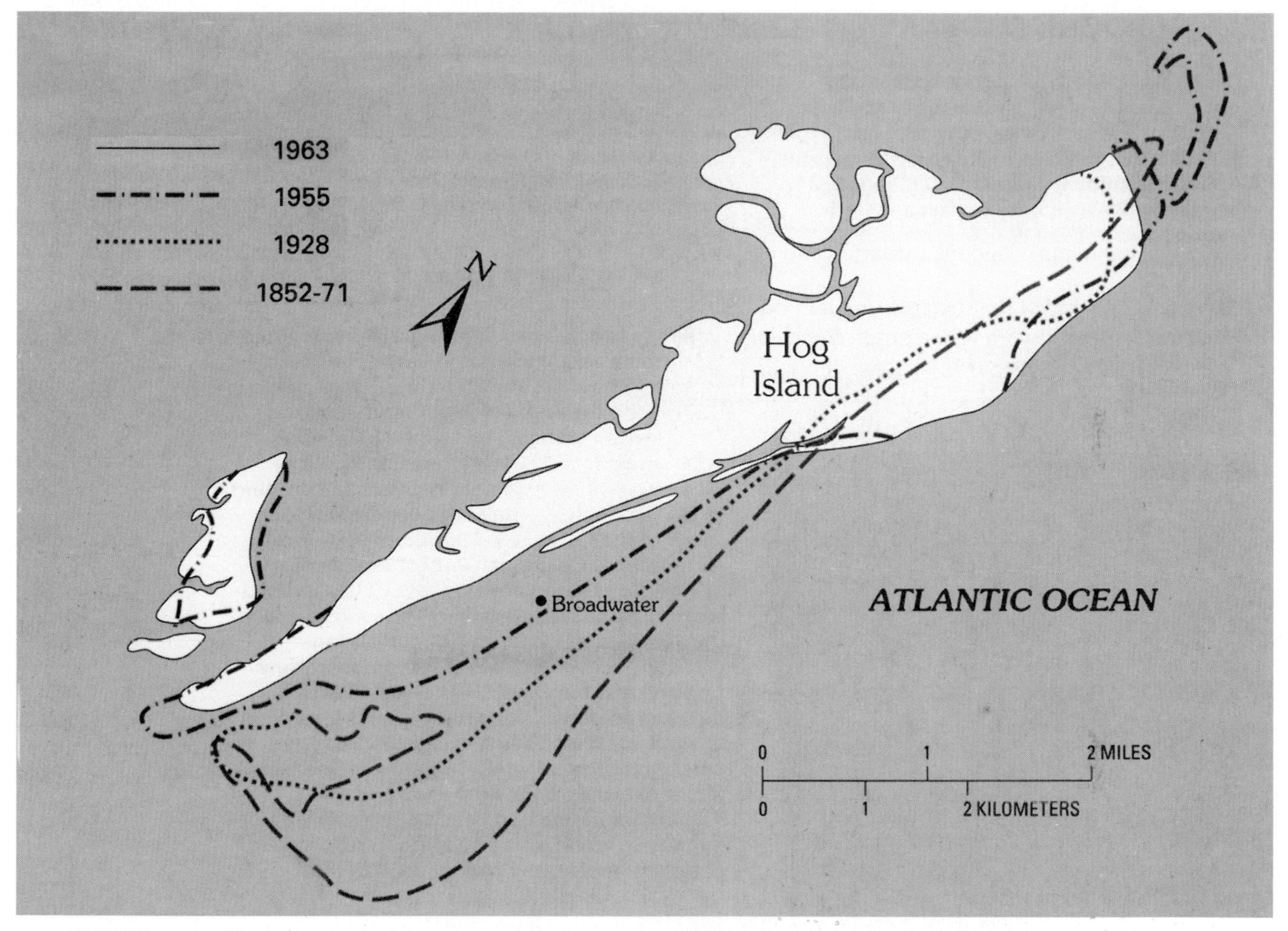

FIGURE 1-18 Hog Island, Virginia—A changing barrier island. (Modified from *Ecosystem Description,* Vol. 1, part B, of Virginia Coast Reserve Study conducted by the Nature Conservancy, [*Source:* Adapted from Williams, et al., 1990])

These habitats are important to shorebirds, water birds, and other waterfowl for their breeding and migration cycles. Many of our nation's endangered and threatened species, such as the loggerhead sea turtle and the whooping crane, make use of these areas. Coastal barriers are also valuable to humans for recreation purposes (e.g., nature walks, birdwatching, photography) as well as the economic benefits of tourism to local coastal communities.

Human impacts on coastal barriers. Coastal barriers are constantly migrating because they are exposed to severe storms. Consequently, these ever changing "ribbons of sand" are an unwise choice for homesites or commercial building. Yet, humans continue to build on these unstable structures. Many of the coastal barriers along Florida's coast have been so overdeveloped that they are a mere sea of roads, hotels, condominiums, and shopping centers. Someday, in fact, they may be literally "returned to the sea." Even the slightest development on a barrier island can provide a dangerous scenario. A case in point is the barrier island called Hog Island off the coast of Virginia. (See Figure 1-18).

In the 1800s, hunting and fishing clubs were established in the pine forests on Virginia's barrier islands.

One of the largest clubs was at Broadwater near the middle of Hog Island—an island of 300 people, 50 houses, a lighthouse, a school, a church, and a cemetery. By 1930, erosion brought down the lighthouse, and a hurricane in 1933 destroyed the protective pine forest and devastated the town. By 1940, most of the inhabitants gave up and left the island. Today, the former site of Broadwater is hundreds of meters offshore, under several meters of water.

Coastal Inlets—Lagoons, Deltas, and Estuaries

Coastal inlets are another form of coastal formation. A **coastal inlet** can be defined as a small indentation in a coastline, and is usually a relatively narrow channel or pocket of water that leads inland. Davies (1977) identifies three broad classes of coastal inlets—lagoons, estuaries, and deltas.

Lagoons. At one end of the spectrum are **lagoons**—long, shallow bodies of water with very restricted exchange with the sea, minimal tidal flux, and no significant freshwater inflow. Consequently, variations in salinity are due primarily to seasonal fluctuations in precipitation and evaporation. Lagoons form wherever sandbars or barriers separate a section of the sea from the mainland. In terms of a habitat for wildlife, lagoons typify that of a hypersaline environment—few species but large populations (Davis 1994).

Deltas. At the other end of the spectrum are **deltas**—land masses formed from alluvial deposits of sand, silt, mud, and other particles at the mouth of a river. Delta shapes are most often explained and classified according to the relative influence of the three major factors affecting their development: the river, waves, and tides. In 1975, William Galloway—a geology professor at the University of Texas at Austin—originated the triangular classification scheme for river deltas. (See Figure 1-19).

River-dominated deltas, such as the Mississippi River delta on the Gulf of Mexico, characteristically have (a) a good freshwater and sediment supply from the drainage

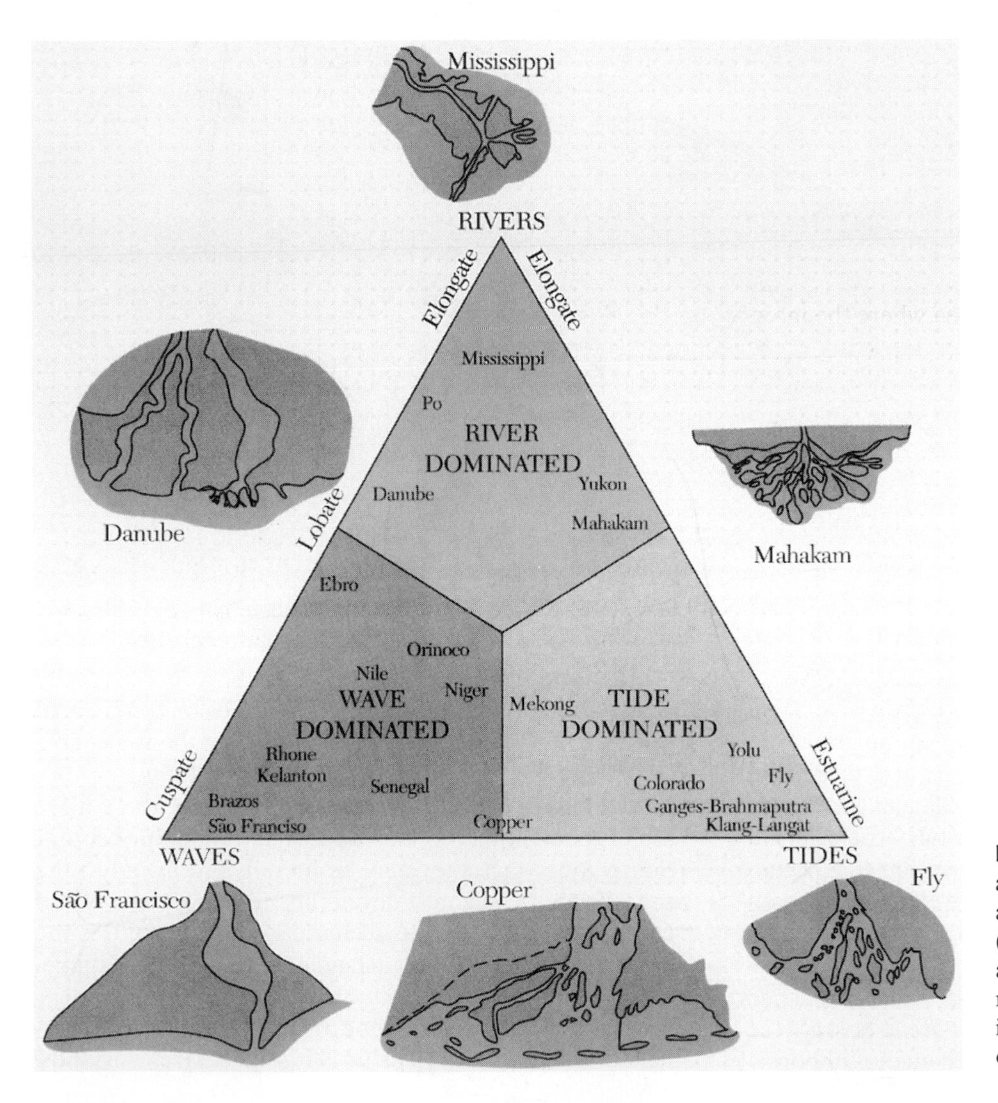

FIGURE 1-19 River delta systems according to William Galloway's triangular diagram classification system. (*Source:* Adapted from L. D. Wright and J. M. Coleman, AAPG ©1973, reprinted by permission of the American Association of Petroleum Geologist)

basin; (b) a well-developed (elongated) delta plain; (c) several tributaries projecting seaward, giving them a "bird's-foot" configuration; and (d) a placid receiving basin—one that has small waves, a small tidal range, and a tectonically stable coast. Although the Mississippi delta is in the path of hurricanes and storm surges that cause accelerated erosion, this activity does not impact overall delta configuration (Davis 1994).

Tide-dominated deltas, such as the Fly River in Papua, New Guinea, generally have (a) strong tidal currents that overpower the freshwater discharge; (b) weak longshore currents; (c) more sediments carried inland by strong tidal currents than river sediment discharge; and (d) salt marshes that rim the intertidal flats. Because of their tidal flats and salt marshes, tide-dominated deltas have a tendency to resemble estuaries.

Wave-dominated deltas, such as the São Francisco River delta on the southern coast of Brazil, characteristically have (a) only one river channel, or just a few distributaries; (b) river and sediment flow is only moderate compared to the stronger distributing power of the waves; (c) weak tides; and (d) well-developed beaches and dunes on the outer delta. Wave-dominated deltas are generally smaller than the other two types, because the waves striking the delta front are more powerful than the river's ability to replace the area with sediment.

Estuaries. Estuaries are a third type of coastal inlet. Whereas lagoons are quiet saline bays with no regular freshwater influx, an **estuary** is the area where the mouth of a river enters the sea and freshwater and seawater intermix. Since it is a place where tides ebb and flow, it is often more turbulent than a lagoon. Another way of thinking about estuaries is that they are "arms of the ocean" that have been thrust into the mouth and lower course of rivers as far as the tide will take them. All estuaries have three parts: (1) the *head* (the inland areas where the river enters); (2) the *main estuary* (the middle or fully estuarine area); and (3) the *mouth* (the seaward end or indentation of the coastline). (See Figure 1-20).

Estuaries are generally classified according to geologic origin and geomorphology, or according to characteristic circulation patterns. When classified according to origin, there are four types: coastal plains estuaries; fjord estuaries; bar-built estuaries; and tectonic estuaries. *Coastal plains estuaries* (also commonly known as "drowned river valley estuaries") are the result of the drowning of a river valley. Many of the estuaries of the eastern United States are of this type, including the Chesapeake and Delaware bays, and the estuaries of the Mississippi, Hudson, and Savannah rivers. *Fjord estuaries* are the result of the drowning of a valley gouged out by glaciers. These estuaries are generally deep, narrow, and u-shaped in cross section. The largest fjord estuaries can be found in Greenland, British Columbia, Scandinavia, and Chile. Estuaries may also result from the development of an offshore barrier, such as a line of barrier islands, a reef formation, a beach strand, or a line of marine debris. Examples of these bar-built estuaries include: Laguna Madre, along the Texas coast; Barataria Bay and Mississippi Sound, along the Gulf of Mexico; and Pamlico and Albemarle sounds in North Carolina. Tectonic processes may also result in the origin of *tectonic estuaries,* such as San Francisco Bay, that resulted from slippage and land subsidence along fault lines.

The above classification of estuaries by geologic origin is useful, especially to physical geographers and geologists, but physical oceanographers choose to classify estuaries according to circulation patterns. According to this classification system, there are three types of estuaries: salt-wedge estuaries; partially mixed estuaries; and fully mixed estuaries. (See Figure 1-21).

Salt-wedge estuaries are referred by some as "stratified estuaries" since the saltwater and freshwater masses are almost completely separate, with no significant amount of mixing occurring. These estuaries are river dominated, with a weak marine inflow due to a small tidal range. The Mississippi River estuary is a good example of

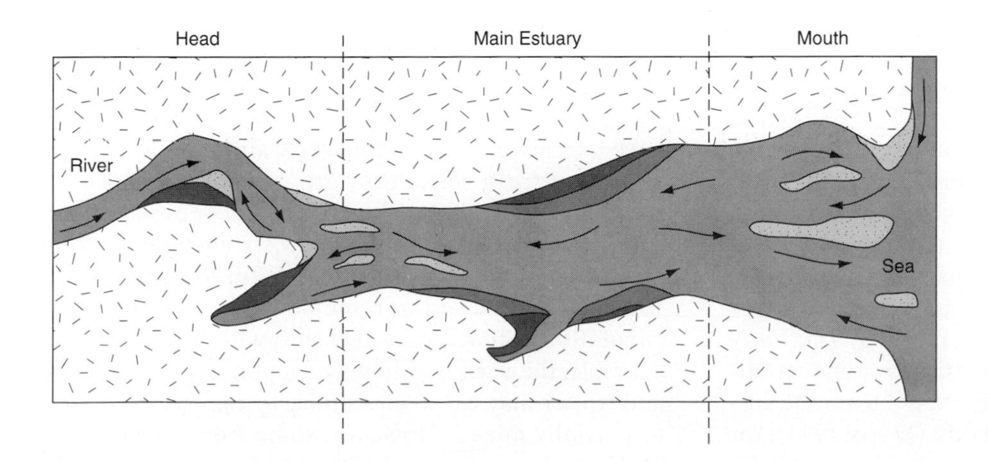

FIGURE 1-20 Three major parts of a typical estuary—Head, Main Estuary, Mouth. (*Source:* Diagram by Heather Theurer.)

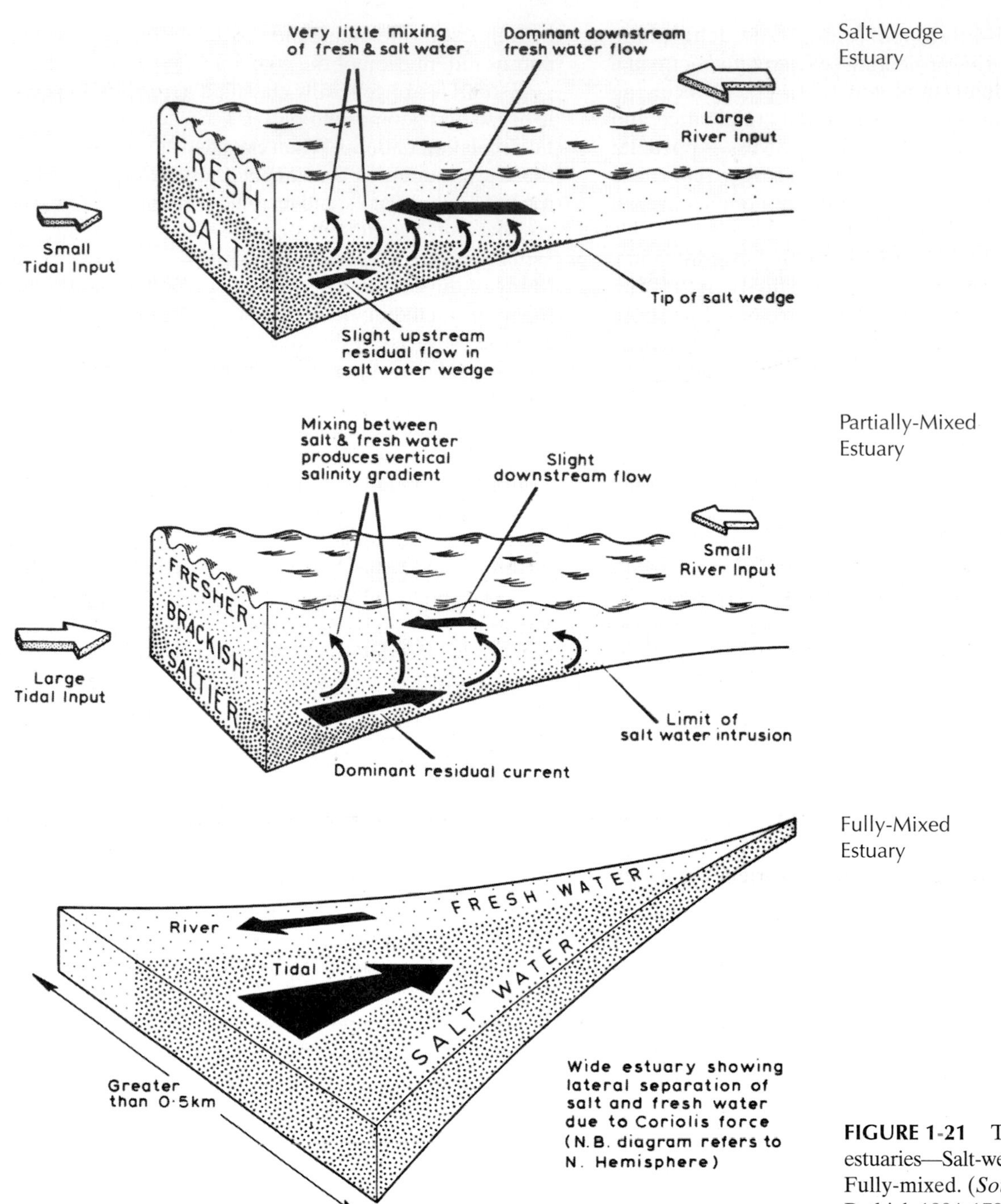

Salt-Wedge Estuary

Partially-Mixed Estuary

Fully-Mixed Estuary

FIGURE 1-21 Three major types of estuaries—Salt-wedge; Partially-mixed; Fully-mixed. (*Source:* Adapted from Pethick 1984, 179–181)

a salt-wedge estuary. On the other hand, the Chesapeake Bay is a good example of a *partially mixed estuary.* In this type of estuary, tidal current, not a river, is the dominant factor in circulation. Consequently, a mixing of salt water and fresh water occurs. Salinities in partially mixed estuaries may range from 0 for fresh water to 35 parts per thousand for undiluted seawater (Davis 1994). *Fully mixed estuaries* may have either vertically homogenous water profiles [i.e., the salinity is the same from the surface to the bottom of the water body (Davis 1994)], or have lateral flow and salinity separation [i.e., the net salt water and fresh water flow are moving in opposite directions (Pethick 1984)]. In any case, these estuaries characteristically have weak river flow, strong tidal flows, and are generally wider than about 0.5 km (0.3 mi) (Pethick 1984). It is important to remember, however, that these circulation patterns may also shift with the seasons. For example, the variation in wave action between summer and winter may cause an estuary to vary from stratified to partially mixed. However, some funnel-shaped resonating estuaries like the Bay of Fundy in Nova Scotia, Canada, have such strong tidal currents that they remain fully mixed throughout the year (Davis 1994).

Coastal Wetlands

Coastal wetlands are another type of coastal phenomenon. Coastal wetlands can be simply defined as wet vegetated areas along the coast. Wetlands have three components: (1) presence of water at the surface or within the root zone; (2) soil conditions that are different from adjacent uplands; and (3) vegetation adapted to wet conditions (*hydrophytes*) (Mitsch and Gosselink 1993). These components are all a part of the major habitats associated with estuaries: salt marshes, freshwater marshes, and mangrove swamps.

Salt marshes. A salt marsh is a "natural or semi-natural halophytic grassland and dwarf brushwood on the alluvial sediments bordering saline water bodies whose water level fluctuates either tidally or non-tidally" (Beeftink 1977). Salt marshes predominate in mid to high latitudes along intertidal shores throughout the world. In the lower tropical and subtropical regions (between 25 degrees N and 25 degrees S), they are replaced by mangrove swamps along coastlines. A cross-section of a salt marsh shows the subenvironments and distribution of different marsh grasses. (See Figure 1-22). Generally, cordgrass (*Spartina*) can be found closest to the tidal creek, while the higher marsh is covered in various varieties of needle rush (*Juncus*).

Freshwater marshes. Freshwater coastal marshes combine many of the features of saline coastal marshes and freshwater inland marshes, yet they remain unique ecosystems. Freshwater marshes reflect greater biotic diversity, as a result of the reduction of salt stress found in salt marshes. Because plant diversity is higher, more birds use this type of marsh than any other marsh type. Freshwater marshes require adequate river flow or rainfall to maintain fresh conditions, a relatively flat gradient from the shoreline inland, and a substantial tidal range. This set of conditions occurs along the middle and south Atlantic shores of the United States. There is not always a clear-cut distinction between tidal and inland freshwater marshes since on the coast they form a continuum. (See Figure 1-23). It is safe to say, however, that tidal freshwater marshes experience tides but are above the salt boundary, whereas inland freshwater marshes experience neither salt nor tides. Since urban centers are generally located further inland from the more saline portions of the estuary, these freshwater coastal marshes are usually more susceptible to human impact than coastal salt marshes.

Mangrove swamps. In tropical and subtropical estuaries and bays throughout the world, the mangrove wetland replaces the salt marsh as the dominant coastal ecosystem. **Mangroves,** commonly called swamps but actually *mangles,* are thick tangles of woody shrub and tree roots of various taxonomic groups. (See Figure 1-24). All the dominant plant species are known for their ability to adapt to the saline wetland environment (e.g., salt exclusion, salt excretion, pneumatophores, and the production of viviparous seedlings). Mangroves are most noted for their spectacular display of prop (above ground) roots, especially the red mangrove (*Rhizophora mangle*) and the black mangrove (*Avicennia germinans*). These massive root systems provide physical stability to shorelines by creating dense sediment-stabilizing mazes. Scientists have also established that mangroves serve as sinks for nutrients and carbon, protect inland areas from severe damage during tidal waves and hurricanes, and export organic matter to adjacent coastal food chains.

These "tropical buffer zones" are limited geographically to the frost free zone of 25 degrees N to 25 degrees S latitude. (See Figure 1-25). In the United States, mangrove wetlands are limited to the southern extremes of Florida and to Puerto Rico. In Florida, the best development of mangroves is along the southwest coast, where the Everglades and the Big Cypress Swamp drain to the sea. This is the area of Florida's Ten Thousand Islands—one of the largest mangrove swamps in the world. The predominant species of mangrove here is the white mangrove (*Laguncularia racemosa*). Development pressures, however, have eliminated a significant portion of these original mangroves. For example, Patterson (1986) estimates that 24 percent of the mangroves were removed from Marco Island—one of the most developed islands in the region. Today, Florida's mangroves are protected against such devastation.

Importance of estuaries and coastal wetlands. In the past, estuaries and their associated wetlands (saltwater marshes, freshwater marshes, mangroves) were considered swampy wastelands. Today, because of scientific findings and changing public perception, these so-called wastelands are taking on new value. We now know that estuaries and wetlands can function as:

- *Wildlife habitats.* They provide shelter, migration rest stops, breeding sites, and food for millions of waterfowl, fishes, invertebrates, and fur-bearing animals. In fact, several endangered species live only in estuaries and their associated wetlands.
- *Fish nurseries.* They provide a breeding ground for large numbers of fishes, including commercially important species. According to Owen and Chiras (1995), 60 percent of the marine fish harvested by America's fishing

a. Southern New England

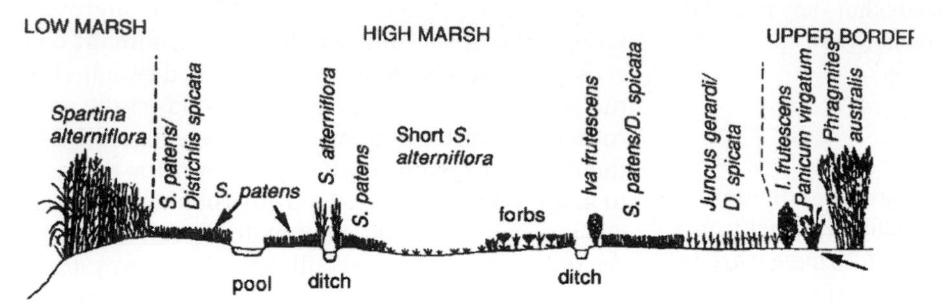

b. South Atlantic Coast

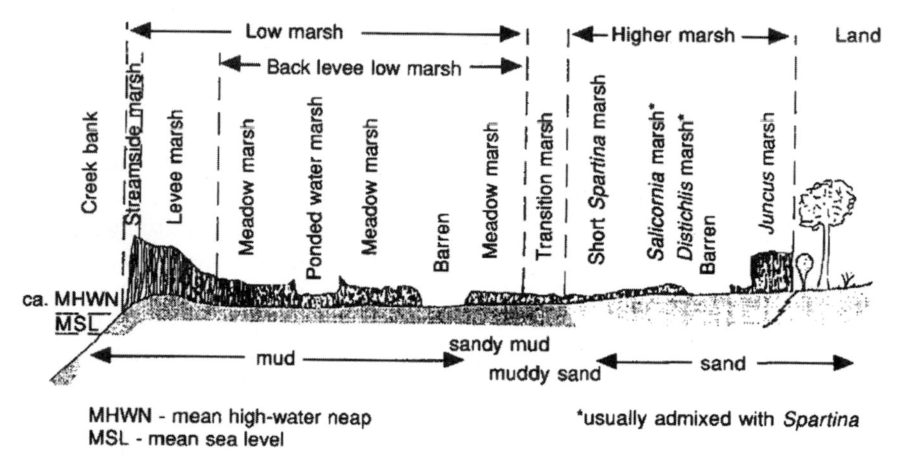

c. Northeastern Gulf of Mexico Coast

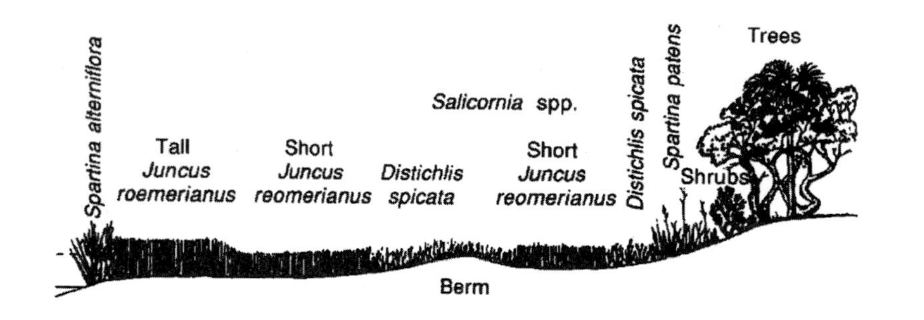

FIGURE 1-22 Zonation of vegetation in typical North American salt marshes. (a. after Niering and Warren, 1980; b. after Wiegert and Freeman, 1990; c. from Montague and Wiegert, 1990. [*Source: Wetlands,* 2/e by Mitsch and Gosselink, © 1993. Reprinted with permission of John Wiley & Sons, Inc.])

industry spend part of their life cycle in estuaries. In the Gulf of Mexico, for example, 98 of every 100 fish that are taken are estuary and salt marsh dependent.

- *Fisheries.* Estuaries provide fish and shell fish as a valuable food resource.
- *Natural farms.* They grow green material (more than our best-managed farms) that feeds a complex food web.
- *Flood and erosion controls.* They provide an important role in flood control, by absorbing the shock of storm-driven waves before they rush inland and cause destruction of property and human life.
- *Natural pollution-filtering systems.* They help cleanse the water of industrial and domestic sewage delivered to the estuaries by rivers. For example, just 5.6 hectares (14 acres) of estuary has the same pollution-reducing ability as a $1 million waste treatment plant (Owen and Chiras 1995).
- *Air purification systems.* They, like any large expanse of green plants, absorb carbon dioxide from the air and release oxygen.
- *Environmental education centers.* They provide "living laboratories" for schools, recreation centers, and mental health retreats.

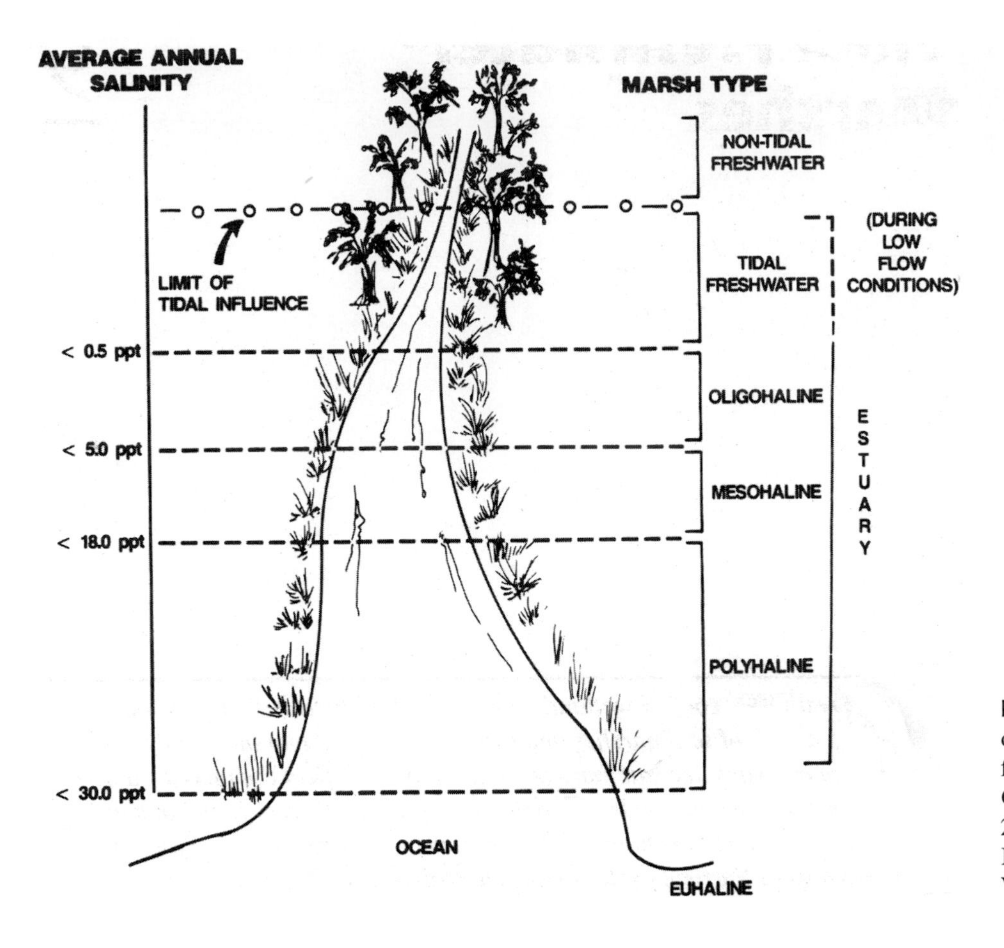

FIGURE 1-23 Coastal marshes lie on gradients of decreasing salinity from the ocean inland. (From W.E. Odum et al., 1984. [*Source: Wetlands,* 2/e by Mitsch and Gosselink, © 1993. Reprinted by permission of John Wiley & Sons, Inc.])

- *Job centers.* They provide jobs for fishermen, wetlands ecologists, marsh managers, tour guides, and nature or recreational store owners.
- *Tourist attractions.* They attract millions of visitors, such as birdwatchers, anglers, hunters, and boaters. The local economy is also supported when these same people "support their sport" by spending money in local nearby towns.

If you add up all the values of a particular estuary and/or its associated wetland, the site that may have once been considered a "muddy wasteland" may be more valuable to society in the long run than if the site were developed for short-term profit.

Destruction of estuaries and coastal wetlands. Despite their importance, our nation's estuaries and associated wetlands are annually dredged to deepen channels for navigation and filled to form solid land for agriculture or construction sites. The economic value of wetland development is an important force causing this change. In the early 1980s, more than 180,000 hectares (450,000 acres) of U.S. wetlands (much of it in estuaries) was destroyed annually (Owen and Chiras 1995). Most of this occurred on the coasts of New Jersey, Florida, Louisiana, Texas, and California. However, efforts are beginning to reverse some of this destruction. There is a growing interest in estuarine and coastal wetland restoration, though most mitigation projects to date have not been very successful. For example, the U.S. Army Corps of Engineers is now beginning to use sediment dredged from rivers to create new marshes. But compared to the current rate of destruction, their efforts are still minuscule.

Nearshore Waters and Open Ocean (including Marine Sanctuaries)

The final major coastal habitat is a combination of the nearshore zone (swash zone, surf zone, and breaker zone) and the offshore zone. Here, on the gently sloping continental shelf, one finds vast numbers of marine organisms, algae, invertebrates, fish, seabirds, and mammals that inhabit these shallow waters. It is also here that one finds tide pools (in the intertidal), kelp forests, sea mounts, and submarine canyons that provide a diversity of ecological habitats for wildlife. It is also within this zone that our nation has its national marine sanctuaries—the oceanic equivalent to our terrestrial national park system. An entire chapter, Chapter 3, will be devoted to discussing these marine and coastal protected areas.

FIGURE 1-24 Mangrove swamps help stabilize sediment. (Upper) Mangroves have dense root systems that act as sediment stabilizing mazes; (Lower) Fast growing mangroves can eventually close navigation inlets to coastal villages, such as here in Belau [Palau], Micronesia. (*Source:* Photos by author)

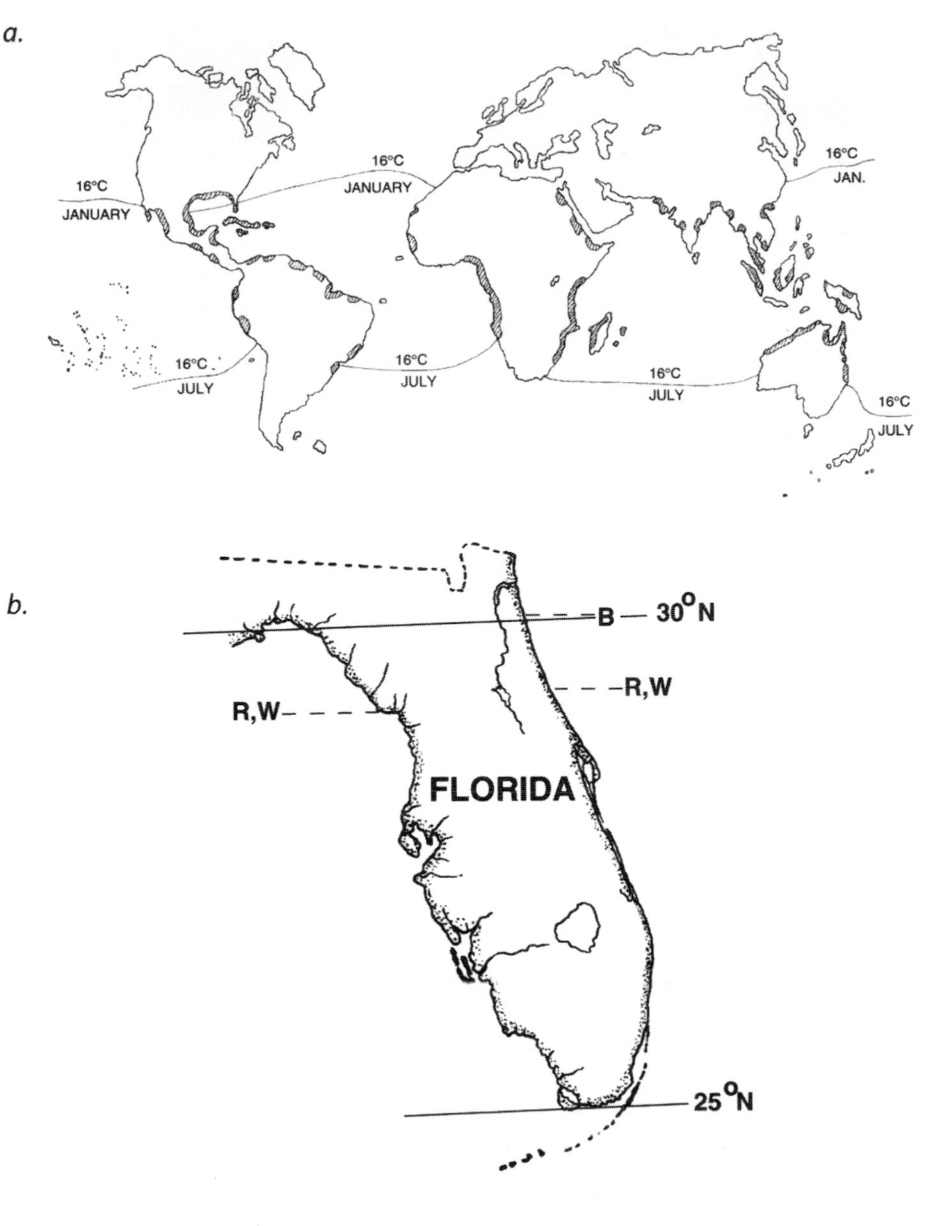

FIGURE 1-25 Distribution of mangrove wetlands in the (a) world, (b) and Florida. (a. After Chapman, 1977; b. From W.E. Odum et al., 1982 and W. E. Odum and McIvor, 1990; by permission, University of Central Florida Press, Orlando. [*Source: Wetlands* 2/e by Mitsch and Gosselink, © 1993. Reprinted by permission of John Wiley & Sons, Inc.])

CONCLUSION

This chapter provided a general introduction to the coastal environment. *Section I, The Coastal Zone,* began with a working definition of the coast, then identified five major coastal issues facing America's coastal areas. *Section II, The Coastal System,* discussed major ways to classify and subdivide coasts. Natural and human factors affecting coastal elements were then discussed. *Section III, Coastal Formations,* then introduced the primary physical formations that can be found in the coastal zone, ranging from inland coastal mountains to open ocean habitats. It should now be clear that the coastal zone is both highly dynamic and ecologically productive, and that *any attempt to manage this complex zone will first require a greater understanding of the physical coastal environment than presented in this brief chapter.* One is encouraged to pursue many of the scientific studies that are listed in this chapter's "Further Reading" section as well as elsewhere. Furthermore, anyone serious about pursuing a career in coastal science or coastal resource management should take university courses such as coastal geology, coastal geomorphology, coastal geography, and coastal meteorology to get better acquainted with the physical characteristics and natural processes present in the coastal environment.

Managing coastal phenomena, including their associated economic and cultural resources, also requires an understanding of the existing political framework for coastal management—the principal players (agencies), programs, and policies that are already in place. We now turn to this subject in the following chapter.

REFERENCES

Association of Bay Area Governments. 1992. *State of the Estuary: San Francisco Estuary Project.* Oakland, California: Association of Bay Area Governments.

Beeftink, W. G. 1977. "Salt Marshes." In *The Coastline,* edited by R. S. K. Barnes, pp. 93–121. New York: Wiley.

Bird, Eric C. F. 1984. *Coasts: An Introduction to Coastal Geomorphology.* 3rd ed. New York: Basil Blackwell.

California Coastal Commission. 1987. *California Coastal Resource Guide.* Berkeley: University of California Press.

Carson, Rachel. 1983. *The Edge of the Sea.* (Originally published in 1955.) Boston: Houghton Mifflin, p. 11.

Carter, R. W. G. 1988. *Coastal Environments: An Introduction to the Physical, Ecological and Cultural Systems of Coastlines.* New York: Academic Press.

Chapman, V. J. 1977. *Wet Coastal Ecosystems.* Amsterdam: Elsevier.

Christopherson, Robert W. 1997. *Geosystems: An Introduction to Physical Geography.* 3rd ed. Englewood Cliffs, New Jersey: Prentice Hall.

Clark, John R. 1977. *Coastal Ecosystem Management.* New York. John Wiley and Sons.

Clark, John R. 1996. *Coastal Zone Management Handbook.* Boca Raton, Florida: CRC Lewis Publishers.

Coastal Zone Management Act. 1994. United States Code. Title 16: Conservation; Chapter 33: Coastal Zone Management; Section 1453: Definitions.

Davies, J. L. 1972, 1977, & 1980. *Geographical Variation in Coastal Development.* London: Longman Group Ltd.

Davis, Richard A., Jr. 1994. *The Evolving Coast.* New York: Scientific American Library.

Dolan, Robert, Bruce Hayden, and Harry Lins. 1980. "Barrier Islands." *American Scientist,* Vol. 68, No. 1 (January–February), pp. 16–25.

Ehler, Charles N., and Daniel J. Basta. 1993. "Integrated Management of Coastal Areas and Marine Sanctuaries: A New Paradigm." *Oceanus,* Vol. 36, No.3 Fall, pp. 6–13.

Griggs, Gary, and Lauret Savoy, ed. 1985. *Living with the California Coast.* Durham, North Carolina: Duke University Press.

Hansom, J. D. 1988. *Coasts.* Cambridge: Cambridge University Press.

Inman, D. L., and B. M. Brush. 1974. "The Coastal Challenge," *Science,* Vol. 181, No. 4094, pp. 20–32.

Inman, D. L., and C. E. Nordstrom. 1971. "On the Tectonic and Morphologic Classification of Coasts." *Journal of Geology,* Vol. 79, No. 1, pp. 1–21.

King, C. A. M. 1972. *Beaches and Coasts.* 2nd ed. New York: St. Martin's Press.

Levin, Harold L. 1994. *The Earth Through Time.* 4th ed. Fortworth, Texas: Saunders College Publishers.

Mitsch, William J., and James G. Gosselink. 1993. *Wetlands.* 2nd ed. New York: John Wiley & Sons.

Montague, C. L., and R. G. Wiegert. 1990. "Salt Marshes," in *Ecosystems of Florida,* R. L. Myers and J. J. Ewel, eds., Orlando, Florida: University of Central Florida Press, pp. 481–516.

Niering, W. A., and R. S. Warren. 1980. "Vegetation Patterns and Processes in New England Salt Marshes," *Bioscience* 30: 301–307.

Odum, W. E., C. C. McIvor, and T. J. Smith, III. 1982. *The Ecology of the Mangroves of South Florida: A Community Profile,* U.S. Fish and Wildlife Service, Office of Biological Services, Technical Report FWS/OBS 81-24, Washington, D.C.

Odum, W. E., and C. C. McIvor. 1990. "Mangroves", in *Ecosystems of Florida,* R. L. Myers and J. J. Ewel, eds. Orlando, FL: University of Central Florida Press, pp. 517–548.

Odum, W. E., T. J. Smith, III, J. K. Hoover, and C. C. McIvor. 1984. *The Ecology of Tidal Freshwater Marshes of the United States East Coast: A Community Profile.* Washington, D.C.: U.S. Fish and Wildlife Service. FWS/OBS-87/17, 177 p.

Owen, Oliver S., and Daniel D. Chiras. 1995. *Natural Resource Conservation: Management for a Sustainable Future.* 6th ed. Englewood Cliffs: Prentice Hall.

Patterson, S. G. 1986. *Mangrove Community Boundary Interpretation and Detection of Areal Changes in Marco Island, Florida: Application of Digital Image Processing and Remote Sensing Techniques.* Washington, D.C.: U.S. Fish and Wildlife Service, Office of Biological Services. Report 86 (10).

Pethick, John. 1984. *An Introduction to Coastal Geomorphology.* London: Edward Arnold.

Schwartz, Maurice L. 1982. *The Encyclopedia of Beaches and Coastal Environments.* Stroudsburg, PA: Hutchinson Ross.

Viles, Heather, and Tom Spencer. 1995. *Coastal Problems: Geomorphology, Ecology, and Society at the Coast.* London: Edward Arnold.

Weigert, R. G., and B. J. Freeman, 1990. *Tidal Salt Marshes of the Southeast Atlantic Coast: A Community Profile.* U.S. Department of Interior, Fish and Wildlife Service, Washington D.C., Biological Report 85 (7.29).

Wells, John T., and Charles H. Peterson. Undated. *Ribbons of Sand: Atlantic and Gulf Coasts Barriers.* Washington, D.C.: U.S. Department of the Interior.

Williams, S. Jeffress, Kurt Dodd, and Kathleen Krafft Gohn. 1990. *Coasts in Crisis.* Washington D.C.: U.S. Geological Survey, p. 3.

FURTHER READING

Beatley, Timothy, David J. Brower, and Anne K. Schwab. 1994. *An Introduction to Coastal Zone Management.* Washington, D.C.: Island Press.

Bertness, M. D. 1992. "The Ecology of a New England Salt Marsh." *American Scientist,* Vol. 80, pp. 260–268.

Chapman, V. J. 1964. *Coastal Vegetation.* New York: The Macmillan Company.

Coates, Donald R., ed. 1980. *Coastal Geomorphology.* London: George Allen & Unwin.

Inman, Douglas L. 1994. "Types of Coastal Zones: Similarities and Differences," In *Environmental Science in the Coastal Zone: Issues for Further Research:* Proceedings of a retreat held at the J. Erik Jonsson Woods Hole Center, Woods Hole, Massachusetts 25–26 June 1992, Commission on Geosciences, Environment, and Resources, National Research Council, National Academy Press, Washington, D.C.

Josselyn, M. 1983. *The Ecology of San Francisco Bay Tidal Marshes: A Community Profile.* Slidell, La.: U.S. Fish and Wildlife Service. FWS/OBS–83/23, 102 p.

Kauffman, W., and O. H. Pilkey, Jr. 1983. *The Beaches Are Moving: The Drowning of America's Shoreline.* Durham, NC: Duke University Press.

Kenchington, Richard, and David Crawford. 1993. "On the Meaning of Integration in Coastal Zone Management." *Ocean & Coastal Management,* Vol. 21, pp. 109–127.

Knecht, Robert W., and Jack Archer. 1993. "Integration in the U.S. Coastal Zone Management Program." *Ocean & Coastal Management,* Vol. 21, pp. 183–199.

National Research Council. 1994a. *Environmental Science in the Coastal Zone: Issues for Further Research.* Washington, D.C.: National Academy Press.

National Research Council. 1994b. *Priorities for Coastal Ecosystem Science.* Washington, D.C.: National Academy Press.

National Oceanic and Atmospheric Administration. 1990. *Estuaries of the United States: Vital Statistics of a National Resource Base. A Special NOAA 20th Anniversary Report.* October. Rockville, Maryland: National Oceanic and Atmospheric Administration.

Odum, W. E. 1988. "Comparative Ecology of Tidal Freshwater and Salt Marshes." *Annual Review Ecological Systems,* Vol. 19, pp. 147–176.

Trenhaile, A. S. 1987. *The Geomorphology of Rock Coasts.* Oxford: Oxford University Press.

Wiegert, R. G., and B. J. Freeman. 1990. *Tidal Salt Marshes of the Southeast Atlantic Coast: A Community Profile.* Washington, D.C.: U.S. Department of Interior, Fish and Wildlife Service. Biological Report 85 (7.29).

Zedler, J. B. 1982. *The Ecology of Southern California Coastal Salt Marshes: A Community Profile.* Washington, D.C.: U.S. Fish and Wildlife Service Biological Services Program. FWS/OBS–84/54.

2 Coastal Management: Players And Jurisdictions

Coastal management can be defined as *planning for, and reacting to, environmental change within the coastal zone.* The goal of coastal management is to preserve, protect, or enhance coastal zones for humans as well as other species. Generally, coastal management involves three broad areas: (a) *policy* (i.e., the administrative and political framework in which coastal management is regulated); (b) *planning* (i.e., the process of resource allocation); and (c) *implementation* (e.g., practices that range from restoring native dune grass, to constructing jetties, to improving navigation, and to dealing with the military's dumping of antiquated nuclear submarines). Since some of the best intended coastal implementation schemes (e.g., the building of a jetty to improve navigation) can have negative consequences (e.g., the loss of sand on adjacent beaches), further planning, policy, and implementation schemes are sometimes needed to correct the original strategy.

Managing coastal habitats in the United States involves a number of overlapping agencies and jurisdictions. This chapter will briefly identify the key components of the coastal management framework. It will look at both the *public players* (e.g., government agencies, their primary coastal management activities, and key authorizing legislation) and the *private stakeholders* in coastal resource management. The purpose is to help coastal students and interested individuals sort out "the management muddle" (Bulloch 1989, p. 77). For example, for just the issue of marine water quality, there are over 21 federal programs, governed by 8 major statutes, administered by 11 different federal agencies. Of course, the issue becomes even more complicated when the other players (international, state, regional, and local) are considered.

To work effectively within these three broad areas as they relate to coastal resources and environments, however, one must understand their social, cultural, political, legal, economic, as well as technical (scientific) aspects. This also entails understanding and being able to use the evolving management tools and techniques practiced by the coastal scientist and coastal resource manager. Coastal management is *interdisciplinary* in its approach.

Prior to the 1970s, coastal management was practiced under the purview of the disciplinary sciences (e.g., marine geology, marine biology, etc.). However, events during the late 1950s through the early 1970s resulted in a "coming of age" for coastal management. Questions over territorial rights of the sea, resource development, and coastal protection were on the governmental agenda throughout the period. In 1966, Congress passed the Marine Resources and Engineering Act, which resulted in the formation of the Commission on Marine Sciences, Engineering, and Resources (COMSER), ultimately known as the Stratton Commission, to investigate the possibilities of a coordinated governmental effort for coastal resource conservation. While considering the needs of ocean science specifically, the Stratton Commission also made recommendations to Congress for a national coastal zone management program in order to organize a coordinated national effort for coastal management. Congress was agreeable to this concept both as an answer to coastal resource management needs and as a response to public demands precipitated by environmental disasters, particularly along the coast, such as the 1968 Torrey Canyon tanker spill off the coast of England, and the 1969 offshore oil platform blowout near Santa Barbara, California (Knecht, Cicin-Sain, and Archer 1988).

In 1972, the Coastal Zone Management Act was enacted, creating a new framework for coastal management, with the subsequent need for new types of multidisciplinary and interdisciplinary specialists. For example, local offices of the California Coastal Commission have the following job descriptions:

- *District Director.* Attends public hearings as a representative of the coastal commission; facilitates intergovernmental cooperation and works with public agencies and private industry to ensure compliance with commission rules and state laws regarding the coastal zone.
- *Deputy District Director.* Acts as staff liaison officer incorporating the work of the entire staff, pertaining to public relations, regulation, zoning, planning and permits for development, or other activities in the district.
- *Legal Council.* Manages legal issues for the regional office of the coastal commission. Conducts legal review of findings and conditions, and facilitates legal cooperation between government and private concerns.
- *Coastal Planner.* Works with federal, state, and local government to write findings and reports and design planning and policy for coastal commission affairs.
- *Public Access Manager.* Works with private property owners and local government agencies to provide and maintain public accessways to coastal areas. Arranges easements within the coastal zone.

The Marine Protection, Research, and Sanctuaries Act (also enacted in 1972) has required new types of multidisciplinary and interdisciplinary specialists, such as the following job descriptions:

- *Environmental Scientist.* Reviews environmental quality projects, such as coastal development permits, desalination plant proposals, dredging projects.
- *Water Quality Protection Program Director.* Directs interagency efforts between a marine sanctuary and adjacent watersheds.
- *Research Coordinator.* Reviews research permits and proposals with the local research community.
- *Education Outreach Specialist.* Develops coastal education materials and programs for schools and the general public.

With almost 50 percent of the U.S. population now living within the coastal zone, the need for such interdisciplinary coastal scientists and resource managers will no doubt increase.

The majority of this chapter is organized from the international perspective down to the local level where the average citizen participates. *The International Players* gives a short description of the global nature of the problem, citing only a sample of the international organizations that are involved. In *The Federal Players,* principal federal agencies within the United States are identified, as well as their primary coastal management interests and activities. *The State Players* discusses how state governments exercise some degree of control over their coastal zone. Two examples will be given: one from California, and the other from Florida. *The Local Players* gets down to the level of the individual stakeholder—the beachfront cottage owner, the fisherman, and the clam digger—who needs a voice in the management of the resources that they utilize.

The chapter then takes a different slant, focusing on those players who are moving in the direction of integrated coastal management—*The Regional Players.* Here we discuss how special regional governmental bodies are sometimes created to manage coastal bays, estuaries, or other coastal ecosystems that do not neatly fit into international, state, or local jurisdictional boundaries. The chapter concludes with *Integrated Management of Coastal Areas and Marine Sanctuaries* wherein the emerging paradigm of integrated coastal management is discussed, and key concepts for 21st century coastal management are identified.

THE INTERNATIONAL PLAYERS

International agencies are most interested in the fair division, and proper management, of **continental shelf** resources. As one might suspect, however, demarcating nautical boundaries and zones leads to numerous conflicts of interest and concern over accessibility, security, and wealth (Prescott 1985). These agencies and their conventions are increasingly playing an important role in regulating pollution prevention, containment, and cleanup standards. To a lesser extent, they are involved in public education, the training of personnel, inspection activities, and environmental enforcement.

International Maritime Organization (IMO)

The *International Maritime Organization (IMO)* is currently the most important international organization involved in maintaining standards and conditions relating to ship operations in near shore and shelf waters. The IMO's emphasis is on the promotion of maritime safety for the protection of the marine environment. As a technical agency of the United Nations, IMO was originally set up in 1959 to develop response strategies for catastrophic pollution problems and to develop long-term plans to minimize global sea pollution through regulation and technical assistance (Kenchington 1990). The precedence for many of the IMO's recommendations were established in old customary laws of the sea. Dating back centuries in some cases, these generally accepted rights and duties of states regarding use and utilization of oceans became "customary" law around the 19th century. Since its creation in 1959, the IMO has conducted about 30 international conventions regulating international maritime activities. Three of the more important conventions are listed below:

- *United Nations Convention on Law of the Sea (UNLOSC, or LOSC).* This was an umbrella convention that provided for a balanced allocation of ocean resources, nation by nation, and dealt with all areas of marine pollution. It also provided for a more efficient regulatory system than that currently practiced. Although presented for acceptance in 1982 at Montego Bay, the Law of the Sea Convention (LOS Convention) has yet to come into force. When the LOS Convention closed in 1984, only 119 coastal states had signed. The U.S., U.K., and West Germany would not sign. The U.S., for example, objected to the provisions regarding sea-bed mining and accessibility. Although not likely to become effective in the near future, some of the environmental provisions could possibly apply as customary law (e.g., the convention's statements related to the 200-mile exclusive fishing zone; see below) (Frankel 1995).
- *Convention for the Prevention of Marine Pollution by Dumping of Wastes and Other Matter (London Dumping Convention, 1972).* In September 1975, this convention did enter into force, and deals with ocean dumping on a global scale. The convention prohibits the dumping of radioactive waste, industrial wastes, oil, and other substances deemed toxic, persistent and bioaccumulative. This convention is aimed principally at wastes originated on shore. The United States is a signatory. See Chapter 6, *Ocean Dumping*, for further details.
- *International Convention for the Prevention of Pollution from Ships, 1973, as amended in 21978 (MARPOL 73/78).* This convention became effective in 1983, and was the first international convention to regulate all forms of marine pollution from ships, except dumping. In other words, the emphasis of this convention was regulating the disposal (purposeful or accidental) of oil or refuse at sea as a part of the operation of a ship. MARPOL was last updated in March 1992. Many of the tighter regulations were a direct result of the 1989 *Exxon Valdez* tanker oil spill that caused so much damage to Prince William Sound, Alaska. As is discussed in Chapter 6, *Ocean Dumping*, the U.S. Navy is exempted from many of these regulations.

Other Key International Organizations

In addition to the International Maritime Organization, there are a number of other international organizations that are active players in coastal and ocean environmental management. For example, there is the *U.N. Environmental Program (UNEP),* the *U.N. Conference on Trade and Development (UNCTAD),* the *Organization of Economic Cooperation and Development (OECD),* the *World Health Organization (WHO)*, and the *World Bank.* Although all have a hand in some aspect of coastal and ocean protection and management, most are weak when it comes to mechanisms to enforce their own standards. In other words, many would argue that they are better at "drafting standards and agreements" than enforcement mechanisms. Unfortunately, installing "traffic signals" means little without a nearby "cop on a motorcycle."

Coast-Parallel Zones

Coastal-parallel zones have also been established by coastal countries and international organizations. (See Figure 2-1).

Territorial Waters, the inner-most zone, is usually 12 nautical miles wide (22.2 km) as established internationally by the United Nations LOS Convention III. In Territorial Waters, a country has sovereign rights to the economic resources, sea, and airspace. Foreign vessels have the right of "innocent" passage through Territorial Waters but a country's territorial airspace is restricted to flights granted prior permission. In the United States, the Coastal Zone Management Act (CZMA) gives individual states jurisdictional rights out to 3 nautical miles. The remaining 4–12 nautical miles were designated federal territorial sea, in which states have only the right to review activities.

The *Contiguous Zone*, established at a maximum of 24 nautical miles (44.4 km) from a coastal baseline, extends beyond the Territorial Waters, usually up to 12 nau-

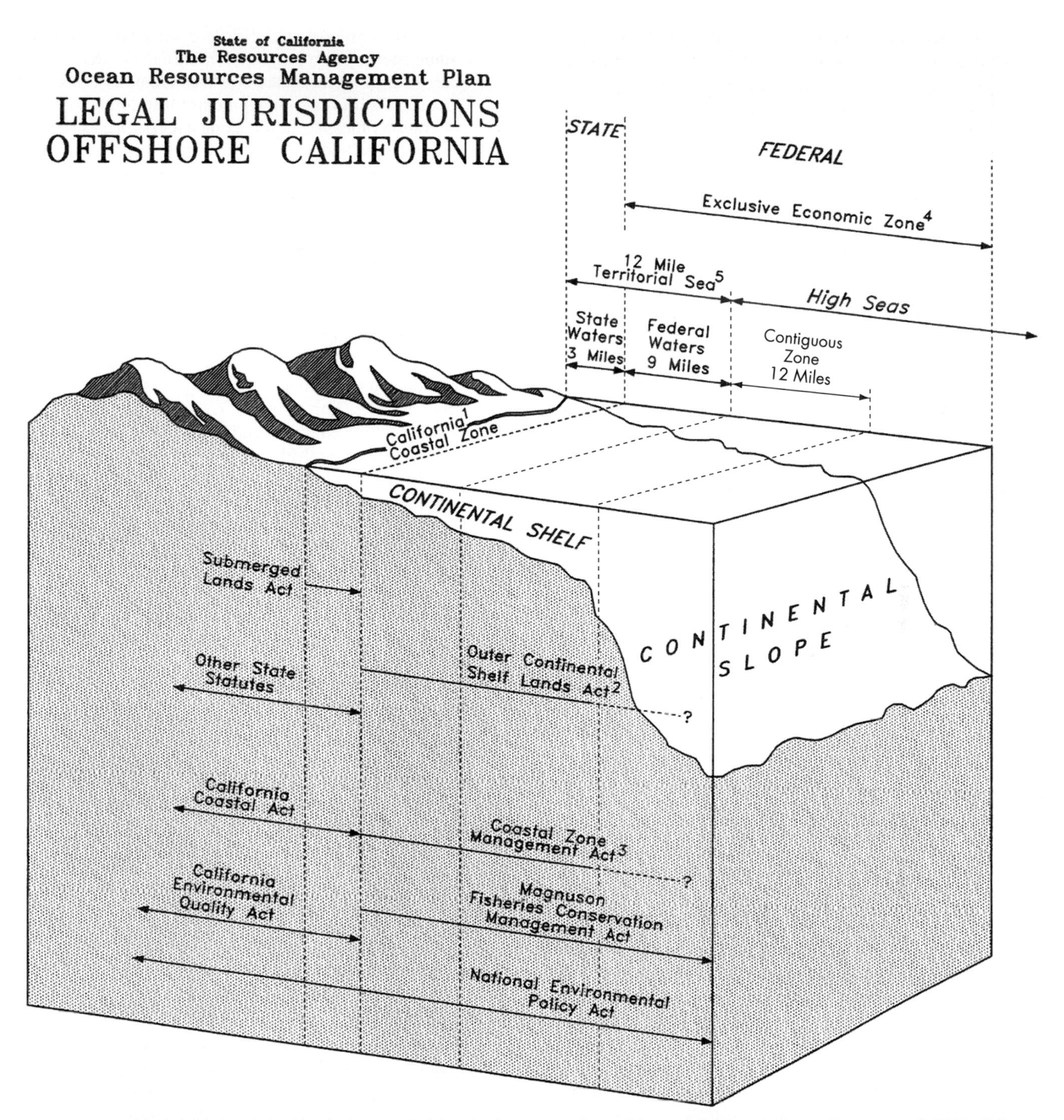

FIGURE 2-1 Legal Jurisdictions Offshore California. (*Source:* Adapted from *California's Ocean Resources* © 1997. California Resources Agency, All Rights Reserved)

tical miles wide. In the Contiguous Zone, a country may exercise the control needed towards foreign vessels to prevent infringement of its customs, fiscal, immigration, or sanitary laws and regulations (Bledsoe and Boczek 1987). Wider limits, usually up to 200 nautical miles (370 km) as measured from the Territorial Sea baselines, were developed in the 1950s through 1980s, with the rush to claim fishing rights *(Exclusive Fishing Zone [EFZ])* and sea-bed mineral resources *(Exclusive Economic Zone [EEZ])*. In 1983 President Reagan proclaimed for the U.S. a 200-mile Exclusive Economic Zone. These extensions came as a result of the 1982 United Nations LOS Convention discussed earlier. As can be easily imagined, this led to many boundary disputes, such as the dispute between France and Britain over the southwest approach to the English Channel (Prescott 1985). Rights of innocent passage, however, are mandated by customary law for foreign vessels traveling in a state's sovereign waters and coastal zones.

The last zone, the *High Seas*, is all waters outside established EEZs and EFZs, seaward of the 200 nautical mile (nm) limit. Though technically the High Seas begins at the outer edge of the Territorial Sea, the coastal state's sovereign rights in the waters out through the EEZ's restrict most international activity. Thus for all practical purposes, the High Seas and its resources begin outside the 200 nm EEZ's and EFZ's (Buchholz 1987). No state may lay claim to sovereignty in the High Seas and aggressive actions against another state's vessels are prohibited here. The High Seas are regulated by a combination of International, National, and State jurisdictions, depending on the nature of an incident. For example, any state may seize a ship suspected of piracy, and if necessary try the pirate crew under the state's own courts. On the other hand, the International Sea-Bed Authority, a U.N.-sanctioned body, has jurisdiction over any mineral resources in the High Seas.

THE FEDERAL PLAYERS

There are a number of federal agencies with regulatory or research responsibilities for America's coasts. In this section, we will first identify these key federal agencies, then later list and discuss the legislation and associated programs that they administer or in which they play a role.

Relevant Federal Agencies

There is no one agency at the federal level that has responsibility for coastal management. However, seven federal agencies are particularly important players, and they are the following:

- *National Oceanic and Atmospheric Administration* [*NOAA* (U.S. Department of Commerce)]. NOAA, more than any other federal agency, implements America's coastal zone management program. The emphasis of this agency is "science for marine solutions." In order to better understand the connection of these coastal habitats to marine ecosystems, NOAA, for example, conducts studies of wetlands and maps watershed and habitat-change analysis. Furthermore, NOAA operates an environmental satellite system called *Coast Watch* that, among other things, helps fisheries managers locate concentrations of endangered sea turtles to minimize conflicts with other fisheries. The agency also prepares nautical charts and geodetic surveys of coastal areas, and administers a grant program for marine research. To help carry out all of its duties, NOAA's Office of Ocean and Coastal Resource Management has a number of field operation offices scattered along America's coasts. In addition to its "science role," it works with states on implementing their individual coastal zone programs. NOAA also manages the National Marine Sanctuary Program (NMSP).
- *U.S. Army Corps of Engineers* [*COE* (*U.S. Department of Defense*)]. Since 1824, the Army Corps of Engineers (COE) has been engaged in applied research and development related to navigational waterways, harbors, and coastal protection. It gets involved in all phases of civil works projects, from design and construction, to operation and maintenance, to technical assistance, to funding activities. For example, river and harbor projects would include channel dredging and inlet stabilization; coastal erosion activities have included the construction of jetties, groin fields, seawalls, and beach nourishment and restoration projects.

  The COE administers laws and issues permits that protect coastal resources and navigable waterways. For instance, the Army Corps of Engineers administers permit programs regulating construction within navigable waters, as well as the discharge of material therein. A number of their projects have been heavily criticized. Part of the controversy is over their historic role in these areas. Major change may be coming in the future.
- *Federal Emergency Management Agency (FEMA, an independent federal organization)*. FEMA is the principal agency of the federal government that deals with riverine and coastal flood hazards. It provides disaster assistance to coastal states and local governments. As is illustrated in Chapter 4, *Coastal Hazards*, it is one of the more controversial federal agencies that deals with coastal resource management.
- *U.S. Environmental Protection Agency (EPA, an independent federal agency)*. The EPA funds scientists who conduct numerous coastal studies, such as contaminant research and, more recently, investigations into sea level change. It also plays an important role in regulating the discharge of coastal pollutants, the disposal of dredged materials, and other forms of ocean dumping. The EPA has ultimate authority over the Army Corps of Engineers regarding certain aspects of dredging related to *Section 404* of the *Water Pollution Control Act*. The EPA is also involved in creating emission standards for airborne pollutants.
- *National Park Service (NPS [U.S. Department of the Interior])*. Two agencies within the Department of the Interior—the National Park Service (NPS) and the Fish and Wildlife Service (FWS)— own and operate extensive land holdings along our nation's coastlines. These agencies manage such areas as coastal barriers, shorelines (including the Great Lakes), estuarine wetlands, and water bodies with migrating fish. In addition to reg-

ular parks, the National Park Service manages National Seashores and National Lakeshores—areas set aside for the protection of their natural environments (See Chapter 8, *Open Space Preservation and Management*, for further details).

- *Fish and Wildlife Service (FWS [U.S. Department of the Interior]).* The FWS has many traditional programs that play an important role in coastal conservation, such as managing its extensive coastal lands as wildlife preserves under its National Wildlife Refuge Program, mapping wetlands, and monitoring migratory birds and endangered species. Since 1992, under the direction of John Turner, the agency has declared the conservation of coastal resources one of its top priorities. A key part of the FWS's coastal emphasis is its newly established Coastal Ecosystems Program.
- *National Marine Fisheries Service (NMFS* [U.S. Department of Commerce]). The seventh and final member of the "Magnificent Seven" is the National Marine Fisheries Service. This federal agency is primarily interested in marine life resources, especially the conservation of fisheries and marine mammals. Claiming to be the nation's oldest conservation organization, it celebrated 125 years of federal marine fisheries research, conservation, and management on February 9, 1996. The organization, however, is not without its critics. There are those that claim it has a very poor fisheries management record. Nevertheless, since it shares a similar interest with the Fish and Wildlife Service (i.e., the conservation of biotic resources), these two agencies generally have a high degree of coordination and cooperation.

In addition to the *major* federal players, there are numerous other federal agencies that are actively involved in some aspect of coastal research, policy, or management. In reading through this book, you will also see mention of such federal players as the U.S. Coast Guard, the U.S. Geological Survey, the U.S. Minerals Management Service, the U.S. Natural Resources Conservation Service (formerly called U.S. Soil Conservation Service), and the National Science Foundation. There is no question that the federal government has played, and will increasingly continue to play, an important role in coastal resource management. We now turn for a brief look at the major pieces of legislation and programs that authorize these agencies and the programs they oversee.

Relevant Federal Legislation

Since the 1930s, Congress has passed legislation developing a variety of programs related to the management of coastal areas. These programs have had a number of objectives, but unfortunately, some have been in direct conflict with each other (e.g., riparian rights v. navigation; protection of fish and wildlife resources v. economic development; public recreation opportunities v. reducing natural hazards, and so on). We will first look at those pieces of legislation and programs that are "coast specific"—legislation specifically designed to address coastal resource management issues. Then, we will briefly identify and discuss "coast/marine" related pieces of legislation, where the focus is perhaps more on marine affairs, but still is relevant to coastal concerns. Finally, we will bring to your attention those "noncoast/marine specific" pieces of legislation that deal with overall environmental quality in the United States, but also bear on coastal resource management.

Coast specific legislation/programs. It helps to simplify "the management muddle" by first looking at the pieces of legislation specifically designed to help protect our nation's coastal resources. Lets start with the most important one of all, the Coastal Zone Management Act.

The Coastal Zone Management Act (CZMA). In 1972, President Nixon signed into public law the Coastal Zone Management Act. The CZMA administered by NOAA's Office of Ocean and Coastal Resource Management (OCRM), focused on coastal resource management more than any previous piece of legislation. Through this act, Congress declared its intention to "preserve, protect, develop, and where possible, to restore or enhance" the coastal areas of the United States (CZMA 1972a). This goal would be accomplished by providing incentives to coastal states to develop individual coastal management programs. If the state developed a coastal management program compatible with the intent of the CZMA, that state would be approved for Federal grants to assist with implementation and operation of the program.

Acceptable state coastal management plans needed to include programs which (a) help protect wetlands, lagoons, reefs, and other coastal habitats; (b) minimize property damage from coastal hazards; (c) improve recreational use of the coast (e.g., public access); and (d) encourage intergovernmental cooperation through procedural and policy standardization. It is important to note that although the CZMA seeks to influence the coastal zone as a region, its management tools are primarily land based, using zoning and regulation to control how land is utilized along the coast. This fact is exemplified by two approved coastal management programs, the Florida Coastal Management Program and the California Coastal Management Program (see "Two State Programs: Florida and California" in this chapter for details).

According to the CZMA, coastal states include all states which border "the Atlantic, Pacific, or Arctic Ocean, the Gulf of Mexico, Long Island Sound, or one or more of the Great Lakes" (CZMA 1972b). Included in this definition are Puerto Rico, Guam, the Virgin Islands, the Commonwealth of the Northern Mariana Islands, and the Trust Territories of American Samoa and the Pacific Islands. In 1998, all but three coastal states had coastal zone management plans—32 were approved; 2 in development (Minnesota and Indiana), and 1 not participating (Illinois). (See Figure 2-2).

Although federal funding provided a strong incentive for state participation, states were also enticed by the **consistency** provisions of the CZMA, designed to aid in coordination and cooperation between federal and state governments with regard to development in the coastal zone. Section 1456(c) of the CZMA states that "each federal activity within or outside the coastal zone that affects any land or water use or natural resource of the coastal zone shall be carried out in a manner which is *consistent* to the maximum extent practicable with the enforceable policies of approved State management programs" (CZMA 1972c). This meant states could control the actions of the Federal government if those actions would interfere with, or were considered inconsistent with, the State's coastal management program. This was especially important to those States concerned with Federal sales of outer continental shelf oil and gas development leases. These consistency provisions also gave the States review rights of other types of Federal activities including larger construction projects such as highways, airports, military installations, navigation and flood control projects, as well as oversight of Section 404 wetland alteration permits and fisheries management (Beatley, Brower, and Schwab 1994). This was a rare opportunity for the States to obtain powers normally reserved for the Federal government.

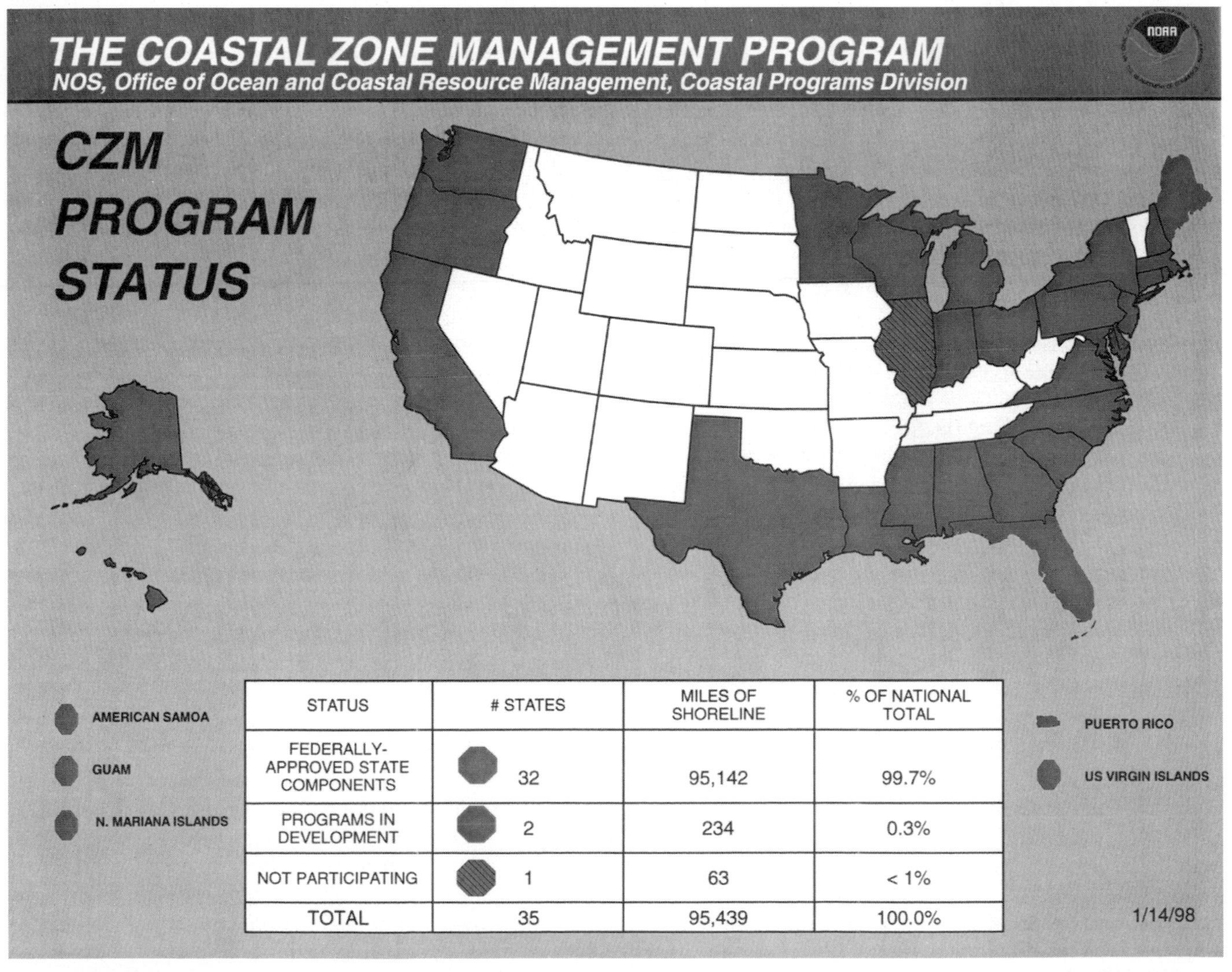

STATUS	# STATES	MILES OF SHORELINE	% OF NATIONAL TOTAL
FEDERALLY-APPROVED STATE COMPONENTS	32	95,142	99.7%
PROGRAMS IN DEVELOPMENT	2	234	0.3%
NOT PARTICIPATING	1	63	< 1%
TOTAL	35	95,439	100.0%

FIGURE 2-2 Figure Program Status of NOAA's Coastal Zone Management Program. [*Source:* Courtesy of NOS, Office of Ocean and Coastal Resource Management, Coastal Programs Division (N/ORM3)]

In addition to financing state coastal management programs and providing a means of state oversight of federal actions, the CZMA also provides funding to conserve specific resources along the coast. For example, Elkhorn Slough on the shores of the Monterey Bay National Marine Sanctuary and Tijuana Estuary in Southern California receive 20 percent of their funding (matched by the state of California) through the National Estuarine Reserve Program, a function of the Federal Coastal Zone Management Act.

The CZMA has been an evolving law—one that has gone under several amendments and reauthorizations over time. Two of the most formative changes occurred during the 1990 reauthorization of the CZMA, in which the *Coastal Zone Enhancement Grants Program* and the *Coastal Nonpoint Pollution Control Program* were established. The *Coastal Zone Enhancement Grants Program* prioritized project funding with the intent of focusing attention on protecting existing and creating new wetlands, addressing natural hazards, increasing public access to the coast, decreasing marine debris, and addressing cumulative and indirect impacts of growth along the coast, among other priorities (Beatley, Brower, and Schwab 1994). Meanwhile, the *Coastal Nonpoint Pollution Control Program* was established to help reduce nonpoint contamination—the single largest source of pollution—of coastal wetlands, coastal bays, and marine waters.

It is hard to ignore the practical success of the CZMA. For example, the *Coastal Nonpoint Runoff Control Program*, has been instrumental in heightening awareness and creating solutions for the nonpoint source pollution problem. However, despite success, the CZMA is not immune to political disfavor, as demonstrated during Reagan era attempts to zero fund the act. Most conservationists agree that if the CZMA is ever weakened or jettisoned, the nation's ability to protect and manage coastal resources will be in peril. (See Chapter 9, *Toward Integrated Coastal Management*, for a brief review of the accomplishments of the CZMA.) For a detailed review of the programs accomplishments, see Mitchell (1982), Brower and Carol (1984), Charis (1985), and Beatley, Brower, and Schwab (1994). For an excellent history of the Coastal Zone Management Act, see Godshalk (1992).

The Coastal Barrier Resources Act (CoBRA). The Coastal Barrier Resources Act of 1982 established the *Coastal Barrier Island Resources System*. The National Park Service (NPS) administers this program to protect 186 undeveloped barrier island units covering 1072 km (666 mi) of shoreline along the Atlantic and Gulf coasts. This Act prohibits most forms of Federal flood insurance under the National Federal Insurance Program (NFIP) or other Federal infrastructure assistance for development of the undeveloped coastal barriers within the system. In other words, the Act forbids Federal expenditures on such development infrastructures as roads, bridges, sewerage hookups, and so on. Consequently, the Act provides a "disincentive" to private development on listed barrier islands.

Marine/coast related legislation and programs. Two federal Acts are designed primarily to protect marine resources, but also have a bearing on coastal resource management: The Marine Protection, Research, and Sanctuaries Act, and the Marine Mammal Protection Act.

The Marine Protection, Research, and Sanctuaries Act (MPRSA). In 1972, Congress passed the MPRSA which established the *National Marine Sanctuary Program (NMSP)*—a system of marine sanctuaries to protect ocean and coastal waters. The program is administered by NOAA. Chapter 3, *Marine and Coastal Protected Areas,* discusses the program in detail. Since the focus of this book is on the need to integrate sanctuary management with what has been traditionally considered coastal resource management, numerous examples of the problems, successes, and needed future directions of this program will be discussed throughout this book.

Marine Mammal Protection Act (MMPA). In the same year that Congress passed the "Sanctuaries Act" in 1972, it also passed the *Marine Mammal Protection Act (MMPA)*. Concern had been mounting over the effect the fishing industry was having on marine mammals, specifically the relationship between purse-seine fishing for yellowfin tuna and the incidental deaths of porpoise. Prior to 1960, the most common method of fishing for yellowfin tuna was the use of pole, line, and live bait. However, fisherman had observed that yellowfin habitually associate with certain species of dolphin (commonly called porpoise), then began to set their nets over porpoise. Speedboats were used to herd the porpoise and tuna into a purse-seine net. Although the fishermen usually tried to free the porpoise, hundreds of thousands of these air breathing mammals were accidentally killed each fishing season. After years of public outcry, Congress reacted with the creating of the Marine Mammal Protection Act. It authorized the National Marine Fisheries Service (NMFS) within the Department of Commerce to promulgate regulations on the basis of the best scientific evidence available for permits for taking of marine mammals. This Act impacts coastal areas in that many marine mammals migrate up and down our coasts and through the marine sanctuaries that are part of the National Marine Sanctuary Program.

Noncoast/marine specific legislation and programs. There are several pieces of general environmental legislation that impact the management of coastal areas. Only the most relevant are highlighted below:

National Environmental Policy Act (NEPA, 1970). The approval of the National Environmental Policy Act in 1970, was a milestone in the planning of coastal projects, though it was not coastal nor marine specific. The Act created the *Environmental Protection Agency (EPA)*. Federally-funded projects henceforth come under the scrutiny of the EPA through the mechanism of the Environmental Impact Assessment (EIA) process. The Act forced the reevaluation of such stock-in-trade techniques as building groins, seawalls, and dredging. The EPA also sets standards for water pollutants (under the Clean Water Act) and establishes emission standards for air pollutants (under the Clean Air Act).

Clean Water Act (CWA). As mentioned above, the EPA oversees the Clean Water Act—a 1972 piece of legislation that has several provisions and programs affecting coastal resource management. (The *Clean Water Act* is an amendment to the *Federal Water Pollution Control Act* of the same year; FWPCA is now an outdated term.) The Clean Water Act regulates discharges into wetlands (Section 404) and places restrictions on point and nonpoint pollution. Under *Section 404 Wetlands Restrictions*, for example, any discharge of dredge and fill materials into coastal waters requires a permit approval from the U.S. Army Corps of Engineers. The EPA, however, has final veto authority over the issuance of these permits. The CWA is the primary federal law governing spills of oil and other hazardous substances into navigable waters.

National Flood Insurance Act (NFIA). In 1968, Congress had several objectives in mind when it passed the National Flood Insurance Act. The *National Flood Insurance Program (NFIP)* that it established was designed to (a) provide reasonably priced Federal insurance in flood hazard areas (prior to this Act, affordable private flood insurance was generally not available); (b) limit unwise development in flood plains (both coastal and riverine); and (c) reduce Federal expenditures for flood control. Criticisms of this program, as well as recommendations for improvement, are discussed in detail in Chapter 4, *Coastal Hazards*.

The Flood Disaster Protection Act (FDPA). In the early years of the National Insurance Program, participation was limited. Consequently, Congress then passed the *Flood Disaster Protection Act* of 1973. This Act *mandated* flood insurance for all mortgages backed by, or associated with, the federal government, such as VA or FHA loans.

The Stafford Disaster Relief and Emergency Assistance Act. In 1988, Congress attempted to revamp its federal disaster assistance framework by passing the Stafford Disaster Relief and Emergency Assistance Act. Its objective was to increase mitigation activities through financial incentives. The Program is administered by FEMA.

Endangered Species Act (ESA). To help protect wildlife resources, Congress passed the Endangered Species Act in 1973. Among other things, the Act established the National Wildlife Refuge System—a system of protected zones for wildlife resources. (See Chapter 8, *Open Space Preservation and Management* for details). The Act also authorizes the enforcement of federal wildlife and endangered species laws, and designs and implements habitat restoration activities. The ESA is administered by the U.S. Fish and Wildlife Service (USFWS).

This section merely touched on the main "non-coast/marine specific" pieces of legislation and associated programs. If space were available, one would also have to include such Acts as the Housing and Community Development Act, the Ocean Resources Management Act, The Outer Continental Shelf Lands Act, and such programs as the Emergency Home Moving Program, the Hazard Mitigation Grants Program, the Upton-Jones Amendment to the Coastal Zone Management Act, and the Great Lakes Program, and many more.

As should now be very clear, there are a great number of federal programs that impact coastal resource management, and that these programs are fragmented and dispersed over several federal agencies and departments. Unfortunately, there is no single federal agency that coordinates all of these programs and laws into a comprehensive, unified national coastal management plan. The problem becomes even more complex when we add the state, regional, and local players to the management muddle. We now turn to a discussion of the state players.

THE STATE PLAYERS

As mentioned in the previous "Federal" section, one of the attractive aspects of the Federal Coastal Zone Management Act is that it gives states cost-share financial assistance, and some control over Federal actions and policies regarding managing their coastal zone. Consequently, this "partnership" (rather than advisory) relationship has resulted in a high degree of state participation in the Federal Coastal Zone Management Program (Refer back to Fig. 2-2). This section of the chapter will briefly discuss (a) the minimum standards that states must meet, and (b) examples from two states: Florida and California.

Minimum Standards

Before implementing their programs, states must address nine national objectives as identified in the CZMA. They are as follows:

1. Public access to shorelines
2. Conservation of natural resources
3. Conservation, planning, and management of living marine resources
4. Coastal development to avoid hazardous areas
5. Redevelopment of urban waterfronts and ports
6. Development priority given to coastal dependent uses and energy facility siting
7. Consultation and coordination with Federal agencies
8. Coordination and streamlining of coastal management governmental procedures
9. Public participation in the coastal decision-making process

Once these minimum requirements have been addressed in their coastal management plans, states can request approval by the Office of Ocean and Coastal Resource Management in NOAA. From that point on, at least in theory, state and Federal actions and policies are consistent with one another.

Two State Programs: Florida and California

As an example of state coastal management frameworks, let's look briefly at two states that have major National Marine Sanctuaries next to them—Florida and California.

Florida. Florida is unusual, in that the entire state is considered a coastal zone. Twenty-seven (27) state laws make up the *Florida Coastal Management Program (FCMP)*, and it is administered by the Department of Environmental Regulation's Office of Coastal Management. Federal consistency reviews are conducted by the Department of Environmental Regulation (DER) with assistance from the Governor's Office of Planning and Budget (OPB). Actual day-to-day program administration is handled by three agencies: the DER, the Department of Natural Resources, and the Department of Community Affairs. These three agencies coordinate state coastal management through a procedural memorandum of understanding. Interagency coordination is further assisted by two committees: the state's Interagency Management Committee, which is made up of heads of all major FCMP agencies, and by the Governor's Citizen Advisory Committee.

Since Federal approval of the FCMP in 1981, it has been cited by NOAA (1990) as having made several accomplishments, including a joint effort by DER and the Department of Education to integrate environmental education programs into the general curricula of public schools, a DER initiative to help protect estuarine habitats, and the passage in 1985 of several Acts—the Coastal Zone Protection Act, the Local Development Regulation Act, the State Comprehensive Planning Act, and the Apalachicola Bay Protection Act. Finnell (1985) and others have indicated that Florida is a state that absolutely attempts to integrate policy building and decision-making at all levels of government.

While Florida has a number of strong, comprehensive and innovative programs as part of its coastal management plan, the very fact that 27 laws and 17 agencies are involved continues to cause problems in implementation and enforcement. The state still faces problems of deteriorating water quality because of continued uncontrolled development on some coastlines; water shortages caused by competition between agriculture and industry for potable water; continued erosion, property loss, and loss of life because of destruction of coastal sand dunes; the continued threat of sea-level a rise in; and the ongoing battle to preserve remaining wetlands and mangrove swamps from extinction.

California. In 1976, the California Legislature passed the *California Coastal Act (CCA)* to conserve (and develop) the state's 1770 km (1,100 mi) coastline. One year later, in 1977, it received Federal approval under the CZMA. The Act followed four years of coastal regulation under *Proposition 20,* the so-called "Coastal Initiative" which was enacted by California voters in November 1972.

The CCA's "wise use" policies guide decisions regarding the conservation and development of the state's coastal zone. These state coastal policies called for the following:

- *Aesthetic preservation:* preserving California's scenic coastal landscape
- *Protection of coastal agriculture:* Maintaining a productive coastal agriculture
- *Wildlife habitat protection:* Protecting wetlands, riparian corridors, tidepools and other important marine and coastal habitats
- *Increased public access:* Increasing public access and recreational use of the coast
- *Energy development:* Locating needed coastal energy and associated industrial facilities where they have the least adverse impact
- *Control of urban sprawl:* Encouraging new housing in existing urbanized areas rather than allowing scattered, sprawling development along the coast.

To implement these policies, the CCA established the *California Coastal Management Program (CCMP).* It is comprised of two segments: (1) a San Francisco Bay segment, administered by the San Francisco Bay Conservation and Development Commission (BCDC), and (2) the remainder of the coast, administered by a commission also established by the CCA—the California Coastal Commission (CCC). The CCC is the lead agency when interfacing with the federal CZMA.

Jurisdictional power is as follows: The CCC administers a coastal zone from the mean high tide line to approximately 1,000 yards inland, or up to 8 km (5 mi) in areas of significant coastal resources, and seaward to the limit of the territorial sea. BCDC's jurisdiction only extends inland generally 30 m (100 ft) from marshes and tidal waters of San Francisco Bay. The BCDC operates under the McAteer-Petris Act. Any proposed development involving dredging, fill, or substantial changes in shoreline use within the designated San Francisco Bay Area requires a BCDC permit.

California's Coastal Act encourages "vertical" power sharing between state and local governments, as well as "horizontal" coordination between government agencies. For example, local governments along the coast (15 counties and 54 cities) are required to incorporate the policies of the CCA into their own **Local Coastal Program (LCP)**—the specific long term management plan for a county or city's own section of the coast. LCPs consist of (a) land use plans, (b) zoning ordinances, (c) zoning district maps, and (d) within sensitive coastal resource areas, other implementation strategies. The CCC exercises permit control over all new development within the coastal zone until a region's LCP has been approved by the Commission. Once approved, the local area regains most permit control over new development. Hence, the "power" or authority over coastal development is "shared" between state and local governments. The CCA recognizes the need for horizontal or lateral coordination and attempts, to some degree, to integrate coastal program objectives with other state agency needs (e.g., energy development, highway construction, sewage disposal).

However, the Commission maintains permanent permit jurisdiction over the immediate shoreline, such as submerged lands, tidelands, and public trust lands. In addition, the Commission monitors local programs, hears appeals from certain local decisions, and reviews and advises on proposed LCP amendments.

California also has another state agency specifically designed to deal with coastal resource management issues—the *California State Coastal Conservancy (CSCC).* In 1976, the California legislature created the CSCC. Whereas the CCC uses "regulation" as its principal tool to conserve coastal resources, the CSCC uses "acquisition" to restore, protect, and manage the coast. The conservation strategy of the Conservancy is to use innovative acquisition techniques, such as development rights transfers, and less than fee interests (these and other acquisition strategies are explained in Chapter 8, *Open Space Preservation and Management).* The Conservancy has had many successes because it is innovative, flexible, and adaptable. For a history of the Conservancy's first 18 years, see Grenell (1994).

Non coast-specific state agencies also try to influence coastal protection. In July 1995, for example, California's Resources Agency issued a report calling for stronger ocean and coastal laws. The report noted that businesses tied to California ocean and coastal areas contribute $17.3 billion a year to the economy, and the state's coastal resources needed additional protection. The report called for:

- Promoting environmentally sound ocean and coastal tourism
- Better programs to protect fisheries and fish farming
- Improving ship traffic safety along the coast, including studying whether to move tanker traffic farther from shore
- Prohibiting new oil and gas exploration leases in tidelands, and opposing oil and gas leases on the Outer Continental Shelf off the coast of California
- Better coordination of federal, state and local pollution-control efforts
- Better maps regarding erosion of coastlines.

The report also recommended creating an ocean-management council of state agency officials charged with coordinating efforts to protect ocean and coastal resources, especially inland waters, bays and estuaries.

California has other state noncoast-specific agencies and legislation which parallel federal agencies, such as California Environmental Quality Act (CEQA), the California Department of Boating and Waterways, the State Department of Parks and Recreation, and the California Division of Mines and Geology. All are important players and have a stake in influencing the management of America's coastlines.

Summary

Overall, coastal states are becoming progressively more involved in coastal resource management. Prior to the Federal Coastal Zone Management Act, many states had little to no coastal planning and management. Today, as illustrated by Figure 2-2, most states have developed coastal management plans. For background information, program accomplishments, major grant tasks, significant program changes, federal consistency, and evaluation findings for each state's Coastal Zone Management Program, see NOAA (1990).

THE LOCAL PLAYERS

A large number of private and public local players have input in decisions about how coasts are used. Coastal resource managers must recognize the diversity of interests and the varied roles of these relevant players. Local players exist in both the private and public sector.

The Private Sector

In the private sector, the types of local interested parties range from developers to neighborhood associations, and to environmental organizations. The National Research Council (1990) has identified some of the more important local players from the private sector:

- *Professional developers and builders*. When the word "developer" comes up in conversation, many environmentalists are immediately put on alert. However, modern-day society needs developers and builders; we could not cope with our burgeoning populations without them. Without question, professional developers often find it highly profitable to build in high-demand coastal areas. They may hold an ownership interest in the site in question, or they may be merely working for the actual owner of the property. Regardless, they are liable professionally for some forms of unwise construction.
- *Private property owners*. It is the property owner that generally initiates change to a piece of property. Although private property owners may "own" the land, they do not have exclusive authority to do whatever they want to their property. The public at large has several protective mechanisms. For example, *nuisance laws* restrict one property owner from putting a stone groin perpendicular to its beach front property if it will negatively affect a neighbor's property; *covenants and deed restrictions* may also restrict future land use at a site by retaining certain rights by a prior seller.
- *Lenders*. A large group of local players are the lenders—the banks, savings and loan associations, and insurance companies that help finance land development. Many are under federal regulations requiring that their borrowers purchase flood or erosion insurance.
- *Realtors*. Real estate professionals have an interest in selling property on the coast, as well as elsewhere. Like developers, they can be liable for concealing certain information, such as erosion or flood hazards.
- *Homeowners associations*. Large development complexes often have "homeowner associations" set up by the developer to enforce subdivision deed restrictions. The members of the association are the owners of the lots within the complex. Homeowners associations especially play a role in coastal management when they own and manage beach and shoreline property on behalf of the subdivision.
- *Neighbors and other residents*. The individual citizen can also play an important role at the local level, since changes in local zoning require mandatory public notice.
- *Private Foundations or Conservation Groups*. These groups also play an important role at the local level, since they are responsible for purchasing and preserving much coastal land.

The Public Sector

At the local level, the public sector includes incorporated municipalities (e.g., cities, towns, villages, and boroughs), counties, and special districts (e.g., harbor districts and school districts). A variety of management tools are used at this level. Only the most often used tools are identified and briefly explained below:

- *Traditional zoning practices*. The most widely used local management tool is **zoning,** in which various parcels of land are assigned certain uses. Typical categories include industrial, utilities, transport, residential (various categories), recreation (e.g., forest or park reserves), wildlife preserves, and floodplains. Zoning can be used to help control growth and to protect coastal areas, but it is also easily influenced or modified by developers. *E-zones* (erosion hazard zones) and *V-zones* (flood hazard zones) are discussed in Chapter 4, *Coastal Hazards*.
- *Urban growth boundaries (UGBs)*. Another management tool used by the local players is the **urban growth boundary**—a line surrounding an urban area beyond which new development is not allowed. Portland, Oregon has used this management approach with great success. In fact, the state of Oregon requires that all incorporated communities adopt the UGB concept. Overall, few coastal communities have adopted UBGs as a management tool. Two that have, however, are Canon Beach, Oregon, and Baltimore's Harborplace (Beatley, Brower, and Schwab 1994).
- *Cluster development*. Some local communities have also required developers to adopt a concept known as **cluster development**—a development strategy which preserves areas of open space within housing complexes. This strategy allows coastal developers to protect highly sensitive areas (e.g., dunes, wetlands, or riparian habitats), while locating and orienting buildings away from more hazardous areas (e.g., typical storm paths; highly erodable cliffs).
- *Building codes*. Local communities have also set standards on how coastal buildings are designed and constructed. Examples include requiring houses be built on stilts so that surge water can easily pass underneath,

or requiring that roofs have as few projections as possible so they are less vulnerable to being blown away by strong winds. See Chapter 4, *Coastal Hazards*, for more details.

- *Setback requirements.* **Setbacks** have been mandated by some local communities to establish an exclusion zone adjacent to hazardous or sensitive areas (e.g., eroding sea cliffs). Within these "forbidden" zones, no building or structures are allowed. For example, Florida has established a "coastal construction control line" based on the estimated inland reach of a 100-year storm event. Construction is not permitted seaward of this line (Shows 1978). See Chapter 4, *Coastal Hazards*, for further details.

In addition to the above, local players use a variety of other management strategies such as subdivision ordinances, geologic hazard ordinances, planned unit developments, bonus or incentive zoning, performance zoning, fee-simple acquisition, transfer of development rights, community rating systems, and taxation and fiscal incentives (Beatley, Brower, and Schwab 1994). Most of these techniques are discussed in this book as they apply to specific subfields of coastal resource management (e.g., Chapter 8, *Open Space Preservation and Management*, or Chapter 4, *Coastal Hazards)*.

THE REGIONAL PLAYERS

In addition to federal, state, local agencies and programs, there are also the regional coastal players—those agencies and advisory committees that *transcend* political boundaries. We will now identify the need for a regional approach, then briefly discuss the most used regional strategies.

The Need for a Regional Approach

The basic reason that regional administrative bodies have been created is to deal with the dilemma that federal, state, and local jurisdictions are political and do not conform to natural ecosystems (e.g., bays, estuaries, lakes), nor do they always include all of the areas that impact the ecosystems (e.g., point and nonpoint sources of pollutants within a watershed). See Chapter 5, *Coastal Pollution*, for a discussion of how the citizens of Chesapeake Bay have established a regional management consortium to coastal resource management.

Problems with Federal Jurisdictions.

- *Too broad in scale.* Federal policy is often too broad, unable to protect natural resources of specific areas. For example, the National Environmental Policy Act (NEPA) and the Clean Water Act (CWA) had to have broad goals in order to be applicable nationwide.
- *Not site specific.* Sensitive resources are often in need of specific management policies for that site or region. For example, the topography of the land can affect the quality of air and the types of controls and regulations that must be enforced; the geology and geomorphology of a coastline can affect the rate of erosion and the types of construction standards that need to be followed.
- *No assurance of Inter-agency cooperation.* A federal policy can establish regulations and require state and local compliance, but it cannot demand (nor enforce) cooperation between state and local agencies. Since there is often disagreement between agencies as to which has ultimate control in environmental planning, a new federal compliance requirement may do nothing more than start a series of interagency conflicts—and the federal government does not serve as a mitigator.

Problems with State Jurisdictions.

- *Differences in interpretation.* A sensitive habitat may be bordered by different states. Unfortunately, states often interpret federal guidelines differently. As a result, a habitat may have different states practicing different (and sometimes) conflicting management strategies.
- *Too broad in scale.* Like Federal policy, state guidelines can also be too broad to address specific local problems. This is particular true of larger states like California that must draft state environmental policy that encompass a great variety of sensitive habitats.

Problems with Local Jurisdictions.

- *Too small in scale.* While federal and state jurisdictions can be too broad in scale, local jurisdictions can suffer from the other extreme; the local political boundaries are probably too small to include the watershed, airshed, or viewshed that impacts the habitat to be protected.
- *Too provincial in perspective.* Local jurisdictions can also be too provincial in viewpoint, being only concerned with increasing jobs, raising the tax base, attracting tourists, and so on. This local narrowness of thought, interest, or activity may not see (or want to see) wider community and national goals. If this is the case, immediate short term benefits (e.g., hotel construction) may be chosen over long term benefits for future generations (e.g., preservation of wetlands).
- *Too many, and too fragmented.* At the local level, there are just too many agencies and special districts. They each have their mini jurisdictions, and there is little incentive to cooperate to solve a bigger problem. Yet, the environmental problems (e.g., air pollution in the Los Angeles Basin; water pollution in the Chesapeake Bay) are regional in scope.

Regional Management Strategies

Cognizant of these traditional jurisdictional problems, there is an increasing number of players interested in trying a regional approach. The five most common regional strategies center around a watershed, estuary, lake, special area, or area of concern. These are briefly discussed below:

The watershed approach. In recent years, there has been an increasing interest in using an area's watershed as the principal planning boundary. Natural drainage patterns within a region make up a drainage or catchment basin called a **watershed**—the entire land surface or gathering ground of a single river system.

Watershed management deals with a particular watershed—its use, regulation, and treatment of water. Though truly multipurpose (e.g., water supply, flood management, water quality), the primary objective of watershed management is to improve the ability of the land *to hold water in place*. Techniques for accomplishing this objective range from reforestation (including the reseeding of denuded areas), to the terracing of slopes, and to altering farming techniques (e.g., promoting contour plowing, strip cropping, and the use of cover crops). To date, the Stillaguamish Watershed Protection Project in Washington State has had some success. It was formed in 1988 to address the livestock waste and other nonpoint source pollutants that feed into Puget Sound. For information on this project, and an overview of other national efforts at watershed management, see EPA (1991).

The estuary approach. Estuaries have also been used as a regional focal point. In 1986, for example, the Regional Administrator of the U.S. Environmental Protection Agency's office in San Francisco began to realize that a regional approach was needed to solve the numerous environmental problems associated with San Francisco Bay (really an estuary). The following year, the U.S. Congress established the *National Estuary Program*—a mandate of the federal Water Quality Act of 1987 (See Chapter 8, *Open Space Preservation and Management*, for details). Among other things, the National Estuary Program provides $60 million in Federal funding over a five-year period for developing comprehensive plans to address the environmental problems facing estuaries.

The *San Francisco Estuary Project* was established in 1988. It has the following goals:

- *Build consensus*. Public agencies, policy makers, and a variety of economic, social, and environmental interests had to agree on management priorities for the Estuary.
- *Identify gaps in technical knowledge.* Comprehensive research projects needed to be established and funded that would cover those areas where technical knowledge was lacking.
- *Conduct public outreach programs*. The general public needed to be informed about the Estuary through meetings, hearings, workshops, publications, and curriculum development.
- *Develop a comprehensive conservation and management plan.* The plan needed to address five major issues: (a) the decline of biological resources; (b) increased pollutants; (c) freshwater diversion and altered flow regime; (d) increased waterway modification; and (e) intensified land use.

The San Francisco Estuary Project finished its Comprehensive Conservation Management Plan in 1992. Over the next twenty years this plan will be enacted through a "Formal Implementation Committee" which has divided the estuary into three geographical subgroups. Initially these groups are placing priority on research and monitoring the health of the estuary, wetlands planning and acquisition, scientific studies to determine the quantity of wetlands needed to maintain the health of the estuary, and monitoring of pollution and fish quality as they relate to public health and subsistence fishermen (Ankrum 1995).

The lake approach. The regional approach can also be beneficial in the management of lakes, or complexes of lakes like the Great Lakes that are on the border between the United States and Canada. The five interconnected Great Lakes are a tremendous resource to these two nations. These lakes contain at least 95 percent of the surface fresh water in the United States and 20 percent of the world's fresh surface water. Some 35 million people depend on the Great Lakes for drinking water, and about 40 million people live within the Great Lakes basin—all using some aspect of this great resource (Miller 1994).

The Great Lakes are too large and complex to directly "manage," but there is a cooperative agreement between Canada and the United States to address its many environmental problems. The *Great Lakes Program (GLP)* was established in the 1970s. Although its focus is on the lakes themselves, this interjurisdictional effort encompasses the entire watershed of the Great Lakes. The International Joint Commission (IJC) monitors the progress of the implementation of the GLP. In the United States, the GLP is administered by the Great Lakes National Program Office (GLNPO), which is an office within the U.S. Environmental Protection Agency.

In 1992 the GLNPO began a five-year cooperative program, the Great Lakes strategy, to coordinate the activities of federal, state, and local governments, including the EPA, NOAA, the U.S. Natural Resources

Conservation Service, the Army Corps of Engineers, the U.S. Fish and Wildlife Service, the U.S. Coast Guard, and state and local counterparts. This program strives to guard and enhance wildlife habitat and associated wildlife, reduce toxic inputs to the lakes, protect human health, and foster cooperation between government entities and citizen groups, with special emphasis on keeping the public informed on the state of the Great Lakes ecosystem. In order to assist with project management, the strategy divides the lake region into geographical management units (Special Geographic Initiatives), such as Remedial Action Plans (RAPs) based on Areas of Concern, and Lakewide Management Plans (LAMPs), which will help focus efforts on local and regional problems. Remedial Action Plans are designed to address specific, localized problems in small geographical areas, such as rivers and harbors, through identification of problems and solution options, while measuring progress against the goals established by the Great Lakes strategy. Meanwhile, Lakewide Management Plans guide efforts within larger units defined by the lakes themselves, such as Lake Superior. Notable program achievements include the collection of 11,000 pounds of "Suspended or canceled" pesticides through a "clean sweep" program, commitment from utility companies in the Great Lakes region to speed up phase out of PCB dependent equipment, and successful creation of RAP and LAMP programs (EPA 1994).

The special area management approach. Another regional approach that stresses cooperation and negotiation is *Special Area Management Planning (SAMP)*. This approach is a function of the Federal Coastal Zone Management Act, designed to address problems associated with cohesive ecological regions. These regions may be as small as a few thousand acres (e.g., San Bruno Mountain in San Francisco, California) to several million acres (e.g., Adirondack State Park in the state of New York), with boundaries defined by the geographic characteristic which makes that region an interdependent unit (e.g., mountain range, watershed, etc.). This planning tool has had varying degrees of success in the United States.

One example of a successfully developing SAMP may be found in the Chesapeake Bay Program. In 1975, the Environmental Protection Agency recognized the Chesapeake Bay as a damaged ecosystem, initiating a study to outline the problems and develop solutions. This study culminated in the first Chesapeake Bay Agreement signed in 1983 by the EPA, the governments of Virginia, Pennsylvania, Maryland, and the District of Columbia, and by the Chesapeake Bay Commission (an interstate legislative coordinating entity), thus initiating the Chesapeake Bay Program. This program was designed to coordinate the research and restoration efforts of governmental and nongovernmental organizations with the states in the 165,747 sq km (64,000 sq mi) Chesapeake Bay watershed. The goals of the program are to address concerns with nutrient enrichment, toxic substances, declines in submerged aquatic vegetation, wetlands alteration, shoreline erosion, hydrologic modification, fisheries modification, shellfish bed closures, dredging and dredge materials disposal, and the effects of boating and shipping on water quality (Environmental Protection Agency 1993).

In 1992, amendments to the Chesapeake Bay Agreement reaffirmed cooperative and outreach efforts and restated a commitment (made at the second Chesapeake Bay Agreement in 1987) to reduce point and nonpoint source pollution by 40 percent of 1985 levels by the year 2000. Overall, the program has been successful in creating cooperative efforts which have resulted in dramatic improvements in the health of submerged aquatic vegetation and the striped bass population. Meanwhile, moderate improvements have been made in dissolved oxygen levels and nutrient loading reduction. However, regional population growth and the difficulty of managing the diverse land uses in this huge watershed have been manifested in reductions of important oyster communities. As recognized by the program developers, this effort is a "work in progress." The states of New York, West Virginia, and Delaware still need to be incorporated in the Chesapeake Bay Program, while existing program participants face many challenges in accomplishing the goals they have established (Environmental Protection Agency 1995).

The "areas of environmental concern" approach. Some states, like North Carolina, incorporate "Areas of Environmental Concern" (AECs) as a planning mechanism within their state Coastal Acts. It is an attempt to identify especially sensitive areas within the coastal zone that require special attention, and a regional approach to management. AECs ensure a review process, through permitting, designed to prevent developments which are incompatiable with maintenance of the resource of concern. Examples of possible AECS are estuarine areas (estuarine waters, coastal wetlands, estuarine shorelines), natural hazard areas (beaches, frontal dunes) that are subject to flooding and erosion, water supplying watersheds and aquifers, and sensitive natural and cultural resource areas. All AECs receive a formal description in the state guidelines which describe the resource area and its significance, sets management goals, and creates use standards for the area. Only very valuable and extremely sensitive regions are given AEC status, in order to preserve the integrity of the designation (Beatley, Brower, and Schwab 1994).

Problems with Regional Management Strategies

Unfortunately, some of the reasons a regional approach to coastal management is required are the very reasons this type of approach is difficult to achieve (Beatley, Brower, and Schwab 1994).

- *Public's resistance toward government jurisdiction on private property rights.* Given our form of government, authority is fragmented among different agencies and levels of government rather than focused at a federal level. Therefore, enforcement power is often not a part of a regional management committee, leaving interpretation and enforcement up to several diverse groups. Some individuals feel government at anything but the local level regarding public property is unconstitutional.
- *Public's overall feeling that the community will survive any short-term problems (e.g., flooding every 100 years).* Many communities focus their limited resources on short-term issues surrounding the local economy and personal well-being of their citizens. It is often difficult to get support for more holistic, ecosystem-level management practices that may immediately affect their short term goals. An occasional disaster is required to get public support and attention for regional management issues, but this is a reactive attitude and usually more expensive in the final analysis.
- *Lack of funding to support regional management activities.* Federal funds are often available for initial planning and environmental impact analysis, but implementation and project monitoring costs must be absorbed by local and state governments. Given the limited resources and different local priorities, funding becomes a major issue in the success of regional management programs.
- *Opposition by local business, real estate companies, homeowners, and area developers.* Our free-market economy puts local interest in profit from investments in direct conflict with regional environmental needs. Consensus and negotiation are the only tools available to the regional management committee to obtain agreement and cooperation from all parties who will be affected by regional management programs.
- *No incentives to protect resources of regional significance.* Incentives require cooperation by government at all levels in the form of tax relief, investment insurance, etc., which is often not considered possible, given the already limited local financial resources.
- *Lack of clear regional management goals and priorities, and a lack of trained personnel to implement the program.* Where regional programs fail the most is in implementation. This stems from not getting consensus and agreement on the top-priority goals to be achieved from all areas effected by the proposed management program. Once the program is ready for implementation, the personnel selected to manage it are often not qualified to work with the diverse groups of people and government that will be affected.

This section has provided a few examples of ongoing strategies using the regional approach and some of the obstacles inherent in this type of approach. Clearly, the regional perspective and ecosystem approach, though not completely fail-safe, will play an increasingly important role in the United States and around the world in the future.

INTEGRATED MANAGEMENT OF COASTAL AREAS AND MARINE SANCTUARIES

Our nation's coastal zone (including marine sanctuaries) is facing urgent problems—problems of human encroachment, natural hazards, loss of habitat, deteriorating environmental quality, reduced biodiversity, and diminishing levels of shellfish and fish populations. Managing these problems will require a new paradigm—a new way of thinking.

Integrated Coastal Management—A New Paradigm

In the past, "coastal resource managers" dealt strictly with coastal areas (the land), while marine biologists and oceanographers limited their concerns to nearshore and offshore waters (the sea). Today, scientists and resource managers are beginning to see the need to integrate the management of land and sea. In other words, a more integrated, highly participatory management approach *(holistic thinking)* must replace the traditional segmented approach *(reductionist thinking)* where each agency (e.g., the U.S. Fish and Wildlife Service) concentrates on one of the coastal and sanctuary picture (e.g., the migration of salmon).

This relatively new way of thinking, best known as **Integrated Coastal Management (ICM),** had its beginnings in the 1980s as other nations began to develop their coastal management programs. Since then, there have been numerous definitions of ICM, as nicely summarized by Knecht and Archer (1993). One of the more comprehensive definitions is provided by Bower (1992, p. 3):

> *At minimum, any definition should include the integration of programs and plans for economic development and environmental quality management, and more specifically the integration of cross-sectoral plans for fisheries, energy, transportation, waste disposal, tourism, etc. ICM should also include the vertical integration of responsibilities for management actions among various levels of government—international, national, state, and local—or between the public and private sectors. It should include all the components of management—from the planning tasks of analysis and design, to the implementation of tasks of installation operation and maintenance, monitoring and*

evaluation of strategies over time. ICM should be cross-disciplinary among the sciences, engineering (technology), economics, political science (institutions), and law. In practice, it is all of the above.

Taking this and other definitions into consideration, Ehler and Basta (1993), have identified at least five possible pathways to "integrate" coastal and sanctuary management:

- *across coastal and sanctuary management agencies* (natural resource management, environmental protection, economic development);
- *across levels of authority and jurisdiction* (federal, state, regional, local);
- *across economic sectors* (recreation, agriculture, industry, energy);
- *across traditional academic disciplines* (science, engineering, social science [law, political science, economics]);
- *across the management tasks themselves* (e.g., wetlands restoration and fisheries management).

Knecht and Archer (1993, pp. 187–188), however, have done the best job of reviewing recent articles and reports, and clarifying exactly what is ICM. They have simplified ICM to four dimensions: (a) Intergovernmental, (b) Land-water interface, (c) Intersectoral, and (d) Interdisciplinary. They have written as follows:

(a) Intergovernmental—*This dimension encompasses the necessary integration of various levels of government into coastal management, especially between the national level and regional/local levels. In the US system, three levels are involved: national, state, and local.*

(b) Land-water interface—*Clearly, integration across the land-water boundary is basic to the concept of coastal management. The coastal zone area to be managed is usually defined in terms of both a shoreland area (the uses of which affect the coastal waters) and a water area (the uses and disturbances of which affect the shoreland). Understanding the effects which traverse the land-water boundary (in both directions) is of fundamental importance.*

(c) Intersectoral—*It has become increasingly clear that the rational management of coastal resources requires that all activities affecting such resources (or the coastal environment in which they reside) come within the 'reach' of the management program. For example, dredging to create a deeper harbor and/or safer navigation channels can potentially affect habitats that are of critical importance to coastal fisheries. Hence, such dredging must be within the policy and regulatory purview of the coastal management program.*

(d) Interdisciplinary—*This dimension pertains to the need for a holistic approach in ICM. It reflects the realization that coastal zone issues not only involve the use and protection of natural resources and the coastal environment, but that significant economic and social issues almost always exist as well. Decisions to protect or develop a particular resource usually have significant economic implications. Major social and cultural issues can also be involved.*

The following chapters help illustrate the need for this integrated approach to coastal and marine sanctuary management, and, when possible, provide examples of "success stories" in integrated management. Unfortunately, most agencies are still practicing the traditional disjointed approach, so the number of examples of "how we can make it work" are few. The book will conclude with an analysis of how the United States is doing in terms of moving toward ICM.

Key Concepts for 21st Century Integrated Coastal Management

Most of the concepts that we need for integrated coastal and sanctuary management are founded on the "building blocks" of modern environmentalism. They include such notions as renewable resource use, harmony with nature, resource conservation, carrying capacity, ecological stability, habitat protection, energy efficiency, biological and cultural diversity, wastelessness, sustainability, durability, steady state economy, recycling, appropriate technology, decentralization, conflict resolution, cooperation, human scale, sense of community, and many more. Oddly enough, most of these "modern ideas" have been practiced for hundreds (and in many cases thousands) of years by traditional (indigenous) societies around the world (Appendix 2-1).

Lets take one traditional practice—the concept of *sustainability*—as an example. There is no question that the latest "buzzword" in government, business, and industry today is "sustainability." Yet, this is perhaps the key lesson that traditional societies have taught us. For 99 percent of human history, humans practiced a way of life that anthropologists and cultural geographers refer to as "hunting/gatherers." The literature is full of examples of hunter/gatherers, as well as agriculturalists and/or pastoralists, that had economies that endured for eons of time (Durning 1992; Klee, ed., 1980; McNeely and Pitts, eds., 1984). Today, the United States and other developed nations are beginning to realize that we must redesign our economies so they are more sustainable. To be more specific, city and regional planners in the U.S. are beginning to hold conferences entitled, "Making Cities Sustainable," that ultimately end up discussing other concepts practiced by traditional societies, such as *resource conservation*, *waste recycling*, and *habitat protection*.

There is *a quiet revolution* that is occurring across the globe. The world is moving forward (though to environmentalists it seems at times at a glacier's pace) *at relearning the concepts and lessons that many traditional societies have yet to forget.* As chief "Worldwatcher," Lester Brown has so clearly pointed out, "If this Environmental Revolution succeeds, it will rank with the Agricultural and Industrial Revolutions as one of the great economic and social transformations in human history" (Brown 1992).

As this book will hopefully make crystal clear, *this quiet revolution is drastically needed, and has actually already begun in the minds and actions of many of those that study, plan, and manage our precious coastal resources.* The purpose of this book is to help practicing (and future) coastal resource managers identify those needed key concepts, and more importantly, see their applicability to coastal issues and alternative management strategies.

CONCLUSION

This chapter has provided a *brief overview* of the principal players and associated programs that constitute the framework for coastal resource management. In no way is it meant to be all inclusive. As should be very clear by now, the "framework" is loose, fragmented, and in need of integration. At the international level, there is no unified approach to coastal zone management. In the United States, coastal issues are within a single major legislative and procedural system, but there is no single agency that oversees it. Coastal management occurs at all levels of government—federal, state, regional, and local. For an excellent *in-depth* comparative analysis of these levels of administration as they apply to coastal zone management, see Beatley, Brower and Schwab (1994).

The need for "integrated coastal management" (ICM) has increasingly been articulated in recognition of the ongoing fragmentation of decision making in virtually all areas with respect to the use and management of coastal resources. The ICM concept is not new, but rather a comprehensive and continuous refinement of such ongoing conservation strategies as "regional environmental quality planning," "comprehensive watershed management," "area-wide waste treatment management," "basin-wide planning," or "special area management." Developing an integrated management approach to coastal resource management will not be easy, and it will most likely be done (if at all) by incremental steps over a long period of time. Most would agree, however, that ICM is an excellent concept that must be pursued. The question is, how do we implement the ideas, concepts and principles that it embodies. This book will try to keep the reader thinking about this difficult task as we delve deeper into the many coastal management issues and concerns. For now, let's turn to an often neglected subject relating to coastal studies—marine and coastal protected areas.

APPENDIX 2-1

Relationship between traditional concepts and environmental goals for 21st century living.

Traditional Concept	Traditional Examples	Environmental Goals for 21st Century Living
Environmental Elements		
Renewable resource use	*Amish* (U.S.A.) use of wind power (wind mills) for water transport; *Ladakh* (northern India) use of animal and muscle power in agriculture, as well as grain mills powered by hydropower; *widespread* use of wind and solar power in *African countries* for threshing and drying.	Goal: To have greater reliance on renewable energy resources rather than fossil fuels.
Harmony with nature	Traditional hunter/gatherers made relatively little impact on the landscape (e.g., *Ohlone Indians*, Central California; *San (Bushmen)*, Kalahari Desert, Africa); *Ladakh* saw the interconnectiveness of all things.	Goal: To live as an integral part of the ecosystem—humans are perceived as just one more animal on the land.
Resource Conservation	*Tsembaga* (New Guinea) protected tree seedlings within their swidden gardens; *Lakakh* prevented overgrazing by always moving herds; *Nepalese* (Nepal) terraced their rice fields to help prevent soil erosion.	Goal: To practice good land stewardship in order to preserve the resource base, whether it be grasslands, forests, croplands, or fisheries.

(continued)

Traditional Concept	Traditional Examples	Environmental Goals for 21st Century Living
Environmental Elements (continued)		
Carrying Capacity	*San (Bushmen)* were nomadic, which minimized impact in one locale; Many *Pacific Islanders* relocated people to maintain proper land/people density ratios.	Goal: To not exceed the resource base upon which your society is based.
Ecological stability	Hillside agro-ecosystems in *Tlaxcala (Mexico)* were designed to minimize ecological disruption.	Goal: To foster land, air, and water practices that minimize ecological disruption.
Habitat Protection	*Tsembaga* use of logs to reduce soil erosion on hillsides; *Marshall and Pukapuka islanders* maintained habitats for game.	Goal: To protect the local natural habitat.
Energy Efficiency	*Eskimos* of southern Baffin Island found the use of their traditional harpoon was more "energy efficient" than using a modern rifle in seal hunts.	Goal: To demand that all things be energy efficient so as to obtain and maintain a sustainable energy supply.
Economic Elements		
Diversity	*Tsembaga,* New Guinea, practiced swidden agriculture—an example of polyculture (many crops in one field); *San (Bushmen)* had a diet much more diverse than modern man.	Goal: To preserve, protect, or restore landscape diversity (flora and fauna).
Wastelessness	Traditional *Miskito Indians* (Nicaragua) ate all parts of the turtle; *Ladakh* were extremely frugal in their resource use.	Goal: To waste as little raw materials and energy as possible; to reduce the flow of energy and matter through your life.
Sustainablility	For 99 percent of human history, humans have been hunter/gatherers—a very enduring (sustainable) way of life; traditional *Ladakh* way of life lasted for centuries.	Goal: To redesign economies so as to emphasize endurance (growth that is sustainable), not uncontrolled growth and unnecessary materialism.
Durability	*Amish* made durable tools, buggies, and clothing.	Goal: To move away from the "throwaway mentality;" to strive for high quality engineering and design as opposed to planned obsolescence.
Steady State Economy	Many *Pacific island societies* limited their economies to fit their small land space.	Goal: To strive for an economy that is based on qualitative growth, not undifferentiated economic growth.
Recycling, Reuse, Waste conversion	*Southern Baffin Island Eskimos* recycled fat from marine animals for fuel; recycling of waste products was a *widespread* practice in traditional *African societies.*	Goal: To increase the share of raw materials that are recycled or reused (e.g., dairies that convert animal manure into methane gas which turns a generator for production of electricity).
Technocratic Elements		
Value of the Generalist	Hunters/Gatherers (e.g., *Ohlone* Indians, Central California; *San (Bushmen),* Kalahari Desert, Africa) were generalists; most *Ladakh* had knowledge and skills to build a house	Goal: To think in broad, holistic, multidimensional terms, rather than as a reductionist or technocrat.
Appropriate Technology	Wisdom of the *Palauans* (Micronesia) when they rejected U.S. desires for an oil super-tanker facility on one of their atolls.	Goal: To always assess the best scale or level of technology for a community, not necessarily the most advanced or complex.
Controlled Technology	*Amish* use of tractors without rubber tires.	Goal: To control the types and rate of cultural change, particularly technological advances.
Simplicity	*Amish* made simple homemade toys for children; simplicity in house design & clothing that they wear.	Goal: To choose, when possible, the simpler, less consumptive lifestyle. Avoid fancy new cars, keeping up with the fashion industry, and insisting upon having the fastest, most powerful computer.
Selective Adaptation	*Amish* would use a combine only if it was drawn by a horse, not if it was pulled by a tractor.	Goal: To favor and adopt new ways of doing things that support an enduring society.

(continued)

Traditional Concept	Traditional Examples	Environmental Goals for 21st Century Living
Political Elements		
Self sufficiency	*Amish* and *Ladakh* constructed houses from local materials; *Tsembaga* practiced a form of swidden gardening and pig husbandry that allowed for self sufficiency.	Goal: To reduce reliance on outside/distant resources.
Decentralization	*Amish* would not allow their homes to be connected to central power sources outside of their community; telephones were used, but not allowed in the home.	Goal: To "unplug" as much as possible from the large centralized infrastructure, such as fossil-fuel or nuclear power plant complexes.
Conflict resolution	*Tsembaga* used pig cycles to help reduce societal conflicts.	Goal: To establish systems of management that help resolve environmental and other societal conflicts.
Cooperation	*Amish* and *Ladakh* both practiced cooperation, not competitiveness, when it came to labor, sharing tools, and farming; the *Dene Indians* of Canada and the *Eskimos* of Alaska survived the harsh winters by practicing cooperation.	Goal: To improve cooperation and coordination at the international level (despite the need to decentralize within national borders); to work with the rest of nature, not to dominate it.
Stratification Elements		
Wealth leveling devices	Uniform dress code in *Amish* culture; *Palauans* had self-imposed social controls that forbade the display of wealth.	Goal: To lesson the gap between the "Haves" and "Have-nots"; to work towards "fixed or reduced wants" within society; to strive for less, not more; concept that "less is more."
Social Security for the Elderly	*Ladakh* families belonged to groups that helped each other out during birth, old age, death; *Palauans* (Micronesia) took care of their elderly within their own homes.	Goal: To work for the inclusion, not exclusion, of the elderly from the main stream of society.
Value Elements		
Humility	Numerous *hunter/gatherer societies* believed they were one with their fellow creatures.	Goal: To have "animal humility" (i.e., to move away from modern industrial society's collective arrogance to dominate nature; to realize that we share the earth with other creatures; we do not own the earth).
Population Control	Traditional Oceanic islanders (e.g., *Pukapuka, Yap, Truk, Kapingamaragi*) practiced various forms of birth control, including abortion.	Goal: To control the rate of population growth so as not to destroy the quality of life of the entire society; childbearing should be optional, not automatic.
Human Scale	Most hunters/gatherers, traditional agriculturists and pastoralists lived in hamlets, villages, or small towns; *Ladakh* maintained that small scale of community helped build community ties.	Goal: To appreciate that bigger is not always better, that smaller may actually be better in some cases (e.g., living in a village may provide more sense of well-being and security than living in a large city).
Quality over Quantity	Life in traditional *Ladakh* was considered rich—much beyond subsistence; traditional *San (Bushmen)* can be viewed as having led an "affluent life;" traditional *Palauans* never thought of themselves as poor until outside Western news reporters told them they were poor.	Goal: To strive for human fulfillment through "life quality" (clean air and water, healthy food, interesting endeavors, love, caring) not "life quantity"—the number of material possessions.
Ecological Religion	Traditional *Taos Indians* of New Mexico believed the trees and lakes were living spirits; *native Hawaiians* believed that volcanoes, lava, streams, and the land itself were godlike sacred places; the *Ladakh* practiced Buddhism—a interdependence and relationships of humans, animal, plants, soil.	Goal: To have a deep, spiritual connection with the earth; to worship a religion (institutional or self-designed) that is earth oriented—one that has an ingrained conservation or environmental ethic.

(continued)

Traditional Concept	Traditional Examples	Environmental Goals for 21st Century Living
Interpersonal Elements		
Sense of Community	*Amish* barn raising activity where community comes together to help the Amish family build a barn; *Ladakh* people worked together on building and repairing irrigation projects.	Goal: To feel a responsibility towards the welfare and well being of the entire community.
Reciprocity	*San (Bushmen)* of Kalahari Desert shared food that was hunted and gathered; *Ladakh* shared in the distribution of water; *Palauans* reciprocated with fish to meet certain debt or obligations.	Goal: To have mutual sharing within society.
Environmental Awareness	Traditional *Palauans* were aware of cycles of nature (sun, moon, stars, fish and bird migrations); *Inuit* of the Arctic regions had an intimate knowledge of the environment, based on ancient teachings.	Goal: To be able to "read the landscape" once again; to feel a "sense of place" with your locale.
Cyclic Time	Traditional *Palauan* calendar and time reckoning based on lunar and other cycles within nature; *South Asians* believed that time moved in cycles within cycles.	Goal: To understand that time and causation are cyclic, not just linear.
Work Elements		
Work is Play	*Ohlone* Indian food gathering was a time for work as well as social activity; the *Ladakh* also saw work & leisure as one; hunting was as much challenging fun as work for the *San (Bushmen)*.	Goal: To strive for a career that is as enjoyable as "play." When you can say to yourself, "Boy, I can't believe they pay me for this," you know you have reached that point.
Labor has a Place	The *Amish* saw value in work—that it strengthened character and resolve.	Goal: To see and appreciate the value of intensive labor v. inappropriate technologies.

Source: Adapted from Klee, Gary A. 1996. "Traditional Concepts for 21st Century Living," *Etnoecológica* (Ethnoecology), vol. III, No. 4–5, August, pp. 5–21.

REFERENCES

Ankrum, Kathryn. 1995. Public affairs officer with the San Francisco Estuary Project. Personal communication, 28 September, Oakland, California.

Beatley, Timothy, David J. Brower, and Anne K. Schwab. 1994. *An Introduction to Coastal Zone Management*. Washington, D.C.: Island Press.

Bledsoe, Robert L., and Boleslaw A. Boczek. 1987. *The International Law Dictionary*. Santa Barbara, CA: ABC-Clio, Inc.

Bower, B. T. 1992. *Producing Information for Integrated Coastal Management Decisions: An Annotated Seminar Outline.* Office of Ocean Resources. Conservation and Assessment, National Ocean Service. November. Washington D.C.: National Oceanic and Atmospheric Administration (unpublished manuscript).

Brower, D. J. and D. S. Carol. 1984. *Coastal Zone Management and Land Planning, Report 205*. Chapel Hill, North Carolina: National Planning Association.

Brown, L. R. 1992. "Revolution in the Making." *Worldwatch*. Vol. 5, No. 1, p. 2.

Buchholz, Hanns J. 1987. *Law of the Sea Zones in the Pacific Ocean*. Singapore: Institute of Southeast Asian Studies.

Bulloch, David K. 1989. *The Wasted Ocean*. New York: Lyons & Burford Publishers.

Charis, S. 1985. "The Coastal Zone Management Act: A Protective Mandate." *Natural Resources Journal*. Vol. 25, No.1, pp. 21–30.

Coastal Zone Management Act. 1972a. U.S.C. Title 16, Chapter 33, Section 1452.

Coastal Zone Management Act. 1972b. U.S.C. Title 16, Chapter 33, Section 1453.

Coastal Zone Management Act. 1972c. U.S.C. Title 16. Chapter 33, Section 1456.

Durning, A. T. 1992. *Guardians of the Land: Indigenous Peoples and the Health of the Earth.* Worldwatch Paper 112. Washington D.C.: Worldwatch Institute.

Ehler, Charles N. and Daniel J. Basta. 1993. "Integrated Management of Coastal Areas and Marine Sanctuaries: A New Paradigm." *Oceanus*. Fall. Vol. 36, No. 3, pp. 6–13.

Environmental Protection Agency. 1991. *The Watershed Protection Approach: An Overview*. Washington, DC: Environmental Protection Agency.

Environmental Protection Agency. 1993. *Chesapeake Bay program a work in progress: A retrospective on the first decade of the Chesapeake Bay restoration*. Annapolis: Chesapeake Bay Program.

Environmental Protection Agency. 1994. *A Report to Congress on the Great Lakes Ecosystem.* Chicago: Great Lakes National Program Office, Environmental Protection Agency.

Environmental Protection Agency. 1995. *The State of the Chesapeake Bay 1995.* Annapolis: Chesapeake Bay Program.

Finnell, Gilbert L., Jr. 1985. "Intergovernmental Relationships in Coastal Land Management." *Natural Resources Journal*. Vol. 25, No. 1, pp. 31–60.

Frankel, Ernst. 1995. *Ocean Environmental Management*. Englewood Cliffs, New Jersey: Prentice Hall.

Godschalk, David R. 1992. "Implementing Coastal Zone Management: 1972-1990." *Coastal Management*. Vol. 20, pp. 93–116.

Grenell, Peter. 1994. "Looking Back in Light of the Future: The Coastal Conservancy's First 18 Years." *California Coast & Ocean*. Vol. 10, No. 1, Winter/Spring, pp. 36–43.

Kenchington, Richard A. 1990. *Managing Marine Environments*. New York: Taylor & Francis.

Klee, Gary A. 1996. "Traditional Concepts for 21st Century Living," *Etnoecológica*. Vol. III, No. 4–5, August, pp. 5–21.

Klee, Gary A. (ed). 1980. *World Systems of Traditional Resource Management*. London: Edward Arnold.

Knecht, Robert W., and Jack Archer. 1993. "'Integration' in the U.S. Coastal Zone Management Program." *Ocean & Coastal Management*. Vol. 21, pp. 183–199.

Knecht, Robert W., Biliana Cicin-Sain, and Jack H. Archer. 1988. "National Ocean Policy: A Window of Opportunity". *Ocean Development and International Law*. Vol. 19, pp. 113–142.

McNeely, J. A., and D. Pitts (eds). 1984. *Culture and Conservation: The Human Dimension of Environmental Planning*. London: Croom Helm.

Miller, G. Tyler, Jr. 1994. *Living in the Environment*. Belmont, California: Wadsworth Publishing Company.

Mitchell, J. K. 1982. "Community Response to Coastal Protection," In *Ocean Yearbook* 3, Chicago, Illinois: University of Chicago Press, pp. 358–319.

National Research Council. 1990. *Managing Coastal Erosion*. Washington D.C.: National Academy Press.

National Oceanic and Atmospheric Administration. 1990. *Biennial Report to the Congress on Coastal Zone Management: Fiscal Years 1988 and 1989*. April. Washington D.C.: National Oceanic and Atmospheric Administration.

Prescott, J. R. V. 1985. *The Maritime Political Boundaries of The World*. London: Methuen.

Resources Agency (The). 1995. *California's Ocean Resources: An Agenda for the Future (Draft)*. Sacramento, CA: The Resources Agency of California.

Shows, E. W. 1978. "Florida's Coastal Setback Line—An Effort to Regulate Beach Front Development." *Coastal Zone Management*. Vol. 4, pp. 151–164.

FURTHER READING

Brower, David. *1991. Evaluation of the National Coastal Zone Management Program.* NCRI-W-91-003. Washington, D.C.: National Oceanic and Atmospheric Administration, U.S. Department of Commerce.

Carter, R. W. G. 1988. *Coastal Environments: An Introduction to the Physical, Ecological and Cultural Systems of Coastlines*. New York: Academic Press.

Cicin-Sain, Biliana. 1990. "California and Ocean Management: Problems and Opportunities." *Coastal Management.* Vol. 18, No. 3, pp. 311–335.

Clark, John R. 1996. *Coastal Zone Management Handbook*. Boca Raton: CRC Lewis Publishers.

Delaney, Richard F., David W. Owens, and James F. Ross. 1991. *Florida's Coastal Management Program: An Independent Assessment*. Prepared for the Governor's Office of Planning and Budget. July.

Fischer, Michael L. 1985. "California's Coastal Program: Larger than Local Interests Built into Local Plans." *Journal of the American Planning Association*. Summer, Vol.51, No.3, pp. 312–321.

Kenchington, Richard and David Crawford. 1993. "On the Meaning of Integration in Coastal Zone Management." *Ocean & Coastal Management.* Vol. 21, pp. 109–127.

Lima, James T. 1990. *Ocean and Coastal Management: The Role and Activities of California Government in Spring 1988.* California Sea Grant College, Working Paper No. P-T-52.

Platt, R. H. 1985. "Congress and the Coast." *Environment*. Vol. 27, No. 6, pp. 12–17, 34–39.

Wenzel, Lauren, and Donald Scavia. 1993. "NOAA's Coastal Ocean Program: Science for Solutions." *Oceanus*. Vol. 36, No. 1, pp. 85–92.

3 Marine And Coastal Protected Areas

Scientists believe that off the coasts of the United States, from the frigid ice-scoured waters of the Arctic Ocean to the tropical reefs of the Florida Keys, the West Indies, and the Pacific Islands, there are more kinds of marine plants and animals in more kinds of marine habitats than are found off any other country in the world.

—*A Nation of Oceans*, descriptive document on the U.S. National Marine Sanctuary Program, 1986.

Placing marine and coastal areas under some form of protection is a conservation strategy that is practiced more and more. In 1989, for example, there were 977 marine and coastal protected areas around the globe, covering some 211,406,000 hectares. One hundred seven (107) of those protected sites, covering 54,317,000 hectares, were associated with the United States (The World Resources Institute 1994). More than 100 protected sites are in the Caribbean region (Sobel 1993). Such worldwide interest in MPAs has led United Kingdom environmental consultant Susan Gubbay to claim that "Marine protected areas have become the flagships of marine conservation programmes in many parts of the world" (Gubbay 1995, 1). (See Figure 3-1).

The World Resources Institute defines a **marine and coastal protected area** as a site over 1,000 hectares (2471 acres) with littoral, coral, island, marine, or estuarine components. Other international organizations, such as the International Union for the Conservation of Nature and Natural Resources (IUCN), have more broadly defined marine protected areas from small, highly protected reserves to large, multiple-use areas and biosphere reserves. For a useful review of the IUCN's ten categories of protected areas, see Sobel (1993).

The focus of this chapter will be on marine and coastal protected areas as a conservation technique, with a particular emphasis on the U.S. National Marine Sanctuary Program. As indicated in Chapter 2, any true effort at Integrated Coastal Management will require the integration of coastal issues (traditionally land oriented) and sanctuary issues (traditionally marine oriented) conservation strategies. Consequently, it is necessary that we include a discussion on marine and coastal protected areas.

The section entitled *Marine and Coastal Protected Areas (MPAs) as a Conservation Technique* begins with a look at the fishery, nonfishery, and overall economic benefits of establishing a marine protected area. This section also discusses the challenges facing those that must try to "integrate" coastal and sanctuary management. *The U.S. National Marine Sanctuary Program* looks at the "marine equivalent" of the U.S. National Park Service. The section on *Program Description,* answers such questions as . . "What is a national marine sanctuary?" "What are the mission and goals of the National Marine Sanctuary Program?" "What are the benefits of designation?" "What is the Site Evaluation List?" "How are sites chosen?" With this background, *Number and Location of Sanctuaries* introduces the reader to the uniqueness of each sanctuary. Conservation challenges for each particular site will be identified. *Problems of the U.S. National Marine Sanctuary Program* will identify the overall prob-

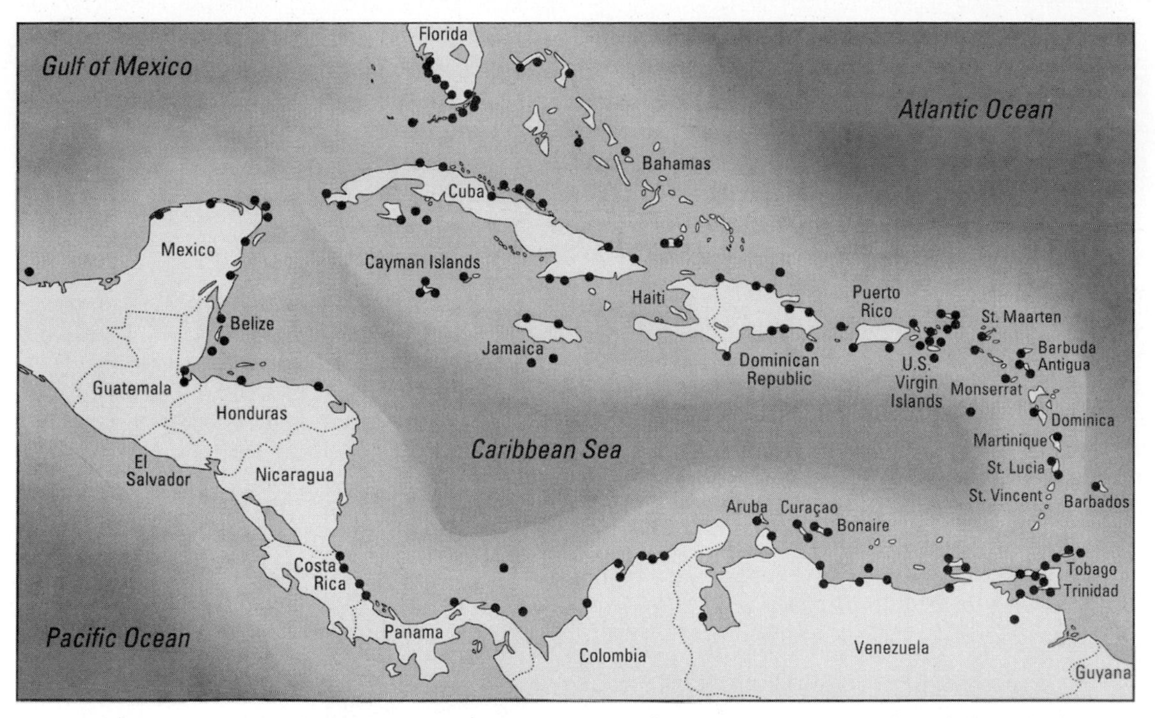

FIGURE 3-1 Marine protected areas in the Caribbean. (*Source:* From "Conserving Biological Diversity Through Marine Protected Areas," by Jack Sobel in *Oceanus*, Vol. 36, No. 3, ©1993 Woods Hole Oceanographic Institute)

lems of this management strategy, and the concluding section, *The NMS Panel—Recommendations for Improvement* introduces the reader to the history of this panel, as well as states their latest recommendations for improving the U.S. National Marine Sanctuary Program.

MARINE AND COASTAL PROTECTED AREAS AS A CONSERVATION TECHNIQUE

There are many benefits (e.g., fishery, nonfishery, overall economic) as well as management challenges associated with marine protected areas and coastlines. But first it is important to note the distinction between the two major types of marine protected areas: (1) marine reserves, and (2) marine sanctuaries. A **marine reserve** generally refers to a **replenishment zone** where it is against the law to catch or handle fishes (i.e., a non-exploitation sanctuary), whereas a **marine sanctuary** is a body of marine water where fishing is allowed, but the area is under other prohibitions, such as the banning of offshore oil development. Furthermore, one can have a marine reserve within a marine sanctuary, but not vice versa. For example, in 1993, several marine reserves were being planned as an essential part of the Florida Keys National Marine Sanctuary, as well as several other areas around the country (Bohnsack 1993). What follows is a brief introduction to the benefits and challenges facing managers of marine protected areas (MPA's).

Benefits Of Marine Protected Areas

Fishery benefits. An exploiting fishing operation can have serious effects on fish populations. For example, intensive fishing operations can alter the natural selection process, resulting in a fishing population that matures earlier, has a shorter life span, and ultimately a smaller adult size (Bohnsack 1993). (See Figure 3-2).

Consequently, fisheries are generally better conserved in marine protected areas that include replenishment zones where fishing is prohibited. If an area is not fished, the total abundance of fishes, their average size, and their overall egg production increases over that of an area that is fished. Once produced, ocean currents can then spread the eggs to other protected areas, as well as to areas that are heavily fished. Additional benefits of protection to fisheries can include (a) the elimination of incidental bycatch mortality, (b) an insurance against stock collapse, and (c) indirect benefits, such as the opportunity for scientific studies of behavior, social organization, and dynamics of certain species for the development of useful fishery management models.

Nonfishery benefits. Marine protected areas (MPAs) offer a variety of other benefits in addition to greater protection of fisheries. Different, though closely related, is the protection of overall marine **biodiversity.** Endangered biotic communities, such as mangrove swamps that fringe

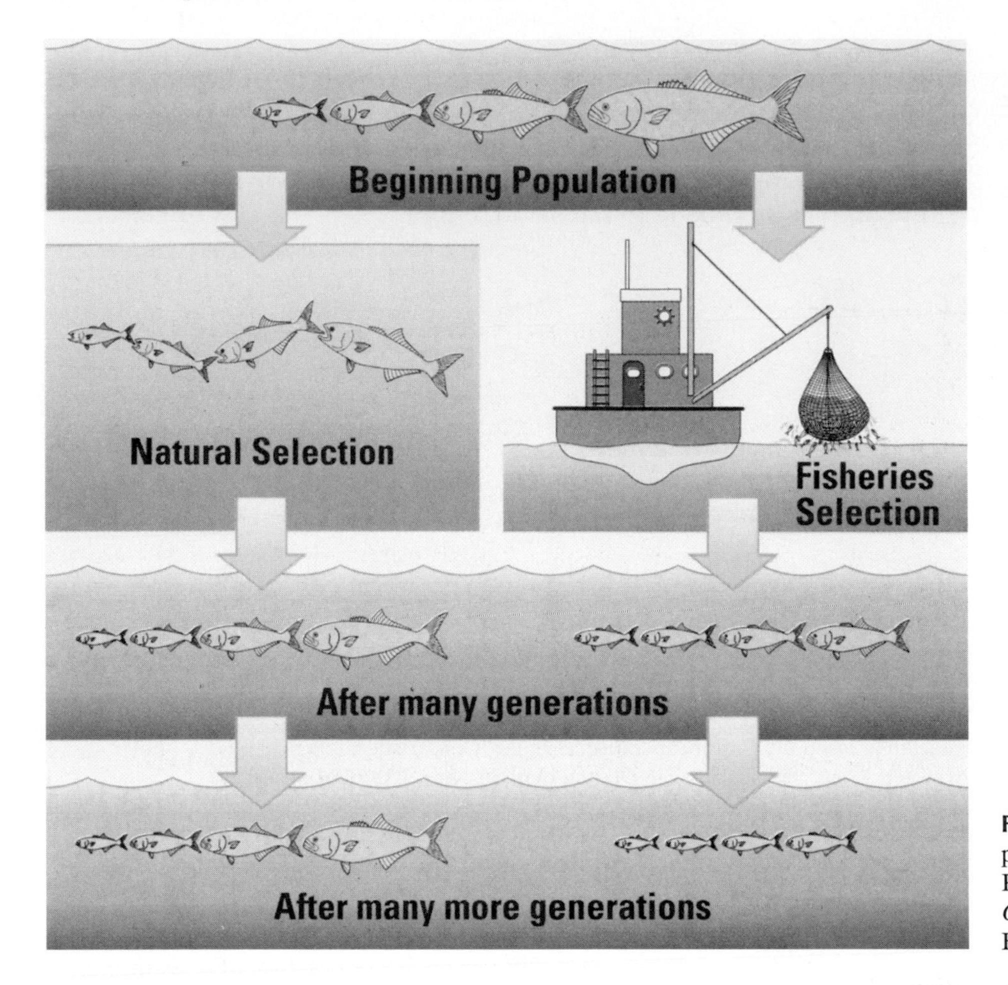

FIGURE 3-2 Effects of fishing on fish populations. (*Source:* From "Marine Reserves" by James Bohnsack in *Oceanus* Vol. 36, No. 3, ©1993 Woods Hole Oceanographic Institute)

the Florida Keys, can be better protected in marine protected areas. To a genetic engineer, such protected communities become priceless "resource banks" crammed with precious biodiversity that reflects millions of years of Earth evolution. To environmental restorationists, environmental engineers, and others interested in restoring damaged ecosystems, the biodiversity preserved in marine protected areas provides a "yardstick" in which to gauge a healthy ecosystem. Without such yardsticks (minimally disturbed areas, sometimes called "reference areas"), it is difficult to understand the impacts of human activities on natural systems. Scientists, therefore, see marine protected areas as *natural laboratories* that provide precious study sites and opportunities for monitoring and research.

These same marine protected areas are also seen as *an insurance policy* to recreationists that their beloved sport will have a place in the future. Scuba divers, for example, will know that destinations still exist in the United States where they can dive among healthy stands of elkhorn coral (e.g., in the Florida Keys) or frolic with seals and sea otters in giant kelp forests (e.g., in Monterey Bay, California). Recreational sailors see marine protected areas as a place of solitude where a person can try to regain a sense of oneness with earth, wind, and sea. (See Cover photo).

Marine protected areas can also be used as a teaching tool to help improve public environmental awareness of natural systems and how humans can alter those systems. Teachers, university professors, naturalists, above sea and underwater photographers, and aquarium operators use marine protected areas to heighten public environmental awareness.

Overall economic benefits. Marine protected areas also benefit the businessmen, whether they be motel operators on Key Largo, Florida, or ecotourist agencies in Monterey, California. A clean, healthy coastal area is attractive to visitors, and lots of visitors means "money in the bank." Marine protected areas (MPAs) also provide jobs through harvest of renewable and nonrenewable resources such as fish and shells. There are also economic benefits that are difficult to express in monetary terms, such as the economic value of healthy reefs that buffer the coast against coastal storms. For a detailed discussion of the overall economic benefits of marine protected areas, see Dixon (1993).

Finally, to a coastal resource manager, large protected areas, such as the relatively new sanctuaries off the coasts of the Florida Keys and Central California, provide an opportunity to develop state-of-the-art programs in integrated coastal resource management (Sobel 1993). The need for such an approach is illustrated in the following section.

Planning and Management Concerns

Each type of marine protected area, whether it be a marine reserve, marine sanctuary, or some other form of MPA, has its unique set of challenges to overcome. However, there are at least five common challenges to the creation and maintenance of marine protected areas (Bohnsack 1993).

Lack of scientific data. First, there is the problem of lack of scientific data about the habits and needs of many of the fishes and other species within a potential MPA. This lack of information leads to difficulty in determining the proper size of an MPA. The sanctuary must be large enough to allow "biological integrity" of the ecosystem (e.g., the normal movements of fishes and other organisms to be protected). However, if the MPA is too large, particularly if we are talking about the creation of a marine reserve, it could limit fisheries production. According to Bohnsack (1993), current information suggests that 10–20 percent of the U.S. continental shelf should be protected in marine reserves for optimum benefit.

Controversy and initial public opposition. Because of the lack of scientific data to adequately determine the ideal number, location, and size of MPAs, a second problem arises—controversy and initial public opposition. It is understandable that most fishermen, for example, do not want their favorite fishing spot put in a *marine reserve,* because it would make fishing there off limits. They are also generally afraid that a *marine sanctuary* would restrict their fishing activities. Consequently many fishermen succumb to the NIMBY syndrome—"not in my backyard." However, this is not always the case. For example, when the Monterey Bay National Marine Sanctuary was proposed, it was the oil industry, not fishermen, that opposed sanctuary status.

Public education and awareness. Consequently, the third major challenge to the creation and maintenance of marine protected areas is public information and awareness. The general public, and particularly fishermen, are likely to initially oppose the creation of a marine protected area (especially a marine reserve) until the function and importance of the specific kind of MPA is adequately explained.

Surveillance and enforcement. Once established, there is also the fourth challenge of marine protected areas—the budget issue of surveillance and enforcement. Without adequate surveillance and enforcement, marine protected areas become nothing more than "paper reserves"—something that looks good on a map, but is ineffective as a conservation strategy. Despite these and other problems, marine protected areas are on the increase around the world.

Integrating coastal and marine sanctuary management. Finally, there is the last and ultimate challenge—integrating the management of marine protected areas with coastal resource management strategies. As Ehler and Basta (1993) have indicated, it will require developing *a whole new paradigm—a whole new way of thinking.* We have to change our management approaches, adjust to the same or fewer resources, and forge new partnerships with governmental, academic, and other institutions. Examples of these management approaches, and how they might be integrated, are illustrated throughout the book.

Before we look at U.S. efforts to better integrate the management of coastal and marine protected areas, it would be useful to have an overview of the National Marine Sanctuary Program (NMSP)—the only U.S. program specifically designed to provide comprehensive protection of the nation's extraordinary marine ecosystems. What follows is an introduction to the program as well as a brief overview of each of the sanctuaries within that program.

THE U.S. NATIONAL MARINE SANCTUARY PROGRAM

Program Description

Definition of a national marine sanctuary. Americans have a rich conservation history. We started protecting "special places" back in 1872, when Congress designated Yellowstone National Park as the first park within a federal system of *terrestrial* parks. At the time, it was considered one of the best ideas Americans ever had. The National Park System now includes more than 360 units and more than 32.4 million ha (80 million acres) of mountains, seashores, battlefields, and other historic sites. This system of federal parks is managed by the National Park Service, a bureau of the U.S. Department of Interior. Other conservation programs were soon to follow. The National Forest System was established in 1891 and today has more than 156 units encompassing more than 75.7 million ha (187 million acres). (Some would argue, however, that this is merely a windfall for the lumber industry.) In 1920, the National Wildlife Refuge System was established and presently includes 485 units covering more than 35.9 million ha (88.6 million acres).

In 1972, exactly one hundred years after the birth of the National Park System, the U.S. Congress passed the *Marine Protection, Research and Sanctuaries Act (MPRSA),* thereby creating the *National Marine Sanctuaries Program (NMSP)*—a framework for the protection of *oceanic* parks. The actual legislation defines marine sanctuaries as "areas of special national significance due to their resource or human use values" and places an emphasis on "conservation, recreational, ecological, historical, research, educational, or aesthetic value of the sanctuary." The National Marine Sanctuary Program is administered by the National Oceanic and Atmospheric Administration (NOAA), within the Department of Commerce.

In 1997, there were 12 designated and 2 proposed sanctuaries within the National Marine Sanctuary Program. (See Figure 3-3). The Florida Keys, the Monterey Bay National Marine Sanctuaries, and some others, embrace all of the criteria of the official definition of a sanctuary, whereas other designated sanctuaries within the system fit only one or two criteria. For example, the first designated sanctuary, The Monitor National Marine Sanctuary, was chosen solely for its historical significance. The sanctuary was established to protect the location off North Carolina where the *Monitor* sank during the Civil War.

The definition of a marine sanctuary is almost as fluid as the habitat which it attempts to protect. As society's needs and attitudes change, so will the definition of marine sanctuary (Earle 1993). According to today's definition, marine sanctuaries are not areas that are sacrosanct—areas regarded as sacred and inviolable. On the contrary, they are areas that are used, everything from extraction industries (e.g., fisheries), to recreation (e.g., power boating and scuba diving), to education (e.g., nature observation). One common element, however, is that these sanctuaries are "special places." And, the definition of a special place, of course, is in the eyes of the beholder. As society's needs and attitudes regarding these special places change over time, however, there may be a day when some places are perceived as more valuable or more vulnerable than others, and will need special care and extra protection.

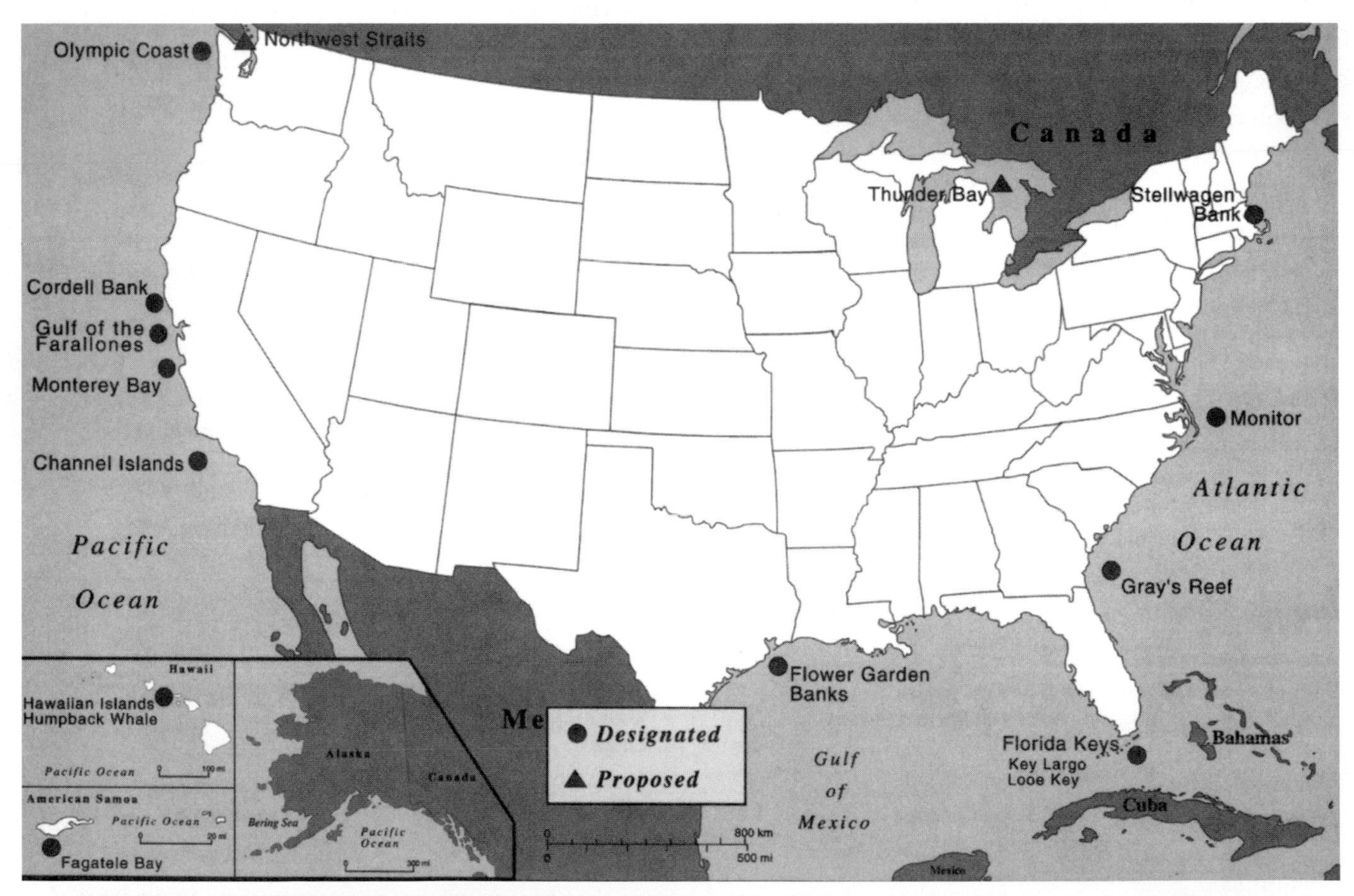

FIGURE 3-3 The National Marine Sanctuary Program. (*Source:* NOAA National Marine Sanctuaries: Accomplishments Report 1997a. Used with permission)

Program mission and goals. The mission of the National Marine Sanctuary Program is to identify, designate, and manage nationally significant marine and Great Lakes waters. Consistent with that mission, the following goals have been targeted:

- *Resource Protection.* To enhance resource protection through comprehensive and coordinated strategies that complement existing regulatory authorities;
- *Scientific Research.* To support, promote and coordinate scientific research and monitoring of the site;
- *Public Awareness.* To enhance public awareness, understanding, and wise use of the site;
- *Multiple Use.* To facilitate multiple use of these marine areas to the extent comparable with the primary goal of resource protection.

On the "public awareness" front, for example, former Channel Islands manager, Francesca Cava, developed a program for middle school children called Los Marineros, which was an educational strategy to bring marine sanctuaries into the classroom. In the Los Marineros program "Captain Cava" presided over two NOAA periodicals, a quarterly newsletter titled Rivers to Reefs, and a magazine dedicated to NOAA's marine sanctuaries and estuarine reserves. These are just a few examples of public outreach by sanctuary managers and or officials.

Benefits of designation. Perhaps the greatest benefit of sanctuary designation is a framework that can be established to administer and manage the area as a complete ecosystem, rather than isolated management of certain resources (e.g., protection of a certain marine species), or isolated regulation of a targeted human activity (e.g., waste water discharge). In other words, a *holistic management plan* can be developed, at least in theory, that attempts to *integrate* and *coordinate* with other agencies. New regulations are only imposed if there is a gap in existing management programs and regulations. Emphasis is on *complimenting* existing programs (e.g., U.S. Fish and Wildlife Service, not duplicating nor adding unnecessary effort or expense).

The site selection process. The process by which NOAA selects sites for potentially new national marine sanctuaries has several steps, is quite complicated, and would take many pages to explain. In brief, however, the process has three major stages: (1) Site Evaluation List; (2) Active Candidate; and (3) Designation. Getting on the *Site Evaluation List (SEL)* is the first step towards national marine sanctuary designation. In conjunction with knowledgeable, interested and affected persons, NOAA identifies sites that have a high degree of natural resource or human-use value. In 1992, there were 25 natural resource based and 10 cultural resource based SEL sites. Being on the SEL is the necessary first step, but it does not guarantee a site will become a sanctuary.

To become an *Active Candidate,* the second stage, the NOAA staff evaluates such things as the biogeographic and resource representation, the availability of staff and resources, and the relative costs and benefits of designation. If the site passes this test, then it is advanced to candidacy which activates the two to three-year process that is required for Designation. This stage is particularly lengthy because it requires public meetings, consultations with government agencies, the preparation of an Environmental Impact Statement and Draft Management Plan, the approval of Congress, and the concurrence of the Governor for sites that include state waters. If the candidate passes all these hurdles, the Secretary of Commerce designates the site as a national marine sanctuary—*designation* being the third and final stage.

Number and Location of Sanctuaries.

Introduction. America's national marine sanctuaries represent a number of distinct marine environments, including nearshore coral reefs, open water, and **benthic** (bottom dwelling life) ecosystems. The sanctuaries range in size from less than 0.64 sq km (0.25 sq mi [Fagetelle Bay NMS]) to over 13,798 sq km (5,328 sq mi [Monterey Bay NMS]), and are located in tropical as well as temperate waters. Furthermore, this system of sanctuaries harbors a fascinating array of plants, animals, significant shipwrecks, and prehistoric artifacts.

The National Marine Sanctuary Program has had an irregular pattern of growth. The first marine sanctuary, Monitor NMS off the North Carolina coast, was established in 1975, three years after the NMSP was established. Less than a year later, Key Largo NMS off the shore of Florida was designated. The NMSP got a real boost in 1979 when President Jimmy Carter began devoting special attention to marine sanctuaries. By 1981, four new sanctuaries had been designated—Channel Islands, Gulf of the Farallones, Gray's Reef, and Looe Key. The program atrophied during the late 1980's with the Reagan administration. Under President Reagan and his controversial Secretary of Interior, James Watt, only one sanctuary was designated—Fagatele Bay NMS in American Samoa. Not surprisingly, Fagatele Bay is the smallest national marine sanctuary so far created. Of the current 14 designated marine sanctuaries, half were established between the years of 1987 and 1995. There was a period when NOAA felt more ambitious and had some 70 sites as candidates for marine sanctuaries. By 1997, the agency had toned down its goals, apparently feeling the winds of political change, and now has only two officially proposed sanctuaries—Northwest Straits, on the northern reaches of Washington State's Puget Sound, and Thunder Bay, located along the northeast coast of Michigan.

Each sanctuary with the National Marine Sanctuary Program will now be briefly introduced, with special emphasis on location, unique characteristics, and threats to its natural resources. (See Table 3-1).

Florida's three national marine sanctuaries: Key Largo, Looe Key, and Florida Keys. Florida is the only state in the continental United States to have a tropical rainforest biome on its southern tip and extensive reef-building coral formations near its shores. The previously designated Key Largo and Looe Key NMS have now been folded into the management of the Florida Keys National Marine Sanctuary as Existing Management Areas. Because their regulations remain much the same as they were, except no-take areas were established on some of the shallow reefs, Key Largo and Looe Key will be discussed below as separate national marine sanctuaries. (See Figure 3-4).

Key Largo National Marine Sanctuary. In 1975, Key Largo National Marine Sanctuary (KLNMS) became the first national marine sanctuary to protect a part of Florida's coral reefs. The sanctuary is located southeast of the Florida peninsula, beginning some 4.8 km (3 mi) seaward of the city of Key Largo, and extending approximately 32 km (20 mi) parallel to the coastline. One hundred (100) square nautical miles of Florida's Reef Tract is protected under this sanctuary.

Key Largo NMS has an abundance of natural, cultural, and historical resources. Distinct natural habitats include sand flats, seagrass beds, bank reefs, patch reefs, hard bottom, and over 100 species of hard and soft corals. The warm clear waters at Key Largo allow for an array of over 500 species of tropical fishes, as well as endangered and protected sea turtles, spotted eagle rays, moray eels, and numerous other reef inhabitants. Cultural and historical resources include the 1852 Carysfort Lighthouse—the oldest functioning lighthouse of its kind in the United States—and numerous shipwrecks, including a Spanish galleon, a British warship, and a World War II freighter.

Management Challenges. Because of its luxuriant reef development, and the fact that metropolitan Miami is only 97 km (60 mi) away, Key Largo NMS is invaded by over 1.5 million visitors per year. To help protect the reefs, the sanctuary has prohibited anchoring in a manner that

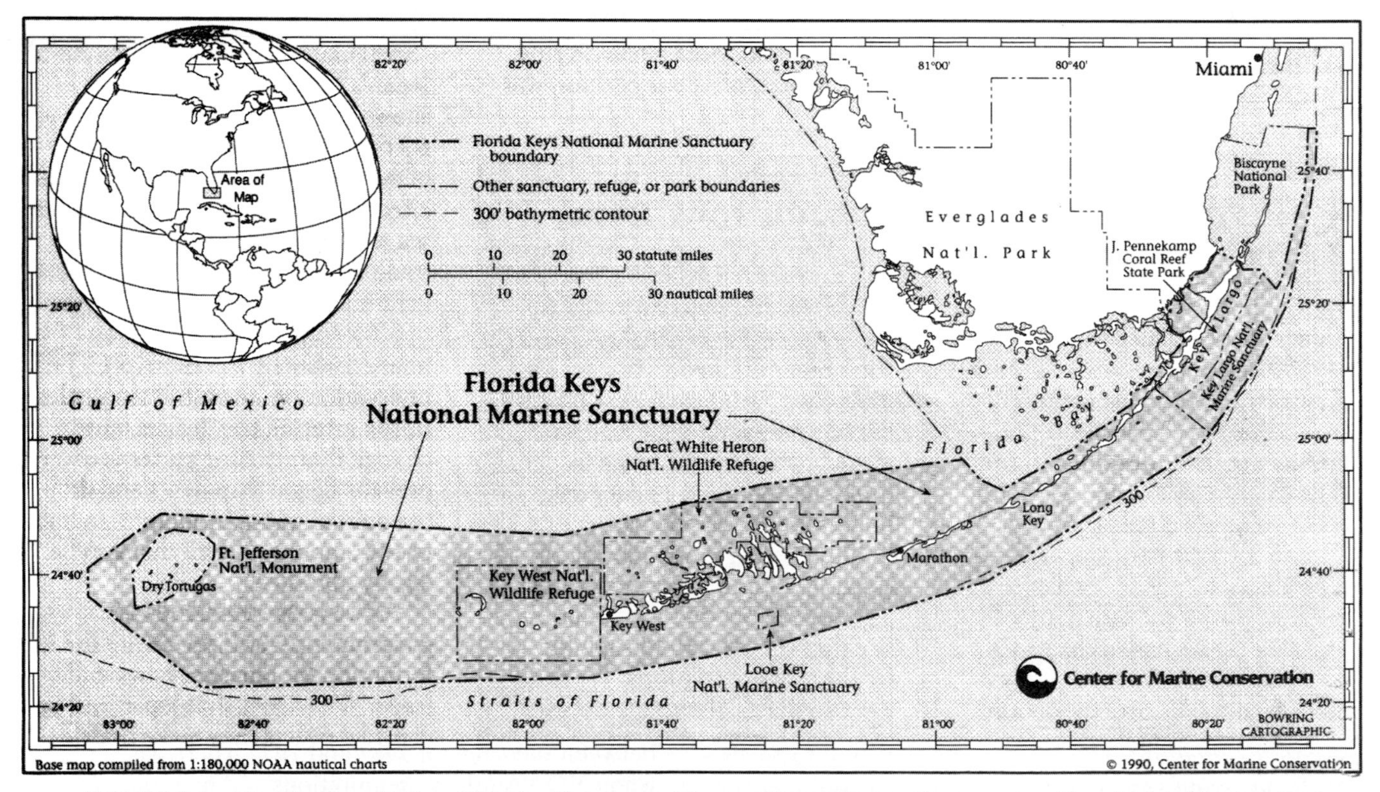

FIGURE 3-4 Florida Keys National Marine Sanctuary. (*Source:* From *Marine Conservation News*, Vol. 3, No. 1, p. 8. ©Center for Marine Conservation. Used with permission)

TABLE 3-1

National Marine Sanctuaries

National Marine Sanctuary	*Habitats*	*Key Species*	*Protected Area*	*Designation*
Channel Islands (California)	Kelp Forests, Rocky Shores, Sandy Beaches, Seagrass Meadows, Pelagic, Open Ocean, Deep Rocky Reefs	California Sea Lion, Elephant & Harbor Seals, Blue & Gray Whales, Dolphins, Blue Shark, Brown Pelican, Western Gull	1,658 square miles	September 1980
Cordell Bank (California)	Rocky Subtidal, Pelagic, Open Ocean, Soft Sediment, Continental Shelf and Slope, Seamount	Krill, Pacific Salmon, Blue Whale, Humpback Whale, Dall's Porpoise, Shearwater, Albatross, Rockfish	526 square miles	May 1989
Fagatele Bay (American Samoa)	Tropical Coral Reef	Crown-of-Thorns Starfish, Blacktip Reef Shark, Surgeon Fish, Hawksbill Turtle, Parrot Fish, Giant Clam	0.25 square mile	April 1986
Florida Keys (Florida)	Coral Reefs, Patch & Bank Reefs, Mangrove-Fringed Shorelines & Islands, Sand Flats, Seagrass Meadows	Brain & Star Coral, Sea Fan, Loggerhead Sponge, Turtle Grass, Angelfish, Spiny Lobster, Stone Crab, Grouper, Tarpon	3,674 square miles	November 1990
Flower Garden Banks (Texas/Louisiana)	Coral Reefs, Artificial Reef, Algal-Sponge Communities, Brine Seep, Pelagic, Open Ocean	Brian & Star Coral, Manta Ray, Loggerhead Turtle, Hammerhead Shark	56 square miles	January 1992
Gray's Reef (Georgia)	Calcareous Sandstone, Sand Bottom Communities, Tropical/Temperate Reef	Northern Right Whale, Loggerhead Sea Turtle, Barrel Sponge, Angelfish, Ivory Bush Coral, Grouper, Black Sea Bass	23 square miles	January 1981
Gulf of the Farallones (California)	Coastal Beaches, Rocky Shores, Mud & Tidal Flats, Salt Marsh, Esteros, Deep Benthos, Continental Shelf and Slope	Dungeness Crab, Gray Whale, Stellar Sea Lion, Common Murre, Ashy Storm Petrel	1,255 square miles	January 1981
Hawaiian Islands Humpback Whale (Hawaii)	Humpback Whales Breeding, Calving, & Nursing Grounds, Coral Reefs, Sandy Beaches	Humpback Whale, Pilot Whale, Hawaiian Monk Seal, Spinner Dolphin, Green Sea Turtle, Trigger Fish, Limu, Cauliflower Coral	1,300 square miles	November 1992
Monitor (North Carolina)	Pelagic, Open Ocean, Artificial Reef	Amberjack, Black Sea Bass, Red Barbier, Scad, Corals, Sea Anemones, Dolphin, Sand Tiger Shark, Sea Urchins	1 square mile	January 1975
Monterey Bay (California)	Pelagic, Open Ocean, Sandy Beaches, Rocky Shores, Kelp Forests, Wetlands, Submarine Canyon	Sea Otter, Gray Whale, Market Squid, Brown Pelican, Rockfish, Giant Kelp	5,328 square miles	September 1992
Olympic Coast (Washington)	Pelagic, Open Ocean, Sandy & Rocky Shores, Kelp Forests, Seastacks & Islands	Tufted Puffin, Bald Eagle, Northern Sea Otter, Gray Whale, Humpback Whale, Pacific Salmon, Dolphin	3,310 square miles	July 1994
Stellwagen Bank (Massachusetts)	Sand & Gravel Bank, Muddy Basins, Boulder Fields, Rocky Ledges	Northern Right Whale, Humpback Whale, Storm Petrel, White-Sided Dolphin, Bluefin Tuna, Sea Scallop, Northern Lobster	842 square miles	November 1992

Source: NOAA, 1997a, 68–69.

damages corals, discharging toxic substances, using certain types of fishing techniques (e.g., spear guns, explosives, wire fish traps), and removing or damaging natural features, marine life, and cultural or historical resources.

Enforcement is often done by "reef rangers"—NMS scuba divers that prowl the depths, armed with grease pencils and plastic slates, ticketing anyone who abuses the fragile coral heads. Divers who touch coral, thereby disobeying the "*Key-mandment*"—"*Don't touch the coral, after all, you don't even know them*"—are instructed to surface by the reef rangers with messages written on slates with grease pencils. Once above water, the divers usually receive a minimum $25 ticket. Heavy fines ($150 and up) are given to divers that cut the coral for souvenirs, and to boat owners that drop and drag anchors, breaking off coral branches and topping heads of coral that have taken hundreds of years to mature.

To help alleviate the anchoring problem, some permanent mooring buoys have been strategically placed around popular diving areas. This allows people to tie their boats to moorings rather than drop anchors. Key Largo sanctuary managers would like to see regulations that would allow only buoy moorings, thus limiting the numbers of visitors to an area, and allowing the damaged areas to recover undisturbed. The same managers would also like to see some kind of boat operators' license and education since boat groundings on the reef are a major problem—the operators simply cannot recognize shallow water. To the surprise of most people, anyone can operate a boat anywhere. Yet, like automobiles, high-powered boats can cause environmental damage and can kill.

A more dramatic form of coral reef damage occurs when an ocean going vessel strays off course and plows into a reef. On August 4, 1984, for example, the 122 m (400 ft) *M/V Wellwood* grounded on Molasses Reef in the Key Largo National Marine Sanctuary. A survey revealed that a 1,282 sq m (13,799 sq ft) area of the reef had sustained a 70 to 100 percent loss of live coral cover as a result of the grounding (Hudson and Diaz 1988). The ship was fined $6.1 million. All the money went into an account set up as a special fund for monitoring and restoration of the damaged site.

Over 5,000 oil tankers alone pass along Florida's coast each year, and southbound ships tend to hug the coastline close to the reef as they try to avoid strong northerly currents. In 1990, sanctuary managers called for a 243 km (150 mi) buffer zone around the Florida reef to force large ships farther out to sea. Ships from other nations are asked to abide by these rules, and for the most part they do.

All the sanctuaries, including Key Largo, are facing the bigger environmental problems of pollution, global warming, and rising sea levels. According to many scientists, these factors combined with others are further retarding the normally slow coral reef growth. Furthermore, such natural phenomena as black-band disease, and the loss of the long-spined urchins that once cleaned the coral of algae, have dealt further blows to the reef track of 6,000 small barrier reefs that extend from Miami south along the Florida Keys all the way to Dry Tortugas. As you can see, it will take more than reef rangers and local buffer zones to address these global environmental problems.

Looe Key National Marine Sanctuary. Just 121 km (75 mi) southwest of Key Largo NMS is Looe Key National Marine Sanctuary (LKNMS). Looe Key, taking its name from a British warship that went aground on the reef in 1744, was designated a national marine sanctuary in 1981 in order to protect 13.7 sq km (5.32 sq mi) of fragile coral reef resources. The area is known for its classic "spur-and-grove" reef system where deep sand grooves separate high profile coral spurs. This natural habitat, together with clear water and moderate sea conditions provides excellent shelter for hundreds of reef inhabitants including branching elkhorn and staghorn corals, huge brain corals, delicate sea fans, numerous brightly colored fish, sponges, lobsters, crabs, and many more.

Management Challenges. Looe Key NMS faces a number of natural and human impacts. Natural impacts include hurricanes and heavy storms. The coral reef ecosystem is also vulnerable to coral bleaching or coral diseases, such as black-band that proliferate during warm summer temperatures. Additional natural impacts are "fish grazing" on corals and "bioerosion." There is little that sanctuary managers can do (nor should they) to protect the coral ecosystem from these types of natural processes.

Looe Key sanctuary managers, consequently, put most of their efforts on such human impact problems as divers that stand on, touch, or destroy corals and boat operators that improperly anchor their boats. As at Key Largo, fines are prescribed against those that do not practice good "reef etiquette." In addition to establishing and enforcing conservation regulations for the area, the Looe Key sanctuary staff are also involved in various research projects (e.g., studies on restoration at boat grounding sites; black-band coral disease; coral bleaching) and public education (e.g., strategic placement of interpretative signage and information brochures; descriptive video tapes, on-site "Coral Reef Classrooms"). Both Key Largo and Looe Key National Marine Sanctuaries are managed through a cooperative agreement with the Florida Department of Natural Resources (FDNR) and NOAA. NOAA provides FDNR a grant to operate both sanctuaries.

Florida Keys National Marine Sanctuary.

> *I doubt that anyone can travel the length of the Florida Keys without having communicated to his mind a sense of the uniqueness of this land of sky and water and scattered mangrove covered islands. The atmosphere of the Keys is strongly and peculiarly their own . . . This world of the Keys has no counterpart elsewhere in the United States, and indeed few coasts of the Earth are like it.*
>
> —Rachel Carson, *The Edge of the Sea, p. 191*

In 1990, President Bush signed into law the Florida Keys National Marine Sanctuary which encompasses over 9515 sq km (3,674 sq mi) of the Atlantic and Gulf of

Mexico waters surrounding the Keys. It is currently the second largest sanctuary (Monterey Bay NMS being the largest) and the third largest protected barrier reef system in the world.

Management Challenges. Florida Keys NMS suffers from many of the same natural stresses and human impacts as Key Largo and Looe Key National Marine Sanctuaries, especially because this newer sanctuary encompasses the previously designated Florida sanctuaries, as well as Ft. Jefferson National Monument and Key West National Wildlife Refuge. Just prior to designation, visual signs of environmental deterioration at the Florida Keys NMS included small-boat groundings, algae growth on and around corals, coral bleaching and diseases, fish kills and decreases in certain fisheries. To top matters off, 1989 was a disastrous year for the reefs. Within a 16 day period, three large freighters ran aground, pulverizing 16,824 sq m (181,097 sq ft) of the reefs. There were also serious proposals for oil exploration and drilling offshore of the Keys. With the Keys two principal industries (tourism and commercial fishing) dependent on a healthy ecosystem, the people of the Keys called for federal protection of their coastal waters.

Florida Keys NMS was the first sanctuary to attempt integrated coastal resource management. Why? The shear size and location of the sanctuary, plus what it encompassed, required a holistic management strategy. The sanctuary encompasses the entire Florida Reef Tract, mangrove, nearshore, and "back country" habitats, such as Florida Bay and Gulf of Mexico waters. It also includes other administrative units, such as previously designated sanctuaries, a national monument, and a national wildlife refuge. Consequently, it was necessary to integrate a multitude of authorities with jurisdiction in the Florida Keys, from the federal level (e.g., NOAA's National Marine Sanctuary Program, U.S. Environmental Protection Agency, U.S. Fish and Wildlife Service, two fishery management councils), state level (e.g., Florida Department of Environmental Protection, undersea park management agencies), to the local level (e.g., Monroe County officials), to mention just a few. User groups, such as divers, jet skiers, fishermen, treasure hunters, ocean freighter operators, and general boaters, also have a voice, and their particular concerns addressed.

Florida Keys' integrated coastal resource management process is only in its initial stages. While some already consider it a "success story" (Barley 1993), others say it is too early to tell (Ehler and Basta 1993). Florida Keys' NMS managers can say, however, that oil and gas exploration is now prohibited in the area; a holistic water quality management program has been established (See Chapter 5, *Coastal Pollution,* for details); boundaries have been established that prohibit large ships from going through the FKNMS waters; an Advisory Council has been created to develop and implement a comprehensive management plan for the sanctuary; and a group of environmental education specialists has developed a program to educate the public about the sanctuary.

This draft management plan, known as *Strategy for Stewardship,* which was released by the Ocean Management Branch of NOAA in April 1995, presents about 100 management strategies, including *ocean area zoning* to create "Sanctuary Preservation Areas" to protect specific reefs and reduce conflicts between consumptive and nonconsumptive uses, as well as "Replenishment Reserves"—areas set aside to sustain important marine species by providing spawning, nursery, and permanent residence areas for marine life. The plan also suggests how to integrate federal, state, and local agencies in their continuous management efforts of the region (NOAA 1995a). The public had until December 31, 1995 to comment upon the draft management plan. A preliminary review of public comments has shown widespread support from the public, as well as government agencies, for the plan. According to Benjamin Haskell, Public Comment Coordinator for the Florida Keys National Marine Sanctuary, 80 percent of the plan will probably be retained. In 1997, Florida Governor Lawton Chiles and his Trustees, unanimously voted to adopt the final management plan for the Florida Keys National Marine Sanctuary.

Central California's three national marine sanctuaries: Monterey Bay, Gulf of the Farallones, and Cordell Bank. In addition to the efforts at Florida Keys NMS, other opportunities exist for integrated management of coastal areas and marine sanctuaries. The complex of three marine sanctuaries on the shores of Central California offer one such opportunity. (See Figure 3-5).

Monterey Bay National Marine Sanctuary. Designated in 1992, Monterey Bay National Marine Sanctuary (MBNMS) is the largest (and deepest) sanctuary in the nation, covering some 13,798 sq km (5328 sq mi)—an area approximately the size of the state of Connecticut. The Monterey Bay NMS is the second largest sanctuary in the world after the Great Barrier Reef in Australia.[1] Prior to the designation of the Monterey Bay NMS, sanctuaries within the NMSP tended to be relatively small areas of the ocean at a distance from large urban centers. The designation of Monterey Bay was a major change in the direction of the NMSP. The 300 nautical miles of urbanized California coastline—with all its

[1]MBNMS is now the third largest marine sanctuary in the world. Australia recently adopted a sanctuary in south Australia that is slightly larger than MBNMS, approved May 1998.

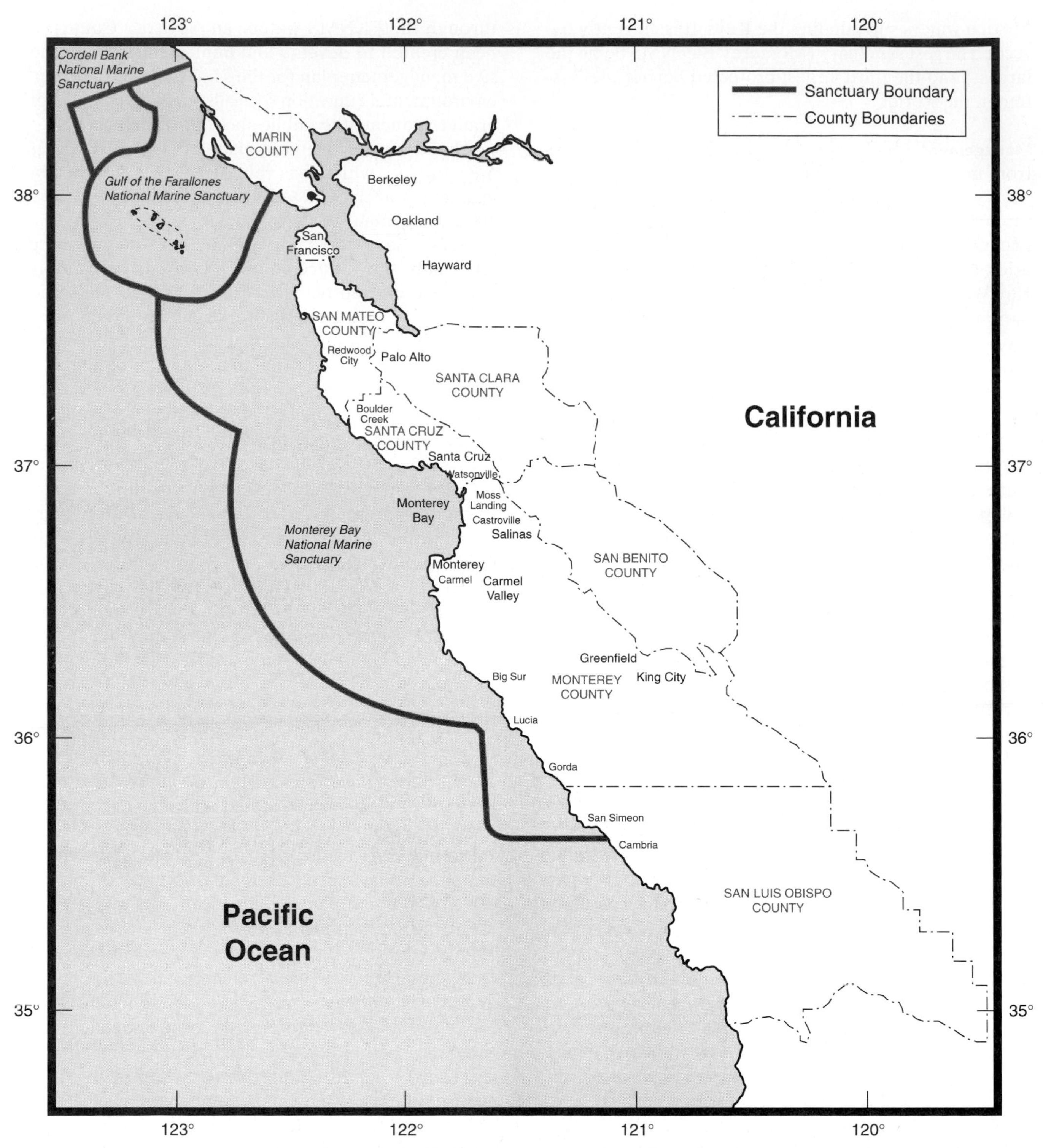

FIGURE 3-5 Monterey Bay National Marine Sanctuary. Note that the Gulf of the Farallones National Marine Sanctuary butts directly against the MBNMS, and Cordell Bank National Marine Sanctuary is directly against the GFNMS. (*Source:* Adapted from Center for Marine Conservation, 1992)

potential multiple use conflicts—demanded an integrative policy approach to coastal and sanctuary management.

At the heart of the Monterey Bay National Marine Sanctuary is the bay itself—a crescent shaped bay that sweeps 64 km (40 mi) of coastline, from Santa Cruz in the north, past farmland famed for artichokes and other truck crops, to the shore of the jutting Monterey Peninsula in the south, which holds the historic towns of Monterey and Pacific Grove. Due to the area's relatively small bay-area population—less than 600,000 people—and the lack of heavy industry, the waters of the Monterey Bay remain relatively unpolluted. From spring to fall, ocean upwelling brings cold, nutrient-rich water close to shore, and fog is a near-daily phenomenon. (See Figure 3-6).

At the center of the Bay is a *submarine canyon* like no other on the continental U.S., nor in the National Marine Sanctuary Program. This spectacular submarine chasm plunges to 90 m (295 ft) less than a kilometer off Moss Landing—making it the closest-to-shore deep ocean environment in the continental United States. The canyon then meanders over 175 km (109 mi) out to sea, reaching a depth of 3048 m (10,000 feet)—twice the depth of the Grand Canyon.

The Monterey Bay NMS also holds other treasures. For example, an exceptional variety of kelp thrive here, including the nation's most expansive forests of *giant kelp and bull kelp* that provide food and shelter to marine animals, fishes, and invertebrates, and a playground for scuba divers and kayakers. The kelp forests are harvested on a

FIGURE 3-6 The crescent bay of Monterey Bay National Marine Sanctuary. (*Source:* Courtesy of U.S. Geological Survey)

sustainable basis for commercially valuable products. Sea stars, anemones, limpets, sea cucumbers, and numerous other invertebrates abound in the tide pools on Monterey's famous *rocky shores.* These rocky shores have long attracted scientists, including the legendary Edward "Doc" Ricketts who based his pioneering ecological study, *Between Pacific Tides,* on what is now in the sanctuary.

Other natural resources within the sanctuary include 26 species of marine mammals, many of which are endangered or threatened. The list of vulnerable marine mammals include the southern sea otter, the Guadalupe fur seal, the Steller sea lion, three species of marine turtle (leatherneck, loggerhead, Pacific ridley), and seven species of whale (gray, humpback, fin, blue, right, sperm, and sei). One of the more interesting marine mammals is the northern elephant seal which breeds at Point Año and in the past decade at Pt. Piedras Blancas to the south. In fact, Año Nuevo State Reserve is considered one of the more important pinniped rookeries in California. The sanctuary's rich and varied habitats also attract 94 species of *migratory and nesting seabirds,* including Forster's terns, tufted puffins, and snowy plovers. Four of the bird species within the area (California brown pelican, American peregrine falcon, California least tern, and short-tailed albatross) are considered either endangered or threatened [There are also California state threatened and endangered species as well]. The area's *sloughs and wetlands* provide an abundant source of food for millions of migratory water birds, as nurseries for many fish species, and permanent habitats for a host of mud-dwelling creatures (NOAA 1992). Elkhorn Slough and Pescadero Marsh are two important wetlands within the sanctuary.

This great diversity of natural resources in the Sanctuary provides a "living laboratory" for Hopkins Marine Station (the oldest marine research center on the West Coast) and 17 other public and private marine research centers. These facilities support over 1,600 scientists and staff that work on projects ranging from basic marine science that tries to understand how the region functions as a system, to monitoring regional environmental changes for the protection of marine resources. The facilities provide information needed to resolve management issues and conflicts within and adjacent to the sanctuary.

At the Monterey Bay Aquarium Research Institute (MBARI), for example, new technologies are being developed that are applicable for sanctuary research. MBARI has designed surface vessels that are highly stable, undersea vehicles that can survey the depths visually, buoys and moorings that can monitor environmental changes, and "bottom stations" that relay information about changes on the seafloor (Robison 1993). These and other new technologies, (e.g., undersea navigation systems, chemical sensors, biotechnology satellite links, new data management systems) will provide new kinds of information that bring new perspectives (and open new doors) to old problems.

Monterey Bay is also internationally known for its history, culture, and local economies that depend on a pristine coast. *Historical and cultural resources* within the area include: native American middens (shell mounds) along the coast adjacent to the sanctuary; lighthouses that supported commercial navigation to the Central California Coast; submerged wrecks of ships; and Monterey's Cannery Row that has now been turned into tourist shops. Cannery Row is the location of the Monterey Bay Aquarium—a first rate center for environmental education about the bay.

Preserving the pristine quality of Monterey's coastline also helps the local economy. (See Table 3-2). Since the Ohlone Indians first harvested shellfish along the shore, the Monterey Bay has attracted and sustained *fisheries.* Despite some boom and crash fisheries (e.g., the sardine industry in the 20th century), the Monterey Bay still provides valuable commercial and recreational fisheries. With over 340 species in the area, including salmon, rockfish, halibut, squid and anchovy, commercial fishing supplies a wide selection to restaurants and fresh fish markets locally, across the nation, and around the world. Over 5,800 coastal farms in the six counties adjacent to the sanctuary, make a major contribution to the local economy.

If you have ever traveled along the shores of the Monterey Bay, you also know that this region is a mecca for *recreation and tourism.* On almost any day, you can see recreational sailors, surfers, scuba divers, kayakers, windsurfers, tidepoolers, swimmers, beachwalkers, birders, and others practicing what they like most.

Management Challenges. Monterey Bay NMS is part of a complex of sanctuaries, since it is adjacent to the Gulf of the Farallones National Marine Sanctuary 3,250 sq km (1,255 sq mi), which in turn is adjacent to the Cordell Bank National Marine Sanctuary—1,362 sq km (526 sq mi). As Ehler and Basta (1993, p. 12) have rightly indicated, "This sanctuary 'complex' and adjacent coastal watersheds of

TABLE 3-2

Economics of Seven Ocean-Dependent Industries in California

Tourism and Recreation	$9.9 Billion/Yr
Water Transportation of Freight and Passengers	$3.4 Billion/Yr
Ship and Boat Building	$2.6 Billion/Yr
Offshore Oil and Gas Production	$852 Million/Yr
Commercial Fishing, Mariculture, and Kelp Harvesting	$554 Million/Yr
Mineral Production	$10 Million/Yr
Total Market Value of Industries	**$17.3 Billion/Yr**

Source: California Resources Agency. 1995. *California's Ocean Resources: An Agenda for the Future.* Sacramento, California: State Printing Office.

Central California provide another opportunity to demonstrate the benefits of using an integrated approach to coastal and sanctuary management."

Sanctuary status brings a number of prohibitions and regulations to the area. There are three major prohibitions: (1) the exploring for, developing or producing of oil, gas or mineral resources; (2) the designation of new dredged-material disposal sites within the area; and (3) discharging of primary treated sewage. There are also activities that are regulated, including the discharging or depositing of any materials; moving, removing, or injuring sanctuary historical resources; altering the seabed or constructing any structures on the seabed; disturbing marine mammals, sea turtles, and birds; and flying motorized aircraft below 305 m (1,000 ft). When the sanctuary was first designated in 1992, there was also a regulation against operating motorized personal water craft (e.g., jet skis) within certain sections of the sanctuary. However, this was so controversial that the regulation was temporarily dropped, then later reinstated (See Chapter 8, *Open Space Preservation and Management,* for further details).

Threats remain to the sanctuary. Inadequately treated sewage, storm drain overflows, and pesticide runoff could degrade the water quality and compromise the sanctuary. Sanctuary staff are concerned about the catastrophic effects that could happen from an oil spill from a passing tanker or barge, and they worry about the possible harmful effects of dredge dumping in waters adjacent to the sanctuary. The sanctuary is working with other federal, state, and local agencies to manage these controversial issues. These concerns are discussed in greater detail in the appropriate chapters throughout the book.

Gulf of the Farallones National Marine Sanctuary. The Gulf of the Farallones National Marine Sanctuary lies along the coast of California just north, south, and west of San Francisco. (See Figure 3-7). Designated in 1981, this sanctuary encompasses 3,250 sq km (1,255 sq mi) of ocean that is characterized by the rugged Farallones Islands, strong currents, and gusty winds. (See Figure 3-8). The sanctuary extends from Bodega Bay in the north to just south of Half Moon Bay, and from the coast to 64 km (40 mi) offshore (NOAA 1987).

A characteristic within the Gulf of the Farallones NMS is the natural phenomena known as **upwelling.** In the spring, northwest gales and southerly ocean currents combine with the earth's rotation to move warm surface water offshore, drawing cold water up from the depths to the coast. (See Figure 3-9).

These "upwelled" waters contain vast quantities of nutrients. The nutrients, in turn, fertilize microscopic plant and animal life which quickly develop into a major marine food source. Upwelling, thus, is the core of a rich food source for schools of squid, young fish, and shrimp, which in turn, get eaten by larger creatures. This rich and resilient food web has made this region one of the most dynamic and productive marine regions in the world. It is a region where whales (including the endangered blue whale) live and migrate, and millions of seabirds feed on the abundant sea life. California's largest breeding population of harbor seals live here, as well as elephant seals and California sea lions. The sanctuary is also a home, nursery and/or spawning ground for commercially valuable species such as Pacific herring, rockfish, and Dungeness crab. It is understandable why it was once heavily harvested by U.S. fishermen.

Management Challenges. Since the sanctuary boundaries include the coastline up to mean high tide, proper management of the sanctuary requires an integrated strategy in cooperation with other federal (e.g., Point Reyes National Seashore), state and locally managed parks, beaches, and surrounding watersheds.[2]

Sanctuary status has brought many safeguards. Oil exploration and drilling, or any other form of alteration of the seabed, is prohibited. There are also regulations against discharging substances, operating vessels within 3.2 km (2 mi) of specific biologically sensitive areas; disturbing marine mammals by flying motorized aircraft at less than 305 m (1000 ft) over the waters within one nautical mile of biologically sensitive areas, and damaging or removing any historical or cultural resources.

Despite these regulations to protect the sanctuary, the Gulf of the Farallones NMS faces many environmental threats. Located at the entrance to San Francisco Bay, tankers and barges must regularly pass through its waters. The sanctuary faces threats not only from future oil spills from tankers and barges but also from past oil spills, such as the *Apex Houston* and *Puerto Rican* (See Chapter 7, *Offshore Oil Development,* for further details).

Because this marine sanctuary is so close to the urbanized San Francisco Bay area, with shoreline boundaries in Marin and Sonoma counties, it faces many coastal pollution problems such as sewage disposal from Santa Rosa and other neighboring areas. There is also the threat from past ocean dumping of radioactive materials and the potential threat of future dumping of dredge spoils adjacent to the sanctuary (See Chapter 6, *Ocean Dumping,* for details).

Cordell Bank National Marine Sanctuary. Cordell Bank National Marine Sanctuary is managed jointly with the Gulf of the Farallones NMS and the northern section of Monterey Bay NMS, with staff headquarters at Crissy Field, Golden Gate National Recreation Area. Designated in 1989, Cordell Bank NMS is 50 nautical miles northwest of

[2] The GFNMS manages the waters within GFNMS boundaries. The islands, themselves, are managed by the state.

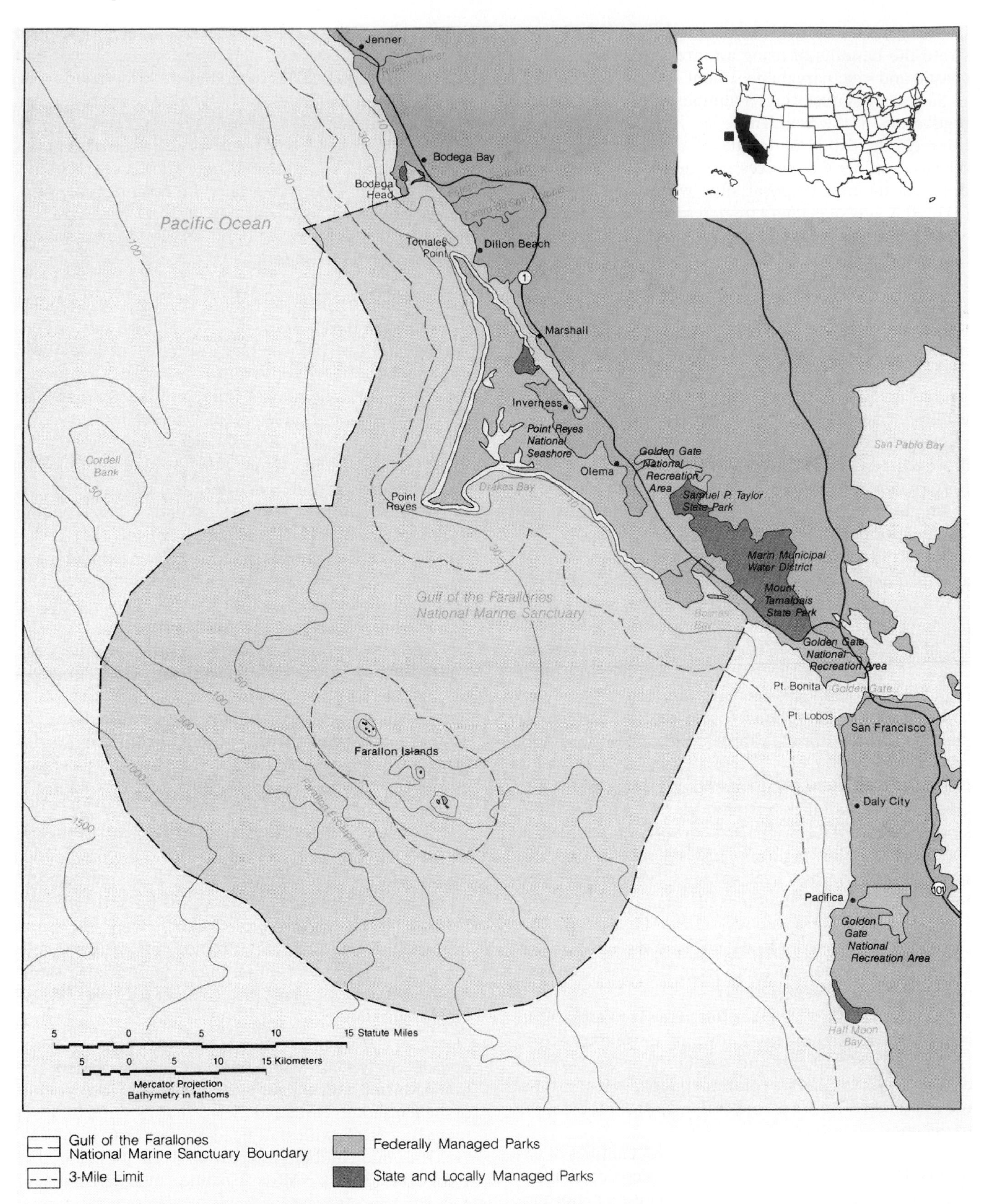

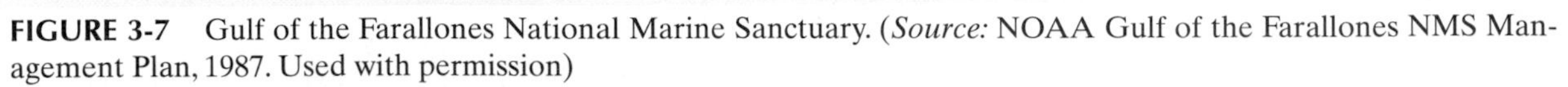

FIGURE 3-7 Gulf of the Farallones National Marine Sanctuary. (*Source:* NOAA Gulf of the Farallones NMS Management Plan, 1987. Used with permission)

FIGURE 3-8 The Gulf of the Farallones National Marine Sanctuary protects the productive waters surrounding the Farallon Islands. (*Source:* Photo by Jan Roletto, GFNMS)

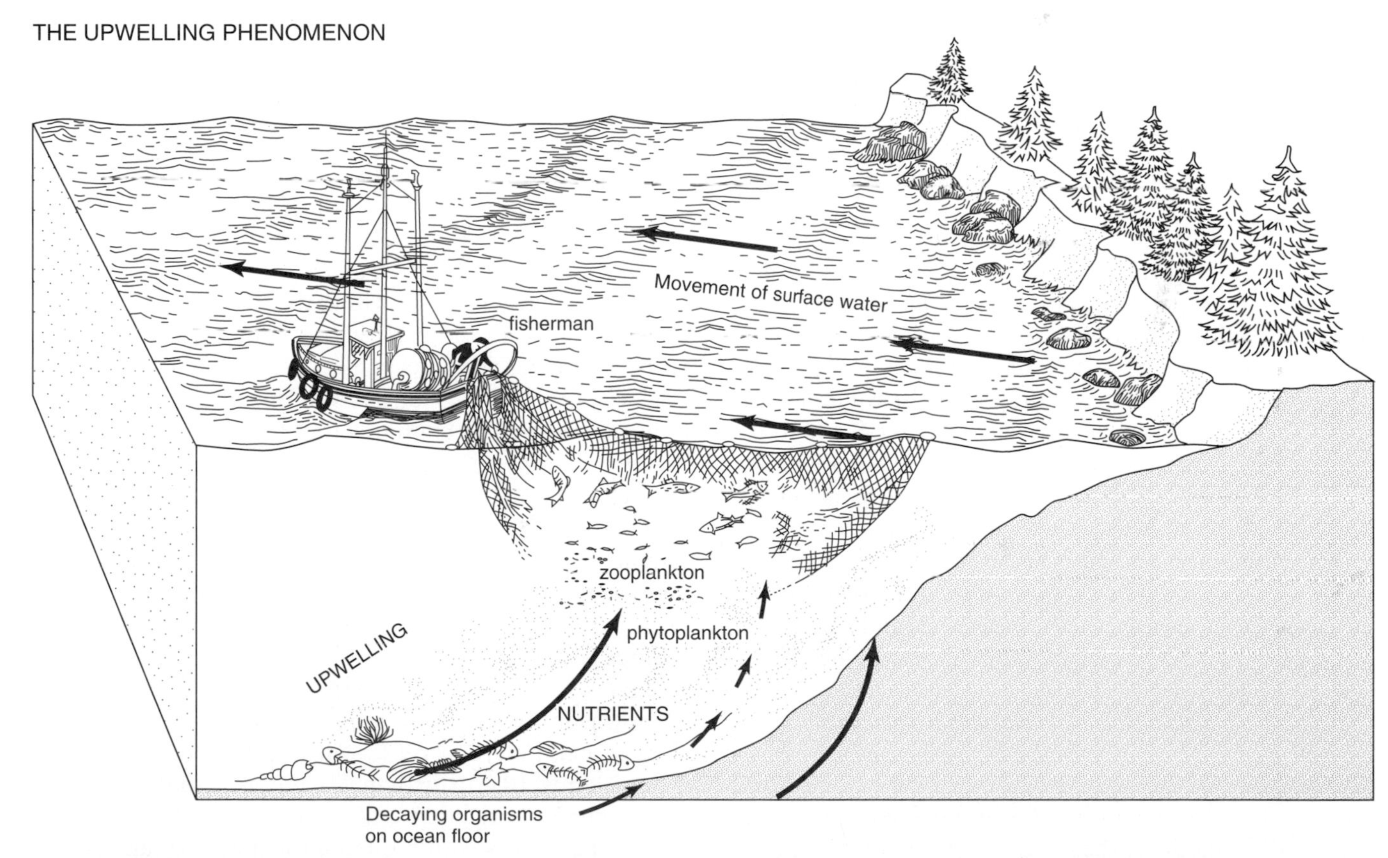

FIGURE 3-9 Diagram illustrating upwelling which occurs at the Gulf of the Farallones NMS. (*Source:* Adapted from *Natural Resource Conservation*, 7/e by Owen and Chiras 1998. Reprinted by permission of Prentice Hall, Inc.)

San Francisco and encompasses 1362 sq km (526 sq mi) of ocean waters (NOAA 1989). This sanctuary is unique because it has the northernmost seamount on the California Continental Shelf. A **seamount** is an isolated submerged mountain of volcanic origin which may rise to 3700 m (12,139 ft) above the deep sea plain. Cordell Bank rises only to 35 m (115 ft) above the ocean floor, but only a few miles away there are water depths of 1,829 m (6,000 ft). Like the Gulf of the Farallones, Cordell bank is blessed with the upwelling of nutrient rich deep-ocean waters that stimulate growth of planktonic organisms. As a result, the sanctuary has a great diversity of algae, invertebrates, fishes, and seabirds, and is called "Albatross Central" by many.

Management Challenges. Many of the marine mammals here, however, are classified as either endangered or threatened; the list includes whales (e.g., blue, gray, right, fin, humpback, sei, and sperm), turtles (e.g., Pacific ridley, leatherback, and loggerhead), and the Guadalupe fur seal.

Like the Gulf of the Farallones, Cordell Bank NMS has regulations against offshore oil exploration and drilling, removing or injuring natural resources; or depositing or discharging materials.

Three island sanctuaries: Channel Islands, Hawaiian Islands Humpback Whale, and Fagetele Bay. In addition to the Gulf of the Farallones, there are three other national marine sanctuaries that are associated with islands—the Channel Islands (Southern California), Humpback Whale (Hawaii), and Fagetele Bay (American Samoa).

Channel Islands National Marine Sanctuary. The Channel Islands National Marine Sanctuary (CINMS) is located approximately 32 km (20 mi) off the coast of Santa Barbara, California, and was designated in 1980 to protect 4294 sq km (1658 sq mi) of unique marine resources. The sanctuary encompasses the waters surrounding five [of the eight] Channel Islands (San Miguel, Santa Rosa, Santa Cruz, Anacapa, and Santa Barbara islands) from mean high tide to 6 nautical miles offshore surrounding each of the islands. (See Figure 3-10).

The sanctuary helps protect a wealth of natural and cultural resources. The combination of cool and warm currents provides a fertile breeding ground for a great number of marine plants and animals. Large nearshore forests of giant kelp provide a nutrient-rich environment and playground for a variety of invertebrates, fishes, and five species of pinnipeds (seals and sea lions). At one time of the year or another, one can also find twenty-eight species of whales and dolphins and sixty species of birds. Since these waters are relatively secluded and undisturbed, the sanctuary provides critical feeding ground for several endangered species, including whales (e.g., blue, gray, humpback, sei), seals and sea lions (e.g., California sea lion; northern elephant seal), and birds (e.g., California least tern; California brown pelican). In terms of cultural resources, the ocean floor within the sanctuary is riddled with prehistoric artifacts from the Chumash Indians, as well as the remains of over 100 historic shipwrecks.

Management Challenges. Administrating this sanctuary is no simple task, since it must be done in coordination with seven major agencies. One of the things that makes Channel Islands NMS unique is that it completely surrounds the Channel Islands National Park (CINP), which was also designated in 1980. The U.S. National Park Service oversees the five islands and the waters for one nautical mile around each island. Consequently, the National Park Service (NPS) and the National Marine Sanctuary Program overlap in the 1.6 km (1 mi) of ocean surrounding the islands. To further complicate matters, the NPS only oversees use and resource protection on four of the five islands. The Nature Conservancy (a national, private, nonprofit conservation organization) owns nine-tenths of Santa Cruz Island, and, therefore has an equal input on how that island is administered.

The Channel Islands NMS manager and staff must also integrate their efforts with the *United States Coast Guard* (e.g., regulations regarding ocean dumping, foreign fishing vessel safety), the *National Marine Fisheries Service* (e.g., regulations regarding the protection of marine mammals), the *California State Water Resources Control Board* (e.g., regulations regarding sewage disposal), the *State Lands Commission* (e.g., protection of shipwrecks and other cultural resources), and the *California Department of Fish and Game* (e.g., protection of state ecological reserves within the sanctuary). The United Nations is even involved in the management of this area, since this site is recognized as a Biosphere Reserve, which is part of the Man and the Biosphere Program administered by the United Nations Education, Scientific and Cultural Organization (UNESCO).

The Channel Islands NMS has a management plan for protecting the natural and cultural resources within its domain. There are the standard prohibitions against (a) discharging or depositing substances (with the exception of fish or chumming materials and biodegradable effluents from vessel use), (b) altering or constructing on the seabed (e.g., no drilling or dredging within two nautical miles of the islands), (c) operating a commercial vessel within one nautical mile of the islands; (d) disturbing marine mammals and birds by flying motorized aircraft at less than 305 m (1000 ft) over the waters within one nautical mile of any island; and (e) removal or damaging historical or cultural resources. To learn the details (and the exceptions) of

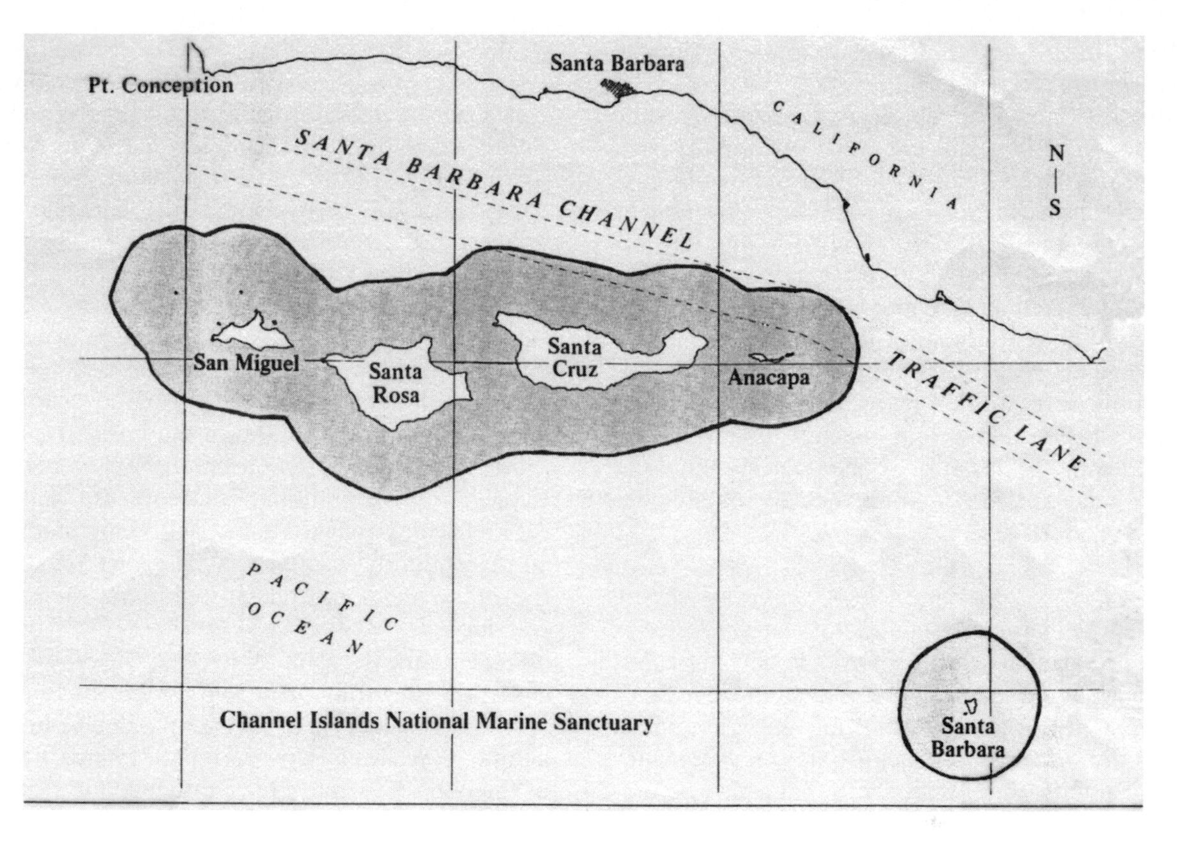

FIGURE 3-10 Channel Islands National Marine Sanctuary. (*Source:* NOAA fold out map 1991. Used with permission)

these prohibitions, one must consult the management plan for each sanctuary.

Water quality is generally good around the islands, primarily because of the absence of municipal or industrial point sources and the relative isolation from the mainland. Sanctuary managers keep note of any observed problems such as unusual accumulations of tar or oil and monitor tar cover in the intertidal plots, though it rarely occurs. Most of the tar on the islands is assumed to come from natural seeps and drilling in the Santa Barbara Channel. The U.S. Coast Guard has jurisdiction over the federal waters regarding pollution and the California Department of Fish and Game has jurisdiction over the marine life in state waters.

The sanctuary management is concerned about oil and tar pollution from non-point sources outside the sanctuary. It is an especially important concern along the north shore of the islands (California State Water Resources Control Board, 1979a, 1979b; Bureau of Land Management 1981). Although offshore oil and gas activities are prohibited within the sanctuary, hydrocarbon development has occurred in the vicinity of the sanctuary area since the late 1800's. Therefore, it is not surprising that sanctuary management is concerned about the lack of information on the effect of offshore oil development outside the sanctuary or in sanctuary waters, not to mention what would be the impact on the sanctuary if there was a major oil spill outside the sanctuary (For further details, see Chapter 7, *Offshore Oil Development*).

Another major environmental concern at the sanctuary is the discovery of significant accumulations of heavy metals within Anacapa, Santa Barbara, and San Miguel islands' areas of special biological significance (Bureau of Land Management 1981). To date, it is not known whether these levels of pollutants are moving up the food chain thereby affecting individuals or populations. The sanctuary staff does not monitor nonpoint source pollution at the islands. However, it does have long-term ecological monitoring studies being done in the rocky intertidal, kelp forest, and sand beach communities observing indicator species at permanent sites around the islands. The sanctuary also monitors nesting seabird phenology and productivity.

Hawaiian Islands Humpback Whale National Marine Sanctuary. The Hawaiian Islands Humpback Whale National Marine Sanctuary (HIHWNMS) was the thirteenth national marine sanctuary to be designated. Signed into law by President Bush in 1992, this sanctuary recognized the importance of the humpback whale (*Megaptera novaeangliae*) and its winter habitat. The sanctuary boundary is the 100-fathom isobath around the main Hawaiian Islands with a protected area of 3367 sq km (1,300 sq mi). The purpose of the sanctuary is to promote comprehensive and coordinated protection, research, education, and monitoring of the humpback whale.

Management Challenges. Protecting the humpback whale, of course, implies practicing wise use of the marine environment which constitutes its habitat. In September 1995, the Hawaiian Islands Humpback Whale National Marine Sanctuary published its Draft Environmental Impact Statement and Management Plan (NOAA 1995b). In addition to protecting humpback whales and their habitat, management challenges include educating the public related to humpback whales and their conservation, managing selected uses of the sanctuary, and providing for identification of marine resources and ecosystems of national significance for inclusion in the Sanctuary (NOAA 1997b).

Fagetele Bay National Marine Sanctuary. Fagetele (Fangh-a-t´eh-leh) Bay National Marine Sanctuary (FBNMS) is unique in many ways. It is the farthest national marine sanctuary from the U.S. mainland, being located off the southwest shore of Tutuila Island in American Samoa, some 14 degrees south of the equator and just east of the International Dateline (NOAA 1984). Fagatele Bay NMS is also the only one to be nestled in a broken crater of an ancient volcano, surrounded by lush green tropical rainforest. The coral here are the most diverse in the NMS program—nearly 200 species representing over 50 genera. These reefs, in turn, support over 200 species of fishes, over 100 mollusks, and a variety of crustaceans, echinoderms, decapods and marine plants. Endangered or threatened whales (e.g., humpback, sperm) and sea turtles (e.g., hawksbill, green) frequent the sanctuary. As mentioned earlier, Fagetele Bay is the smallest of the National Marine Sanctuaries, protecting an area less than 0.64 sq km (0.25 sq mi).

Management Challenges. One of the reasons for sanctuary designation in 1986 was to allow the coral reefs of American Samoa to recover after a devastating attack by a catastrophic population bloom of the crown-of-thorn starfish (*Ancanthaster planci*). (See Figure 3-11). In the late 1970s, millions of these starfish swarmed over the reefs, consuming the coral as they migrated. Less than 10 percent of the corals survived. There are still scars of the attack, and it may take a century for the reef to fully recover.

Despite the devastation to coral, the crown-of-thorns starfish is now fully protected under sanctuary regulations. The sanctuary staff remains puzzled by the rationale behind those regulations. Fortunately, since the outbreak of the late 1970s, there have only been low numbers of *Ancanthaster* found in Fagatele Bay (and elsewhere around Tutuila).

Samoans still use their reefs for subsistence fishing, and this often becomes a problem because some traditional fishing gear techniques (e.g., pole spears, spear guns, bow and arrow, seines, or fixed nets) are no longer allowed in sanctuary waters. There are also prohibitions against using fishing poles or hand-lines near shore. This is also one sanctuary where commercial fishing is prohibited. One of the related management concerns is trying to stop such destructive fishing methods as using dynamite and fish poison. Though both are now illegal in sanctuary waters, the practice has been difficult to stop. According to sanctuary personnel, there have been no studies on the impact of traditional fishing in Fagatele Bay. Such studies would help the sanctuary coordinator in addressing enforcement questions and reviewing current and prospective regulations.

Fagatele Bay NMS is run through a cooperative agreement with the American Samoa Government, the coordinators employer (hence the title, coordinator, rather than manager). The staff works with the Department of Marine and Wildlife Resources (regarding office space and technical support) and the Department of Education (e.g., the sanctuary staff cooperates in a summer camp that is run by DOE personnel). According to sanctuary personnel, the 1984 management plan is in great need of revision, but limited funds have prevented that activity.

Although information about Fagatele Bay NMS is limited, one could start with Berkeland, et al. (1987); and Berkeland, Randall, and Amesbury (1988).

FIGURE 3-11 Crown-of-Thorns Starfish. (*Source:* From *Conservation Ecology* by G. Cox © 1992. Use with permission of the McGraw-Hill Companies)

Flower Garden Banks National Marine Sanctuary. In 1992, Flower Garden Banks National Marine Sanctuary was designated to protect 145 sq km (56 sq mi) of coral reef in the Gulf of Mexico—approximately 120 nautical miles south southwest of Galveston, Texas. The reefs are located some 483 km (300 mi) north of the Tropic of Cancer, thus making it the northernmost living coral reefs on the U.S. continental shelf. By comparison, the reefs of the Florida Keys lie just 97 km (60 mi) north of the Tropic of Cancer. Both the Flower Garden Banks and the Keys' reefs are unusual, since we generally associate coral reefs with the tropics—the equatorial band below the Tropic of Cancer. Massive coral colonies approaching heights of 4.6 m (15 ft) or more are characteristic of the reefs at Flower Garden Banks NMS. Whereas sea fans, sea whips or branching elkhorn and staghorn corals are common in the Keys, massive brain and star corals are more characteristic of the reefs at Flower Banks. (See Figure 3–12.) Over 500 species of plants, invertebrates, and fish are located within the Flower Gardens Banks NMS.

Management Challenges. Unlike the Florida Keys NMS, this reef is located some distance from land and is not inundated by millions of visitors each year. Only the more serious scuba divers or researchers visit this area. However, the area does suffer from some human impacts. For example, anchor damage is a major management problem. Sport divers and sport fishermen from Louisiana and Texas, and commercial fishermen from as far away as Florida visit the Banks and drop and/or drag their anchors on the delicate coral reefs. However, the principal human activity—and potentially greatest threat to the sanctuary—*is the exploration and development of oil and gas within and adjacent to the sanctuary. This is the only sanctuary that actually has a pre-existing gas production platform located directly in the sanctuary.*

While the prevalence of oil and gas production makes it a logical target as the most significant threat to sanctuary resources, 20 years of monitoring have not revealed any discernible impacts from oil and gas production. More significant has been the physical destruction of coral habitat from anchoring, and possibly towing cables, used by large shipping vessels and barges. There is a major shipping fairway just south of the Flower Gardens. FGNMS has not yet managed to get on all of the international maps as a protected area, so the banks show up simply as a topographic high on many maps—a convenient shallower place to anchor when a vessel has engine trouble or needs to hang around waiting for instructions from the home office before heading into port. FGNMS is still investigating the most recent incident, but the other two it knows about did not involve the oil and gas industry in the Gulf of Mexico—they were foreign vessels.

Gray's Reef National Marine Sanctuary. Gray's Reef is a highly productive marine habitat that lies 15–2 m (50–75 ft) deep, 17.5 nautical miles due east of Sapelo Island, Georgia. (See Figure 3-13). Designated in 1981, this 60 sq km

FIGURE 3-12 Large boulder-shaped corals (brain coral in foreground) at the Flower Gardens. (*Source:* Photo by Dr. Steve Gittings.)

(23 sq mi) sanctuary protects the largest inner-shelf "live-bottom" reef found off the southeastern coast of the United States. The term "live-bottom" refers to the habitat provided by a limestone platform with diverse sessile (attached) benthic invertebrates. (See Figure 3-14).

The live bottom, with its accompanying 1.8 m (6 ft) in height ledges and sandy, flat-bottom troughs, provides a habitat that attracts numerous species of fish (e.g., snapper, grouper, black sea bass, mackerel) and threatened or endangered sea turtles (e.g., loggerhead, leatherback, green, and Kemp's ridley). The most endangered of all the great whales—the North Atlantic right whale (*Eubalean glacialis*)—also migrates into Gray's Reef NMS. The whale cows give birth to a single calf within the sanctuary before returning to their feeding grounds off New England in the early spring.

Management Challenges. As with many of the other sanctuaries, the management plan prohibits altering the seabed, damaging or removing bottom formations, discharging substances, and using fishing techniques that could damage or over exploit the reef, such as using trawls, wire fish traps, and explosives (NOAA 1983).

Human activities, and consequently human impacts at Gray's Reef NMS, are minimal. Although Gray's Reef NMS is the closest natural reef offshore of Georgia, the reef is still 18 nautical miles out at sea and requires a seaworthy vessel, electronic navigation equipment, and offshore experience. A boat trip to Gray's Reef from the closest departure point may take between 1-3 hours, depending on the type of vessel. Consequently, this sanctuary only attracts the most serious recreational fisherman or scuba diver. Furthermore, there is currently little or no interest in the Gray's Reef area for commercial fishing, marine minerals development, military activities, nor ocean dumping (including dredge material disposal). There is interest, however, in the reef as a natural laboratory for research and educational programs. For further information about Gray's Reef NMS, see NOAA (1980, 1983).

Monitor National Marine Sanctuary. In 1975, NOAA designated the very first national marine sanctuary—Monitor National Marine Sanctuary. Interestingly enough, it was not established to protect a unique coral reef or other form of biological habitat, but rather solely to protect the remains of the Civil War ironclad *U.S.S. MONITOR*—a Civil War warship that lay in 73 m (240 ft) of water 26 km (16 mi) off Cape Hatteras, North Carolina. Monitor NMS is one of the smallest national marine sanctuaries, since it is a mere 1 nautical mile diameter [2.6 sq. km., or .79 sq. mi.] in area over the wreckage of the *U.S.S. MONITOR*.

The U.S.S. MONITOR is historically significant because it was the first exclusively iron clad U.S. warship, and thus marked the end of wooden sailing vessels as implements of war. (See Figure 3-15). Consequently maritime historians consider the *U.S.S. MONITOR* one of the most significant vessels in American history. The Monitor has been designated a National Historical Landmark. In 1862, the *U.S.S. MONITOR* sank while being towed to Beaufort, North Carolina. The main purpose of the sanctuary is to preserve the remains of the wreckage and educate the public of its historical significance. Since most of the public cannot get to the site, the Mariners' Museum

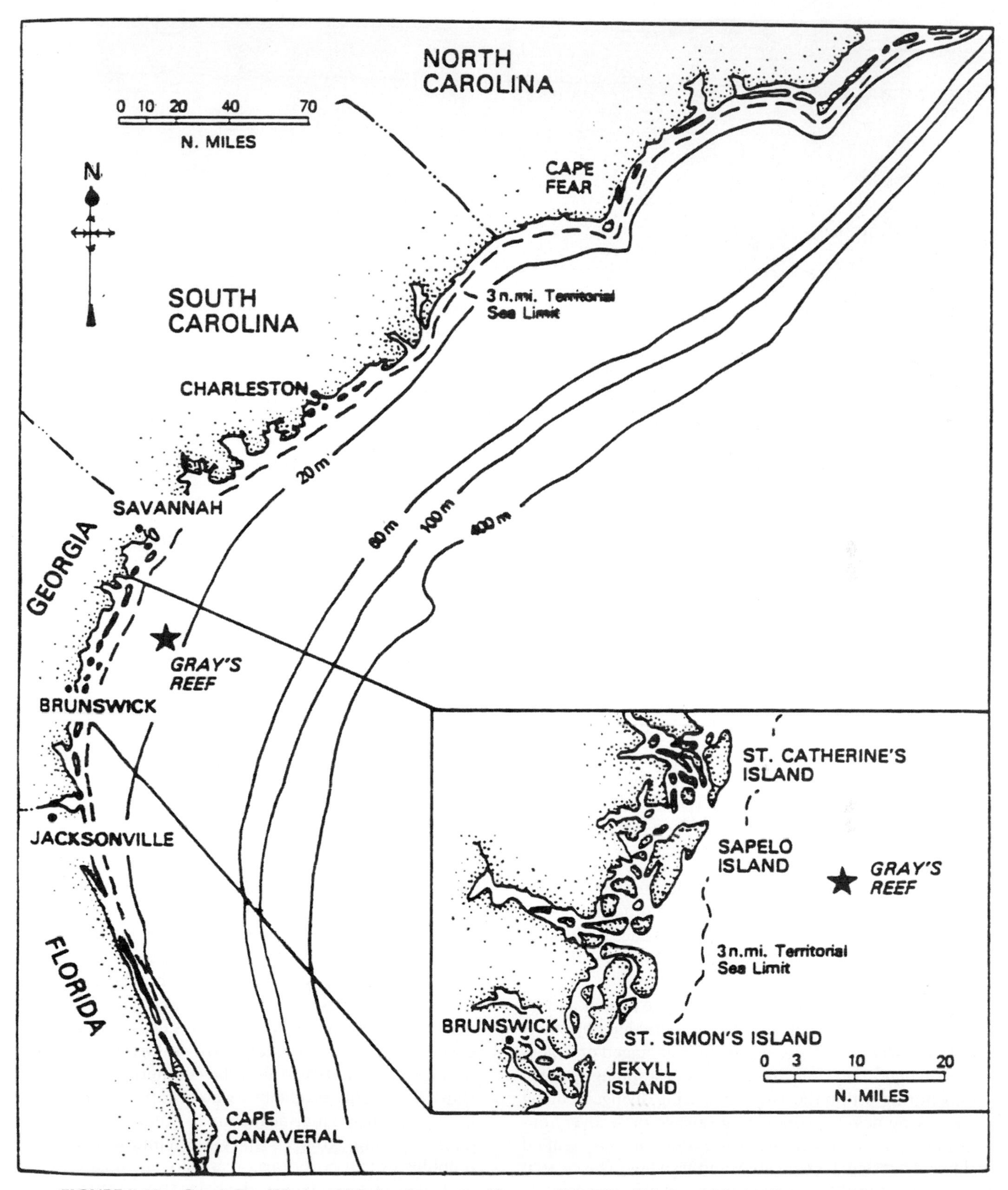

FIGURE 3-13 Gray's Reef National Marine Sanctuary. (*Source: Gray's Reef National Marine Sanctuary Management Plan.* NOAA 1983. Used with permission)

FIGURE 3-14 Limestone rock (photo) and reef habitats (diagram) at Gray's Reef NMS. (*Source:* Courtesy Gray's Reef NMS.)

in Newport News, Virginia has been designated the principal museum for the sanctuary. Among other things, interested persons can view underwater scenes of the ship in the sanctuary's video, "Down to the MONITOR."

Management Challenges. As might be expected, a long-term management problem is making sure that objects on the *U.S.S. MONITOR* are not further damaged or stolen. The sanctuary, therefore, has prohibited stopping, drifting, or anchoring within the sanctuary; using diving, dredging or wrecking devices; conducting salvage or recovery operations; conducting underwater detonations; seabed drilling or coring; laying cable; and trawling. The current management problem is to preserve as much as possible of the Monitor's hull, which is rapidly disintegrating.

Stellwagen Bank National Marine Sanctuary. In 1992, Congress designated New England's first sanctuary—Stellwagen Bank National Marine Sanctuary. The sanctuary is located 9.7 km (6 mi) off the northern end of Cape Cod, Massachusetts in the southwestern Gulf of Maine. (See Figure 3-16). Stellwagen Bank gets its name from hydrographer Henry S. Stellwagen who first discovered the glacially-deposited submerged sand bank located 30 nautical miles east of Boston. The sanctuary boundary encompasses approximately 2,181 sq km (842 sq mi) of ocean waters that surround the submerged Stellwagen Bank. The highly productive waters support large populations of fish and marine mammals. Cetaceans are abundant, including Atlantic white-sided and white-beaked dolphins, harbor porpoises, orcas, pilot and minke whales, as well as three endangered whales (fin, humpback, and northern right whales). So many whales use the area as a nursery and feeding grounds that the whale-watching industry brings more than one million visitors to the Bank each year. The great number of recreational boaters that

FIGURE 3-15 Monitor National Marine Sanctuary. (Left) Painting of the *U.S.S. MONITOR*; (Right) Diver measures the sunken Civil War ironclad—the *U.S.S. MONITOR*. (*Source:* NOAA *Marine Sanctuary,* Spring/Summer 1993, p. 15. Used with permission)

use Stellwagen Bank makes it one of the busiest places in the Gulf of Maine. Recreational fishermen from Gloucester to Provincetown troll for giant bluefin tuna or other highly valued gamefish.

Management Challenges. The sanctuary has regulations against ocean dumping or discharging, sand and gravel mining, alteration of or construction on the seabed, placing submerged pipelines or cables, and vessel lightering (transfer of fuel at sea). The taking of marine mammals, reptiles, and seabirds is also prohibited. NOAA has empowered the sanctuary manager to take action against any outside source of pollution that enters the sanctuary and injures its resources.

Stellwagen Bank NMS is another excellent example that coastal and sanctuary management must be integrated if sanctuary resources are to be protected. Stellwagen Bank NMS faces intense urban pressures from Boston and the surrounding areas (Eldredge 1993). Although ocean dumping is prohibited within the sanctuary, the Environmental Protection Agency has designated the Massachusetts Bay Disposal Site for dredge materials less than 1 km (0.6 mi) from the Sanctuary. There are also potential impacts upon the sanctuary from moving Boston's sewage effluent outfall out of Boston Harbor into Massachusetts Bay. Stellwagen's unique thriving temperate ecosystem and Massachusetts Bay support the legendary New England fishing industry.

Since NOAA chose a multiple use approach for the Bank, fishing activities are not regulated by the sanctuary management plan. In addition to the hundreds of recreational boats that visit the area on any given day, commercial fishermen venture to the Bank to drag for flounder, tuna, haddock, cod, or other popular New England fish. Consequently, the sanctuary staff is concerned with prevention of accidental marine mammal entanglement in fishing gear.

There are no rules addressing the important issue of vessel traffic in the sanctuary. Huge container ships, oil tankers, and cruise ships maneuver through designated shipping channels to and from the Port of Boston. Unfortunately, collision with vessels is one of the major causes of whale mortality in the region. Sanctuary officials are currently evaluating the feasibility of vessel speed limits, shifts in traffic lanes, requiring onboard lookouts, and other strategies.

In other words, sanctuary designation for Stellwagen Bank provides an administrative framework for coordinating coastal and marine management. In order for these efforts to be successful, existing marine management initiatives will also have to be incorporated, such as the Massachusetts Coastal Zone Management Program, the Massachusetts Ocean Sanctuaries Program, the Massachusetts Bays/National Estuary Program, and the Gulf of Maine Initiative.

For the final environmental impact statement/management plan for Stellwagen Bank NMS, see NOAA (1993a).

Olympic Coast National Marine Sanctuary. Olympic Coast National Marine Sanctuary (OCNMS) is the newest, and twelfth member of the national sanctuary program. Designated in 1994, this 8,572 sq km (3,310 sq mi) sanctuary is twice the size of Yosemite National Park and lies off the northern coast of Washington State. (See Figure 3-17). Its 217 km (135 mi) coastline extends along Washington's Olympic Peninsula from the Strait of Juan de Fuca to Copalis Beach south of Point Grenville. Here, one can see rocky outcrops, tidepools, sand and cobble

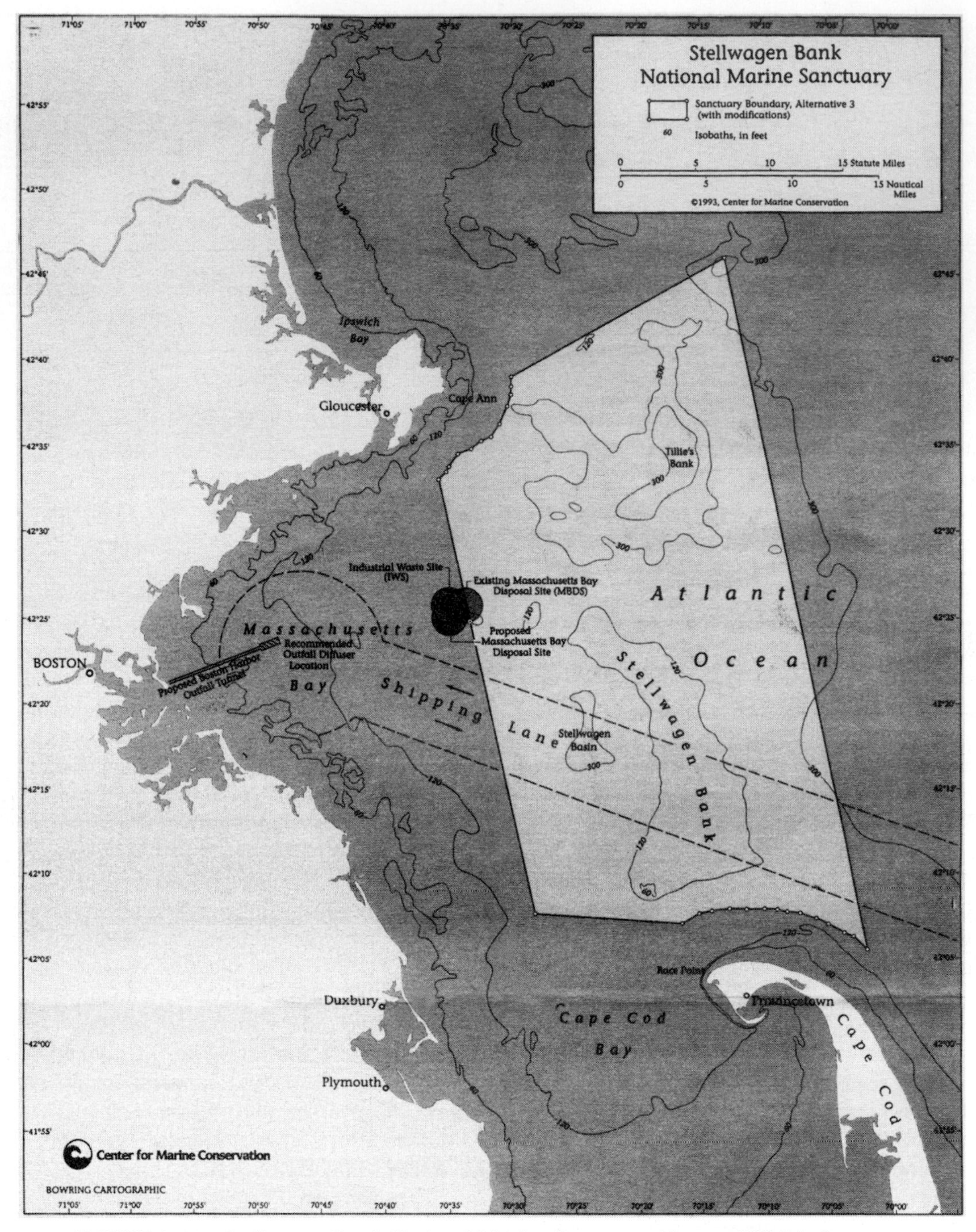

FIGURE 3-16 Stellwagen Bank National Marine Sanctuary. (*Source:* NOAA *Sanctuary Currents* Winter/Spring 1993, p. 3. Used with permission)

beaches, and small offshore islands. The offshore waters are one of the world's most productive areas due to the upwelling of nutrient-rich waters from undersea canyons. Like the Monterey Bay NMS, this area has one of the greatest diversity of kelp and species of whales, dolphins and porpoises in the world. The sanctuary has some of the largest colonies of seabirds in the continental United States, and its coastline is home to one of the largest populations of bald eagles in the lower 48 states.

Management Challenges. An integrated ecosystem management approach is clearly needed in managing the offshore and nearshore resources that come under the protection of this sanctuary. It must compliment, and integrate, the management of the sanctuary with preexisting protected areas along its coastline, including national wildlife refuges, wilderness areas, Olympic National Park, and a Biosphere Reserve and World Heritage Site. Due to the sanctuary's rich diversity of fishes (five species of

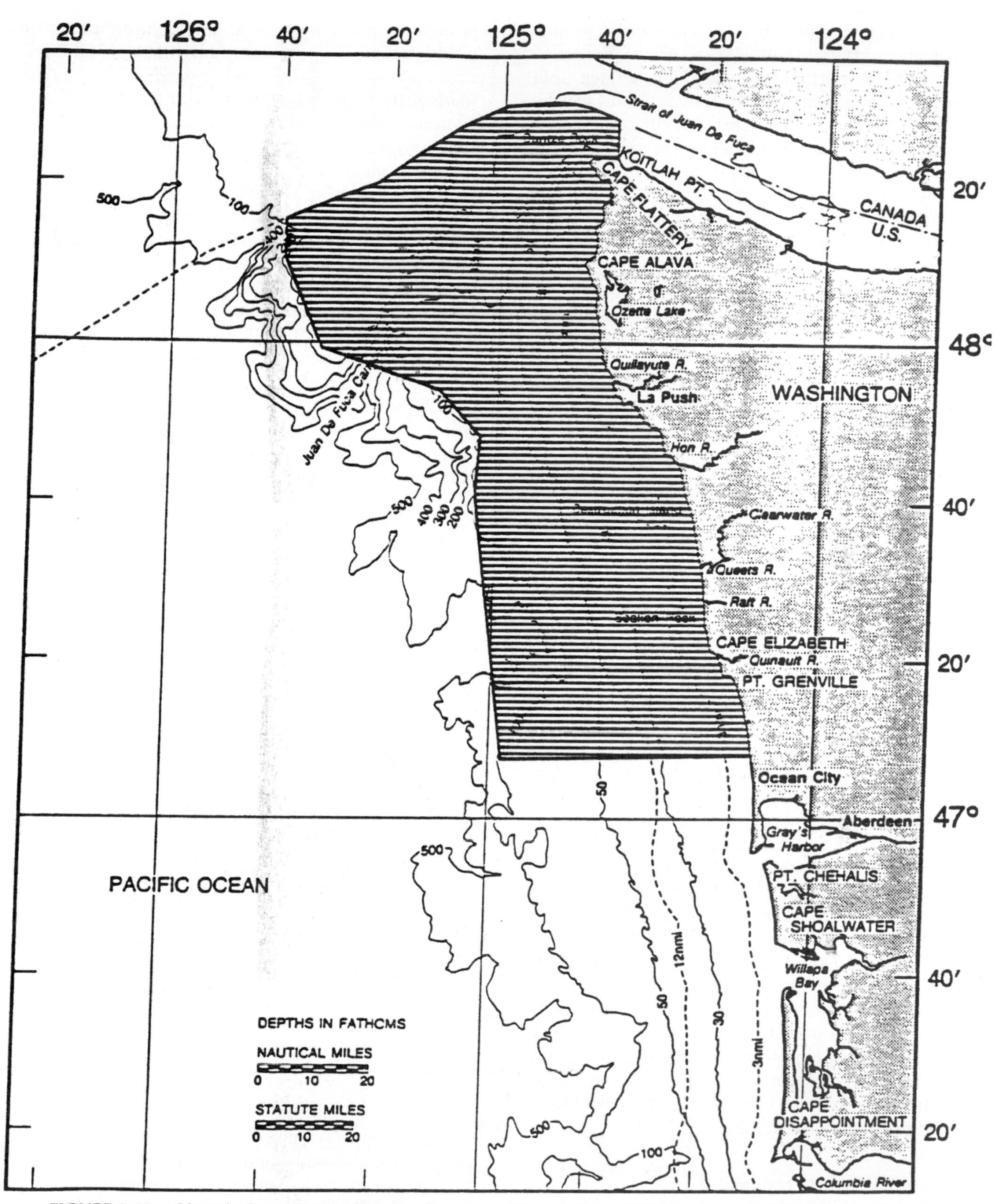

FIGURE 3-17 Olympic Coast National Marine Sanctuary. Olympic Coast NMS borders tribal lands, national parks, and other protected lands. (*Source: Olympic Coast NMS Final Environmental Impact Statement Management Plan,* Vol. 1, NOAA 1993b p. II-6. Used with permission)

salmon, numerous bottomfish, such as rockfish, halibut, flatfish, and cod) and shellfish (e.g., crabs), the area is heavily used by sport and commercial fishermen. Sport fishing, for example supports the local economy, with annual catches of over 9 million fish. One unique aspect of this sanctuary, compared to other sanctuaries within the NMSP, is the integration of adjacent Indian reservation concerns. The reservations of the Makah, Quileute, Quinault, and Hoh Tribes adjoin the sanctuary.

Sanctuary regulations now keep the area free of oil and gas drilling activities, and the Navy no longer uses Sea Lion Rock, within sanctuary waters, for practice bombing raids. Now that "oil rigs and bombs" are kept out of the Olympic Coast NMS, environmentalists and sanctuary officials have turned to another major threat to the sanctuary—large vessels carrying hazardous cargo. Before the sanctuary was officially designated, NOAA developed a proposal to establish an "Area to be Avoided" off the coast. Vessels carrying hazardous cargo would be requested to stay out of waters adjacent to the coast, thereby increasing vessel traffic safety and increasing time for response in the event of an oil or other toxic spill (See Chapter 7, *Offshore Oil Development,* for further details on vessel traffic).

For the final environmental impact statement/management plan for Olympic Coast NMS, see NOAA (1993b).

Additional Sanctuaries Under Consideration

Two proposed sanctuaries have "Active Candidate" status. A *Northwest Straits National Marine Sanctuary* is being considered for the waters north of Puget Sound, Washington. It would encompass the waters surrounding the San Juan Islands, north to the Canadian border. The area is unique for its 12 foot tidal range, its variety of natural habitats, and having the single largest concentration in the continental United States of bald eagles. Management challenges would include dealing with an area that has multiple recreational uses, extensive commercial fishing, and intense research activity.

The proposed *Thunder Bay National Marine Sanctuary* is in Lake Huron off the coast of Alpena, Michigan. In addition to an array of natural features, the site is unique for having over 100 shipwrecks. Management challenges would range from protecting these cultural resources from recreational divers, to protecting wildlife on Thunder Bay Island which is part of the Michigan Islands National Wildlife Refuge.

Challenges to the U.S. National Marine Sanctuary Program

All 12 national marine sanctuaries face five common problems: (1) inadequate funding levels; (2) understaffing; (3) bureaucratic interference; (4) unnecessary fights over prohibiting incompatible uses; and (5) the difficulty of patrolling a three-dimensional realm with fluid boundaries.

Inadequate funding levels. Since it began in 1972, the National Marine Sanctuary Program has become increasingly popular—as we have seen above, there are now 12 marine sanctuaries around the country. Unfortunately, the money per square mile of marine sanctuary is actually decreasing. In 1992, three new sanctuaries were designated: Monterey Bay in California, Stellwagen Bank off of Massachusetts, and the Hawaiian Islands Humpback Whale. Yet in 1993, under the Clinton Administration, the National Marine Sanctuary Program received a 2 percent decrease to 7 million dollars. To further stretch this already thin budget, the 12th National Marine Sanctuary (Olympic Coast) was added in 1994.

Without adequate funding, the NMS program can become nothing more than a series of "paper parks." The Center of Marine Conservation has some interesting comparisons that help illustrate the problem: In 1993, the National Park Service received $1.4 billion, whereas the National Marine Sanctuary Program received only $7.1 million. Yet, one sanctuary alone—Monterey Bay NMS—is twice the size of Yellowstone National Park. Despite the fact that both Yellowstone National Park and the Florida Keys NMS have the same number of visitors each year (several million), the Florida Keys sanctuary staff must get by with 10 percent of Yellowstone's budget (Sobel and Merow 1993).

In 1991, an independent review panel representing diverse interests (fishing, conservation, petroleum, research, education, diving, and government) studied the budget problem and estimated that it required a minimum of $30 million for the National Marine Sanctuary Program to carry out its mandate (Potter 1991).

Should we continue designating sanctuaries if the budget does not keep pace? There are far more reasons than not to continue designating areas as part of the National Marine Sanctuary Program, despite the fact that the budget may not follow suit. Paul Pritchard (1993, p. 4) nicely sums it up: "Even if no specific 'resource protection' measures are taken, there are important and valuable psychological and emotional connections associated with sanctuaries. Designating an area a national park or monument elicits respect, as should naming a site a sanctuary."

Understaffing. Without an adequate budget, most sanctuaries face a second common problem—understaffing. In 1993, for example, Yellowstone National Park had a budget of $18,247,000 and employed a staff of 508 employees, whereas Monterey Bay NMS had a budget of $548,000 and only two employees. Of course, without much of a staff, public outreach, enforcement, and other aspects of operating a sanctuary suffer.

Bureaucratic uncertainty and interference. There are so many federal, state, and local agencies with "their hand in the pie" when it comes to managing coastal and marine ecosystems, it is often difficult to know which agency assumes what role for the protection of a particular resource. Furthermore, NOAA's National Marine Sanctuary Program is small and relatively unknown, as compared to say, the NPS' National Park System, the USFS' National Forests, or the USFW's National Wildlife Refuge System. To some representatives of these larger more established agencies, NOAA sanctuary staff are an "irrelevant irritant" that they must now contend with in the already complex bureaucratic maze. Most seem to agree that the only answer to this problem is what might be called cooperative outreach. Only by outreach to other existing agencies, and integrating NOAA's desires with their management strategies, can the National Marine Sanctuaries Program succeed.

Unnecessary fights over prohibiting incompatible uses. Managing a piece of property, whether it be land or sea, is always difficult when a "multiple use" approach to management is used. All too often, sanctuary staff must contend with resolving differences that are seemingly incompatible. Debates rage over whether jet skiers, for example, should have free reign within a sanctuary when biotic communities might be harassed. Researchers that want to perform various scientific tests within a sanctuary often rile the public with schemes that, at least at first, sound like outrageous activities for a marine "sanctuary." Protecting a marine sanctuary, while at the same time satisfying the diverse interest of a general public, is no easy task.

Patrolling a three-dimensional realm with fluid boundaries. A marine sanctuary is a block of ocean space that is three dimensional. Protecting such a special ocean place is not quite the same as protecting a parcel of land, such as Yosemite National Park. Much of the ocean habitat—the water itself—flows in currents through the political boundary of the sanctuary, as do many of the biotic species the sanctuary was designed to protect. Patrolling "a liquid sanctuary" means enforcing laws at depths (e.g., scuba divers impacting coral reefs) as well as patrolling what happens on the surface of the sanctuary.

CONCLUSION

The National Marine Sanctuary Program was established in 1972 to provide authority for comprehensive and coordinated conservation and management of marine areas that would compliment existing regulatory authority. In 1990, NOAA called for a 12-member panel review of the NMSP to make recommendations on ways to improve it. This so-called "Potter Committee" released its report on February 22, 1991, calling for —among other things—greater visibility, aggressive program leadership, and focus (Potter 1991). Furthermore, the Potter report called for making the Florida Keys and the sanctuaries on the Central California coast the "centerpieces" of this renewed effort.

Since this 1991 report (but not solely due to it), the NMSP *has* undergone a radical change in direction. No longer are the sanctuaries within the program just small, isolated areas where specific resources are to be protected. Many are now within the sight, mind, and impacts of large urban areas. With the designation of Monterey Bay NMS (1992) and Florida Keys NMS (1993), the NMSP has moved into a new realm where it must integrate the diverse and often conflicting policies of the federal, state, regional, and local governments.

As shown above, the National Marine Sanctuary Program has many challenges ahead, especially if it hopes to integrate the management of coastlines and coastal hinterlands (e.g., watersheds) that directly impact sanctuary waters (e.g., nonpoint pollution). One of the areas that this new integrative coastal and sanctuary management paradigm (way of thinking) must address is how to mitigate the numerous problems that come under the category of *coastal hazards*—the subject of our next chapter.

REFERENCES

Baggett, L. S. and T. J. Bright. 1985. *Coral Recruitment at the East Flower Garden Reef. Proc. 5th Int. Coral Reef Congress,* Tahiti. Vol. 4, pp. 379–384.

Barley, George. 1993. "Integrated Coastal Management: The Florida Keys Example From an Activist Citizen's Point of View." *Oceanus.* Vol. 36, No. 3, pp. 15–18.

Berkeland, Charles E., et al. 1987. *Biological Resource Assessment of the Fagatele Bay National Marine Sanctuary. NOAA Technical Memorandum NOS MEMD 3.* Washington, D.C.: National Oceanic and Atmospheric Administration

Berkeland, Charles, Richard S. Randall, and Steven S. Amesbury. 1988. *Coral and Reef-Fish Assessment of the Fagatele Bay National Marine Sanctuary.* Report to the National Oceanic and Atmospheric Administration, Mangilao, Guam: University of Guam Marine Laboratory.

Bohnsack, James A. 1993. "Marine Reserves: They Enhance Fisheries, Reduce Conflicts, and Protect Resources." *Oceanus.* Vol. 36, No. 3, pp. 63–71.

Bureau of Land Management. 1981. *Final Environmental Impact Statement, Proposed, 1982 Outer Continental Shelf Oil and Gas Lease Sale Offshore Southern California.* OCS Sale No. 68.

California Resources Agency. 1995. *California's Ocean Resources: An Agenda for the Future.* Sacramento, California: State Printing Office.

California State Water Resources Control Board. 1979a. *California Marine Waters, Areas of Special Biological Significance Survey Report—Santa Cruz Island.* Water Quality Monitoring Report No. 79-8.

California State Water Resources Control Board. 1979b. *California Marine Waters, Areas of Special Biological Significance*

Survey Report—Anacapa Island. Water Quality Monitoring Report No. 79-9.

Cox, George W. 1993. *Conservation Ecology.* Dubuque, Iowa: Wm. C. Brown Publishers.

Dixon, John A. 1993. "Economic Benefits of Marine Protected Areas." *Oceanus.* Vol. 36, No. 3, pp. 35–40.

Earle, Sylvia. 1993. "Sanc'-tu-ar'-y." *Marine Sanctuary.* Vol. 1, No. 1 Spring/Summer, pp. 4–7.

Ehler, Charles N. and Daniel J. Basta. 1993. "Integrated Management of Coastal Areas & Marine Sanctuaries." *Oceanus.* Vol. 36, No. 3 (Fall), pp. 6–14.

Eldredge, Maureen. 1993. "Stellwagen Bank: New England's First Sanctuary." *Oceanus.* Vol. 36, No. 3, pp. 72–74.

Gittings, S. R., G. S. Boland, K. J. P. Deslarzes, C. L. Combs, B. S. Holland, and T. J. Bright. 1992. "Mass Spawning and Reproductive Viability of Reef Corals at the East Flower Garden Bank, Northwest Gulf of Mexico." *Bulletin of Marine Science.* Vol. 51, No. 3, pp. 420-428.

Gittings, S. R., G. S. Boland, C. R. B. Merritt, J. J. Kendall, K. J. P. Deslarzes, and J. Hart. 1994. *Mass Spawning by Reef Corals in the Gulf of Mexico and Caribbean Sea: A Report on Project Reef Spawn '94.* Flower Gardens Fund Technical Series Report No. 94-03. Corpus Christi, Texas: The Flower Gardens Fund/Gulf of Mexico Foundation.

Gubbay, Susan, ed., 1995. *Marine Protected Areas: Principles and Techniques for Management.* London: Chapman and Hall.

Hudson, J. H., and R. Diaz. 1988. "Damage Survey and Restoration of *M/V Wellwood* Grounding Site, Molasses Reef, Key Largo National Marine Sanctuary, Florida." *Proceedings of the 6th International Coral Reef Symposium, Australia, Vol. 2.* Miami Beach, Florida: U.S. Geological Survey, pp. 231–236.

National Oceanic and Atmospheric Administration (NOAA). 1980. *Final Environmental Impact Statement on the Proposed Gray's Reef Marine Sanctuary.* Washington D.C.: National Oceanic and Atmospheric Association.

National Oceanic and Atmospheric Administration (NOAA). 1983. *Gray's Reef National Marine Sanctuary Management Plan.* Washington, D.C.: National Oceanic and Atmospheric Association.

National Oceanic and Atmospheric Administration (NOAA). 1984. *Final Environmental Impact Statement and Management Plan for the Proposed Fagatele Bay National Marine Sanctuary.* Washington, D.C.: National Oceanic and Atmospheric Association.

National Oceanic and Atmospheric Administration (NOAA). 1987. *Gulf of the Farallones National Marine Sanctuary Management Plan.* Washington, D.C.: National Oceanic and Atmospheric Association.

National Oceanic and Atmospheric Administration (NOAA). 1989. *Cordell Bank National Marine Sanctuary: Final Environmental Impact Statement/Management Plan.* Washington, D.C.: National Oceanic and Atmospheric Association.

National Oceanic and Atmospheric Administration (NOAA). 1991. *Flower Garden Banks National Marine Sanctuary: Final Environmental Impact Statement/Management Plan.* Washington, D.C.: National Oceanic and Atmospheric Association.

National Oceanic and Atmospheric Administration (NOAA). 1992. *Monterey Bay National Marine Sanctuary: Final Environmental Impact Statement/Management Plan. Vol. I&II.* Washington, D.C.: National Oceanic and Atmospheric Association.

National Oceanic and Atmospheric Administration (NOAA). 1993a. *Stellwagen Bank National Marine Sanctuary: Final Environmental Impact Statement/Management Plan. Vol. I & II.* Washington, D.C.: National Oceanic and Atmospheric Association.

National Oceanic and Atmospheric Association (NOAA). 1993b. *Olympic Coast National Marine Sanctuary: Final Environmental Impact Statement/Management Plan. Vol. I and II.* Washington, D.C.: National Oceanic and Atmospheric Association.

National Oceanic and Atmospheric Administration (NOAA). 1995a. *Florida Keys National Marine Sanctuary: Strategy for Stewardship. Draft Management Plan/Environmental Impact Statement. Vol. I, II, III.* Silver Springs, Maryland: National Oceanic and Atmospheric Association.

National Oceanic and Atmospheric Administration (NOAA). 1995b. *Hawaiian Islands Humpback Whale National Marine Sanctuary.* Draft Environmental Impact Statement/Management Plan. Silver Springs, Maryland: National Oceanic and Atmospheric Association.

National Oceanic and Atmospheric Association (NOAA). 1997a. *Accomplishments Report: 25th Anniversary/National Marine Sanctuaries—1972–1997.* Silver Spring, Maryland: National Oceanic and Atmospheric Association.

National Oceanic and Atmospheric Administration (NOAA). 1997b. *Hawaiian Islands Humpback Whale National Marine Sanctuary: Final Environmental Impact Statement/Management Plan.* Silver Springs, Maryland: National Oceanic and Atmospheric Association.

Owen, Oliver S. and Daniel D. Chiras. 1995. *Natural Resource Conservation: Management for a Sustainable Future.* 6th ed. Englewood Cliffs, New Jersey: Prentice Hall.

Potter, Frank. 1991. *National Marine Sanctuaries: Challenge and Opportunity.* A Report to the National Oceanic and Atmospheric Administration. Washington, D.C.: National Oceanic and Atmospheric Association.

Pritchard, Paul. 1993. "Undiscovered Diamonds for the Crown Jewels." *Oceanus.* Vol. 36, No. 3, pp. 3–5.

Robison, Bruce H. 1993. "New Technologies for Sanctuary Research." *Oceanus.* Vol. 36, No. 3, pp. 75, 80.

Sobel, Jack. 1993. "Conserving Biological Diversity Through Marine Protected Areas: A Global Challenge." *Oceanus.* Vol. 36, No. 3, pp. 19–26.

Sobel, Jack and Alison Merow. 1993. "Budget Leaves Ocean Programs Dry." *Sanctuary Currents.* Winter/Spring, pp. 1–2.

World Resources Institute. 1994. *World Resources 1994–95: A Guide to the Global Environment.* New York: Oxford University Press.

FURTHER READING

Clark, John R. 1996. *Coastal Zone Management Handbook.* Boca Raton, Florida: CRC Lewis Publishers.

Seaborn, Charles. 1996. *Underwater Wilderness: Life In America's National Marine Sanctuaries and Reserves.* Boulder, Colorado: Roberts Rinehard Publishers.

Ticco, Paul C. 1995. "The Use of Marine Protected Areas to Preserve and Enhance Marine Biological Diversity: A Case Study Approach." *Coastal Management.* Vol. 23, No. 4, pp. 309–314.

4 Coastal Hazards

Coastal hazards are created where destructive natural processes may interact with the human created environment. The coastal resource manager's job is to understand how the human environment and natural hazards interact and use this information to design plans and actions which mitigate negative interactions between these two elements. People who choose to live along coastlines are susceptible to a number of risks from natural hazards, whether the hazards are classified as cataclysmic natural disasters, gradual erosion, or steadily rising sea levels. Storms, for example, can erode beaches, crumble cliffs, flood buildings and roadways, and bring general havoc to a coastal community. As the demand for coastal recreation increases, so does the pressure to develop harbors, resort areas, and single family vacation houses along the shore. The consequences of such development are increased exposure to natural hazards and the potential loss of natural resources, property, and even life, as experienced when Hurricane Emily made landfall on September 1, 1993. After a five day meander across the southeastern Atlantic, Hurricane Emily blew ashore with winds of up to 115 mph, and sent thousands of locals and vacationers scurrying as the storm bashed dunes and cottages on Virginia and North Carolina beaches.

A year earlier, two major hurricanes hit U.S. coastlines within a two-month period. On August 24, 1992, Hurricane Andrew—one of the strongest Atlantic hurricanes of the century—slammed its 140 mph winds on the Southern Florida coast, flipping over huge commercial fishing boats, peeling off rooftops, totally destroying hundreds of houses and other buildings, and all but annihilating Homestead Air Force Base, which is located about 50 km (31 mi) south of Miami Beach. Hurricane Andrew left over $20 billion in damage in south Florida alone before it moved on to unleash its fury on the Louisiana coast. Off the Louisiana coast, Hurricane Andrew not only leveled houses, it destroyed islands. Andrew's furious storm surge washed away so much sediment that Raccoon Island lost half of its 4.8 km (3 mi) length. Geologists predict that the last vestige of Raccoon Island will be washed out to sea within the next 10 years. All in all, Hurricane Andrew is considered by many as the costliest hurricane in U.S. history.

In September 11, 1992, the 145 mph winds of Hurricane Iniki strafed the rural island of Kauai, Hawaii, causing $1.6 billion in damage and claiming four lives. The roaring gale knocked out power and telephone service island-wide, and tourists fled as residents struggled for basics like water and shelter. This was the worst storm on record for Hawaii, and its up to 9.1 m (30 ft) waves and 4.0 to 5.5 m (13 to 18 ft) **storm surge** utterly transformed hardest hit areas, such as popular Poipu Beach on the southern shore of Kauai (Fletcher et al. 1995).

In this century, hurricanes and other ocean storms have also caused severe damage and loss of life in Charleston, South Carolina (1989), along the coasts of Louisiana and Mississippi (1969), along the Northeast coast (1938), and along Miami Beach (1926). As long as development continues to flourish on the coast lands and the barrier islands of this nation (and any other country), property damage and loss of life from storms and hurricanes are inevitable.

There are a variety of coastal hazards along America's shores, though not all are as dramatic as hurricanes. In 1989, for example, the United States Geological Survey identified and mapped such hazards as shoreline erosion, overwash penetration, storm surge, storm/wave damage, and earth movements. Combined, these factors determine an "overall hazard assessment" for various segments of the coast. (See Figure 4-1). Although shoreline erosion is the primary focus of this chapter, it should be clear that the above categories of coastal hazards are highly intertwined (i.e., storms erode beaches; without beaches, waves erode cliffs; without coastal barriers such as beaches, sand dunes, and cliffs, flooding of inland areas then occurs; and so on).

This chapter, Coastal Hazards, is divided into five sections. The first introduces the reader to the factors that affect shoreline change, with the central theme being coastal erosion. The second section discusses the major strategies used by coastal resource managers to deal with coastal hazards. The third section provides a case study of dune restoration efforts on the shores of the Monterey Bay National Marine Sanctuary. The fourth section provides some general policy recommendations regarding coastal hazards. The chapter concludes with a comment about coastal hazards issues and the need to always question "Institutional Inertia" and "Engineering Mentality".

FACTORS AFFECTING SHORELINE CHANGE

America's shores are rapidly eroding, and the erosion process is accelerating. The coastline of Cape Shoalwater on Washington's Olympic Peninsula has retreated as much as 30 m (100 ft) per year since the turn of the century (Owen and Chiras 1990). Although not nearly so severe, California's coastline is receding at an average rate of 15 cm to 0.3 m (6 in to 1 ft) per year. The state's highest erosion rate occurs at Point Año Nuevo on the Central California coast. Here, the coastline has receded at an average erosion rate of 2.7 m (9 ft) per year for the last 300 years (Griggs and Savoy, eds. 1985).

Although coastal resource managers refer to "average" erosion rates, geologists have found that erosion is usually irregular, if not *episodic.* For example, if we pick one particular site on the shores of the Monterey Bay National Marine Sanctuary (MBNMS), such as the small city of Capitola, the average erosion rate is 0.3 m (1 ft) per year. Yet, one episode—the severe winter storm of 1983—caused large slabs of its cliffs to collapse, thus pushing back the city's shoreline from 1.5 to 3 m (5 to 10 ft) overnight (Griggs and Savoy, eds. 1985). Coastal resource managers, urban planners, and potential coastal dwellers must keep in mind the difference between short term rates versus the possibility of sudden, unexpected episodes that drastically alter those erosion rates. A safe setback for a house, for example, should be based on long-term (30–50 year) erosion averages that include episodic events. The problem, however, is that long-term erosion rate data are seldom available. With this brief introduction to erosion rates, we can now turn to the natural and human-induced factors that affect shoreline change.

Natural Factors

From sand size to tsunamis. There are a variety of complex *natural* processes that affect erosion rates. They include such variables as the geologic characteristics of the shore; sand size, shape, and density; effects of longshore currents, waves, tides, and other fluctuations in sea level (Garrett and Maas 1993; Giese and Chapman 1993); winds, tropical storms, and hurricanes (Williams, Doehring, and Duedall 1993); the bathymetry of the offshore sea bottom, tectonic instability (e.g., earthquakes), and tsunamis. (See Table 4.1).

Regarding geologic characteristics of the shore, for example, steeper leading-edge coasts characteristically have narrow continental shelves, dropping off abruptly into the sea. This type of coast is therefore subject to higher wave energy (due to the deeper water near shore) resulting in smaller beaches and steep coastal cliffs. Leading-edge coasts also typically have high nearshore mountains, resulting in rivers which dump quickly into the sea, transporting sediments a shorter distance, which allows the sand to remain larger in size, and rougher in shape. In other words, the longer distance a sediment particle is transported, the smaller and more rounded it becomes.

Trailing-edge and marginal coasts characteristically have wider continental shelves, dropping off gradually into the sea. This type of coast is subject to smaller, less energetic wave action due to the shallow or protected water near shore. Hills tend to be low and drainage systems tend to be long and wide, resulting in tremendous sediment deposition at the shores. Sediments tend to be smaller in size than leading edge coasts, due to the long distance they must travel, or due to the weak wave action near shore which can only carry smaller sediments.

Tsunamis: seismic sea waves. **Tsunamis,** also called *seismic sea waves* (improperly called *tidal waves*), are

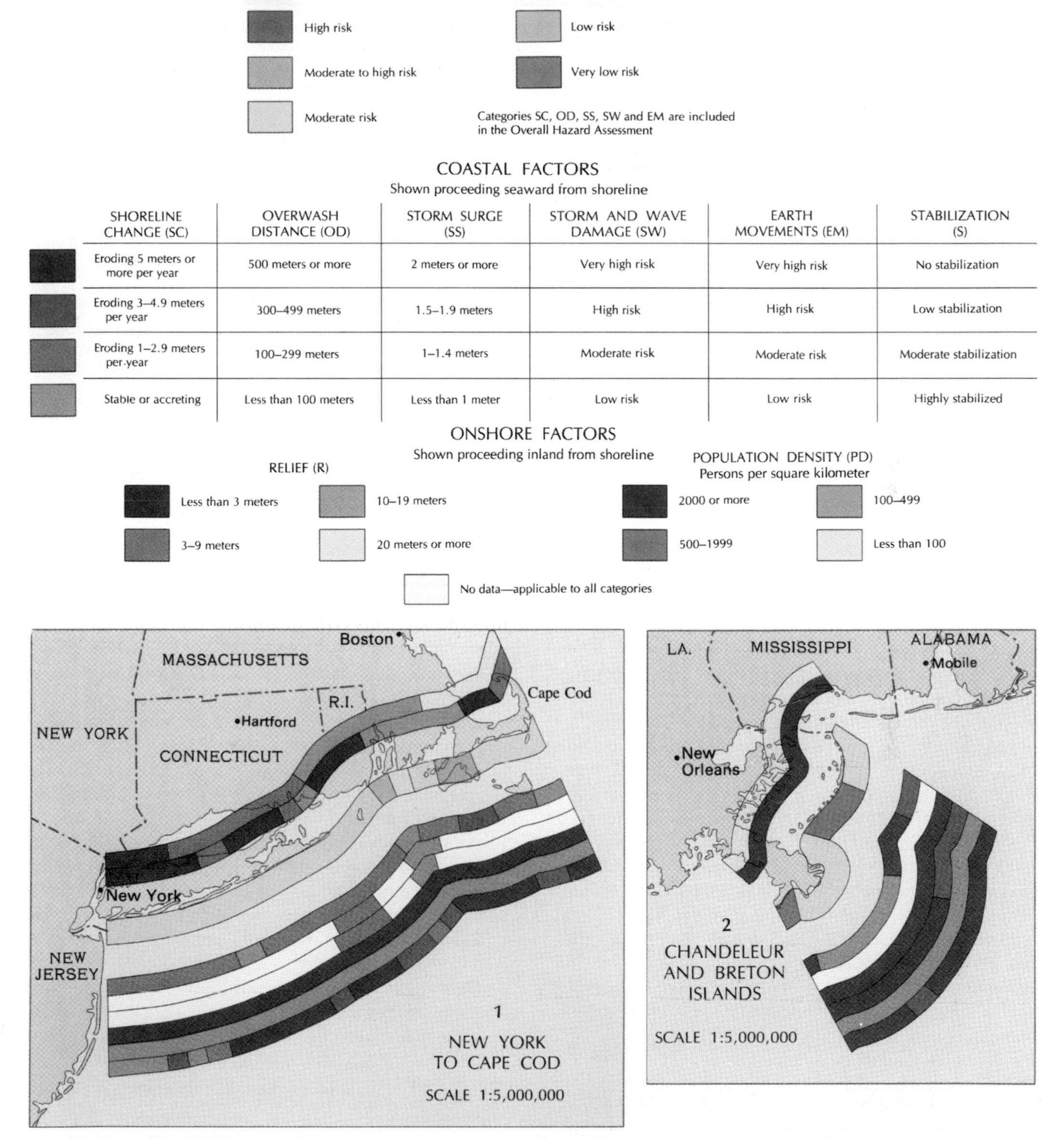

FIGURE 4-1 Overall hazard assessment at two selected sites. (*Source:* U.S. Coastal Hazards map, 1985, Courtesy of U.S. Geological Survey)

TABLE 4-1
Summary of Natural Factors Affecting Shoreline Change

Factor	*Effect*	*Time Scale*	*Comments*
Sediment supply (sources and sinks)	Accretion/ erosion	Decades to millennia	Natural supply from inland (e.g., river flood, cliff erosion) or shoreface and inner shelf sources can contribute to shoreline stability or accretion
Sea level rise	Erosion	Centuries to millennia	Relative sea level rise, including effects of land subsidence, is important
Sea level change	Erosion (for increases in sea level)	Months to years	Causes poorly understood, interannual variations that may exceed 40 years of trend (e.g., El Nino)
Storm surge	Erosion	Hours to days	Very critical to erosion magnitude
Large wave height	Erosion	Hours to months	Individual storms or seasonal effects
Short wave period	Erosion	Hours to months	Individual storms or seasonal effects
Waves of small steepness	Accretion	Hours to months	Summer conditons
Alongshore currents	Accretion, no change, or erosion	Hours to millennia	Discontinuities (updrift ≠ downdrift) and nodal points
Rip current	Erosion	Hours to months	Narrow seaward-flowing, near bottom currents may transport significant quantities of sediment during coastal storms
Underflow	Erosion	Hours to days	Seaward-flowing, near-bottom currents may transport significant quantities of sediment during coastal storms
Inlet presence	Net erosion; high instability	Years to centuries	Inlet-adjacent shorelines tend to be unstable because of fluctuations or migration in inlet position; net effect of inlets is erosional owing to stroage in tidal shoals sand
Overwash	Erosional	Hours to days	High tides and waves cause sand transprot over barrier beaches
Wind	Erosional	Hours to centuries	Sand blown inland from beach
Subsidence			
Compaction	Erosion	Years to millennia	Natural or human-induced withdrawl of subsurface fluids
Tectonic	Erosion/accretion	Instantaneous	Earthquakes
	Erosion/accretion	Centuries to millennnia	Elevation or subsidence of plates

Source: National Resource Council. 1990. *Managing Coastal Erosion.* Washington, D. C.: National Academy Press, p. 22.

major oceanic wave systems triggered by undersea tectonic or volcanic events. In a thrust earthquake, for example, one block of earth is heaved up over another along a fault line, thus pushing the water above it into a hump a few yards high. The resulting tsunami wave then rolls through the ocean at great speeds. On the open sea, tsunamis can be inconspicuous because they have low heights and long wavelengths. A passenger on a ship might be unaware that a tsunamis is traveling underneath. The series of waves could be 145 km (90 mi) long, with intervals of 15–200 minutes, traveling at speeds of more than 435 miles per hour. Although harmless in the deep open ocean, tsunamis cause devastation when they hit coastlines. As these long, high-speed waves approach shallow coastal zones, they bunch up and can rise to heights of 30 m (100 ft)—the height of a ten-story building.

The coastlines most vulnerable to the enormous destructive power of tsunamis are in and along the Pacific Rim where, due to subduction zones along the earth's plate boundaries, worldwide earthquake activity is most concentrated (thus commonly referred to as "the ring of fire"). Japan, for example, has had more than 80 major tsunamis caused by submarine fault movements since 1891 (Abe 1979).[1]

On July 12, 1993, for example, a 7.8 magnitude earthquake (as measured on the Richter scale) struck northern Japan at 10:17 p.m. The associated fault activity triggered two separate tsunamis, caused significant **liquefaction** in **soft soils** around coastal areas, and caused huge fires which completely destroyed the small coastal town of Aonae on the island of Okushiri. There was a significant amount of death and destruction. Okushiri, a small island 48 km (30 mi) south of the epicenter, was hardest hit by the larger second tsunami which measured 30.5 m (100 ft) in height. (See Figure 4-2).

Despite the fact that Japan is known to have the best tsunami warning system in the world, the close proximity of the quake epicenter allowed only 5 minutes before the first tsunami arrived, with the second tsunami following only 7 minutes behind. This short amount of time

[1]Earthquakes do not cause tsunamis, as sometimes stated. Rather, earthquakes and associated tsunamis both result from the same fault movement. Furthermore, tsunamis may also be caused by terrestrial or submarine landslides, or by volcanic action.

FIGURE 4-2 Scene of tsunami damage at Aonae, Okushiri—the Japanese island community most devastated by a July 12, 1993 tsunami. (*Source:* Courtesy of Michael Blackford, Acting Director, International Tsunami Information Center, Honolulu, Hawaii)

was not enough to warn local residents adequately. As each wave struck land, hundreds of people were seen running from their neighborhoods with houses riding the giant waves of the tsunami in their pursuit. Damage included 196 lives lost (120 due to the tsunami itself), 540 homes destroyed (340 by fire), 154 homes significantly damaged, 1826 homes partially damaged, 31 public buildings damaged or destroyed, and numerous disruptions to railway and highway systems. The harbor systems were disrupted for weeks by the floating debris (e.g., boats, cars, and broken structures) which were pulled into the harbor with the returning sea water (Yanev and Scawthorn 1993).

Tsunamis also occasionally hit U.S. coastlines, particularly Hawaii and California. According to reports on West Coast tsunamis by the National Oceanic and Atmospheric Administration (NOAA), there have been at least 18 tsunamis since 1812 along the West Coast triggered by underwater landslides (Lander, Lockridge, and Kozuch 1993). Examples include a 6 m (20 ft) wave that washed up against the Cliff House restaurant in San Francisco in 1868, triggered by a moderate earthquake on the Hayward fault; a 6 m (20 ft) wave that swept into Santa Monica in Southern California after a 5.2 magnitude earthquake in 1930; and a 9.2 magnitude earthquake that struck Alaska on Good Friday in 1964, killing 131 people, and spawned tidal waves as far south as Crescent City, California.

Even the shores of the Monterey Bay National Marine Sanctuary experience occasional tsunamis—the 1989 Loma Prieta quake in the Santa Cruz Mountains set off a landslide in the steep, mile-deep canyon beneath Monterey Bay which resulted in a tsunami wave, though it was only 0.3 m (1 ft) high. Along the California coast, Southern California faces the greatest danger from tsunamis because it has some of the roughest offshore terrain in the world, which marine geologists have labeled the Southern California Borderland. In addition to California's other notorious problems (drought, riots, wildfires, mudslides), marine geologists are now warning

Southern Californians that they may face another threat: tsunamis. Unlike a 4.6 m (15 ft) wave, a 4.6 m tsunami can undermine foundations and wash entire buildings away. Tsunamis do not crash onto shore like ordinary waves. They simply raise sea level so fast that people cannot outrun the onrushing waters.

Earthquakes and cliff erosion. It was mentioned earlier that the 1983 storm off the California coast severely affected the cliff erosion rate at Capitola on the shores of the Monterey Bay National Marine Sanctuary. Subsequently, the October 17, 1989 Loma Prieta earthquake further loosened the already unstable cliffs. Six of the 25 apartments perched on the Depot Hill bluffs in Capitola had to be demolished so they would not fall into the Bay. The same earthquake also severely damaged roadways around Santa Cruz harbor in northern Monterey Bay. Sections of the roadways subsided as much as 0.6m (2 ft).

Coastal communities in California are also regularly impacted by nonearthquake related cliff erosions, such as those along Highway 1—one of America's most scenic roadways. In September 1990 between Stinson and Muir Beaches, about 24 km (15 mi) north of San Francisco, a section of Highway 1 of approximately 182 m (600 ft) began sinking toward the ocean below at a rate of 0.3 to 0.6 m (1 to 2 ft) per month—a result of general erosive forces. (See Figure 4-3).

The slide, known locally as the "Lone Tree Slide," presented the coastal community with a major management problem. Repairing the roadway required recarving its path 61 m (200 ft) deeper into the hillside to place it behind the slide plane. This resulted in an anticipated 500,000 cubic yards of excess soil. The question then became, what does a coastal community do with excess soil? This question initiated a three year environmental study. Some local residents argued that the excess soil should be just pushed over the side into the ocean, since the road was "marching" in that direction anyway. To support their argument, they cited the fact that a major landslide in 1982 on Highway 1 south of Big Sur dumped 3.5 million cubic yards of soil in the Pacific alone, with no obvious environmental impact. Certainly, pushing the excess soil into the ocean was the most expedient and least expensive alternative. However, the National Marine Fisheries Service, the California Coastal Commission, and the League of Marine Conservation reminded those that argued for the "dump it into the ocean" alternative that the Gulf of the Farallones National Marine Sanctuary is located just 1,500 yards from the coast, and its animal and plant residents might suffer from increased sediment loads. Consequently, other alternatives were considered in the three year environmental study. They included:

- *trucking the soil to local canyons and gulches.* The problem with this alternative was that all adjacent land is in Mount Tamalpias State Park;
- *trucking it to distant ranches, where it could be used for grading.* This would require 30,000 truckloads of soil with an anticipated cost of $10 million. Plus, the trucks would all have to pass through the once quiet coastal community of Stinson Beach;
- *abandoning the road, and letting nature take its course.* The merchants and commuters of Stinson Beach opposed this alternative, since their livelihoods are heavily dependent on that stretch of Highway 1. Caltrans (California Department of Transportation) favored this last alternative, since the land would end up in the sea anyway, regardless of whatever temporary improvements they made in the roadway.

Fearing economic troubles from isolation created by the collapse of Highway 1, residents of Stinson Beach pressured politicians to resolve the issues quickly. This meant immediate funding, rapid environmental law compliance, and an expeditious engineering solution to soil removal and road repair, were required. Funding was secured through emergency disaster relief. Meanwhile, Caltrans decided that pushing excess soil into the ocean was the optimal engineering solution to rapid project implementation. Environmental laws regarding this option were satisfied by declaring no adverse environmental impact would occur as long as Caltrans committed to restoration of a degraded wetland area and monitoring of the 5.4 acre marine soil disposal site (McDonnell 1995). The solutions allowed Highway 1 to reopen in the summer of 1991.

What to do with excess soil from landslides (earthquake-induced or otherwise) will certainly be a future environmental issue for other coastal communities across the country, particularly those that have a national marine sanctuary on their shore.

Human-induced Factors

Human intervention alters the above natural processes through such actions as the dredging of lagoons to create harbors, the periodic dredging of harbor berths to maintain proper water levels, the construction of jetties to improve navigation, and several other actions that will now be discussed.

Inlets, dredging, and jetties. Inlets are passageways between inland harbors and the open sea. Humans have long channeled (straightened and dredged) these inlets to ease the passage of ships. This action, of course, disturbs the natural flow and mixing of water, sediments, nutrients, and organisms between the terrestrial and marine

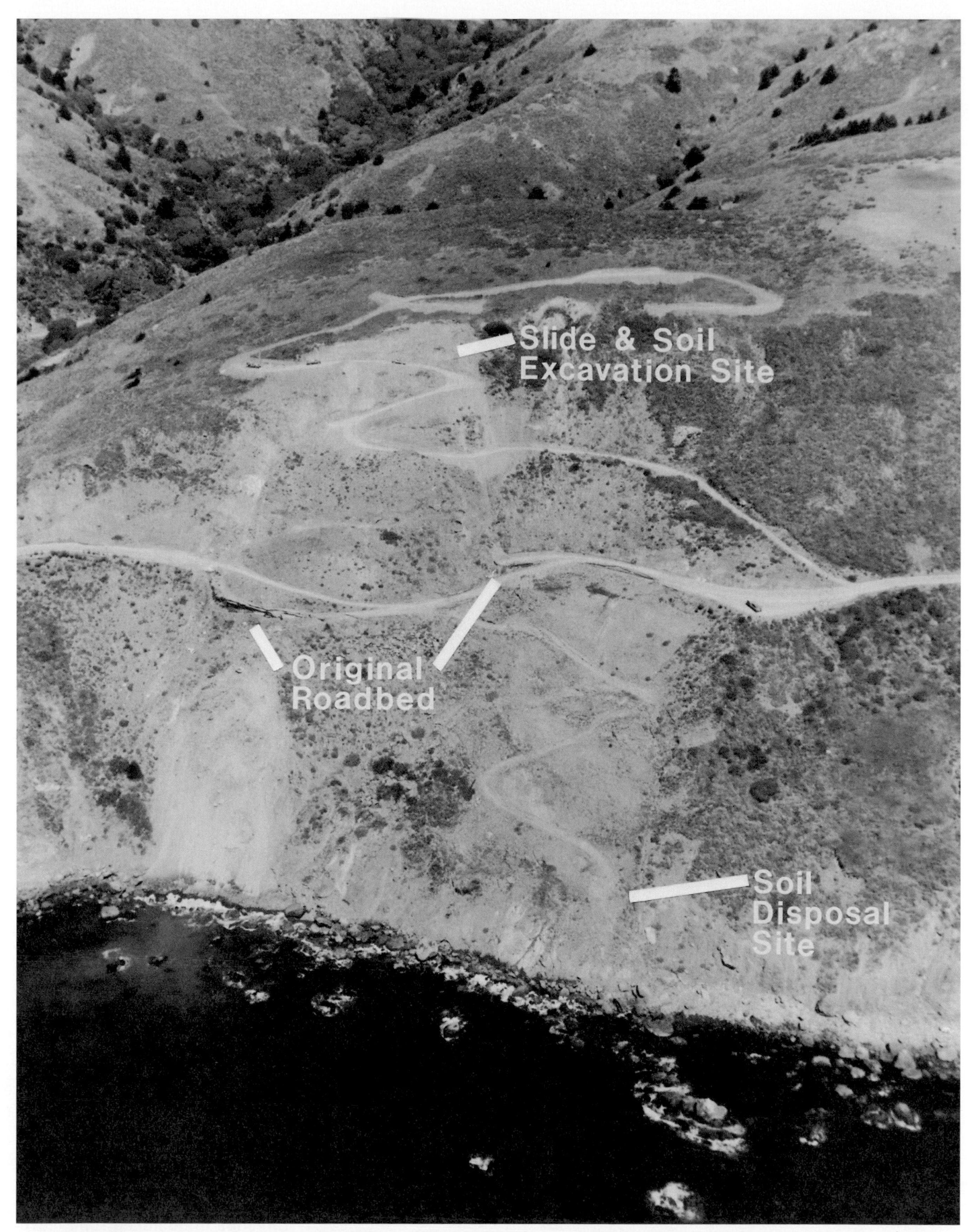

FIGURE 4-3 Stinson Beach coastal Highway 1 1990 landslide, Marin County, California. (*Source:* Courtesy of Lynn G. Harrison, California Department of Transportation, Photo file # 90-03032-5; Photo taken 9/12/90)

environments. In an effort to further improve the navigability of an inlet, humans have also constructed jetties. A **jetty** is an elongate structure that extends into a body of water. Jetties are built in pairs for the purpose of directing the flow of water. They, too, have a pronounced effect on coastal processes and erosion, both upcoast and downcoast of the jetty.

The jetties of the Santa Cruz Small Craft Harbor on the shores of the Monterey Bay National Marine Sanctuary can serve as one case in point. In 1965, the U.S. Army Corps of Engineers completed the construction of the two jetties of the Santa Cruz Small Craft Harbor. (See Figure 4-4). The purpose of the jetties was to create an artificial harbor. The jetties succeeded as planned. There is now a permanent structured inlet (channel) for boats to enter the harbor. A positive side effect of this jetty construction was the widening of an upcoast beach (Seabright Beach) that now protects a stretch of urbanized cliffs.

The U.S. Army Corps of Engineers only partially anticipated the accelerated erosion of downcoast cliffs from this jetty construction, and did not foresee at all two other negative side effects: (1) shoaling of the harbor entrance, and (2) the loss of Capitola's beach farther downcoast.

FIGURE 4-4a Jetties at Santa Cruz Small Craft Harbor, California, illustrating how jetties interrupt the direction of dominant sand movement. (*Source:* Photos courtesy of Gary Griggs)

Littoral drift is the movement of beach sand along the coast due to wave action. Because the winds are predominately northwesterly outside the harbor entrance, the predominant movement of sand is southward, or downcoast. The engineers did not accurately predict the amount of littoral drift within the northern Bay.

When littoral drift brings sands into a narrow inlet such as the Santa Cruz harbor entrance shoaling can occur. (See Figure 4-5). **Shoaling** is defined as the gradual build up of sand that makes shallow a once deeper harbor inlet, river, or lake. Santa Cruz harbor must now be dredged at an annual expense of approximately $450,000. Even with regular dredging, shoaling still occurs at Santa Cruz harbor, though it is not as severe. Nevertheless, this minor form of shoaling creates another problem that was unforeseen by the Army Corps of Engineers—the buildup of waves and navigation hazards at the harbor entrance. With the proper tidal condition and even moderate seas, waves can be pushed up by the shoaled area that it passes over—somewhat like a car passing over a speed bump on the road. Many a boater, including the author, has "surfed" a 2.4–3.0 m (8–10 ft) wave into the harbor, like a surfer rides the crest of a wave. This situation can be dangerous, since captains can lose control of their boats under such perilous conditions. (See Figure 4-6).

In addition to shoaling of the harbor entrance and accelerated erosion of downcoast cliffs, the creation of the Santa Cruz harbor jetty also caused an interruption in the natural migration of sand down the coast. This resulted in a partial loss of sand at the 26th Avenue site 1 km (0.6 mi) down the coast, and loss of the entire Capitola beach 3.5 km (2 mi) further south. Although Capitola's beach naturally disappears during winter storms, it always returns during the early summer months just in time for the summer tourists. After the construction of Santa Cruz harbor jetties, however, the beach did not return. The sand that would normally have replenished the beach was trapped by the Santa Cruz jetty. Capitola partially solved this problem by constructing a groin to trap the littoral drift. However, there is still a winter beach loss problem if waves come from the West or Southwest. (See "groin" under "engineering strategies" below for further details.)

Morris Island Lighthouse, a well known local landmark for the port city of Charleston, South Carolina, also illustrates some of the complexities involved with jetty construction. The 50 m (164 ft) tall Morris Island Lighthouse was first lit in 1876. Today, the lighthouse is not on land, nor near the shore, but 500 m (1640 ft) offshore—the victim of a rapidly retreating shoreline due to jetty construction. Twenty years after the lighthouse was lit, the Charleston Harbor jetties were completed to improve navigation into the harbor. From that point on, Morris Island and its lighthouse were affected. Sand that is predominantly transported southward along the shore is interrupted by the jetties, resulting in severe erosion south of these structures. Sand is now

FIGURE 4-4b (Continued)

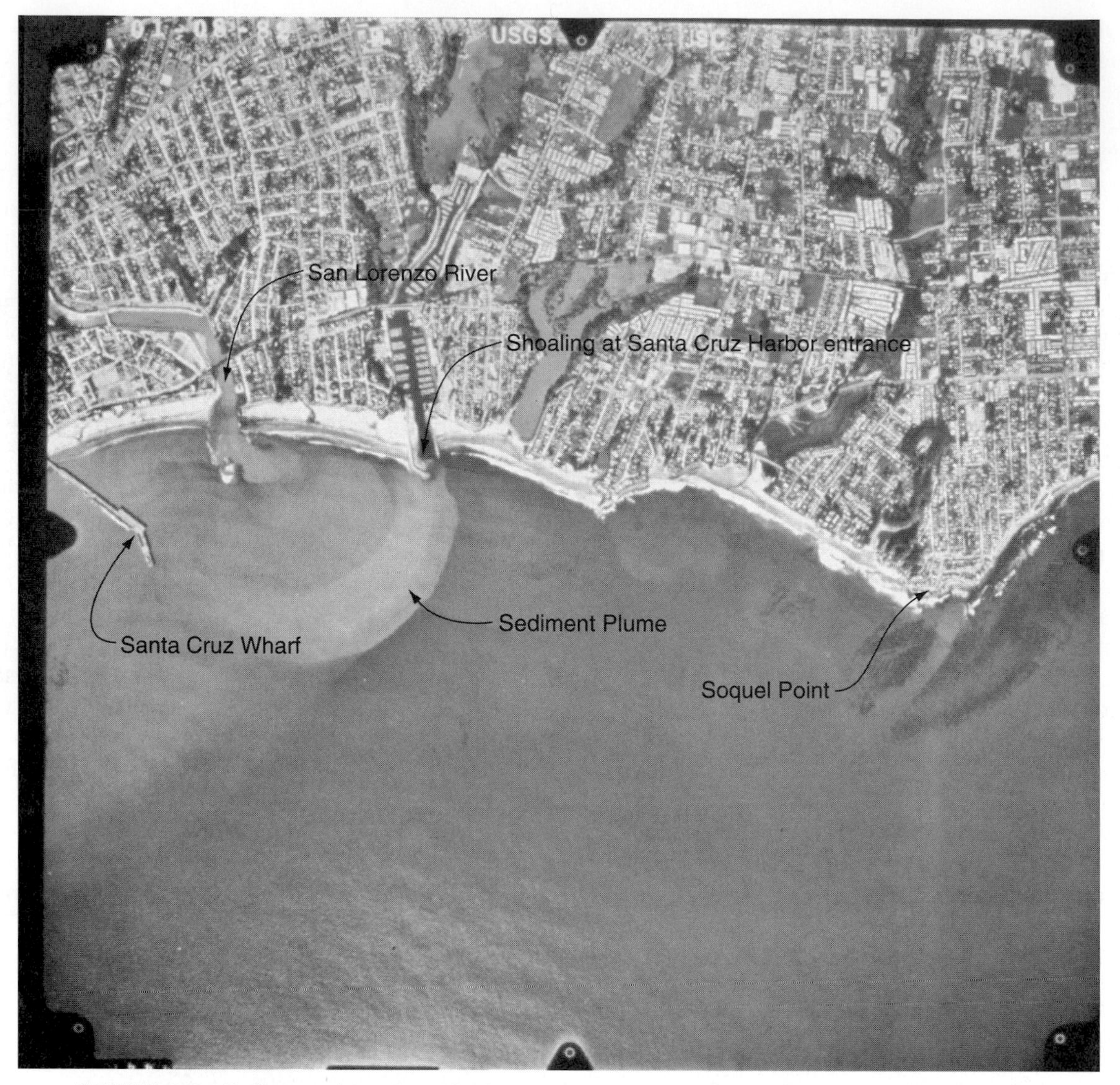

FIGURE 4-5 Sediment plume from the San Lorenzo River as it enters Northern Monterey Bay. Note how it leads to shoaling of the mouth of the Santa Cruz Small Craft Harbor, Santa Cruz, California. (*Source:* Courtesy of U.S.G.S.)

building up the beaches of Sullivan's Island (north of the jetties) while destroying Morris Island and its lighthouse (south of the jetties).

Sand disposal offshore. If beach-quality sand is dredged from inlets and then indiscriminately dumped offshore, rather than back onto nearby beaches, shoreline erosion can be the only result. For example, Florida dumps over 56 million cubic yards of inlet-dredged material offshore. Consequently, many of Florida's beaches are experiencing accelerated erosion rates. However, if Florida transferred that 56 million cubic yards of dredged sand to its 603 km (375 mi) east coast, it would be sufficient to advance those beaches 7.6 m (25 ft) seaward (National Resource Council 1990).

Sand mining. A sand deficit along beaches can also occur if sand is quarried. Although sand mining is not itself a coastal hazard, it can lead to shoreline erosion which, when extensive enough, can become a coastal hazard. One extreme example can be found in Northern Ireland where 80 percent of the area's coastal erosion can be accounted for by sand loss via extraction (Carter et al. 1992). Though to a lesser degree, environmental impacts resulting from sediment removal from beaches and dune landscapes exist around the world, including along the shores of the Monterey Bay National Marine Sanctuary.

Four sand mining sites once operated in the Monterey Bay region. More than 9 million tons have been removed from the dune landscape at the south end of Monterey

FIGURE 4-6 Boat out of control resulting from shoaling at harbor entrance, Santa Cruz Small Craft Harbor, California. Note a passenger even fell into the water. (*Source:* Photo courtesy of Keith Angell)

Bay, between Sand City and the city of Marina (Oradiwe 1986). (See Figure 4-7). and approximately 350,000 cubic yards that were removed annually from one site led directly to shoreline erosion (Griggs et al. 1992). The sands of the Monterey Bay region were highly valued for their industrial uses, such as sandblasting, stucco manufacture, and water filtration. The white sands of Carmel and other areas on the Monterey Peninsula are world renowned for their beauty. However, these same sands were highly valued for their utility as well. Some of the southern Monterey Bay beach and dune sand was also used for **beach nourishment,** even in distant locations. In the mid-1970's, for example, sand excavated from the surf zone near Marina was transported north to three beaches in the San Francisco Bay Area (Gordon 1996).

Whether sand is mined from riverbeds (which could be transported to beaches during rainy periods) or directly from dunes and beaches, the result is open pit scars, shifting dunes, and ultimately accelerated beach erosion along the coast. The sand mining operations in southern Monterey Bay were active for nearly a century before being curtailed in 1989 (Griggs et al. 1992). Today, beach mining in the area has been eliminated altogether.

Human-induced subsidence. Shorelines are also changed by the extraction of oil from coastal regions and the mining of water from coastal aquifers. As oil or water is extracted from underground pore spaces, the soil compacts and the land subsides. Soil compaction destroys an aquifers' water-holding capacity, while **subsidence,** caused by oil and water extraction, can damage coastal infrastructure. Between 1943 and 1964, the Galveston Bay-Houston, Texas area subsided over 1.5 m (5 ft) as a result of combined water and hydrocarbon extraction (Gabrysch 1969). In the Terminal Island-Long Beach region in California, so much oil has been extracted that the land around the extraction site subsided nearly 9.1 m (30 ft) in 27 years (National Research Council 1987). Subsiding land and accompanying sea encroachment has caused $100 million damage to pumping, transportation, and harbor facilities in the area (Monroe and Wicander 1994).

Dams. Since many coastal areas are dependent upon sand nourishment from rivers, dams also cause a loss of sand from beaches by (a) trapping sediment behind the dam, and (b) reducing peak flows that carry suspended sediment to the shore. This can lead to intensified shoreline erosion. For example, numerous dams have been built on the Brazos River in Texas, resulting in a 71 percent decrease in the suspended load of the river. The sand discharge at the coast where the river empties into the Gulf of Mexico is five to nine times less than was normal, without impoundment (Coates 1981). Lowered sediment discharge is a particular problem in the Pacific coastal states of California, Oregon, and Washington, where it has intensified problems of beach erosion.

Geologists estimate that 75 to 95 percent of California's beach sand is originally derived from streams (Griggs and Savoy, eds. 1985). The southern half of the state suffers the greatest sediment deprivation resulting from dam construction. For example, dams on the Santa Clara River (southeast of Santa Barbara) have reduced the river's sediment flow by 37 percent (Griggs and Savoy, eds. 1985). Northern California is more fortunate, since there are fewer dams to trap sediment flow.

FIGURE 4-7 Sand-mining operation near the city of Marina on Monterey Bay, California. (a) Impact of sand mining operation looks relatively insignificant from the main highway. (Photo by author) (b) An aerial view of same sand mining operation better illustrates the impact of this kind of activity to dunes. (*Source:* Photo courtesy of Christine Kook)

Groins, seawalls, gunnite, and bulkheads. Groins, seawalls, gunnite, and bulkheads are all "engineering strategies" *intended* to *prevent* coastal erosion. Unfortunately, they can also cause the very problem they are trying to remedy. (See Figure 4-8).

A **groin** is an elongate structure constructed of either boulders, concrete, wood pilings or steel sheeting. The device is usually erected perpendicular to the coast. Although its purpose is to stabilize a beach by trapping the littoral, a longshore drift of sand, it usually also causes downcoast erosion. This is what happened to the City of Capitola in northern Monterey Bay. In 1969, Capitola built a single groin along its shore as a counter measure to recover its beach after Santa Cruz Harbor constructed its "beach-stealing jetties" in 1963. As long as the winds came from their traditional northwest direction, Capitola's groin trapped sand and a beach developed. The groin saved Capitola City Beach (west of the groin) but, some have argued, at the expense of eroding Capitola Bluffs (east of the groin). Others claim that the Capitola Bluffs area has always had coastal erosion problems because of the lack of a protective beach. The question remains whether the placement of the Capitola groin has *intensified* the shoreline erosion problem at Capitola

FIGURE 4-8 Groins can trap sediment flow (littoral drift) and stabilize a beach (upcoast of a groin), but may intensify erosion in an area downcoast of the groin that was already prone to severe erosion because of the lack of a protective beach, such as at the City of Capitola on Monterey Bay, California. (*Source:* Courtesy of Gary Griggs)

Bluffs. Regardless, the City of Capitola is now discussing the construction of a seawall to protect the apartment complexes that are atop Capitola Bluffs.

A **seawall** is a concrete structure, of which there are three types. The "curved" type with toe protection resembles a large median strip divider for a freeway. (See Figure 4-9). The two other types are the "vertical" seawall constructed of resistant interlocking blocks, and the "curved and stepped" seawall which is secured by pilings (Carter 1988). (See Figure 4-10).

In addition to being extremely expensive and aesthetically unattractive, seawalls have their environmental problems as well. For example, *undercutting* (sediment loss at the base) and *overslumping* (sediment flow over the top of the seawall) can occur if there is poor drainage, falling beach levels, and/or impeded sediment exchange (Carter 1988). If normal sediment supply is interrupted, *flanking* can occur at the end of the wall. This means that the cliffs at the ends of the seawall can slump, which would eventually allow scouring behind the wall, causing it to collapse.

Despite these environmental limitations, residents who live on the towering bluffs between Capitola Village and New Brighton State Beach decided in 1992 that they could no longer afford to lose a foot of their property to erosion each year. They requested the city's help in building a seawall to save 853 m (2,800 ft) of shoreline from being further carved away. If the so-called "Grand Avenue Seawall" is built, the wall will affect one-third of Capitola's shoreline and rank as one of the largest public works projects in the city's history. It is estimated that this 3 m (10 ft) high concrete structure would cost over $2 million—most of it borne by the 28 property owners.

In an attempt to reduce cliff erosion, engineers have even resorted to building structures or frameworks over cliffs then *spraying concrete* in a slurry form onto it—a technique known as **gunnite.** From the shore, a gunnite structure has the appearance of a small dam which is visually unattractive. (See Figure 4-11). This engineering technique, too, has its limitations. Erosion still occurs at the margins of the structure, ultimately

FIGURE 4-9 A "curved" type seawall at Seascape Beach on Monterey Bay, California. (*Source:* Photo by author)

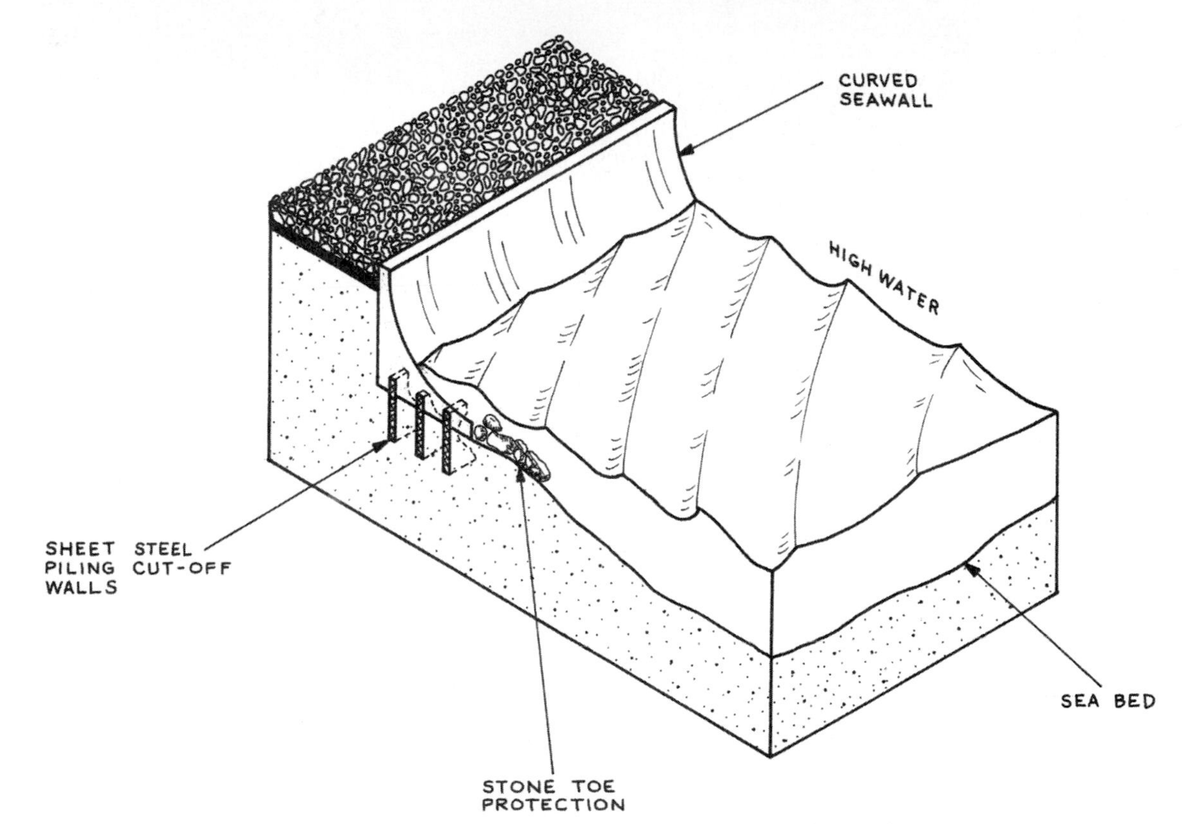

FIGURE 4-10 Three types of seawalls: (Above) Curved sea wall. (Continued)

undermining it. Inadequate drainage behind the coating causes additional problems.

Bulkhead, a term sometimes used interchangeably with seawall, is actually a vertical wall constructed mainly for the purpose of retaining loose fill. (See Figure 4-12). Bulkheads are not designed to protect against flooding from high tides and surges like seawalls, and consequently are constructed of less expensive and less durable materials, such as wood. Some steel and concrete, however, may also be used.

Riprap and revetments. **Riprap** is a pile of rock, boulders, or prefab concrete pieces (e.g., **tetrapods**) dumped along a shoreline to intercept waves. The rock or prefab

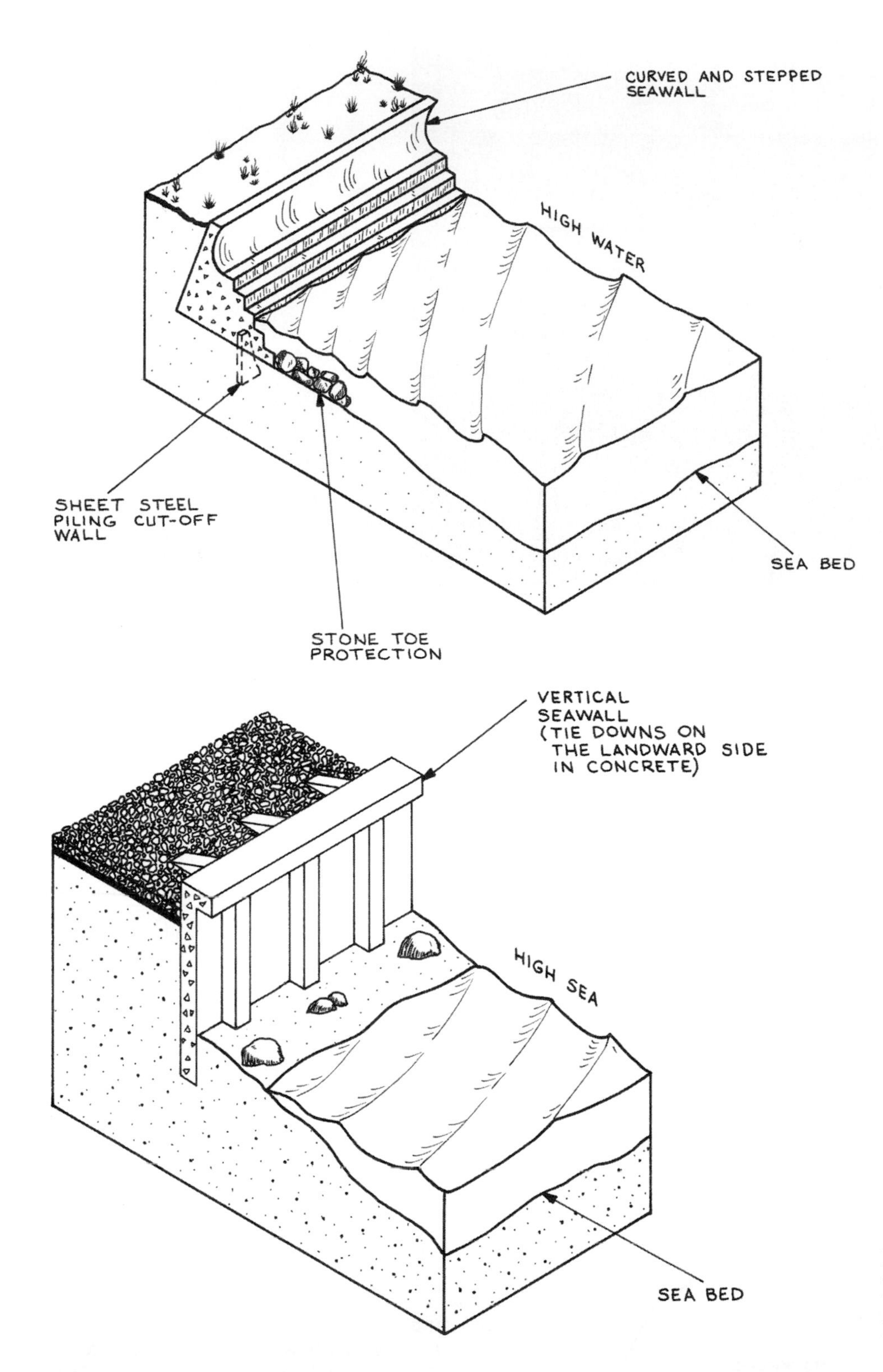

FIGURE 4-10 (Continued) (Upper) Curved and stepped wall secured by piling; (Lower) Vertical seawall constructed of resistant interlocking blocks. (*Source:* Diagrams by Char Holforty)

must be large in size (3 to 5 tons), durable (e.g., granite or marble), and available within a cost-effective transportation distance. In the past, the placement of riprap was done in an "emergency" to protect a building, road, or other threatened structure. Very little design was required, and riprap was simply dumped from trucks over the shoreline cliff. Today, in states like California that have very strict coastal guidelines, better design is mandatory, and emergency permits are difficult to obtain.

The cost of riprap depends on the quantity and location. For instance, if you opt for 3-ton boulders, you must also rent a crane and a qualified crane operator. Along the Central California coast, for example, granite boulders are approximately $54.00 per ton (1994 dollars).

FIGURE 4-11 Gunnite as a shoreline protective technique; Fitzgerald Marine Wildlife Preserve, Northern California. (*Source:* Photo by author)

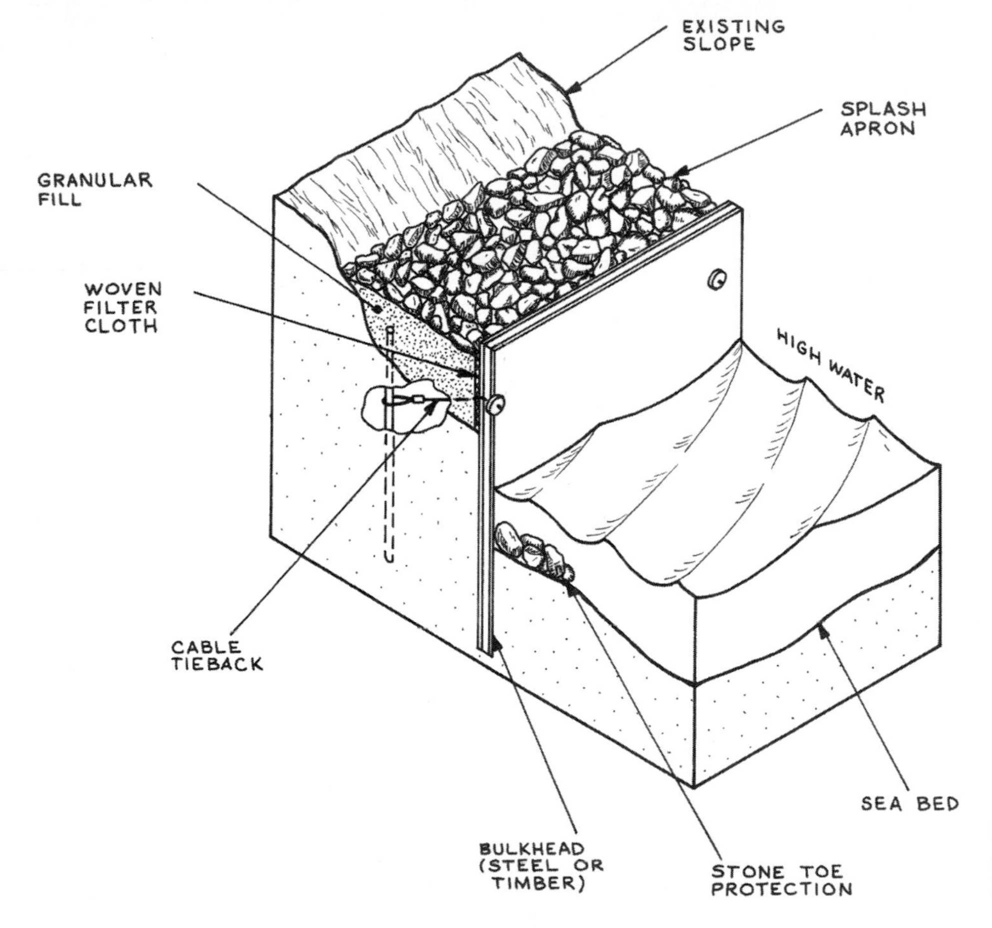

FIGURE 4-12 Bulkhead of wood or steel. (*Source:* Diagram by Char Holforty)

One property owner along the shoreline of northern Monterey Bay had to put riprap before his property on 15 occasions since 1953—a total of 7200 tons of rock at a cost of $185,000 (Griggs et al. 1992). Like most hard structural approaches, riprap can ruin the natural aesthetics of the shore and cause more problems than it solves.

In the Monterey Bay area, riprap has been used for all the traditional reasons—to protect low marine ter-

FIGURE 4-13 Riprap to protect low marine terraces on West Cliff Drive; Santa Cruz, California. (*Source:* Photo by author)

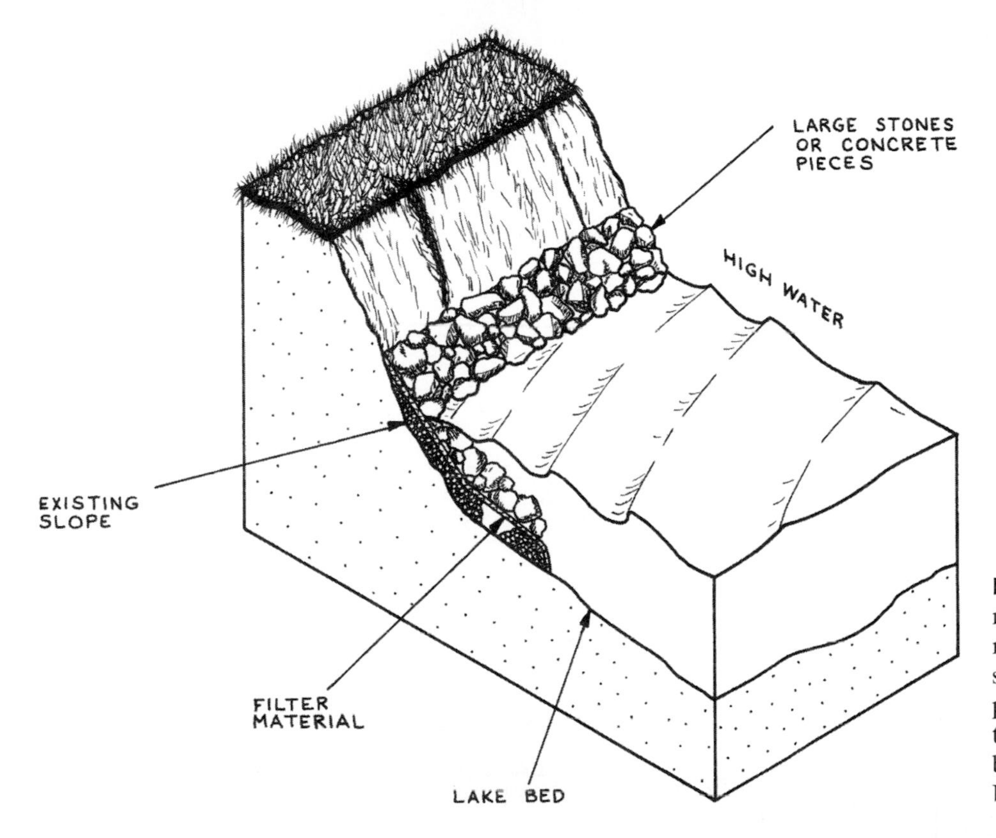

FIGURE 4-14 A revetment—a more refined design of riprap. Here, rock is layered with smaller core stone and larger cap stone or concrete pieces on top of a filter cloth—all for the purpose of minimizing sand scour by wave action or currents. (*Source:* Diagram by Char Holforty)

races that are occasionally overtopped during storms, (See Figure 4-13) to protect highly erodable cliffs of siltstone and sandstone (e.g., between S. C. harbor and Soquel Point [Pleasure Point]), and to protect active dunes subject to undercutting and erosion during severe wave and tidal conditions (see case study on Asilomar Dunes within this chapter).

A **revetment** can be thought of as merely a "better designed riprap." It employs the concept of "layering" and "rock gradation"—small core stones are laid at the base and large cap boulders (2–5 tons per rock) armor the surface (Griggs and Savoy, eds. 1985). (See Figure 4-14). A porous basal filter cloth is used to help maintain the layers within the revetment. Revetments are designed to

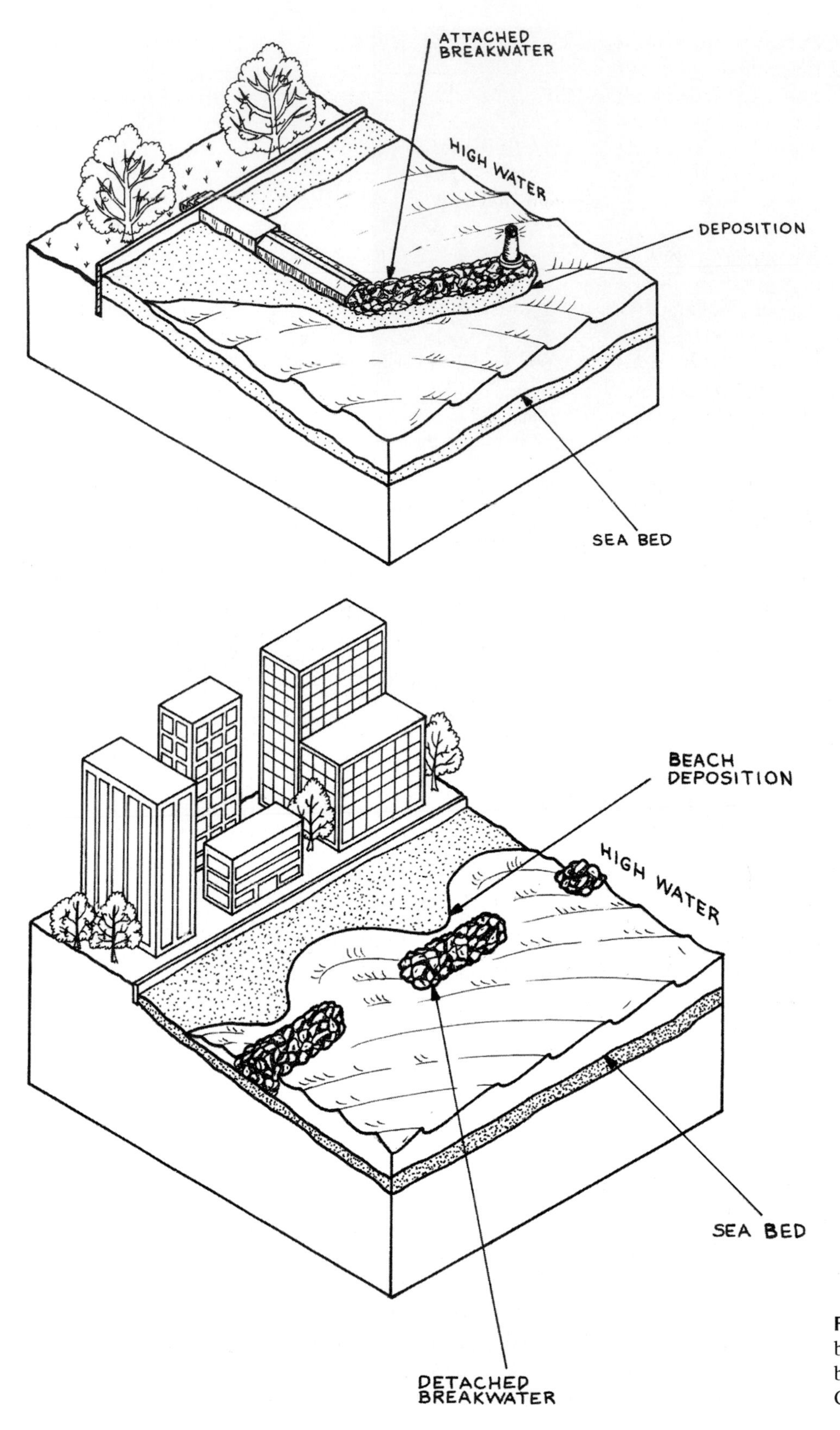

FIGURE 4-15a (Upper) An attached breakwater. (Lower) A detached breakwater. (*Source:* Diagrams by Char Holforty)

minimize several factors—wave reflection, scour beneath the structure, and overtopping. Furthermore, revetments are designed to better absorb wave energy. Revetments are usually less expensive to install than concrete structures, they do not require special drainage systems, and they are easier to maintain and modify (Fulton-Bennett and Griggs 1985).

Breakwaters. Whereas most structures engineered for coastal erosion control are *attached* to the shore, **breakwaters** are often *detached* from shore, in the sense that they are strategically located *within the wave zone* to protect an anchorage or harbor inlet. (See Figure 4-15A). The idea is to reduce the incident wave energy or deflect currents. However, they can also change the shape of the coastline or cause shoaling problems, as can attached breakwaters. In California, for example, the cities of Crescent City and Half Moon Bay on the north coast and Santa Barbara and Long Beach on the south coast all have *attached* breakwaters protecting their harbors. When the Army Corps of Engineers built Santa Barbara's breakwater over 70 years ago, they had no idea that it would eventually form a sand trap that regularly needs dredging at an average annual cost of $496,000 (1992 dollars). (See Figure 4-15B).

Sea level rise. Rising sea levels also accelerate coastal erosion. Since the last ice age some 20,000–11,000 years ago, glacial ice stored on continents has been slowly melting and causing the seas to rise. The seas reached their present level about 5,000 years ago.

FIGURE 4-15b The attached breakwater at Santa Barbara, California, created a major sand trap and has been dredged on a continuous basis for over 70 years. (*Source:* Courtesty of Pacific Western Aerial Surveys, 329 South Salinas, P.O. Box 1588-93102, Santa Barbara, CA, 93103)

Many scientists maintain that the average global sea level is once again rising, at an average of about one-tenth of an inch per year. But researchers still do not agree among themselves as to how much is due to (a) thermal expansion of the water, (b) shifts in the shape of the sea floor, or (c) addition of water to the world's oceans from global climatic temperature rise (due to human-induced "Greenhouse" gases), and consequent glacial melting. Increases in the surface temperature of the Pacific Ocean off the coast of Southern California have swelled the volume of the water enough to raise sea level there more than an inch since 1950.

The term *greenhouse effect* refers to the accelerated levels of carbon dioxide being added to the atmosphere. An increase in carbon dioxide levels is one byproduct of the burning of fossil fuels (coal, oil, and natural gas). Analogous to the way a "greenhouse" traps heat, carbon dioxide "holds in heat from the sun," thus causing the atmosphere to warm and eventually causing alpine and polar glaciers and ice caps to melt, which eventually leads to rising sea levels. Scientists with the United States Environmental Protection Agency estimate that greenhouse gases are likely to cause a sea level rise of 34 cm (13.34 in) by 2100, due to a temperature rise in the range of 1.0 to 2.5 degrees Celsius (Titus 1995). Other estimates range from a rise of 0.5 to 2.0 m (1.6 to 6.6 ft) by the year 2100 caused by temperature rises in the range of 0.5 to 4.5 degrees Celsius (Beatley, Brower, and Schwab 1994). This, of course, comes at the same time as coastal settlement is intensifying.

Figure 4-16 illustrates what the San Francisco Bay region would look like with a 1 meter rise in sea level. How should coastal resource managers and community leaders plan for a possible long-term rise in sea level? Should they try to drastically curtail greenhouse emissions, as environmentalists maintain, or should they try to "hold back the sea" with seawalls and other engineering devices? For an interesting discussion of the sea-level rise dilemma facing coastal resource managers, see Aubrey (1993). A more comprehensive analysis of the effects of a rising sea level on coastal environments is provided by Bird (1993).

Rapid development of U.S. coastlines. All of the existing threats to coastlines (e.g., hurricanes, tsunamis, sea-level rise) are exacerbated by the rapid rate at which America's coastlines are being developed. The most rapid population growth and land development in the United States is occurring near the coasts. According to the 1990 census, 50 percent of all Americans live within 80 km (50 mi) of a coast, with the number likely to increase to 75 percent by 2010 (Morris 1992). Rapid coastal development intensifies shoreline erosion, generally degrades water quality, and increases the likelihood of greater physical damage and human suffering from natural disasters such as hurricanes.

MANAGEMENT STRATEGIES

Various management strategies exist to reduce coastal hazards. They include five major categories of approaches: (1) *Engineering strategies*—both *hard structural* approaches (e.g., groins, seawalls, revetments, breakwaters) and *soft structural* approaches (e.g., beach nourishment); (2) *Building and land use controls*, such as regulations, construction requirements, acquisition, relocation, economic incentives and disincentives; (3) *Evacuation planning and educating the public*; (4) *Mapping and monitoring shoreline change*; and (5) *Nonresponse: development abandonment*. We will now briefly look at these various strategies.

Engineering Strategies

Many of the causes of human-induced erosion (e.g., the construction of jetties, groins, seawalls, and breakwaters) are the same devices used to control coastal erosion. Since many of these engineering strategies have already been discussed under "human-induced causes" of erosion, they will not be listed and discussed again. Just keep in mind that communities usually turn to engineering strategies to "solve" their coastal hazards.

Beach nourishment. Beaches are dynamic—they are always in motion. Various forces either deposit sand or take it away. Beach nourishment refers to replenishment of sand on a beach. It is either brought about naturally (e.g., a longshore transport) or artificially. On the shores of the Monterey Bay National Marine Sanctuary, for example, the city of Capitola trucked in sand from local quarries to charge the groin following its construction in 1969–70. (See Chapter 5, *Coastal Pollution,* for further details and photos on Santa Cruz Harbor's beach nourishment program).

There are many other examples of successful and unsuccessful beach nourishment projects. On the shores of the Florida Keys National Marine Sanctuary, Miami Beach is often cited as a successful project. From 1976–1981, a total of 14 million cubic yards of sand was placed over a 16 km (10 mi) stretch of its beach at a cost of $64 million. In 1987, the beach had to be "renourished" with only 300,000 cubic yards—a loss rate of less than 0.3 percent per year. In the same state, however, one can also find a prime example of unsuccessful beach nourishment. For example, 500,000 cubic yards of sand was placed on a 3.2 km (2 mi) stretch of Indialantic Beach in Florida. One year later, little sand volume remained (Na-

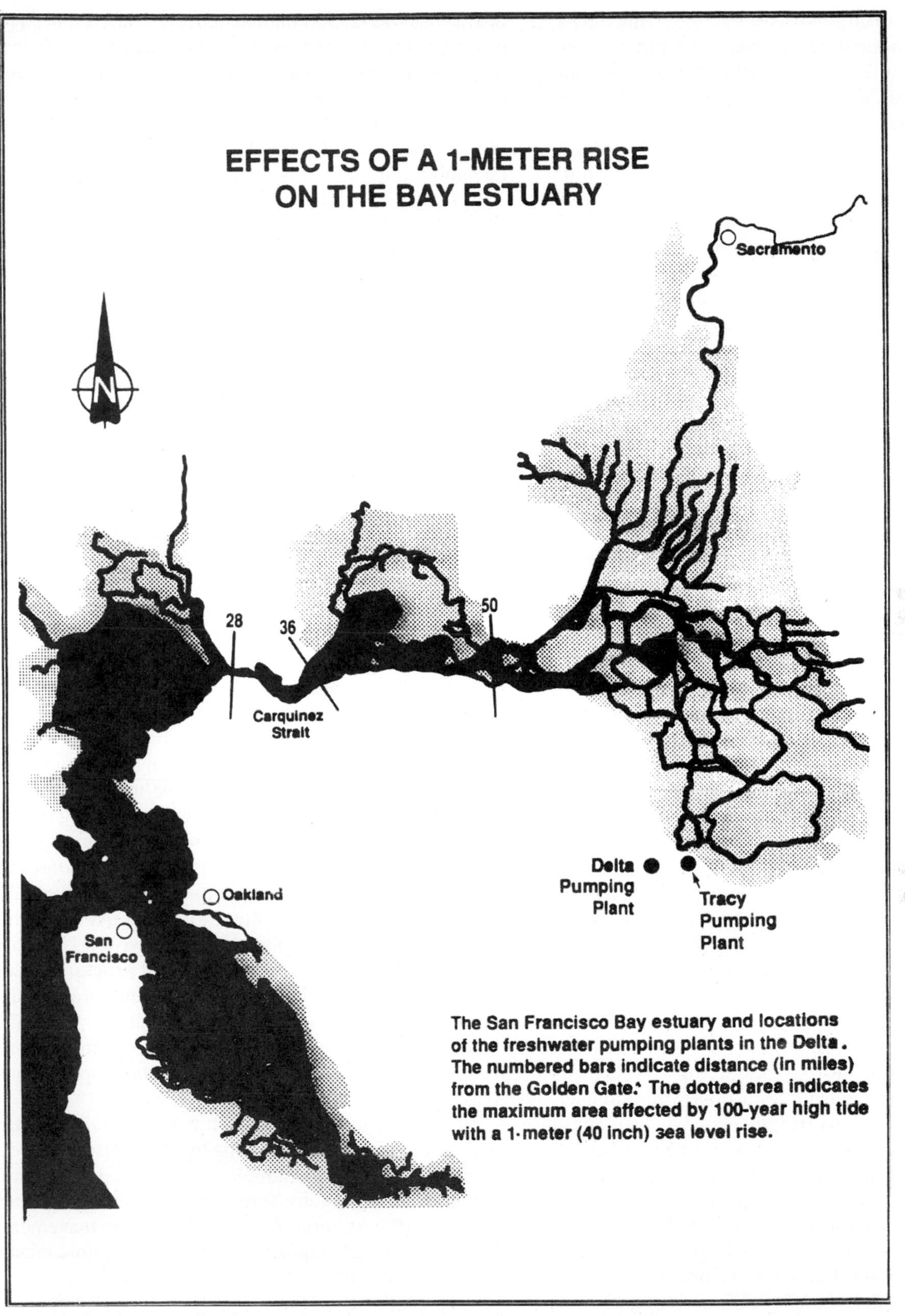

FIGURE 4-16 San Francisco Bay as it would look with a 1 meter (40 in) rise in sea level. (*Source:* California Energy Commission. 1991. *Global Climate Change: Potential Impacts & Policy Recommendations. Vol. II, Sacramento, CA. p. 2–13)*

tional Resource Council 1990). By 1992, the Army Corps of Engineers had nourished 118 miles of United States beaches with 117 million cubic yards of sand at a total cost of $306 million, with local funding accounting for $67.4 million of this total (Griggs et al. 1992). Coastal geologic features (e.g., trailing-edge vs. leading edge) along with ongoing second- and third-order processes of wind, waves, and weather need to be considered when developing beach nourishment projects. The feasibility of a project depends on current and future beach forming processes, sea level, and sediment supply (grain size is particularly important), such that an equilibrium can be obtained between the erosional and depositional forces for that area.

Dune building. *Natural sand dunes* can form when winds blow onshore over a beach and transport sand landward. Once grass and small shrubs grow on the dunes, the dunes become an excellent natural barrier against the sea. In addition to acting as a buffer against wave attack, dunes also provide a reservoir of beach sand. So much, in fact, that a post-storm beach in front of a dune may even be wider than the pre-storm beach. Knowing this, humans have attempted in many locations around the world to mimic nature by creating *artificial sand dunes.* In the United States, the largest dune construction projects have occurred in New Jersey, North Carolina, Florida, and Texas (National Resource Council 1990). Restoring or rehabilitating dunes to their former condition is known as *dune restoration.* On the shores of the Monterey Bay National Marine Sanctuary, a major dune reconstruction and revegetation project is occurring at Asilomar (See case study in this chapter).

Vegetative control of erosion. A soft engineering approach to controlling erosive waves is the use of selective plantings. For example, Texas conservationists have planted salt-tolerant cordgrass in the shallow waters near the coastline of Galveston Bay with apparent success.

Energy dissipation devices. Extensive kelp beds and shallow or exposed offshore rock outcrops create drag on waves, thereby decreasing their energy and impact as they approach the shore. Engineers have tried mimicking "nature's way" of reducing shoreline wave energy and erosion by designing artificial energy dissipation devices such as plastic seaweed. Much more work needs to be done in this area, however, since all efforts to date have contrasting degrees of success. Experiments along the eastern U.S. seaboard illustrated a number of problems, such as the artificial holdfasts becoming lost, either to erosion or burial, and when the material suffered severe dislodgment, boat propellers were fouled (Hall 1985). On the positive side, the cost of artificial seaweed operations was far less than a quarter of that required to install alternative devices of similar efficiency, such as a breakwater (Carter 1988). On the other hand, if such energy dissipation devices as artificial holdfasts do not work (and some coastal geologists would argue that there is no good evidence that they do), then cost may be immaterial.

One thing is certain, whether we are talking about Miami Beach or Monterey Bay, humans have used a variety of engineering devices and strategies to protect against coastal hazards. Unfortunately, the engineering approach has often led to adverse effects on natural processes (e.g., disruption of beach sand supply; disruption of littoral drift; reduction of bluff stability; and modification of shoreline erosion patterns and rates) as well as to adverse effects on the quality of public trust lands (e.g., adverse visual effects; and constraints on public access [Griggs et al. 1992]). The next section of this chapter now turns to the non-engineering strategies that coastal resource managers have most often used.

Building and Land Use Controls

Humans have lived in the coastal zone for a long time, yet we often ignore our existing knowledge of shoreline processes. For example, although coastal geologists, coastal engineers, physical geographers, and environmental scientists may understand and appreciate the intricate interactions between winds, longshore currents, waves, beach sand, and dune formation, this information is often not adequately considered when planning shoreline development. However, we are slowly learning that the traditional *hard structural* (engineering) responses to coastal hazards are often inadequate, environmentally destructive, and aesthetically unpleasant. At the same time, we are beginning to see how expensive it is for citizens to replace or repair storm-damaged structures. And, we are beginning to admit that public agencies are limited in their ability to evacuate densely developed high-hazard areas. As a result, coastal resource managers are looking more seriously at *soft non-structural* approaches to living with the coast.

One non-structural approach is to use *building and land use controls* to influence the location, design, and elevation of new or substantially redeveloped structures. Within this category are three major building and land use strategies appropriate to coastal hazard management: (1) regulation, (2) acquisition, and (3) economic incentives and disincentives. We will now take a brief look at each one of these strategies.

Regulation. All states have a constitutional right to regulate land use. Unfortunately, as more and more humans flock to the coast, the number of regulations (restrictions on development) will have to increase. Coastal resource managers currently use five major types of regulations to help protect the environment: (a) land division and subdivision restrictions; (b) setback lines; (c) building codes; (d) relocation; and (e) owner-assumed liability.

Land division and subdivision restrictions. One type of regulation is outright restriction of land division and subdivisions in areas along the coast that are deemed hazardous. Mapping hazardous zones according to erosion data is now a possibility. Using ancient charts, old and new aerial photographs, and sophisticated computer techniques, geologists can now more accurately predict erosion rates. In fact, the National Research Council (1990) has suggested that government officials use this information to delineate three types of zones—imminent hazard, intermediate zone, and longer-term hazard—and to limit building types and densities accordingly. (See Figure 4-17). This technique is particularly useful in *undeveloped areas* that have few or no structures.

Setback lines. In *partially developed areas,* an older and less sophisticated concept known as "setback lines" is often used. Florida was the first state to promote this management strategy in the late 1960s. At that time, a setback line was merely a standard linear distance from the most seaward dune crest or the high water mark. In Florida's case, it was usually 10 m (32 ft) (Carter 1988).

Setbacks can generally be divided into two types: "stringline" and "rolling" (Pepper 1985). With a **stringline setback** regulation, imagine a string being drawn between the most seaward portions (including decks) of two existing structures on the coast. According to this regulation, all future development must be behind (landward)

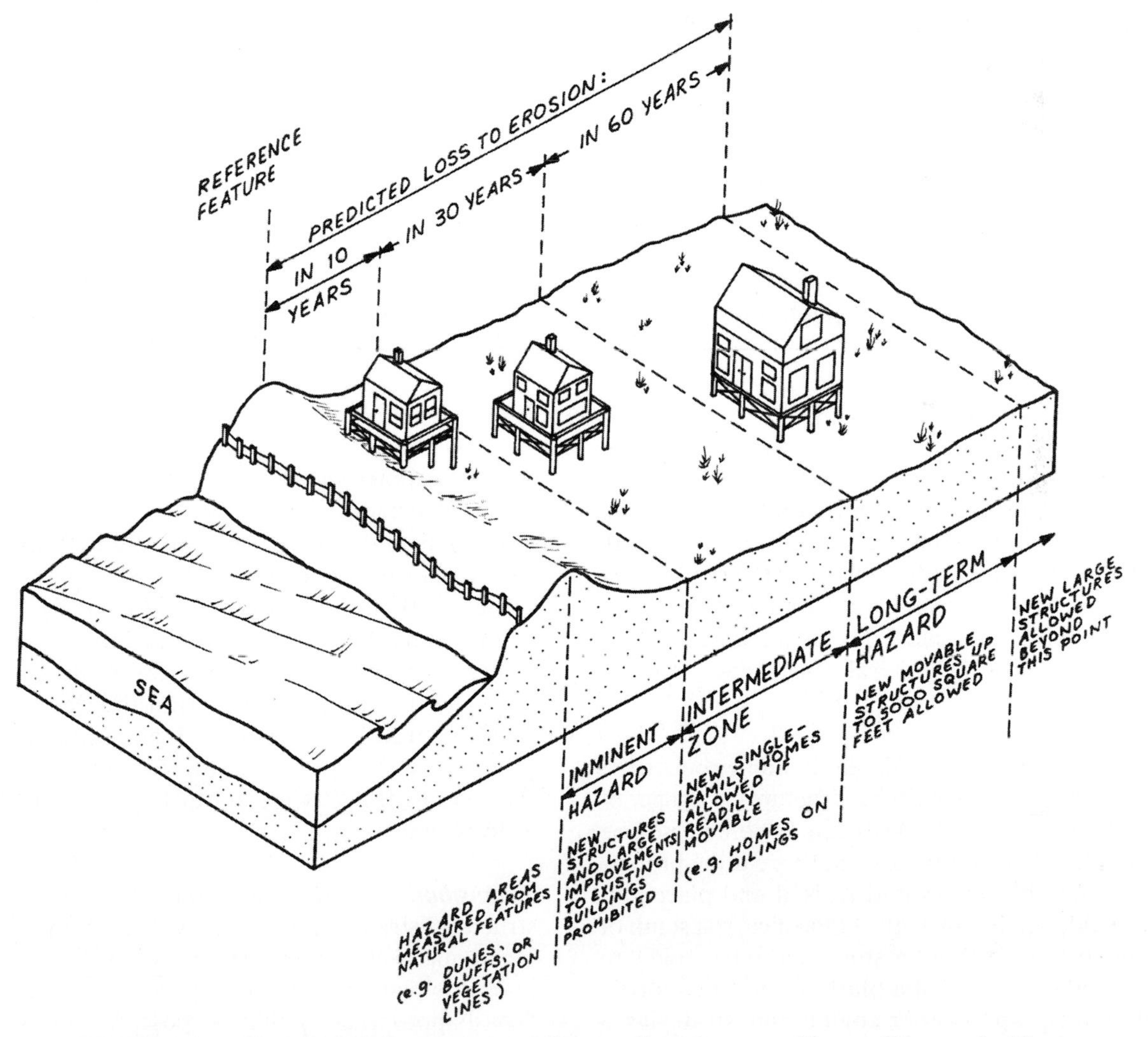

FIGURE 4-17 Limiting building types according to soil erosion rates. (*Source:* Diagram by Char Holforty, adapted from National Research Council, 1990)

of the string. Stringline setbacks are the weakest of the two types, since it allows new construction to occur *at existing setbacks*—even if they occur in hazardous areas. By contrast, a **rolling setback** regulation slowly *moves the "string"* (the seaward-most develop line) landward away from the hazardous zone. For example, if the coastal cliffs at a desirable building spot were known to erode at an average rate of 0.3 m (1 ft) per year, than a hypothetical "100-year structure lifetime" would require that each new building be located 30 m (100 ft) inland from the edge of the cliff. Since the cliff erodes each year, the setback line would likewise move landward each year—keeping a safe distance from the hazardous zone.

Building codes. Coastal hazards can also be reduced through the prudent design and construction of structures. Designs that allow the passage of wind and water around the structure have been found to be the most survivable. Houses built on stilts allow surge water to pass underneath. The pilings can be deeply imbedded, so as to ensure structural integrity during a 100-year storm tide and associated erosion rate. The house floor (lowest horizontal plane) should be elevated above the 100-year wave crest, with the calculations taking into account the annual erosion rates of the site. The January 1983 storm that so severely affected the California coast provided a lesson in house design. At Stinson Beach (Marin County) and Rio Del Beach (northern Monterey Bay), those houses that were built directly on the sand on conventional slabs were destroyed. Some houses built on shallow posts were undermined and collapsed. But the coastal homes that were built on "stilts" (deeply imbedded piles) survived (Griggs and Savoy, eds. 1985).

Structures can also be designed with aerodynamics in mind—the smoother the profile the better. Eaves, balconies, and other protruding objects on a structure are analogous to "sails on a sailboat"—they catch the wind. (See Figure 4-18). Any protruding object should be designed to withstand a 100-year wind loading. For further information on proper coastal design and construction, see Collier et al. 1977; Federal Emergency Agency (FEMA) 1984; and Pilkey et al. 1983.

Relocation and elevation. Whereas building codes have to do with new structures, existing structures in danger of storms and other coastal hazards can, in some cases, be relocated (moved further back from shore if the property is large enough) or elevated (raised and placed on deeply imbedded piles), or both. At Pacifica, just south of San Francisco, the 1983 winter storm caused a rapid 9 to 15 m (30 to 50 ft) retreat of a bluff. Rather than fortify the bluff with riprap or other engineering strategies, a threatened 3-story building was picked up and moved back from its ocean-front site (Griggs and Savoy, eds. 1985). Relocation is generally a wiser, more permanent solution to dealing with coastal hazards. It represents the philosophy of "living with the coast," as opposed to "fortifying against the coast." To move a 457 m (1500 sq ft) house back 30 m (100 ft) from the shore on the Central California coast, in 1992, would cost $10,000 to $18,000, or more. (However, such a management strategy may mean the unacceptable idea of putting the house in the middle of the street or in the neighbor's lot). The cost of building a protective seawall would be several times as expensive over the life of the residence.

Planning for small to average storms is one thing, but if you are talking about the hurricane-prone southeast coast of the U.S., just how many houses and other buildings would have to be relocated, or just how high off the ground would the remaining structures have to be to cope with the fury of a storm surge—the flood of sea water that storms drive onto the coast? In 1992, for example, when Hurricane Andrew swirled toward the southern Florida coast, one of the biggest worries was the potential damage and loss of life that could result from the storm surge. With sustained 140-mph winds, Hurricane Andrew was a Category 4 storm, meaning that the average storm surge ranged from 4 to 5.5 m (13 to 18 ft). If high tides are normally at 1.8 m (6 ft), a 5.2 m (17 ft) surge creates a 7 m (23 ft) **storm tide**—storm surge compensated for tidal condition. Should houses and other structures be allowed in hurricane prone locations? If so, how do you best plan for storm and tidal surges? A number of coastal communities are considering establishing minimum elevations for new oceanfront construction—minimum heights based on the expected levels to be reached by storm waves under extreme conditions.

Owner-assumed liability. Another nonstructural management strategy is to require that home owners who knowingly build in a hazardous area be fully liable for structural repairs resulting from winter storms, wave surge, or other coastal hazards. If the local city or county government warns the potential builder against constructing at a hazardous site, the builder should be required to sign a waiver within the deed restriction saying that he/she cannot seek government aid if the building is damaged or destroyed. This act will force builders to think twice about building within a hazardous zone.

Acquisition. A land acquisition program is another nonstructural strategy for coping with coastal hazards. This technique is especially appropriate if the site to be acquired is planned for open space, recreation, or other "coast appropriate" public purpose. A complete acquisition program includes three avenues: (a) purchase, (b) gifts, and (c) condemnation.

(a)

(b)

FIGURE 4-18 Beach-front houses can be designed to withstand certain levels of storm damage. (a) Although this house was built on stilts to allow surge water to pass underneath, the eaves, deck, and even the floor of the house can act as a giant "sail" for the wind to catch. (Photo by author) (b) This beach-front house, also built on stilts, has a much lower profile and has few protruding objects which could catch the wind. A much better designed house for beach-front property. (*Source:* Photo by author)

Outright state *purchase* of property is an enduring form of land use control. However, state "coffers" are often empty. Land is also acquired through *gifts,* but this low-cost form of acquisition is too random and depends on voluntary donations. Simply put, the site donated is most often not the site needed to be acquired. *Condemnation* (mandated purchase with compensation) is the least popular path to pursue. Public agencies do not want be "heavy-handed" anymore than citizens want to be pressured by government. Not only is condemnation an unpopular management strategy, it is also legally complicated. It is generally considered as a last resort management strategy. However, it is a strategy that may be increasingly common since so many existing developed sites are subject to coastal hazards (Pepper 1985).

Economic incentives and disincentives. There are also national legislative techniques by which money can be used as a carrot for proper coastal development. For example, in 1992, the U.S. House of Representatives passed the first major reform of the nation's flood insurance program in almost 20 years. The measure comprises amendments to the *Federal Flood Insurance Program.* The program covers hundreds of thousands of structures on riverbanks and seacoasts. The changes would restrict insurance coverage to reflect the fact that coastal property is threatened not just by sporadic flooding but also by long-term erosion. The bill limits construction in erosion-prone coastal areas and provides funds for relocation or demolition of buildings in coastal flood zones. It also includes incentives such as grants and insurance premium reductions for communities that take steps to reduce their flood losses.

The insurance program, begun in 1968, was designed to restrict shoreline development by providing coverage only to property in communities with stringent building codes. But, unfortunately, it has had the opposite effect: by making insurance available, it encouraged people to build in otherwise uninsurable areas like barrier islands—the narrow, shifting islands that line much of the East and Gulf coasts. Among other things, the amendments would require communities participating in the program to determine erosion rates on their beaches and calculate 10-year, 30-year, and 60-year erosion hazard areas (refer back to Fig. 4-17). Specific development guidelines would be established for each zone.

Community participation would be *voluntary,* but if an erosion-prone community did not participate, new construction or substantial renovation would be *ineligible* for Federal insurance and the community would be ineligible for Federal development assistance and aid in preventing and mitigating flood damage. Structures in the 10-year zone, for example, would lose their coverage after their next claim.

As might be expected, most regions (e.g., Central California) use all three types of building and land use controls—regulation, acquisition, and economic incentives or disincentives.

Evacuation Planning and Educating the Public

Some coastal communities include evacuation planning and associated public education programs as part of their overall strategies to deal with coastal hazards. This is particularly important in hurricane, flooding, and erosion-prone areas such as Florida, and on islands that rely on causeways or ferry links (e.g., the Florida Keys). Evacuation planners produce flood hazard and emergency maps that illustrate *horizontal evacuation*—the use of roads to move people to higher ground inland. In addition, evacuation planners must consider *vertical evacuation*—the movement of people into storm proofed high-rise buildings. But imagine the difficulty of such planning. For example, in 1992, just before Hurricane Andrew hit the Gulf coast, more than 2 million people in Louisiana, Texas, and Mississippi were asked or ordered to evacuate. Traffic heading north from the Cajun coastline was bumper-to-bumper on U.S. 90. How do you plan for evacuating 2 million people with only a one-or two-day warning? Where will they stay? How will they get there? What about food and water?

Having a plan is only half the battle. Local residents and visiting tourists must be constantly educated about the coastal dangers and what to do if a storm siren is sounded. The success of well-developed evacuation and planning efforts was evident in Florida's 1985 hurricane season. Before Hurricanes Elena and Kate reached the coast, emergency officials evacuated Pinellas County and the Panama City area quickly and safely. All the following factors served to make the evacuation successful: (1) well-marked evacuation routes, (2) easily available shelters, (3) effective communication systems, and (4) a defined chain of command (National Ocean Service 1990). Other states, such as Delaware, have made slide shows of hurricane effects on structural and non-structural erosion control measures, and have used them as public information tools to illustrate how a similar storm might affect their own coastline.

Mapping and Monitoring Shoreline Change

In an effort to better understand how coastal ecosystems work, more and more U.S. states are incorporating extensive mapping and monitoring studies *as part of* their overall coastal management strategy. In itself, mapping and monitoring is not a management strategy, but simply an information base for informing other approaches.

In Duck, North Carolina, for example, scientists ride on the platform of a giant three-legged instrument that wades through the surf like a scene out of a science fiction movie. With this device, as well as a converted Army landing craft, and gauges tethered in the waves, scientists are studying the "nearshore"—the transition zone where wind, water, and land interface. They are convinced that ignorance about the nearshore is why so many beach nourishment projects fail. With these instruments, scientists can trace the movement of water and sand with new precision, including previously invisible wave forms, currents, and patterns of sand transport. Knowing how a sandbar develops, dissipates, and recovers is critical if one wants to understand how a beach is doing, and whether it will come back after a winter storm. However, some coastal geologists maintain that very little if any of the experimental money spent on Duck has helped coastal hazards planning in any significant way.

Remotely sensed images of coastal regions from satellites, such as the Topex-Poseidon, and aircraft, as well as other types of spatial data, are increasingly analyzed with **Geographic Information Systems** (GIS). GIS can be defined as "computer-assisted systems that can input, retrieve, analyze and display geographically referenced information useful for decision making" (Kam, Paw, and Loo 1992). With GIS it is possible to store information regarding topography, land use, conservation and protected areas, fisheries, flora and fauna habitat, oceanography, environmental quality, and socio-economics, calling upon this information to perform an integrated analysis of coastal resource management problems (O'Regan 1996). Resource managers at the National Oceanographic and Atmospheric Administration have used GIS to map the extent of benthic habitat in the Florida Keys National Marine Sanctuary. This information will be utilized in concert with recreational use data, and other types of data as they become available, to properly manage human activity in the marine sanctuary (Clark and Rohmann 1994). Other uses of GIS in coastal resource management include water quality monitoring, coastal erosion management, and flood hazard mapping. Advanced applications include simulations and predictive models for forecasting impacts such as coastal wetland and lowland responses to changing sea levels. Integrated coastal resource management requires open access to large quantities of spatial data from many sources; GIS is an important tool allowing scientists, planners, and resource managers to more easily assimilate and analyze this data (O'Regan 1996).

Nonresponse: Development Abandonment

As with most planning strategies, one must also include the "do nothing" alternative to a potential problem. This approach, of course, has the lowest immediate cost but is potentially the highest risk (to the home owner) in dealing with a situation. One example is Pajaro Dunes—a development project of expensive oceanfront condominiums and homes that stretch for a mile on the shores of the Monterey Bay National Marine Sanctuary (Griggs et al. 1992). Although most environmental scientists would argue that this project should never have been built on a **foredune (primary dune)** in the first place—the dune closest to the beach and most susceptible to erosion—the fact is that the project was built in 1969, so how does one now protect it? Or, should one even attempt to protect it? Despite warnings, such as the winter of 1978 when storm waves came within 3–4.6 m (10–15 ft) of several homes, the homeowners decided to do nothing—the immediate lowest cost alternative. In 1983, however, winter storm waves combined with extremely high tides eliminated up to 12 m (40 ft) of dune width, creating a near-vertical scarp 4.6 to 5.5 m (15 to 18 ft) high that came right up to many dune-front homes. A $1 million emergency rip-rap was installed, and had to be replaced later with a $3 million rip-rap revetment for 1768 m (5800 ft) of protection.

With this brief introduction to coastal hazards and management strategies, let's look at a case study in dune reconstruction and revegetation adjacent to the Monterey Bay National Marine Sanctuary.

CASE STUDY: DUNE RECONSTRUCTION AND REVEGETATION AT ASILOMAR

Dunes, like wetlands, are often wrongfully perceived as marginal or waste land. However, there are many advantages of retaining, restoring, and maintaining coastal dunes. In addition to protecting the land from winter storms and high tides as discussed above, dunes provide important habitat for wildlife, they help concentrate and regulate the flow of fresh water, and they are important aesthetic, open space, agricultural, and industrial resources. One management strategy used by coastal managers is called restoration. **Restoration** (sometimes called *environmental restoration* or *habitat restoration*) has to do with the repair or rehabilitation of a habitat (in this case coastal dunes) for the purpose of restoring its structure and ecological processes. The following is a look at how one community on the shores of the Monterey Bay National Marine Sanctuary has initiated a dune restoration project.

Location

Since 1985, a major dune restoration project has been underway at the Asilomar State Beach and Conference Grounds on the Monterey Peninsula, California. The area is known by locals as "Asilomar"—a Spanish term that suggests "Refuge by the Sea." Today, Asilomar is a 42-hectare (105 acre) state park consisting of stretches of fine-grained white sand beaches (derived from local granite rock); rocky shores; tidepools inhabited by limpets, sponges, and chitons; sand dunes; and a Monterey pine and coast live oak inland area with conference grounds. Asilomar has been the site of many notable events, including the 1971 meeting that resulted in guidelines for Proposition 20 (the Coastal Initiative), which ultimately led to the California Coastal Act (1976) and the creation of the California Coastal Commission. Twenty four hectares (60 acres) of Asilomar's grounds, much of it dune habitat, are undergoing environmental restoration.

Zones of the Dune Habitat

The dunes at Asilomar can be divided into four zones based on predominant vegetation types: the beach and foredune, the bluffs, the middunes, and the reardunes

(a)

(b)

FIGURE 4-19 Four vegetation zones on Asilomar Dunes, Pacific Grove, CA.: (a) Beach and foredune; (b) bluffs; (Continued)

(Goodhue, Haisley, and Hall Consulting 1983). (See Figure 4-19).

The first two zones experience the most severe environmental conditions because of their direct exposure to winds, salt spray, and drifting sand. The *beach and foredunes* (the first zone) naturally have the sparsest of vegetation. Representative species are dune ryegrass (*Elymus mollis),* yellow sand verbena (*Abronia latifolia*), beach bur (*Ambrosia chamissonis*), and beach sagewort (*Artemisia pycnocephala*). The *bluffs* (the second zone) also experience the harshness of the elements. For some unknown reason, there are no plants unique to this zone. However, representative plants include coyote brush (*Baccharis pilularis*), lizard tail (*Eriophyllum staechadifolium*), yarrow (*Achillea borealis var. california*), and even the California poppy (*Eschscholzia californica var maritima*).

The last two zones are inland and more elevated, thus providing greater protection against the marine elements. The *middunes* (the third zone) are sparsely vegetated with plant cover amounting to approximately 35 percent. Common to this zone are Beach sagewort, beach aster (*Lessingia filaginifolia*) and mockheather (*Ericameria eriocoides*). Further inland are the *reardunes*—the fourth

FIGURE 4-19 (Continued) (c) mid-dunes; (d) reardunes. (*Source:* Photos by author)

and final zone. These dunes are large and well stabilized. It is on these dunes that one can find over 50 rustic lodging and meeting facilities scattered amidst a forest of Monterey pines and coast live oak.

Environmental Concerns and Management Strategies

Asilomar is not your typical state park. Although Asilomar is a part of the California Department of Parks and Recreation—a department seemingly always facing budget cutbacks—it fortunately has its own separate source of funding. Asilomar is funded and operated by a concessionaire under contract with the state. It is for this reason that the State Parks Department could take on such an expensive restoration project. To date, over a million dollars have been spent to (1) eradicate exotic plants, (2) reconstruct the dunes, (3) revegetate the dunes, (4) protect the vegetation, (5) protect rare species, (6) interpret the resources, (7) monitor the resources, and (8) make long-term management plans. All of this has been done for the mere sake of preserving the natural heritage of a piece of the California coast. What follows is an explanation of the problem the restoration efforts are intended to solve, including a brief description of the management responses chosen.

(a)

(b)

FIGURE 4-20 Hottentot fig, an exotic plant, Asilomar Dunes, Pacific Grove, CA. (a) area of hottentot fig was sprayed with herbicide; (b) Fencing used to protect sensitive plant area from trampling by "shortcutters." (*Source:* Photos by author)

Exotic plant eradication. In the past, non-native plants were "introduced" to stabilize the dunes and beautify the Conference Grounds. Today, all non-native species are targeted for eradication. To date, the majority of the eradication effort has been against the hottentot fig (*Carpobrotus edulis*), an exotic ice plant from the seacoasts of South Africa. (See Figure 4-20). Hottentot fig was introduced into the United States in the early 20th century. Highway departments in California and other Western states used it widely to landscape along freeways, since the plant requires little irrigation, grows well on slopes, and is easy to propagate. The smaller sea fig (*Mesembryanthemum chilense*) was also eradicated. (According to Kozloff [1983], however, the consensus among most botanists is that *M. chilense* was also introduced.) In areas predominately covered in hottentot fig, the plant was eradicated by using *Round-up*™, which manufacturers claim to be a water-soluble, nonpersistant herbicide. When the hottentot fig was intermixed with native flora, it was extracted by hand.

Another plant invader, beachgrass (*Ammophila arenaria*) has been a more difficult plant to eradicate, since it has roots that can penetrate as much as 1.8 m (6 ft) deep. It was introduced from Europe about a hundred

FIGURE 4-21 Monterey cypress tree on reardunes; Pacific Grove, California. (*Source:* Photo by author)

years ago. (Prior to that time, dunegrass [*Elymus mollis*] was the predominant grass on dunes in Northern California.) Other invasive "weeds" at the Asilomar site include kikuyu grass (*Pennisetum clandestinum*), pampas grass (*Cortaderia atacamensis*), French broom (*Cytisus monspessulanus*), and acacia (*Acacia longifolia* and *A. verticillata*).

Monterey cypress (*Cupressus macrocarpa*) trees are being planted as a short term (one generation) replacement for the Monterey Pines (*Pinus radiata*) which are all (85 percent presently) succumbing to pine pitch canker. (See Figure 4-21).

Dune reconstruction. With approximately 300,000 guests each year, the dunes at the Asilomar conference site were exposed to heavy foot traffic, which resulted in the creation of deep paths and even gullies, and ultimately the loss of the dune-stabilizing vegetation. Some 50 years ago, ice plant and other exotics were introduced to prevent the shifting dunes from burying the conference grounds and parking areas. By 1984, management strategies and policy regarding the use of exotics changed. A program to eradicate ice plant and other exotics began. In 1987, bulldozers were used to reconstruct the dunes, aligning the ridges in the direction of the prevailing wind and pulling sand back from wetland areas.

Once reconstructed, a temporary irrigation system was installed on the dunes so that young seedlings could be easily established. In addition, a layer of mulch (organic material such as straw, leaves, plant residue, or sawdust) was sprayed over the area in a process called **hydromulching**. In the Asilomar dune restoration project, the hydromulch contained a mixture of water, native plant seeds, fertilizer, and wood fiber. According to Tom Moss, State Parks Ecologist, hydromulching is one of the best ways to minimize erosion of dunes that have been temporarily stripped of their vegetation. Riprap was installed to protect Sunset Drive. Guideline (chain) fences and split rail fences surround the dunes and define accessways through the dunes and to the beach. Sections of the bluffs west of Sunset Drive had eroded over the years to bedrock. These areas were also re-built with sand, salvaged from neighborhood developments and restored to their natural condition.

Dune revegetation. The undisturbed dunes at Asilomar provided the model for local natural dune ecology. After an inventory of the plant types, seeds were collected from those pristine sites and added to a hydromulch slurry. In a 1987 application, the slurry contained the seeds of more than 20 native plant species, including yellow sand verbena (*Abronia latifolia*), beach sagewart (*Artemisia pycnocephala*), and beach bur (*Ambrosia chamissonis var. bipinnatisecta*). A greenhouse was constructed on the property so that seeds and young plants would easily be available. Since the project began, 50,000 nursery plants annually have been transplanted to the dunes, with a better than 90 percent success ratio.

Vegetation protection. To keep people from trampling the revegetated dunes, a flat, wood-planked boardwalk with strategically planned benches was constructed to attractively meander throughout the dunes. (See Figure 4-22). Signs were also erected to inform the public

FIGURE 4-22 Boardwalk across dunes helps prevent trampling by the general public, Asilomar Dunes, CA. (*Source:* Photo by author)

not to walk on the dunes, picnic, or create camp fires. Guideline fences using heavy chain through low posts were placed around the foredunes as an added measure of precaution against potential foot traffic. Other than some occasional graffiti on the fences, the public has generally respected the site and what the restoration managers are trying to bring about.

Rare species protection. There are a number of rare species that the Asilomar resource managers are attempting to enhance or reintroduce. For example, they have been successful at increasing the numbers of two California endangered plant species—the Menzies' wallflower (*Erysimum menziesii*) and the Tidestrom's lupine (*Lupinus tidestromii*). Once they successfully improve the density of wild buckwheat (*Eriogonum parvifolium*), the managers have plans to reintroduce the Smith's blue butterfly—a state endangered animal species that uses buckwheat habitats for food and propagation (Monterey Bay Dunes Coalition 1989).

Resource monitoring and public involvement. Regular monitoring and partial reliance on public participation is part of the restoration strategy. Infrared aerial photographs are regularly taken so that native vegetation trends can be recorded and analyzed. Recent surveys indicate that native vegetation on the Asilomar dunes is taking hold.

Revegetating dunes is highly labor intensive, and consequently, dune restoration projects often rely on a variety of sources for the manual labor. At Asilomar, the dune ecologists obtain help from outside sources of labor, such as volunteers from the community (e.g., local school groups) and individuals required to do community service (traffic violators). Advertised "Community Plantings" have attracted numerous workers. For example, on April 6, 1991, over 100 people showed up and planted 15,000 nursery seedlings. With each community planting, the press is invited to cover the story, thus further spreading the word about this highly successful dune restoration project.

GENERAL POLICY RECOMMENDATIONS

In 1988, after an extensive analysis of options for coastal hazards management, the National Research Council (NRC) made a number of detailed recommendations to FEMA's Federal Insurance Administration (FEMA/FIA) so that it could better understand and administer erosion management strategies. What follows is a shortened and abbreviated version that merely highlights some salient aspects of those recommendations. For the complete set of policy recommendations, see the Executive Summary of the NRC's *Managing Coastal Erosion* (National Research Council 1990, pp. 1–15).

Erosion Hazard Reduction

1. Hazard Delineation
   - FEMA's coastal hazard delineations should include data on *erosion hazards.*
   - FEMA should delineate coastlines subject to erosion (*E-zones*).

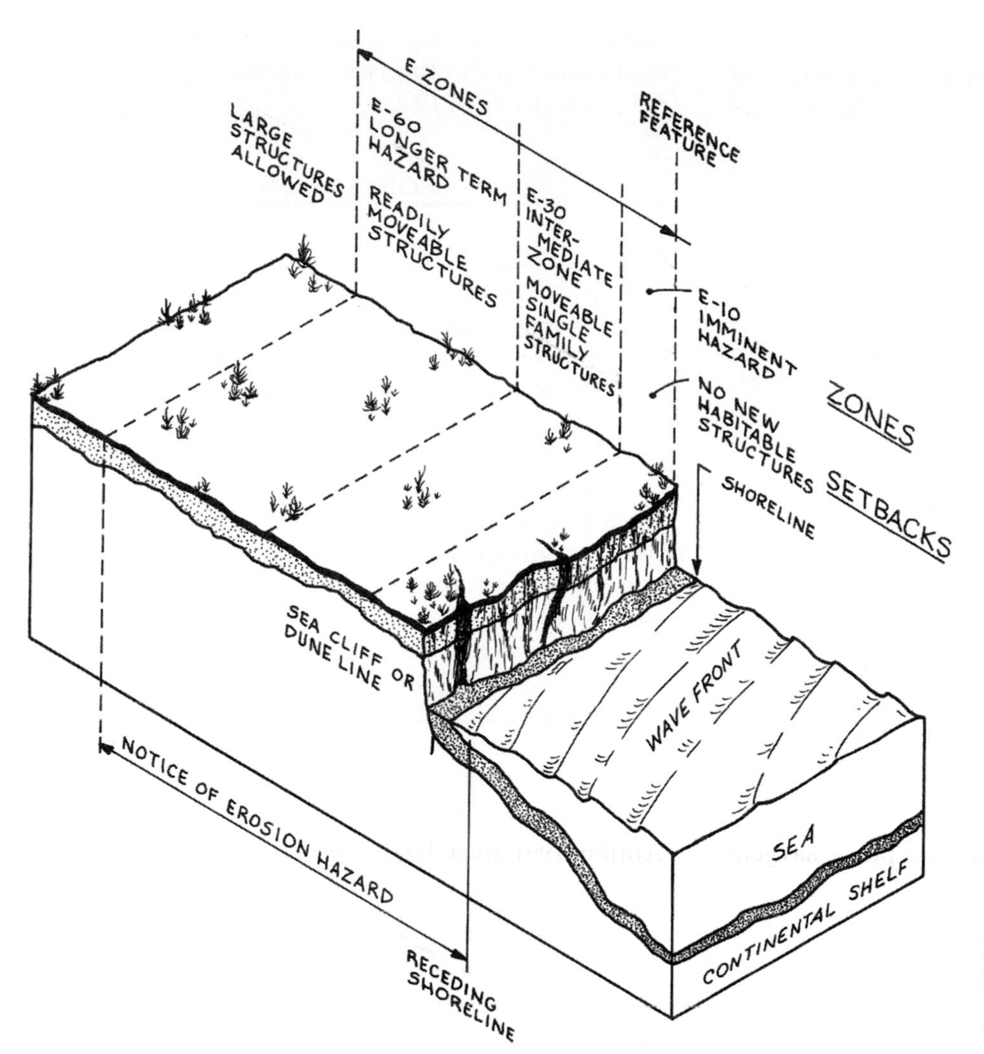

FIGURE 4-23 Summary chart of E-lines and E-zones. (*Source:* Diagram by Char Holforty; adapted from National Resource Council 1990, p. 6)

2. Recommended Methodologies
   - FEMA should, for the short run, start using *historical shoreline change data* to better determine shoreline recession rates.
   - FEMA should, over the long run, change to the more precise methodology based upon *oceanographic data* and statistical techniques.
3. Standards for Development
   - FEMA should establish *minimum standards* for all state or local areas experiencing significant erosion, (e.g., no new development should be permitted seaward of the E-10 line [Imminent Hazard Zone] except such things as docks). (See Figure 4-23).
   - FEMA should establish *insurance rates and availability* based on erosion zones (e.g., no new National Flood Insurance Program [NFIP] policies should be issued for structures in delineated E-10 zones).
   - FEMA should provide benefits for *relocation and demolition* (e.g., existing structures in E-10 zones should be eligible for Section 544 [Upton-Jones] relocation and demolition benefits).
4. Impacts of Navigational and Flood Control Projects on Shore Stability
   - FEMA should develop policies that encourage harbor masters to use high quality dredged sand for *beach nourishment.*
   - FEMA should develop procedures for evaluating the erosion potential of *engineering structures* (e.g., jetties and breakwaters), as well as a means for shifting the cost of erosion from downdrift property owners to the sponsors of the project that caused the erosion.
5. Erosion Control Through Coastal Engineering
   - FEMA should permit engineering structures (e.g., jetties, seawalls) only after an *exhaustive study* of possible adverse environmental impacts.
   - FEMA should *monitor* beach nourishment projects to determine whether they are successful in the long-term.

6. Sand and Gravel Mining
   - FEMA should develop procedures to evaluate the erosion potential of *mining sand and gravel* from beaches and riverbeds.
   - FEMA should develop policies that *shifts the cost of erosion* resulting from mining to the mining operator.
7. Subsidence
   - FEMA should develop procedures for evaluating potential erosion from the *removal of subsurface fluids.*
   - FEMA should develop methods to *mitigate* erosion-caused subsidence.

Education

- FEMA should develop a *public education program* as part of the national policy on coastal erosion.
- FEMA should *notify land owners* in E-zones about the existence and magnitude of erosion (e.g., through notations on deeds, annual flood insurance premiums notice, etc.)

Data Base Development and Research

- FEMA should develop a *shoreline change data base* for use in implementing a national erosion insurance element of the NFIP.
- FEMA should develop *standards* for a national data base; Florida and New Jersey's data base may serve as models for FEMA's data acquisition program.

Unified National Program for Floodplain Management

- FEMA should *revise the Unified National Program for Floodplain Management* to reflect policies and programs concerning erosion zone management.
- FEMA should *convene a national task force of experts* on Coastal Erosion Zone Management to assist FEMA to develop and promulgate nationwide standards for erosion hazards reduction.

These recommendations were incorporated in legislation proposed in 1990. After passing the U.S. House of Representatives by a large majority, the bill was stifled in the Senate when objections were raised by special interest groups, including the National Homebuilders Association and the National Homerenters Association, who were concerned with the negative economic impacts imposed by the legislation. In 1994 less restrictive measures were incorporated in the National Flood Insurance Reform Act, including an economic impact analysis requirement designed to address public concerns about the costs associated with bill implementation. Currently, FEMA is proceeding with the economic impact analysis and will report to Congress with the results which will be used to guide policy formation regarding erosion hazard reduction (Crowell 1996).

CONCLUSION

Finally, a comment about coastal hazards issues and the need to always question "Institutional (e.g., Army Corps) Inertia" and "Engineering Mentality." Coastal hazard management (e.g., beach erosion management) is not a job for amateur or inexperienced practitioners; it requires the best professional expertise available (Clark 1996). But as a citizen, student, or coastal resource manager, it is important to never accept the so-called "expert" opinion that you are "unqualified to comment," regardless of your background. For example, you do not have to be a civil engineer or coastal geologist to question the need, type, or placement of a shoreline device. The following famous case may make the point.

In the 1960s, two young assistant professors, Robert Dolan and Paul Godfrey, decided to take a position regarding one of the long standing debates in coastal resource management: whether the human attempt to stabilize the dunes, beaches, and inlets are essentially futile and injurious to the natural forces that would otherwise govern them. The Army Corps of Engineers proposed a dune stabilization project for the Cape Hatteras National Seashore shoreline. It would cost $50 million initially, and $2.5 million in annual maintenance. The idea was to create a big dune, 6 m (20 ft) high and 152 m (500 ft) wide, all along the Outer Banks. To Dolan and Godfrey, these plans were ludicrous. They began to speak out. Their opposition made statements like, "who are they, merely a couple of assistant professors (botanists), who challenge the authority of the Army Corps of Engineers and the National Park Service." Thus began one of the most potent collaborations in the history of coastal research. Together, Dolan and Godfrey set out to stop the largest long-term coastal management project in history. The odds were formidable. But, as Dolan notes, he and Godfrey were armed with the most powerful weapon ever devised—the *truth.* According to these two individuals, it does not matter how much ammunition the other side has, if you are *right.* Dolan and Godfrey made speeches, attended meetings, and wrote scientific papers. Their ideas were adopted and spread by others. In the end, no other scientist disputed their findings. The National Park Service had to admit that past stabilization policies had not worked. Not only did the Park Service accept their arguments, it acted upon them. Dolan and Godfrey had won. Furthermore, Dolan and Godfrey achieved what few scientists are ever able to achieve: a marriage of science and public policy that reverses a decades-old way of doing business. In sum, "Truth" and "Right" = "Might." For the complete "Dolan and Godfrey story," see Alexander and Lazell (1992).

The battle against the twin evils of "Institutional Inertia" and "Engineering Mentality" has also been taken up by others, such as geologists Orrin Pilkey, Jr. of Duke University, Stan Riggs of East Carolina University, and Gary Griggs of the University of California, Santa Cruz. You, too, and can join in and use "the power of ideas" *to question* (if not topple) those failing (or failed) technologies that have been traditionally used to minimize coastal hazards, as well as *to recommend* softer and more ecologically acceptable approaches to living with the coast.

REFERENCES

Abe, Katsuyuki. 1979. "Size of Great Earthquakes of 1837–1974 Inferred From Tsunami Data." *Journal of Geophysical Research.* Vol. 84, No. B4: pp. 1561–1568.

Alexander, John, and James Lazell. 1992. *Ribbon of Sand: The Amazing Convergence of the Ocean and Outer Banks.* Chapel Hill, North Carolina: Algonquin Books of Chapel Hill.

Aubrey, David G. 1993. "Coastal Erosion's Influencing Factors Include Development, Dams, Wells, and Climate Change." *Oceanus.* Vol. 36, No. 2: pp. 5–9.

Beatley, T., D. J. Brower, and Anna K. Schwab. 1994. *An Introduction to Coastal Zone Management.* Washington, D.C.: Island Press.

Bird, C.F. 1993. *Submerging Coasts: The Effects of a Rising Sea Level on Coastal Environments.* Chichester, U.K.: John Wiley & Sons.

California Energy Commission. 1991. *Global Climate Change: Potential Impacts & Policy Recommendations. Vol. II* Sacramento, California: California Energy Commision.

Carter, R. W. G. 1988. *Coastal Environments.* New York: Academic Press.

Carter, R. W. G., D. A. Eastwood, and P. Bradshaw. 1992. "Small-scale Sediment Removal from Beaches in Northern Ireland: Environmental Impact, Community Perceptions, and Conservation Management." *Aquatic Conservation: Marine and Freshwater Ecosystems.* Vol. 2, pp. 95–113.

Cialone, Mary A. 1994. "The Coastal Modeling System (CMS): A Coastal Processes Software Package." *Journal of Coastal Research.* Vol. 10, No. 3: pp. 576–587.

Clark, Amyu E., and Steven O. Rohmann. 1994. "Mapping and Analyzing Benthic Habitats in the Florida Keys National Marine Sanctuary." In *Proceedings of the Second Thematic Conference on Remote Sensing for Marine and Coastal Environments: Needs, Solutions, and Applications,* 31 January–2 February 1994. New Orleans, Louisiana. Ann Arbor, Michigan: Environmental Research Institute of Michigan.

Clark, John R. 1996. *Coastal Zone Management Handbook.* Boca Raton, Florida: CRC Lewis Publishers.

Coates, Donald R. 1981. *Environmental Geology.* New York: Wiley.

Collier, C. A., K. Eshagi, G. Cooper, and R. S. Wolfe. 1977. "Guidelines for Beachfront Construction with Special Reference to the Coastal Construction Setback Line." *Florida Sea Grant Report. No. 20.* Gainesville, Florida: Sea Grant Program.

Crowell, Mark. 1996. Mitigation Directorate, Federal Emergency Management Agency. Telephone communication, 27 February, Washington, D.C.

Edgerton, Lynne T. 1991. *The Rising Tide: Global Warming and World Sea Level Rise.* Washington, D.C.: Island Press.

Federal Emergency Management Agency (FEMA, USA). 1984. "Design and Construction Manual for Residential Buildings in Coastal High Hazard Areas." *Report FEMA-55.* Washington, D.C.: Federal Emergency Management Agency.

Fletcher, C. H., B. M. Richmond, G. M. Barnes, and T. A. Schroeder. 1995. "Marine Flooding on the Coast of Kauai during Hurricane Iniki: Hindcasting Inundation Components and Delineating Washover." *Journal of Coastal Research.* Vol. 11, No. 1, pp. 188–204.

Fulton-Bennett, Kim, and Gary Griggs. 1985. *Coastal Protection Structures and Their Effectiveness.* Joint Publication of the California Department of Boating and Waterways and the Institute of Marine Sciences. Santa Cruz, California: University of California at Santa Cruz.

Gabrysch, R. K. 1969. "Land-Surface Subsidence in the Houston-Galveston Region, Texas." In *Land Subsidence: Proceedings of the Tokyo Symposium, September 1969, Vol. 1,* by UNESCO. Paris: UNESCO, pp. 43–54.

Garrett, Chris, and Leo R. M. Maas. 1993. "Tides and Their Effects." *Oceanus.* Vol. 36, No. 1: pp. 27–37.

Giese, Graham S., and David C. Chapman. 1993. "Coastal Seiches." *Oceanus.* Vol. 36, No. 1: pp. 38–46.

Good, James W. and Sandra S. Ridlington, eds. 1992. *Coastal Natural Hazards: Science, Engineering, and Public Policy.* Corvallis, Oregon: Oregon State University.

Goodhue, Haisley, and Hall Consulting. 1983. *Asilomar State Beach and Conference Grounds Resource Management Plan and General Development Plan.* Pacific Grove, California: California Department of Parks and Recreation.

Gordon, Burton L. 1996. *Monterey Bay Area: Natural History and Cultural Imprints.* 3rd ed. Pacific Grove, California: The Boxwood Press.

Griggs, Gary B. 1995. "Relocation or Reconstruction of Threatened Coastal Structures; a Second Look." *Shore and Beach.* Vol 63, No. 2, April, pp. 31-37.

Griggs, Gary B., and Lauret Savoy, eds. 1985. *Living with the California Coast.* Durham, North Carolina: Duke University Press.

Griggs, Gary B., James E. Pepper, and Martha E. Jordan. 1992. *California's Coastal Hazards: A Critical Assessment of Existing Land-Use Policies and Practices.* Berkeley, California: California Policy Seminar.

Hall, M.J. 1985. "Seascape Installation: Stone Harbour Point, New Jersey." Rider College, New Jersey.

Kam, S. P., J. N. Paw and M. Loo. 1992. "The Use of Remote Sensing and Geographic Information Systems in Coastal Zone Management." In T. E. Chua and L. F. Scura (eds.). *Integrative Framework and Methods for Coastal Area Management.* ICLARM Conference Proceedings 37. Bandar Seri Begawan, Brunei Darussalam, 28–30, April, pp. 107–131.

Kaplow, David. 1989. *Sand City Dune Restoration Techniques.* San Francisco: California Coastal Commission.

Kozloff, Eugene. 1983. *Seashore Life of the Northern Pacific Coast.* Seattle, Washington: University of Washington Press.

Lander, J., P. Lockridge, and M. Kozuch. 1993. *Tsunanis Affecting the West Coast of the United States, 1806–1992.* Boulder, Colorado: National Oceanic and Atmospheric Administration.

McDonnell, Tom. 1995. Senior environmental planner with the California Department of Transportation, Environmental South Division. Telephone communication, 5 October, Oakland, California.

Monroe, James S. and Reed Wicander. 1994. *The Changing Earth: Exploring Geology and Evolution.* St. Paul, Minnesota: West Publishing Co.

Monterey Bay Dunes Coalition. 1989. *Monterey Bay Dunes: A Strategy for Preservation.* Pacific Grove, California: Monterey Bay Dunes Coalition.

Morris, Marya. 1992. "The Rising Tide: Rapid Development Threatens U.S. Coastal Areas." *EPA Journal.* Vol. 18, No. 4: pp. 39–41.

National Ocean Service. 1990. *Coastal Management Solutions to Natural Hazards.* Technical Assistance Bulletin #103. A prepublication copy. Washington D.C.: NOAA.

National Research Council. 1987. *Responding to Changes in Sea Level: Engineering Implications.* Washington, D.C.: National Academy Press.

National Research Council. 1990. *Managing Coastal Erosion.* Washington, D.C.: National Academy Press.

Oradiwe, E. N. 1986. "*Sediment Budget for Monterey Bay.*" M.S. thesis, U.S. Naval Postgraduate School, Monterey, California.

O'Regan, Peter R. 1996. "The Use of Contemporary Information Technologies for Coastal Research and Management—A Review." *Journal of Coastal Research.* 12 (Winter), pp. 192–204.

Owen, Oliver S., and Daniel D. Chiras. 1990. 5th ed. [1995, 6th ed]. *Natural Resource Conservation.* New York: Macmillan.

Pepper, James. 1985. "Coastal Land-Use Planning and Regulation: Reducing the Risk of Environmental Hazards." In *Living with the California Coast,* edited by Gary Griggs and Lauret Savoy, pp. 69–80. Durham, North Carolina: Duke University Press.

Pilkey, O. H., Sr., W. D. Pilkey, and W. J. Neal. 1983. *Coastal Design: A Guide for Builders, Planners and Homeowners.* New York: Van Nostrand Reinhold.

Titus, James G. 1995. *The Probability of Sea Level Rise.* Washington, D.C.: Environmental Protection Agency.

Williams, John M., Fred Doehring, and Iver W. Duedall. 1993. "Heavy Weather in Florida." *Oceanus.* Vol. 36, No. 1: pp. 19–26.

Yanev, P. I. and C. R. Scawthorn. 1993. *Hokkaido Nansei-oki, Japan Earthquake of July 12, 1993.* Buffalo, New York: National Center for Earthquake Engineering Research, NCEER-93-0023.

FURTHER READING

Bush, David M., and Orrin H. Pilkey Jr. 1993. "A Tale of Two Lighthouses." *Oceanus.* Vol. 36, No. 1: pp. 93–96.

Bush, David M., Orrin H. Pilkey Jr., and William J. Neal. 1996. *Living by the Rules of the Sea.* Durham, North Carolina: Duke University Press.

Chapman, David. 1994. *Natural Hazards.* Melbourne, Australia: Oxford University Press.

Coch, Nichola K. 1995. *Geohazards: Natural and Human.* Englewood Cliffs, New Jersey: Prentice Hall.

Good, James W., and Sandra S. Ridlington, eds., 1992. *Coastal Natural Hazards: Science, Engineering, and Public Policy.* Corvallis, Oregon: Oregon State University.

Griggs, Gary B. 1995. "Relocation or Reconstruction of Threatened Coastal Structures; a Second Look." *Shore and Beach.* Vol. 63, No. 2, pp. 31–47.

Inman, Douglas L., and Patricia M. Masters. 1994. "Status of Research on the Nearshore." *Shore & Beach.* Vol. 62, No. 3, pp. 11–20.

Lewis, Roy R., ed. 1982. *Creation and Restoration of Coastal Plant Communities.* Boca Raton, Florida: CRC Press.

Millemann, Beth. 1993. "The National Flood Insurance Program." *Oceanus.* Vol. 36, No. 1: pp. 6–8.

Platt, Rutherford H., H. Crane Miller, Timothy Beatley, Jennifer Melville, and Brenda G. Mathenia. 1992. *Coastal Erosion: Has Retreat Sounded?* Monograph #53. Boulder, Colorado: Natural Hazards Research and Applications Center.

Salmon, J., D. Henningson, and T. McAlpin. 1982. *Dune Restoration and Revegetation Manual.* Gainesville, Florida: Florida Sea Grant Program.

Steign, R. C., and H. G. Wind. 1993. "Dune Erosion Management: A Cost Benefit Analysis." In *Coastal Zone '93: Proceedings of the Eighth Symposium on Coastal and Ocean Management,* edited by Orville T. Magoon, W. Stanley Wilson, Hugh Converse, and L. Thomas Tobin, pp. 1951–1965. New York: American Society of Civil Engineers.

5 Coastal Pollution

In 1990, scientists at the National Oceanic and Atmospheric Administration (NOAA) reported that pollution was *diminishing* along U.S. coastlines (NOAA 1990a). According to this report, chemical contamination decreased—or at least did not increase—during the previous six years, particularly along the southeast coast and the Gulf of Mexico. Furthermore, it appeared that the significant concentrations of contaminants measured were limited primarily to urbanized estuaries. *Although not quite a clean bill of health, the NOAA prognosis for U.S. coastal waters sounded encouraging.*

By 1995, however, a different team of scientists reached a less encouraging scenario for U.S. coastal waters. Scientists working for the Natural Resource Defense Council (NRDC) compiled data relating to beach closures along the 21 contiguous coastal states, Great Lake states, Hawaii, Guam, Puerto Rico, and the Virgin Islands. Information regarding the number of beach closures along with the sources of the pollution was presented in *Testing the Waters VIII: Has Your Vacation Beach Cleaned Up Its Act?* (Paul 1998). Between 1988 and 1997, the number of beach closures/advisories had gone from **484** beach closures (with 3 permanent closures) in 1988 to **4,153** individual closings and advisories in 1997 (See Figure 5-1). (*Individual* closings and advisories is a single beach for which a closure/advisory has been issued for a single day; *Extended* closings and advisories = 6-12 weeks; *permanent* closings and advisories = over 12 weeks). In 1997, in addition to individual closings, there were also 17 extended and 55 permanent closings. Between 1988 and 1997, there have been **22,819** closings and advisories of one type or another. The major causes of beach closures were stormwater runoff and associated sewage overflows into coastal water bodies. The closures were posted due to resulting high levels of microbial pathogens (microscopic disease-causing organisms) from human and animal wastes. Contrary to the 1990 NOAA report, these findings were not so encouraging.

Regardless of which study the reader believes, something is going afoul on America's beaches. In addition to finding beaches closed because of chemical contamination, pick almost any shore and you will come across either oil slicked rocks, sewage balls (floating chunks of sewage), or a plethora of beach litter, from cigarette butts to discarded plastic Christmas trees. Beach closures have become so frequent in Southern California that residents have a standard reply when asked about beach pollution in their area—"Different day, different beach!" If America's coastlines are becoming less contaminated as suggested by the 1990 NOAA report, why are there more and more beach closures? And, if our coastlines are indeed experiencing greater coastal pollution, what can be done to help remedy the situation?

It helps to remember that there are basically two forms of **coastal pollution**—water pollution, and marine debris (sometimes called *floatable debris*). **Water pollution** can be defined as the addition of *harmful* (e.g., toxic wastes) or *objectionable* (e.g., sediment) materials to water causing an alteration of water quality. Keep in mind that water too polluted for one type of use (e.g., drinking water) may still be suitable for another use (e.g., swimming, industrial production). It is obvious that water is one of the major natural resources of the coastal zone and minimizing water pollution is a major factor in cleaning up our coast. But the task will involve much more than just cleaning up the water. It

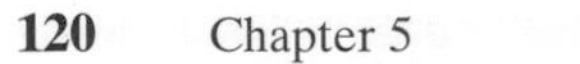

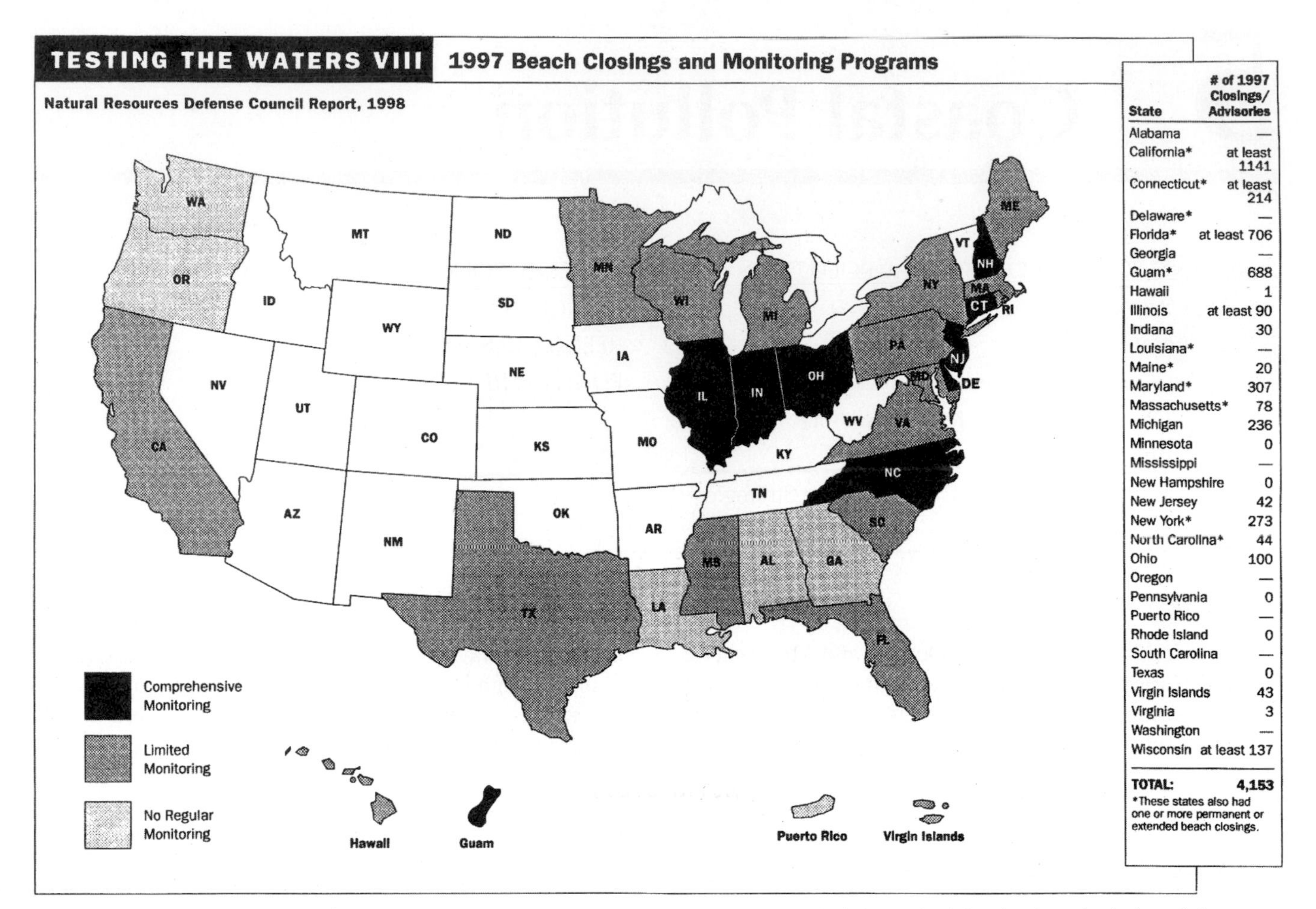

State	# of 1997 Closings/ Advisories
Alabama	—
California*	at least 1141
Connecticut*	at least 214
Delaware*	—
Florida*	at least 706
Georgia	—
Guam*	688
Hawaii	1
Illinois	at least 90
Indiana	30
Louisiana*	—
Maine*	20
Maryland*	307
Massachusetts*	78
Michigan	236
Minnesota	0
Mississippi	—
New Hampshire	0
New Jersey	42
New York*	273
North Carolina*	44
Ohio	100
Oregon	—
Pennsylvania	0
Puerto Rico	—
Rhode Island	0
South Carolina	—
Texas	0
Virgin Islands	43
Virginia	3
Washington	—
Wisconsin	at least 137
TOTAL:	4,153

*These states also had one or more permanent or extended beach closings.

FIGURE 5-1 1997 Beach Closings and Monitoring Programs. *Individual* closings and advisories is a single beach for which a closure/advisory had been issued for a single day. *Extended* closings and advisories = 6–12 weeks; *permanent* closings and advisories = over 12 weeks. (*Source*: Natural Resources Defense Council Report, 1998)

will also involve reducing all the **marine debris,** such as medical waste and garbage.

This chapter on coastal pollution is divided into four parts. The first part introduces the reader to the types and sources of coastal pollution, including point source, nonpoint source, and physical and hydrological modifications. The *Case Study* looks at Soquel Creek—a small creek that feeds into the Monterey Bay National Marine Sanctuary (MBNMS). This case study illustrates how complex coastal watershed management can be, even with a small creek that has had minimal human impact. Relative to much of the United States, the water quality along the Monterey Bay shore is good (Minerals Management Service 1987). This is partially due to the periodic upwelling within the bay and its continual mixing with the open ocean. It also has to due with the relatively low human population around the Bay. However, as this case study illustrates, even the relatively clean Monterey shore has its share of water quality and marine debris problems. The third part provides some general recommendations for reducing pollution on U.S. coasts. The Conclusion, provides some final thoughts on reducing coastal pollution.

TYPES AND SOURCES OF COASTAL POLLUTION

Coastal pollutants come from a wide variety of natural and human sources, which leads to a complex range of water management problems. (See Figure 5-2). To simplify matters, water resource managers generally classify water pollutants as either *point source* or *nonpoint source.* **Point source** pollutants are clearly discernible in terms of origin, such as a municipal sewage outfall (e.g., from a pipe, ditch, or tunnel). Oil tanker spills and

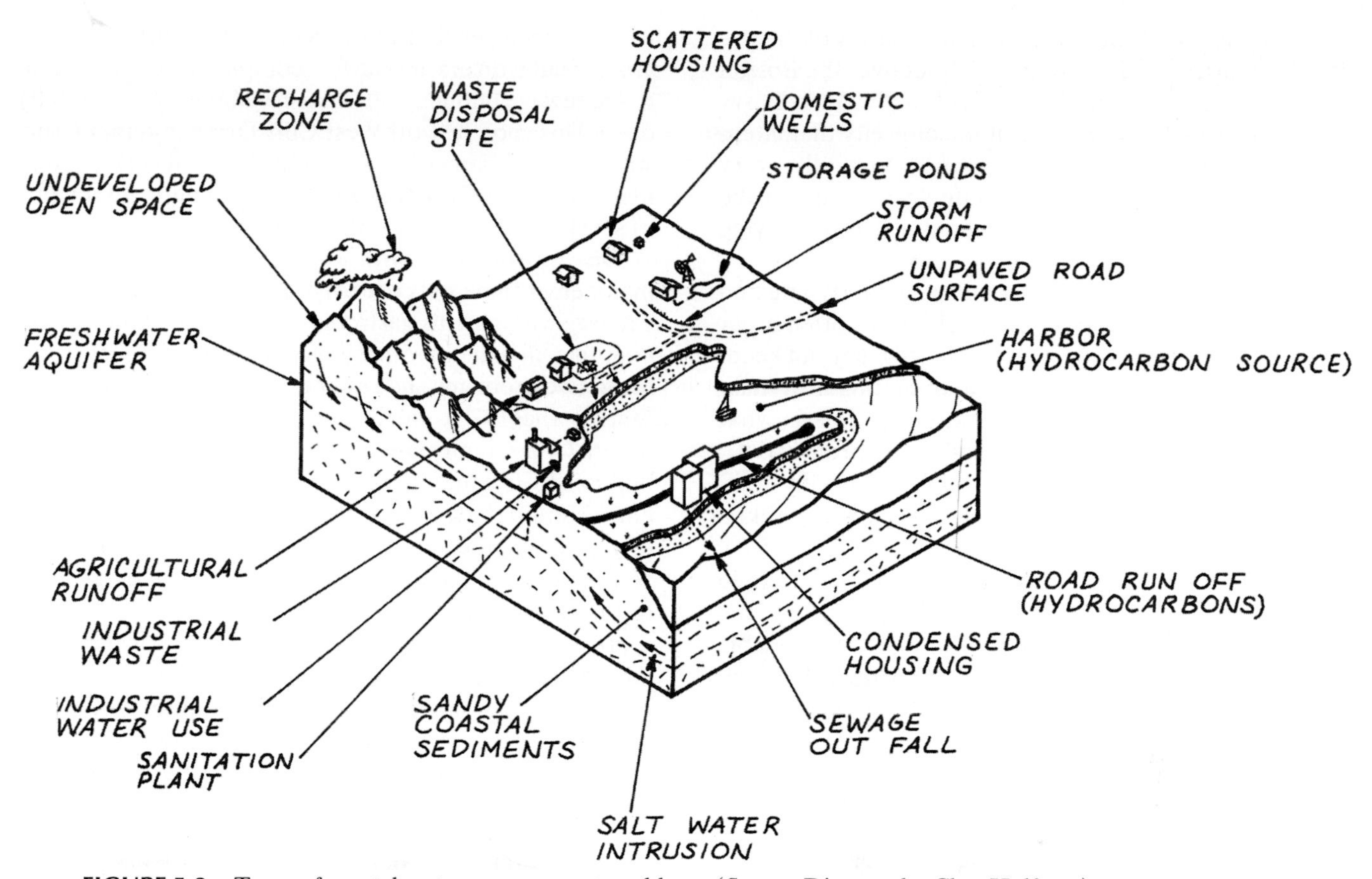

FIGURE 5-2 Types of coastal water management problems. (*Source:* Diagram by Char Holforty)

offshore oil well blowouts are also classified as point source pollutants. (See Chapter 7 for a discussion of coastal pollution resulting from offshore oil development and transportation.) By contrast, **nonpoint source** (or runoff) pollutants arise from ill-defined or diffused sources, such as agricultural fields, city streets and parking lots, or logging operations. Because this source of pollution comes from a large or broad area, rather than a discrete point, it is the more difficult form of water pollution to control.

Physical and hydrological modifications of shorelines also affect water quality. Along many U.S. coastlines, periodic harbor dredging, commercial sand mining, and overdrafting of aquifers resulting in salt water intrusion affect water quality. The construction of storage reservoirs and water diversion canals, of course, also comes under this category.

Point Sources

Oil tanker spills and offshore oil well blowouts are two point source pollutants that are very dramatic and can be devastating to estuaries and bays, fish and wildlife, and human settlements. Consequently, these two point source pollutants receive the most public attention, and in many cases, outrage. In fact, a primary reason the MBNMS was created was to keep oil platforms off the Central California coast. The offshore oil development issue is so important and complex, that a separate chapter, Chapter 7, is devoted to this topic. What follows now is a look at the other forms of point source pollutants that affect water quality on our coastlines.

Sewage outfalls. Coastal waters are often polluted from such point sources as sewage outfalls, industrial outfalls, and solid waste disposal sites. In the past, ocean discharge has been the simplest and cheapest form of waste disposal. In Boston Harbor, for example, 500 million gallons of essentially untreated waste is intentionally (and legally) dumped 9 m (30 ft) offshore into shallow water every day. Boston harbor is so polluted, in fact, that it has been used as a "political football" in the last two U.S. presidential campaigns. In 1988, vice president Bush used the harbor's degradation to help derail Democratic rival and Massachusetts Governor Michael Dukakis' claim that he was an environmentalist. Four years later, in 1992, Bill Clinton's democratic campaign charged that President Bush had failed to request funding to help pay for the $6 billion harbor cleanup in the first and second years of his presidency. (For two recent summaries of the political, historical, and ecological "fallout over the Boston harbor outfall," see Aubrey and Connor [1993] and Levy [1993].) Boston, of course, is not the only city that intentionally dumps its sewage into a bay. Virtually every coastal city in California, the United States, and in the world, discharges untreated to treated sewage of varying levels into the sea.

Coastal waters, however, are also regularly affected by *accidental* sewage spills. Accidental spills often occur because cities operate too close to the edge, by not keeping up with population growth and operating near maximum capacity with no backup storage. This is what caused the San Diego sewage spill in February 1992, when 681 million liters (180 million gal) of partially treated sewage—containing about 50 tons of waste—leaked *daily* into the Pacific Ocean, at a depth of 10.7 m (35 ft), just less than a mile from the shore of Point Loma. (See Figure 5-3).

An outfall located 2 miles from shore is a reasonable distance if the waste flow is adequately treated. However, San Diego with a population of over 12 million people, continues to use a 30-year-old sewage plant designed to treat and pump the waste of only 250,000 people. In February 1992, heavy swells caused the old (and overused) underwater pipe to burst. Before the broken 2.7 m (9 ft) diameter concrete pipe could be repaired, billions of gallons of sewage had spewed closer to the shore than normal. Governor Pete Wilson declared a state of emergency as officials reported high levels of bacteria at several spots along a 6.4 km (4 mi) stretch of the coast. The San Diego story should serve as a warning to the dozens of other cities that continue to operate outdated and overloaded sewage systems.

Faced with a similar problem, though on a smaller scale, the 1989 Loma Prieta earthquake damaged seals in the sewage outfall pipe that services the city of Monterey, located on the shores of the MBNMS. For two months, sewage was dumped closer to the city's shore than under normal conditions. To help upgrade the system, the city of Monterey built the Monterey Regional Waste Water Treatment Facility (MRWWTF) which now replaces the smaller (and less sophisticated) treatment plants at Seaside, Fort Ord, Salinas, Castroville, and Monterey.

Several major problems arise from the discharge of sewage into coastal waters, including loss of aesthetics, disease, deoxygenation, enrichment, and toxicity (Carter 1988). Although *loss of aesthetics* is the least serious environmental problem associated with sewage in coastal waters, there is no doubt that it elicits the greatest public outrage and the most calls for immediate action. This is easy to understand since no one likes to smell obnoxious odors nor swim in the presence of fecal matter. Prior to the 1988 extension of the Santa Cruz outfall, for example, scuba divers and surfers complained of being able to see sewage bubbling offshore in water only 12 m (40 ft) deep. This spot, just off West Cliff Drive in Santa Cruz, was even nicknamed "sewage hill" by local surfers. Residents on West Cliff Drive still occasionally get a whiff of sewage effluent as it is discharged offshore.

Disease associated with sewage-tainted water is also a major environmental concern. Bacteria-tainted water can expose swimmers to gastroenteritis, which can bring on vomiting, diarrhea, stomach pain, nausea, headache, and fever. Eye problems and respiratory infections may occur. During the 1970s, surfers in the Pleasure Point area (Santa Cruz County) regularly complained of rashes and illnesses from contact with sewage in the water. Since scientists still debate about the epidemiological effects of sewage discharge, it is not certain that all of the medical concerns of the Santa Cruz surfers were linked directly to the sewage discharge. Most patients (and many doctors), however, suspect some link between sewage-tainted water and health problems.

Sewage can also cause environmental problems through deoxygenation (taking away oxygen) or enrichment (adding nutrients). *Deoxygenation*—the depletion of dissolved oxygen in water—occurs because most bacteria and microorganisms in sewage consume lots of oxygen; scientists refer to this "need for oxygen" as biological oxygen demand (BOD). Without enough dissolved oxygen in the water, fish can suffocate and die, resulting in mass kills. (Of course, this is more true in restricted bodies of water such as lakes, bays and harbors, than in ocean or coastal waters where mixing and oxygenation occurs.) Sewage also brings phosphates, nitrates, and other nutrients to a body of water. When the water body is "overfertilized," the enrichment may result in algae blooms and other forms of eutrophication. **Eutrophication** is usually considered a symptom of pollution, whereby addition of excess nutrients leads to excess growth of algae and depletion of oxygen, which may in turn lead to unpleasant odors and mortality of aquatic animals and vegetation.

Finally, sewage may also contain toxic wastes—chemicals that are poisonous to humans and aquatic animals (e.g., mercury and other heavy metals and polychlorinated biphenols—commonly known as PCBs). Whether they entered the sewage system by inadvertent or clandestine means, these toxic wastes can be taken up by microorganisms and organisms in the food chain. Eventually, toxins can concentrate in fish and be passed on to humans at the top of the food chain.

Coping with sewage disposal problems can even be problematic in small coastal communities. For example, in 1992, Monterey and Santa Cruz counties—the two counties that surround Monterey Bay—had only

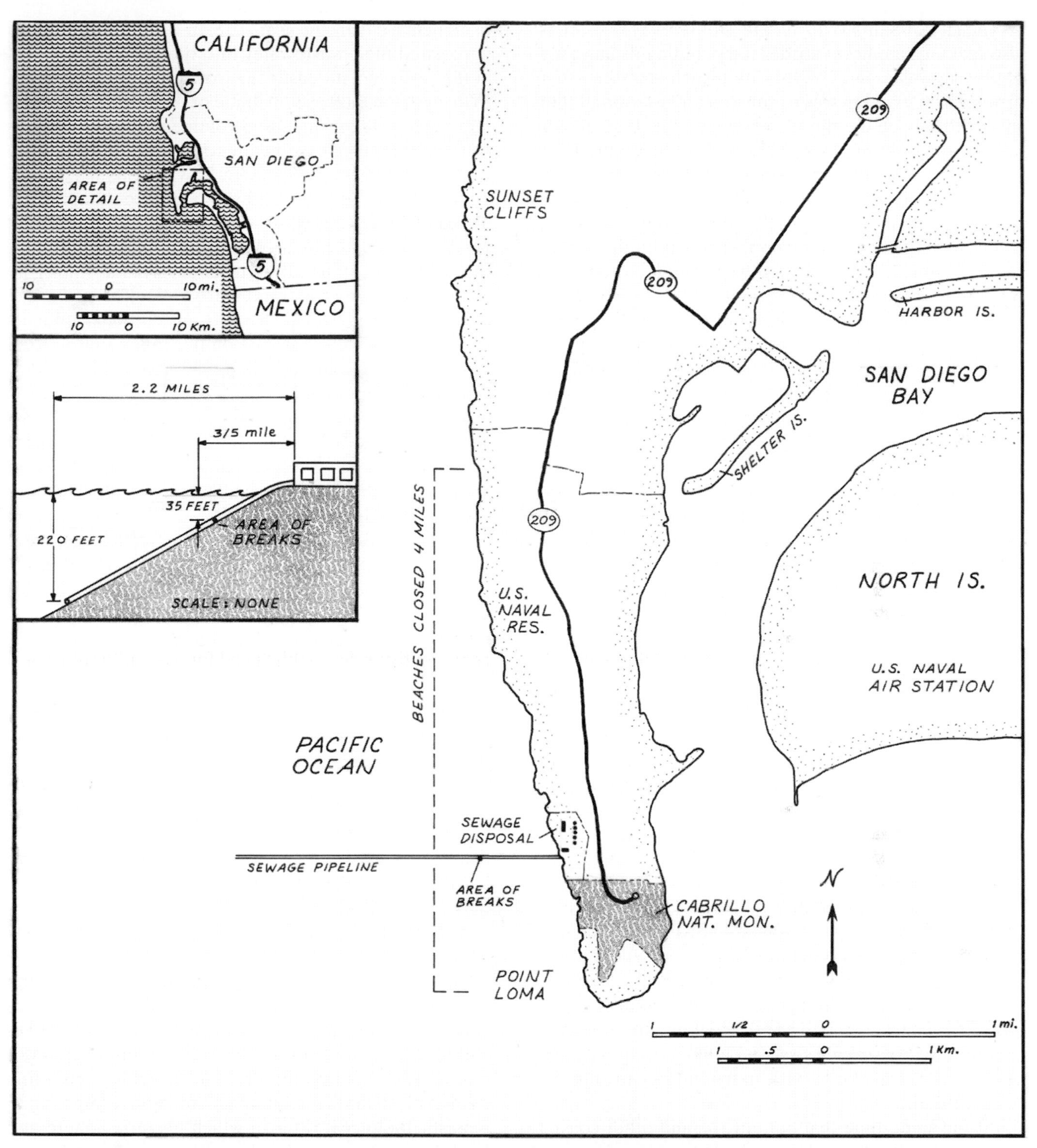

FIGURE 5-3 Sewage spill along San Diego coast. (*Source:* Diagram by Char Holforty)

544,000 people, but the region's population is expected to reach 755,000 by the year 2005 (AMBAG 1987). Before recently upgrading to a secondary treatment plant in November 1997, the Watsonville and Seaside Wastewater Treatment facility flushed sewage waste that received only **primary treatment**—the first stage in wastewater treatment, wherein only floating debris and solids are removed. In primary treatment, the remaining waste stream is often chlorinated before being flushed out to sea.

The city of Santa Cruz on the Monterey Bay shore is also making significant strides in upgrading its handling of sewage waste. In 1988, the city launched a three-step process. The *first step* was to construct a new outfall that was further from the shore and deeper; the original outfall was only 610 m (2,000 ft) from shore and only 12 m (40 ft) deep. The new outfall extends to approximately 2.4 km (1.5 mi) offshore, and discharges waste at a depth of 34 m (110 ft) of water. In 1992, both systems were still in use, with the older system only being used during the rainy season. The *second step* was completed in 1991, when the primary treatment facility at Neary Lagoon in Santa Cruz was upgraded to an "Advanced Primary Wastewater Treatment Plant," which meant that polymers were added to enhance flocculation, hence settling. The **sludge** (remaining solid or semisolid by-product of wastewater treatment) is collected and heated, with the resulting gases used for electricity to power a good portion of the plant. After the gas is extracted, the sludge (approximately 50 tons a day) is hauled to the city landfill. The *final step,* the conversion of the facility to a secondary treatment plant, is projected to be completed by late 1998. The new **secondary treatment** facility will use mechanical means and bacteria that consume organic wastes, thereby increasing the removal of suspended solids up to 95 percent. The treated water is then chlorinated again.

Some scientists and coastal communities are beginning to question the wisdom of using a secondary treatment plant. Clark (1996), for example, argues that a secondary treatment plant is nothing more than a "nutrient refinery" and "delivery system." Though functioning secondary treatment plants may reduce the organic load of wastewaters, Clark maintains that secondary "treatment" also converts wastes from a relatively benign organic state to dissolved phosphate or nitrate—a refined or more active state readily taken up by nuisance plants. (This is a greater problem in restricted bodies of water [e.g., lakes, bays] than in the ocean.) Whether it is increasing doubts about the effectiveness of secondary treatment plants, or whether secondary treatment plants are too expensive to build and maintain, many coastal communities are taking a second look at using only primary treatment plants with extended ocean outfalls (Clark 1996).

At the same time that Santa Cruz and other coastal cities around the United States are trying to solve their local sewage problems, *adjacent inland cities* are looking for ways to get rid of their increasing sewage loads. For example, the inland cities of Morgan Hill and Gilroy in Santa Clara County, California, are in the process of expanding their existing sewage facility to meet projected population growth. They plan to eventually increase the flow capacity of their plant from 6.1 to 15 million gallons of effluent a day by the year 2018. In 1992, these two cities percolated their wastewater into the ground via ponds at the sewage plant east of Gilroy. The question is what to do with the increased volume of wastewater. This system can handle the increased volume during the dry season, but an alternative needs to be developed for the rainy season when the sewage flow is at its maximum. Santa Clara Valley officials have proposed dumping the increased volume of wastewater into the Monterey Bay via the Pajaro River near Watsonville. Concerned about the possible contamination of underground water (the source of drinking water in Pajaro Valley), as well as the concern over the possible harm from additional effluents into Bay waters, the Association of Monterey Bay Area Governments (AMBAG) filed a lawsuit in 1986 to stop the project. A six year legal battle ensued, but Monterey lost. From Monterey's perspective, the only positive aspect of this decision is that perhaps Monterey County farmers will be able to use some of the highly treated water from Gilroy and Morgan Hill for irrigation on their drought-stricken farmlands.

Our nation's beaches are also affected by accidental sewage spills *from other nations.* In 1994, for example, 15 miles of beach from the United States-Mexico border north were polluted after a Tijuana sewage spill. Beaches from Imperial Beach to Coronado were closed, drawing attention to the ongoing problem of pollution from Tijuana, where the population growth has outstripped the ability to handle sewage. The Tijuana facility was shut down temporarily to carry out regular maintenance, thereby diverting over 95 million liters (25 mil gal) of raw sewage into the Tijuana River. The Tijuana River begins on the Mexican side of the border but empties into the Pacific Ocean on the U.S. side. The result: U.S. county health officials had to temporarily close the beaches for fear of an outbreak of hepatitis and gastrointestinal diseases.

Solutions to these sewage treatment problems are being sought. In Tijuana, Mexico, a joint venture between scientists and citizen groups from both Mexico and the United States developed *Ecoparque.* This low-tech, inexpensive, wastewater treatment facility is being tested as a method to reduce sewage problems within the border region. This experimental treatment facility uses a fine stainless steel mesh screening unit to remove large solids, a plastic biological filter to which bacteria adhere and feed on organic matter, and a settling pond where more solids are removed and transferred to a composting unit. The treated water and compost produced with this method may have landscaping and agricultural uses. The result is a facility which treats water, produces fertilizer, and provides wildlife habitat (Luecke and De La Parra 1994). Eventually, this treatment facility may be expanded to include artificial wetlands for additional filtering.

Industrial wastewater. Industrial facilities also create point source pollutants. For example, industrial outfalls regularly pour highly toxic organic chemicals (e.g., industrial solvents, detergents, used oil) and toxic metals (e.g., mercury, lead, chromium) into 32 rivers and streams that flow to the eastern shores of the U.S. (Basta et al., 1985). Not surprisingly, the highest levels of chemical pollutants are found in localized "hot spots" around urban and industrialized areas (Group of Experts 1990). There are numerous examples of hot spots where coastal sediment deposits reflect industrial waste disposal histories. For example, New Bedford Harbor in Massachusetts contains sediment contaminated with PCBs generated in the manufacture of electronic capacitors from the 1950s through the early 1970s.

Chemical contaminants can affect marine organisms at several levels, such as metabolic impairment or damage at the *cellular level,* physiological or behavioral changes at the *organism level,* changes in mortality or biomass at the *population level,* and changes in species distribution or altered trophic interactions at the *community level* (McDowell 1993). Recent studies have shown a clear association between diseased fish and chemically polluted sites (NOAA 1990a). Industrial waste chemicals have caused lesions and tumors in fish (Sindermann 1996), and chemicals have caused outbreaks of fatal viral diseases affecting whales in the St. Lawrence Seaway (Shabecoff 1988) and dolphins in the Atlantic off the United States and Canada coast (Cowell 1991).

Scientists now believe that the effects of global industrialization pose as great a threat to the world's marine mammals as the huge factory whaling fleets of the past. They point to a sharp increase in marine mammal mortalities in recent years. Six mass deaths of dolphins and seals, sometimes numbering in the thousands, were recorded worldwide from 1987 to 1995. There were only four recorded mass deaths in the previous eighty years. Though the immediate cause of these deaths were viruses, pollution is indicated as weakening the creatures to the point where the viruses were able to kill them (Harmstead 1995).

Some coastal communities, such as along the MBNMS, are fortunate because they are not ringed by dozens of heavy industries that pump effluents into their coastal waters. In fact, the MBNMS has only two industrial waste outfalls, both of which are at Moss Landing. The Pacific Gas and Electric fossil fuel power plant uses an average of 962 million gallons per day of sea water for cooling. The plant has an outfall that extends 183 m (600 ft) from the shore and discharges (in 1985) 961 million gallons per day of cooling water at a depth of 14 m (45 ft) (NOAA 1990b). Adjacent to the PG&E plant, the National Refractories and Minerals Corporation (formerly Kaiser Refractories) operates a sea water magnesia plant. After removing magnesium, the plant discharges seawater containing magnesium chloride with an altered ionic composition. The outfall extends 189 m (620 ft) from shore at a depth of 13 m (43 ft) (NOAA 1992). With a few exceptions, current discharges from both plants were in compliance with the guidelines set by the State Water Resources Control Board that monitors water pollution along the California coast.

Solid waste disposal. Some coastal cities, like New York, used to use large barges to transport their municipal garbage for burial at sea. Other cities, like Los Angeles, once used outfalls to discharge sewage sludge into a nearby bay. Whether it's garbage, sludge, or some other form of non-liquid discarded material, these types of solid waste disposal come under the heading of point source pollutants. In the Los Angeles example, the disposal of sewage sludge has smothered and destroyed many indigenous kelp communities, reducing the ecological diversity (Carter 1988). Fortunately, the coastal communities adjacent to the MBNMS use local landfills, not the Pacific, for the disposal of garbage and sewage sludge.

See the following chapter, Chapter 6—*Ocean Dumping,* for further details on the environmental problems and management strategies associated with solid waste disposal along U.S. coastlines.

Nonpoint Sources

Our nation's shorelines also receive pollutants from nonpoint sources, including runoff from harbors and marinas, agriculture lands, forest lands, urban areas, and marine debris. Less obvious, but nevertheless a source of nonpoint pollutants, are air and noise pollution. Along most U.S. coasts, including areas with marine sanctuaries, nonpoint source discharges are the major source of coastal pollution. A brief explanation of these now follows.

Harbors and marinas. Harbors and marinas, of course, are places where boats or ships are berthed and maintained. Where there are people and boats, and especially where there are repair yards, maintenance activities such as engine repair, sanding, varnishing, painting, and cleaning add pollutants into the waste stream and ultimately into coastal waters. Repair yards (also known as boatyards or shipyards) are areas where boat owners have their vessels maintained periodically. Although all kinds of repair work can be done in a boatyard, most small boat owners "haul out" every 1–2 years to have the hull (boat bottom) scraped, sanded, and repainted. Because boats and ships require constant maintenance, there is the constant discharge of waste from these facilities. Despite precautions, the electric sanders and pressurized spray painters disperse toxic particles into the air and on to the ground, and much of it enters nearby waters. (See Figure 5-4).

Even more difficult to regulate is what the individual boat owner does on an almost weekly or monthly basis while the boat sits *in the water* in its slip (berth). Although more and more environmentally friendly maintenance products (e.g., biodegradable boat soap) are coming out on the market, individual boat owners can still be seen using cleansing detergents, oils, varnishes, or acids to maintain their boats (including the author). And, unknown to most non-boaters is the fact that *all* harbor craft (from small recreational fiberglass sailboats to large military ships) have highly toxic antifouling paint on their hulls to ward off barnacles and other hull-loving marine organisms. This antifouling paint slowly releases poisons directly into the water, one of which, **tributyltin (TBT)** can be highly toxic to marine organisms in concentrations of less than 1 part per billion (Cox 1993). TBT has been shown to cause mortality and deformation of oysters and other forms of marine life in bays. Unfortunately, potentially high concentrations of TBT have been detected in bays and harbors in several locations in North America (Goldberg 1986; Champ and Lowenstein 1987). In the United States, the Environmental Protection Agency (EPA) has now restricted the use of TBT paints only to aluminum vessel hulls, or to vessel hulls greater than 25 m (82 ft) in length. According to Clark (1996), the use of antifouling paints is mainly a local problem (e.g., harbors and small bays), not a major coastal problem.

Another wave of controversy regarding harbors, vessels, and coastal pollution has to do with the recent debate over *powerboat pollution.* Until recently, no one thought of the two-cycle outboard motor as a source of great pollution. Speedboats pulling water skiers on lakes, reservoirs, and bays have been synonymous with summer vacation for many Americans. Just how much pollution is created by recreational motorboats each year, and what, if any, harm is done to the environment? According to the EPA's 1991 report entitled *Nonroad Engine and Vehicle Emission Study,* nonroad engines (e.g., boat motors, mowers, chain saws, skimobiles) contribute more than 19 percent of the hydrocarbons, 15 percent of the nitrogen oxides, and 14 percent of carbon monoxide released into the nation's air each year. Furthermore, anyone who owns a powerboat knows that unburned gasoline and lubricating oil can always be seen spreading out on the surface of the water body from the boat's motor when it is in use. Some have estimated that as much as one-third of all the fuel that passes through the motor actually ends up in the water. In his 1993 book, *Polluting for Pleasure,* former boat designer, Andre Mele, estimates that 150 million to 420 million gallons of unburned fuel are exhausted into the environment each year by the nation's 12 million gas-powered pleasure boats. Studies of the environmental impact of this type of pollution are either rare or nonexistent. By comparison, the *Exxon Valdez* spilled about 10 million gallons of crude oil into Prince William Sound in Alaska. One major difference, of course, is that the Valdez spill occurred in a short period of time in one location, not dispersed over time and place.

Beginning in 1998, the EPA will require that new boat engines burn cleaner and emit less pollution. The aim is to cut hydrocarbon emissions by as much as 75 percent. While some marine engine manufacturers are taking a "wait and see" approach, others are conducting research on cleaner burning engines, such as the direct-injection two-cycle engine that is 60 to 80 percent cleaner than today's models. These companies hope to offer a cleaner but more costly engine before the deadline.

Another form of coastal "pollution" that threatens native marine life, particularly in bays and harbors, consist of organisms that invade U.S. waters by hitchhiking a ride in the ballast water of foreign ships. These invasions are taking place on a vast and previously unrecognized scale in all of the world's ports, with San Francisco Bay among the hardest hit. When the foreign ships arrive in port and empty their ballast tanks to make room for cargo, they inadvertently transplant a few species of plants and animals from the distant ports. In the San Francisco area, for example, the Asian clam (*Potamocorbula amurensis*) has taken over the bottom of Suisun Bay since it arrived in 1986. It grows fast and feeds voraciously, sucking virtually all the nutrients out of the water. Other recent invaders in San Francisco Bay linked to ballast water include the hydroid (*Cladonema uchidal*) which arrived from Japan and China in 1979, the Asian clam (*Theora fragilis*) which arrived from Asia in 1982, and the Atlantic green crab (*Carcinus maenas*) which arrived from Europe in 1990. According to local oceanographers, the problem is acute in San Francisco Bay, where the ecological damage has been accumulating for more than a century. At least 225 foreign species are now living in the bay's waters. The problem has been occurring for so long, in fact, that oceanographers have no clear idea of what the natural fauna of San Francisco Bay was like before European settlement.

The same holds true throughout the world. Invasions have taken place on a large scale since the 1880s, when ships began using water as ballast. Researchers will have to reconsider what they think is truly native to an area. The problem is also accelerating, as ships grow bigger and faster and as new trade routes open to places such as China. Ironically, the more pollution is cleaned up, the more hospitable the water becomes to this other form of pollution—foreign marine organisms. To help remedy the problem, the International Maritime Organization, a United Nations group based in London, has issued guidelines calling for ships to flush out their ballast tanks in

FIGURE 5-4 (a) Boat owner does repair work on his sailboat at Breakwater Cove Marina near Monterey Harbor in Monterey Bay, California. (b) More distant photograph of the same boater doing repair work. No matter how careful he may be, contaminated air particles and wastewater from the repair work settles right into the adjacent bay waters. (*Source:* Photos by author)

midocean and refill them with ocean water, which does not contain organisms that would thrive in a coastal port (Carlton and Geller 1993).

Alien species are also accidentally introduced into waters when they escape from an aquaculture or maraculture facility. For example, Louisiana crayfish imported to the Dominican Republic for aquaculture escaped into adjacent waterways by burrowing through the dikes of it's holding pond (Clark 1996). Exotics are even deliberately introduced on occasion, such as when "pet owners" release no longer wanted aquarium fish into local streams. The problem, of course, is that these exotic species may reproduce to the detriment of endemic species.

Agricultural lands. Agricultural operations in coastal watersheds also contribute to the degradation of shoreline waters by introducing sediments, fertilizers, and pesticides. Sediment is the main nonpoint pollutant by volume of agriculture activities (Clark 1996). Although sediments can originate from other sources such as runoff from forest lands, its effect on the environment is the same. Erosion of soils is the single most significant factor to directly affect coastal waters. Cloudy waters reduce light penetration into the water, therefore decreasing photosynthesis. Suspended particles settle on benthic organisms and smother them. Coral reefs are particularly sensitive to sediment pollution.

Fertilizers and pesticides can enter coastal waters via runoff or groundwater infiltration. Farm fertilizers are rich in nitrates and phosphates, and are a significant contributor to nutrient runoff and **cultural eutrophication** (accelerated enrichment or aging of a water body because of human activities, such as farming). Agricultural runoff also carries pesticides that drain into bays and estuaries. Fertilizers and pesticides are particularly hazardous in confined bodies of water, such as in estuaries where there is less flushing out and dilution than in open waters (Crutchfield 1987). Pesticides, of course, can have a deleterious effect on aquatic life. For example, one half of the 128 coastal "fish kills" off the South Carolina coast between 1977 and 1984 were attributed to pesticides (Trim 1987). Regular consumption of fish and shellfish in heavily contaminated areas also poses a serious health risk to humans, particularly fetuses and children (Committee on Evaluation of the Safety of Fishery Products 1991).

Another major contaminant of coastal watersheds is manure that has accumulated outside barnyards. Concentrated nutrients such as the nitrogen that washes from rich heaps of manure can be poisonous to fish and can damage aquatic productivity. Even drinking water supplies can be contaminated. The problem is so acute that Congress for the first time is addressing this issue as well as other forms of agricultural runoff in its revision of the Clean Water Act. Furthermore, new business ventures like Biorecycling Technologies Inc. (BTI) in California are working to help solve the manure problem and turn a profit at the same time.

BTI realized early that California is drowning in cow manure. There are now more than 1.2 million dairy cattle in the state, which in 1993 overtook Wisconsin as the nation's leading producer of dairy products. Those cows are creating so much manure that in some places it is piling up into pungent mountains that must be compacted and transported by heavy trucks. In the meantime, rains come and wash the nutrients into creeks and streams and eventually to coastal bays and ocean waters. BTI launched a "manure-based venture" devoted to transforming animal wastes into a line of trouble-free, environmentally friendly fertilizers. Some nurseries on the Central Coast have experimented with BTI's solid fertilizer, dubbed "NutriTex." They found that not only did the material perform well as a soil ingredient, but it also contained beneficial microbes that controlled fungus and other diseases in potted plants, reducing the need for chemical treatments. As the manure problem literally continues to mount, no doubt other venture capitalists will attempt to turn a waste product into a resource—which, of course, is exactly what is needed.

A wide variety of pesticides, including endosulfan, chlordane, and DDT, have been entering the MBNMS for many years. Of particular interest, and concern, is the fact that undegraded DDT from past legal agricultural use is still appearing in Monterey's waterways. Shaw (1972), for example, found that DDT and its degradation products were in the tissues of eight species of marine fishes in Monterey Bay. Though it was banned in 1972, over 25 years ago, DDT still persists in some wetlands (e.g., Elkhorn Slough) and some harbors (e.g., Moss Landing Harbor) that ring the Monterey Bay National Marine Sanctuary. Unfortunately, the process of **bioaccumulation** occurs when DDT is introduced into an ecosystem. This means that it accumulates in invertebrates and microbial organisms, which are then eaten and accumulated in the organs of the larger organisms and animals. The pesticide can in this way remain in the food web for generations—the reason that DDT can still be found in the mussels of Elkhorn Slough today.

Forest regions. Such forestry practices as cutting trees and building access roads can also impact watersheds, and ultimately, coastal waters. When hillsides are stripped bare of their vegetation, the underlying soil is exposed to wind and rain. The *sediments* are then transported by runoff to nearby streams that feed coastal waters. Coral reefs, seagrass, and other sensitive coastal habitats can be smothered by these sediments (Group of Experts 1990). An interesting study on the relationship between forestry practices as they affect coastal waters and fisheries in Bacuit Bay, the Philippines (Hodgson and Dixon 1988) noted that poor upland logging practices led to increased sedimentation of reefs in Bacuit Bay, thereby reducing live coral cover as much as 50 percent. This, in turn, resulted in further declines in both coral species diversity and fish biomass.

Urban areas. Urban runoff is also an important nonpoint source of plant nutrients and toxins. In this instance, nitrogen, phosphorus, and various toxins are added to water bodies from such sources as gardening fertilizers, pesticides, and herbicides; urine and feces from domesticated pets; leaf litter; and cleaning agents used to clean driveways, automobiles and other surfaces. Adverse effects of nutrient and chemical pollution are numerous. One of the first signs of *nutrient pollution* is the formation of algal blooms—the proliferation of algae due to "overfertilization" on the surface of streams, ponds, or lakes, resulting in eutrophication. (Keep in mind that the coastal and estuarine environment does not just include the sea coast, but streams, sloughs, ponds, lagoons, and lakes as well [see "Point Sources" earlier in this chapter for further explanation.]) Management for this problem may involve herbicide use, which, along with other urban chemical run-off, creates aquatic *chemical pollution.* Chemical pollution, which often destroys aquatic plant and animal life, may exacerbate eutrophication, or become a problem in its own right, by creating losses of dissolved oxygen through the action of decomposition. Of course, the loss of plant and animal life is a problem unto itself. These problems are acute, cumulative, and catalytic, resulting in major water quality, aquatic wildlife habitat maintenance, and watershed management problems.

If left uncontrolled, urban runoff will be an increasing source of nonpoint pollutants to coastal waters in the future for two reasons. First, by the year 2000, demographers predict that one fifth of the world's population—1 billion people—will live within coastal cities (World Resources Institute 1992). This concentration of people on the coast cannot help but degrade coastal waters. Second, researchers have established a correlation between a watershed's human population density and the amount of pollutants (e.g., nitrogen) at the mouth of the river that feed into a particular bay, estuary, or coast (Peierls et al., 1991). This is significant because most of the world's population lives in watershed areas (World Resources Institute 1992).

In Florida, urban and agricultural development have severely impacted the Everglades watershed. It is believed that excess phosphorus in runoff from the sugar industry is degrading marshes and killing native plant species, which in turn, result in native fish and animal die offs. Also, the present configuration of canals to supply the water needs of agriculture and the cities has left the Everglades under supplied with fresh water. Furthermore, massive die-offs of plants, animals, and coral reefs have seriously threatened Florida's leading industry—tourism. It is estimated that 75 to 95 percent of the native wildlife is now gone. Major studies are now underway to assess the impacts and determine what can be done to restore the Everglades to its natural state (*Sun Sentinel* 1995).

Other coastal cities are beginning to look at the larger issue of storm runoff and whether and to what degree they are polluting coastal waters. For example, The Association of Monterey Bay Area Governments (AMBAG) is currently developing an "Urban Runoff Water Quality Management Plan" for the Monterey Bay Region. Project objectives include (a) investigating the impact of storm water runoff from urban areas on water quality in the urban coastal areas from Santa Cruz to Point Lobos (near Monterey), (b) preparing a management plan to reduce potential water quality degradation of Monterey and Carmel Bays, and (c) providing regional coordination to address potential water quality degradation from nonpoint source pollution. See the case study in this chapter for an example of a water quality management program.

Marine debris. Of increasing concern is the problem of marine debris that is appearing in our coastal waters and washing up on our beaches. Until recently, *marine debris*—any human-made object of glass, wood, plastic, cloth, metal, rubber, or paper that is present in the marine environment—was of little concern compared to the other forms of point and nonpoint pollutants. But today, everything from false teeth to hypodermic needles to grenades are washing up on U.S. shores. The Center for Marine Conservation sponsors an annual International Coastal Cleanup Day. In September of 1993, more than 200,000 volunteers from 36 U.S. states and territories and 55 countries scoured over 8047 km (5,000 mi) of beaches and waterways to gather over 2,400 tons of trash as well as record the types and quantities of the debris they collected (Center for Marine Conservation 1994). The "Top Trash"—the 12 most frequent pieces of garbage collected across U.S. beaches in this 1993 CMC Sixth Annual International Coastal Cleanup—was as follows: Cigarette butts; paper pieces; plastic pieces; foam-plastic pieces; broken glass; plastic food bags; plastic caps and lids; metal beverage cans; plastic straws; glass bottles; plastic bottles; and foam plastic cups. Cigarette filter butts easily won out as the greatest number of any one item present.

Plastics are a special case, since a growing proportion of marine debris consists of plastics which deteriorate slowly, can break up into smaller pieces, and can float long distances from their dispersal point. Six-pack plastic rings for beer or soda, styrofoam cups, plastic ice bags, and plastic straws can be especially harmful to sea birds and other marine wildlife. Where does all this litter come from? Most of this litter used to come from boats and ships that dumped their trash overboard, not from the weekend beachgoer. According to the 1994 study mentioned above, ship galley waste declined from more than 5 percent in 1988 to less than 2 percent of the total items collected. This may be a result of the 1988 ban against throwing plastics overboard. For the specific rules and regulations regarding vessel ocean dumping, see Chapter 6, *Ocean Dumping*.

Marine debris also arrives on our coastlines from inland sources—from items purposely discarded or accidentally washed into creeks and rivers within coastal watersheds. For example, in the Monterey Bay area, one merely has to stand at the mouth of the San Lorenzo River in Santa Cruz to see human-created debris floating down the river out to the MBNMS.

Air pollution. Coastal water pollution from air borne contaminants has only been recognized recently. According to the Group of Experts on the Scientific Aspects of Marine Pollution (1990), almost all of the cadmium, copper, lead, zinc, and iron that enters the world's oceans comes from the atmosphere. Much of the nitrogen that enters our coastal waters arrives from the atmosphere in the form of acid rain (Environmental Defense Fund 1988). An average of 27 percent of the total nitrogen inputs into the Chesapeake Bay are due to atmospheric deposition from distant agricultural activities, industrial facilities, mobile sources, and power generating plants.

Although 27 percent is the average for the Chesapeake Bay, some of the Bay's estuaries can reach as high as 40 percent in North Carolina, or as low as 15 percent in Delaware with the difference attributed to varying levels of point-source nitrogen pollution (Linker 1996). Acid rain and airborne toxics have affected other coastal areas and the Great Lakes (Lee 1990). As with many nonpoint pollutants, cleansing our coastal waters of airborne contaminants will require attacking the source. One of the major sources of airborne lead was the burning of gasoline. In the 1970's, lead was phased out of gasoline due to concerns about its toxic effects on the environment and human health. Since the phasing out, water quality studies using mussels and oysters in U.S. coastal waters have shown decreased lead levels, indicating an overall decrease of lead contamination in the marine environment (NOAA 1990a).

Noise pollution. Human-produced noise may be the latest, and least recognized, form of nonpoint coastal pollutant. In 1989, the National Wildlife Federation reported that scientists are beginning to find evidence that sounds from vessel traffic and offshore oil drilling can interfere with the behavior, growth, and reproduction of marine life. For example, one scientist, Arthur A. Myrberg, Jr. of the University of Miami, maintains that bowhead and California gray whales have been known to change their migration routes to avoid drilling ships or other sources of noise. Furthermore, he notes that other studies have shown that noise can damage fish eggs and reduce reproduction rates among several species of fish and shrimp. Professor Myrberg even theorizes that some marine species may be attracted to oil rigs for food and shelter, then later become deaf and unable to function normally in other waters. Additional scientific studies are needed to verify these conclusions.

One of the distinguishing features of the human species is that it is the noisiest organism on the planet. No other organism has produced the noise levels created by roaring jet aircraft, blaring amplifier systems, and the constant background noise of freeway traffic. In March 1994, the citizens on the shores of the MBNMS began to debate whether humans should be allowed to intentionally broadcast loud sounds in a National Marine Sanctuary (NMS). The debate centered around a scientific climate monitoring experiment known as the *Acoustic Thermometry of Ocean Climate (ATOC)* project.

The ATOC project was a federally sponsored international research initiative to measure long-term ocean climate changes on global scales using deep ocean acoustic sound paths to precisely measure synoptic ocean temperatures. Experiments were to be made with several acoustic sources and a network of receivers throughout the Pacific over a two-year period (1994–1995) and to assess the feasibility of an international global ocean climate monitoring effort. To accomplish this goal, ATOC would place two acoustic sound transmitters on the ocean floor to emit low-frequency (60-90 HZ) sound signals at approximately 195 decibels (at 210 watts of power). The transmission signals would have continued for twenty minutes each, and would have been repeated every four hours at a maximum rate of 6 times per day. The rumbling noise would have been generated from two transmitters—one located within the MBNMS, 850–950 m (2788–3116 ft) deep at 42 km (26 mi) west of Pt. Sur; the other at 14 km (8.6 mi) north of Kaihu Point, Kauai, Hawaii. (See Figure 5-5).

The rumblings would have been picked up by 18 receivers scattered 4828–9656 km (3,000–6,000 mi) away. The biggest controversy, of course, is whether or not such a sounding device should be located within one of our nation's newest and largest marine protected areas, the MBNMS. Furthermore, the Kaihu Point location is adjacent to the Hawaiian Islands Humpback Whale National Marine Sanctuary. If this experiment proved successful, the Strategic Environmental Research and Development Program (a program established by Congress to redirect defense spending to environmental matters) would ultimately place transmitters in all the world's oceans as part of a decade-long project to measure whether the Earth—more than 70 percent of which is covered by water—is, indeed, warming.

When the general public learned of this study, the global warming project was immediately attacked. Scientists, hundreds of individuals, and several environmental groups began taking sides. Dozens of questions began to arise: Why wasn't the general public alerted to this

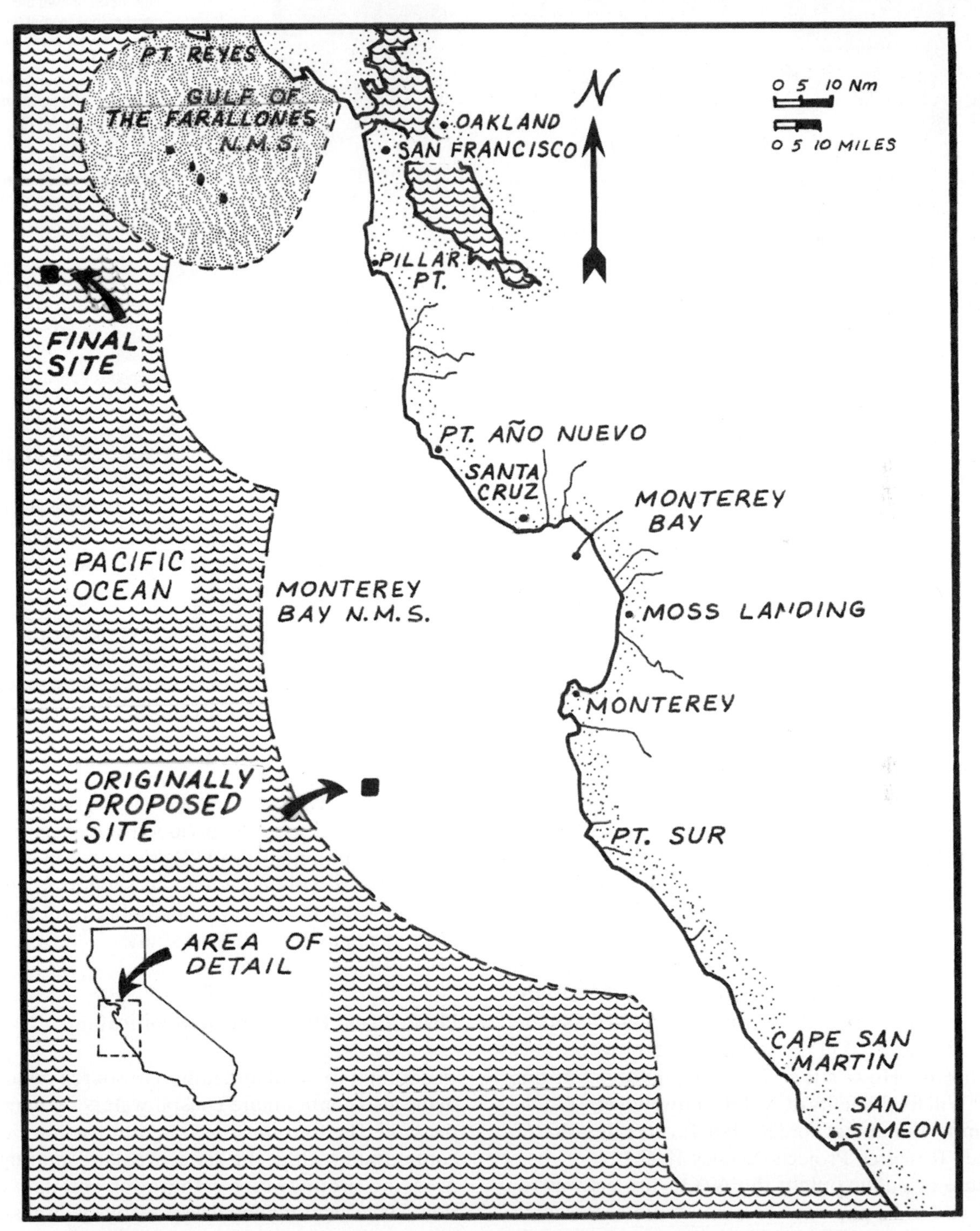

FIGURE 5-5 (a) Map illustrating initial proposed location of sound generator in the Monterey Bay National Marine Sanctuary, as well as the final location outside sanctuary waters. (*Source:* Map by Char Holforty).

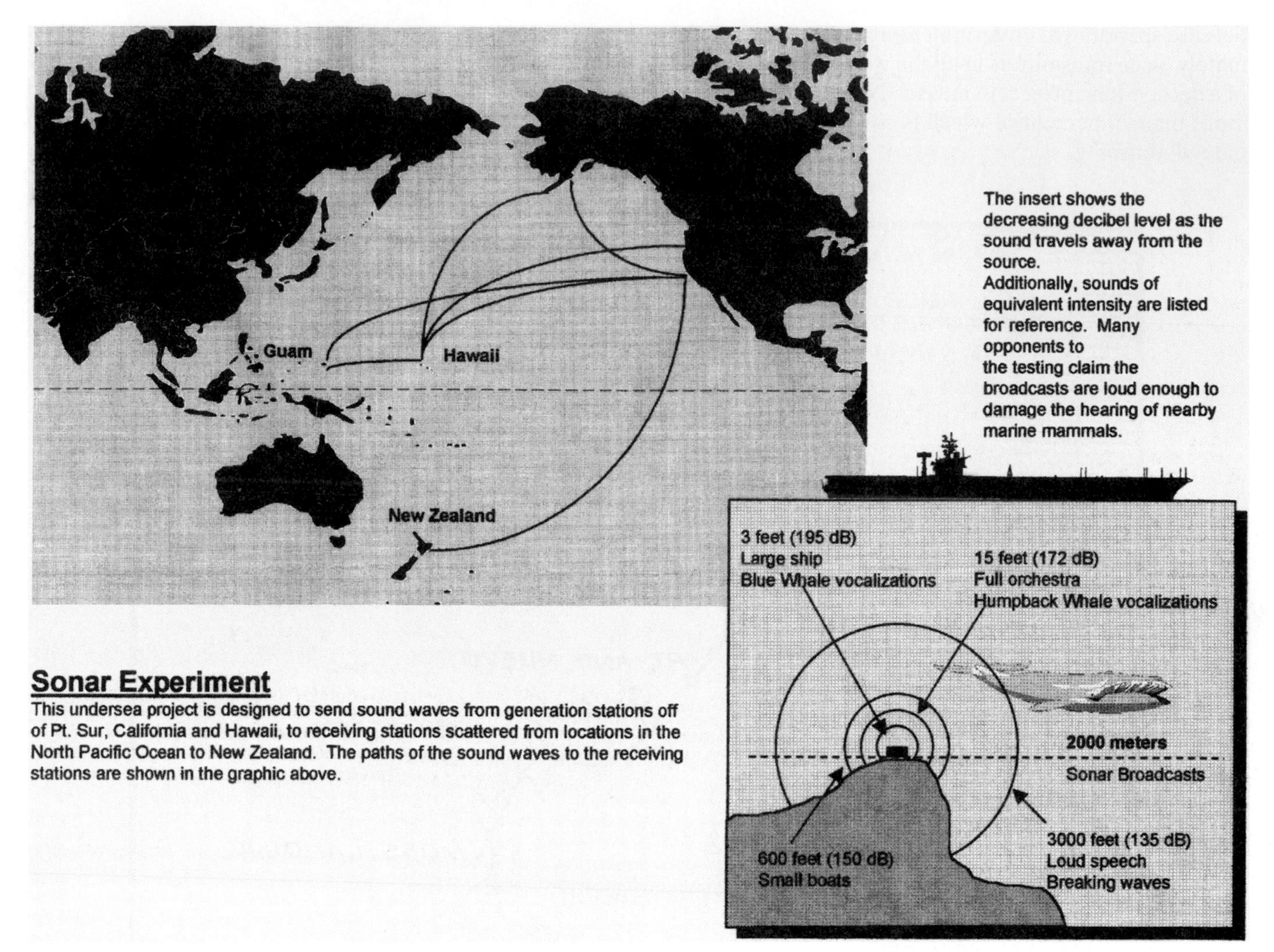

FIGURE 5-5 (Continued) (b) The Sonar Experiment. (*Source:* Map and diagram by Char Holforty)

project prior to its approval? Why use sound to measure ocean warming? Should humans purposely be creating high level sounds within a marine sanctuary? Why must one transmitter be located within a designated marine sanctuary? What are the possible effects on marine mammals? How will scientists monitor effects on marine mammals? Will the sound be a "blast" or a "low-pitched rumbling"? What does the proposed noise level of 195 dB compare to? How will the proposed sound dissipate through the water? Will the sound interfere with whale vocalizations? How is the U.S. Navy involved in this project? Will the benefits of ATOC outweigh any potential harm to the environment? For further details, see Advanced Research Projects Agency 1995.

Because of public outcry, the ATOC project was rescinded, and in its place the Marine Mammal Research Program was initiated. The MMRP is investigating the possible affects of ATOC before continuing with the project. This incident teaches us two lessons: (1) much more needs to be learned about noise pollution as it effects coastal and marine environments; and (2) scientists can no longer do as they please without community support; scientists must seek approval from the general public prior to launching any potentially controversial study.

Physical and Hydrological Modifications

In addition to point and nonpoint source pollutants, humans also contaminate coastal waters by manipulating the land. The four most prevalent ways in which coastal communities alter the land are sand mining, harbor dredging, groundwater withdrawal, and dam construction. Since sand mining was previously discussed under "coastal hazards" in Chapter 4, only the last three categories will be discussed below.

FIGURE 5-6 Beach nourishment. (a) A dredge sucks in bottom sediments at the inlet to the Santa Cruz Small Craft Harbor, Santa Cruz, California; (b) The dredged spoils from the Santa Cruz harbor entrance are then piped to "nourish" nearby Twin Lakes Beach. (*Source:* Photos by author)

Harbor dredging. Since many harbors regularly fill with sediment from streams that feed into them, or by littoral drift, they need to be periodically dredged. Dredging can destroy clam beds, grass beds, or other types of critically important habitat for commercial and recreational fish species (See Chapter 4). But in addition to the physical disruption, dredging can also resuspend heavy metals, pesticides, and other types of toxins that are present in bottom sediment, thus exposing plants, animals, and humans to dangerous contaminants. If the dredge spoils meet environmental requirements, they can be dumped at a disposal site or used for beach nourishment. (See Figure 5-6).

When the dredge spoils do not meet the criteria for beach nourishment they must be taken to an approved toxic waste site (See Chapter 6, *Ocean Dumping,* for information regarding the controversy over dumping dredge spoils from San Francisco Bay). This general practice came into question when Monterey Bay received sanctuary status. The presence of DDT under "A" dock in the South Harbor is the major management problem facing the Moss Landing Harbormaster. What can be done with the DDT laden dredge spoil now that it cannot be dumped into the sanctuary?

In May 1996, Moss Landing Harbor was given a permit by the California Coastal Commission allowing for the dredging of the inner harbor. The main entrance of the harbor is dredged every three years by the U.S. Army Corps of Engineers with the dredge spoils used as beach supplement. The inner harbor though, has a core sampling of 300 parts per billion of DDT requiring the dredge spoils to be removed from the coastal waters and not used as beach supplement, since natural coastal erosion would place these spoils back into the marine environment. The DDT-laden spoils will be "dewatered" at a temporary holding site located at the harbor before they are trucked off to local landfill in the town of Marina. The dredge spoils will be used as earthen cover for the landfill. The dredging project is expected to take two to four years before the inner harbor is completed, by which time it is suspected that the process will have to begin again. Until there is a joint effort by the upland farmers, landowners, city, county, state, and federal agencies to reduce the amount of sedimentation flowing into the harbor, the process of dredging will be only a short-term solution rather than a long-term one.

Groundwater withdrawal. Heavy pumping from aquifers (underground water supplies) for agricultural and urban use increasingly threatens the availability of fresh water in U.S. coastal communities. As farmers and urban populations increase their pumping of underground supplies, the problem of saltwater intrusion intensifies. **Saltwater intrusion** occurs when a body of saltwater invades a body of freshwater, usually resulting from the over-drafting (over-pumping) of aquifers. In Southern California, the groundwater aquifers have been so over-pumped and contaminated by saltwater intrusion that some coastal communities are close to running out of uncontaminated groundwater. In Central California, the problem of saltwater intrusion is also intensifying. In the Monterey Bay area, for example, the problem is especially acute in three locations: the Fort Ord and Marina area, in the fields between Castroville and the mouth of the Salinas River, and along both sides of the Pajaro River. (See Figure 5-7).

In 1991, approximately 6,071 hectares (15,000 acres) of the 80,940 irrigated ha (200,000 acres) under the jurisdiction of the Pajaro Valley Water Management Agency experienced leaching of saltwater into underground aquifers.

To reduce saltwater intrusion caused by agriculture, water officials in many coastal communities are looking at such alternatives as recycling decontaminated wastewater, piping water from other locations, and building new reservoirs. Water officials along the Central California coast are looking at the prospect of *using treated wastewater* from local sewage treatment plants (including the controversial plant expansion proposed by the inland

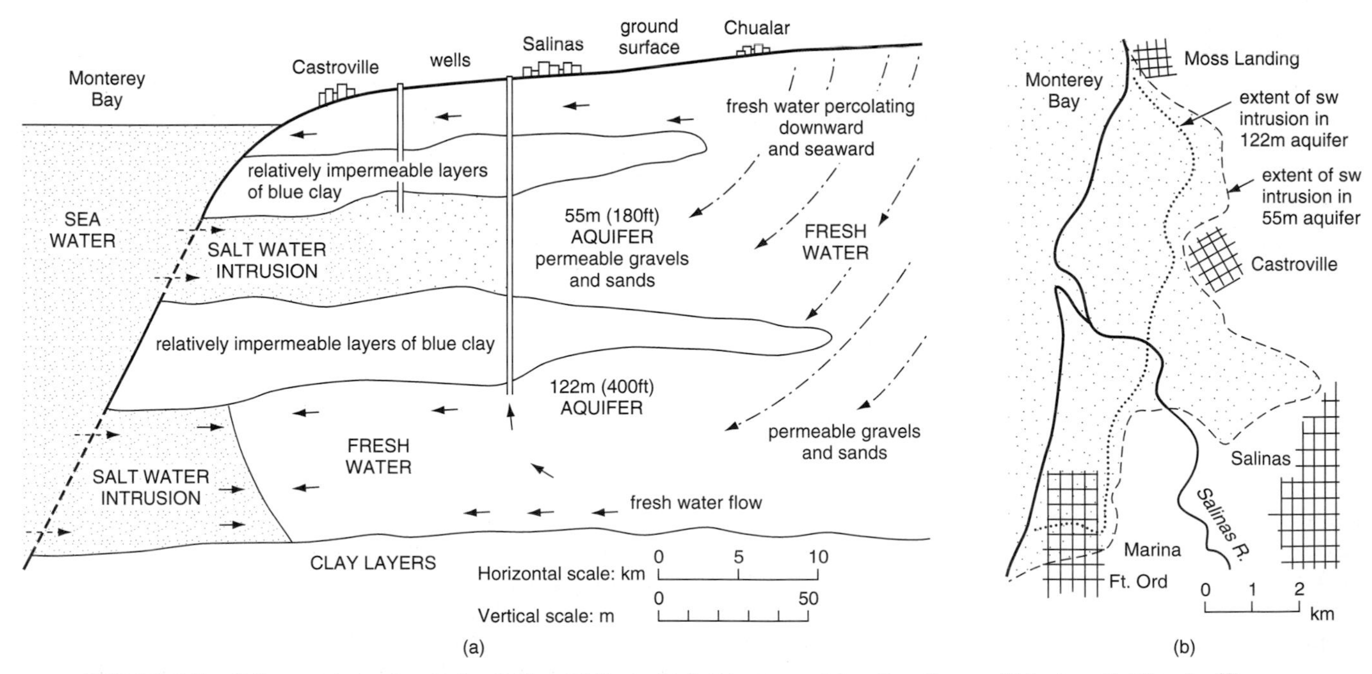

FIGURE 5-7 Saltwater intrusion in the Salinas Valley, which is between Moss Landing and Marina, California. (*Source:* Diagram by Heather Theurer. Adapted from *Monterey Bay Area: Natural History and Cultural Imprints* 3/e, 1996, Boxwood Press)

communities of Morgan Hill and Gilroy) for irrigation. A second alternative being discussed in the region is the feasibility of *tapping the San Felipe Project*—the massive pipeline system that already brings water from San Luis Reservoir to Santa Clara County and the Hollister area. Developing *new reservoirs* in the lake district east of Watsonville is a third alternative being considered.

To reduce saltwater intrusion caused by urban areas, many U.S. coastal communities are working on *water conservation* as a management strategy. In 1992, the City of Watsonville—an agricultural community in the Monterey Bay area—approved a $100,000 toilet retrofit and rebate program. To encourage Watsonville residents to replace their water-guzzling toilets with more efficient low-flow models, a $100 rebate was given to the first 1,000 residents who replaced their water-wasting toilets. Water officials hope this conservation strategy will reduce water use within Watsonville by 16 percent. Other coastal communities have considered such conservation strategies as (a) *creating laws to ban water waste,* (b) *starting conservation education programs in local schools,* and (c) *using water-saving irrigation methods.*

Another area of concern is the combined effect of excessive groundwater withdrawal and rising sea levels. For years, the citizens of Yankeetown, Florida have been struggling to discover what is causing the death of hundreds of Sabal palm and other trees on their coastal properties. Dr. Francis Putz, a botanist with the University of Florida, has found that Sabal palms along a 322 km (200 mi) stretch of Florida's Gulf Coast were dropping their fronds by the hundreds of thousands. Sick trees were also being reported on the Atlantic coast near Jacksonville and on the Cumberland Island National Seashore in Southern Georgia. One by one, diseases such as lethal yellowing were investigated as the possible cause, then eliminated. In 1992, the research was beginning to point to an increase in the salinity of the water that nourishes the trees—increased salinity levels, in part, from rising sea levels and salt water intrusion. The "Sabal palm story" may be a preview of what will happen to much coastal vegetation around the world as sea levels rise in combination with excessive groundwater withdrawal.

Dam construction and irrigation. Perhaps the most controversial structural method of water resource management is the construction of dams. For every advantage (e.g., water storage, flood control, power production), there is often a disadvantage (e.g., breeding ground for diseases, siltation, high capital investment).

From a coastal point of view, however, dam construction can cause loss of sand from beaches, block the migration of anadromous fish, prevent an adequate supply of nutrients to bays and estuaries, and introduce pollution from associated irrigation projects (Dixon et al., 1989). Many of these issues were debated regarding the proposed New Los Padres Reservoir on the Carmel River that flows into the Monterey Bay National Marine Sanctuary. Water district officials and other champions of the proposed dam argue that the new dam would solve three related problems facing the Monterey Peninsula: periodic water shortages, stagnant growth, and low river flow. The dam, they claim, will provide drought security, allow limited growth, and restore year-round river flows to bring back 4,000 steelhead trout. If built, the concrete dam would hold 24,000 acre-feet of water in a 266-acre reservoir on the upper reaches of the Carmel River near the Los Padres National Forest. It would replace an existing earthen dam, built in 1948, that creates a 55-acre lake. As of this writing, a majority of water customers on the Monterey Peninsula must approve the project.

Canal and levee construction. The construction of drainage canals and levees can also bring about drastic physical and hydrological changes. Florida Bay, the great blue outback of the Everglades National Park, is now locally referred to as the "Dead Zone." This once lucrative shrimp and lobster breeding ground has now become stinking muck from dead and dying sea grass. The bay's basic problem, according to local marine biologists, is obvious: 90 percent of the fresh water that once flowed into it from Lake Okeechobee through the Everglades has been diverted into the drainage canals and levees that turned the alligator swamps of Southern Florida into prime real estate. Proposals to solve the problem are costly and controversial (See Chapter 8, *Open Space Preservation and Management* for details). But the cost of doing nothing could be even higher. Spillover from the bay may be killing delicate coral reefs in the tourist-packed Florida Keys to the south. In 1992, Congress authorized a five-year, $9 million program to protect water quality in the Keys National Marine Sanctuary.

The following case study from the Monterey Bay region nicely illustrates the natural and cultural complexities that arise when humans even temporarily reduce the flow of a small creek that flows to the coast.

CASE STUDY: SOQUEL CREEK WATERSHED

Upon designation of the Monterey Bay National Marine Sanctuary in 1992, the National Oceanic and Atmospheric Administration encouraged the development of an Integrated Coastal Management (ICM) program for sanctuary water quality management. The process being implemented in Monterey was first developed in the Florida Keys National Marine Sanctuary, where there was a similar need to resolve water quality issues. The resultant management strategy was comfortable for most users, for the ICM process allows the airing of many divergent viewpoints.

In the case of the Monterey NMS, the first step was to get eight federal, state, and local agencies to sign a *Memorandum of Agreement* to protect and improve the water quality of the sanctuary. The federal agencies included NOAA and the U.S. EPA; state agencies included the California Environmental Protection Agency, California Coastal Commission, and State Water Resources Control Board; local agencies included Central Coastal Regional Water Quality Control Board, San Francisco Regional Water Quality Control Board, and the Association of Monterey Bay Area Governments (AMBAG). Seventeen other resource and regulatory agencies, public and private groups, later joined these eight agencies to form a Water Quality Protection Program Committee to develop and implement a comprehensive program.

In the past 4 years, Monterey Bay NMS has sponsored many "multi-stakeholder" workshops that focused on a series of issues, such as urban runoff, marinas and boating activities, agricultural practices, regional monitoring, educational outreach, and mechanisms for ongoing local-state-federal-public-private collaboration. In order to finalize plans, implementation steps needed to be defined, responsible groups or agencies needed to be identified, and who would oversee the implementation also needed to be agreed upon. In August 1998, three detailed action plans had been completed and implementation begun, and several other draft plans were well underway (Dr. Holly Price, Director, Water Quality Protection Program, MBNMS, personal correspondence).

Both the Florida and Monterey experiences at Integrated Coastal Management illustrate the absolute need for (a) *full participation* by divergent users, and (b) *flexibility*—the ability to understand others and compromise. It was also clear that the ICM process is new, complicated, and will not be learned overnight. What follows is a close look at Soquel Creek that flows into the MBNMS, and the environmental issues and complexities that are involved in managing even a small, relatively undeveloped creek and its watershed.

Location

Flowing through the villages of Soquel and Capitola on the north shore of the MBNMS is a perennial coastal stream known as Soquel Creek. The creek originates at about 305 m (1000 ft) elevation on the western slope of the Santa Cruz Mountains near Loma Prieta, then meanders down for 24 km (15 mi) to Capitola, where it enters the Monterey Bay. The Soquel Creek watershed covers approximately 109 sq km (42 sq mi) and includes about 80 km (50 mi) of streams, making it the third largest watershed in Santa Cruz County. (See Figure 5-8).

Only the watersheds of the Pajaro and San Lorenzo Rivers cover more acreage. The tributaries all feed into Soquel Creek, which funnels into the MBNMS at Capitola Beach. Any pollutants drained out of the watershed with the runoff end up concentrated there as well. A lagoon (sometimes natural, sometimes artificial) periodically appears from the beach to approximately 366 m (1200 ft) upstream. (See Figure 5-9).

Floods, droughts, wildfires, earthquakes, and landslides have impacted the watershed over the past 12 years alone. Furthermore, urbanization and changing land use patterns have put additional pressure on the watershed. Reducing creek pollution and algae growth, maintaining and enhancing fisheries, and revegetating with California native plants are the major watershed management tasks facing this creek-side coastal community.

Environmental Concerns & Management Strategies

There are a number of environmental concerns and management problems associated with Soquel Creek (Alley and Associates 1992), but the one that is unique to this site along the MBNMS is the environmental ramifications of *artificially* closing the mouth of this creek each year. During the summer months when stream flow is minimal, the estuary or tidal mouths of most Central Coast streams become blocked by a sandbar—the *natural* consequence of sand deposited by ocean currents. The result is the formation of a **lagoon,** a body of water that is separated from the sea by a narrow strip of land. With the coming of winter rains, increased stream flow, and storms on the bay, the sandbar is breached (opened up) and the sand is redeposited offshore. For over 70 years, however, the city of Capitola has interfered with this natural process so that they can be assured of the presence of a lagoon for summer bathers and for the popular Begonia Festival every September. (See Figure 5-10).

Mounting public concern about water contamination has now raised questions about this historical practice. These questions can be grouped into four major management issues: (1) creek pollution; (2) algae control; (3) fish and wildlife; and (4) revegetation along the creek. We will now very briefly discuss each one.

Creek pollution. According to the Habitat Restoration Group (a team of researchers that prepared a report on Soquel Creek for the city of Capitola in 1988), the increasingly high fecal coliform count in the creek is coming from the concentration of seagulls, pigeons, ducks, and geese that congregate around and on the artificial lagoon. Capitola has no septic tanks bordering the creek which could be a source, and regular checks of the city's sewage lines have found no leaks. Another source of fecal coliform, however, is the number of dogs that traffic the area. One can find dog excrement almost everywhere, particularly on the dirt paths, parks, and private lawns that surround the creek. One can also regularly see people encouraging their dogs to thrash through the creek by tossing a stick into it.

To minimize these contaminants, the city has discussed relocating some of the geese and domestic ducks and has already erected signs cautioning people not to feed the birds. (See Figure 5-11). Although it will not be popular, a "leash/pooper scooper" law may have to be created and rigorously enforced to resolve the dog problem. Just as we

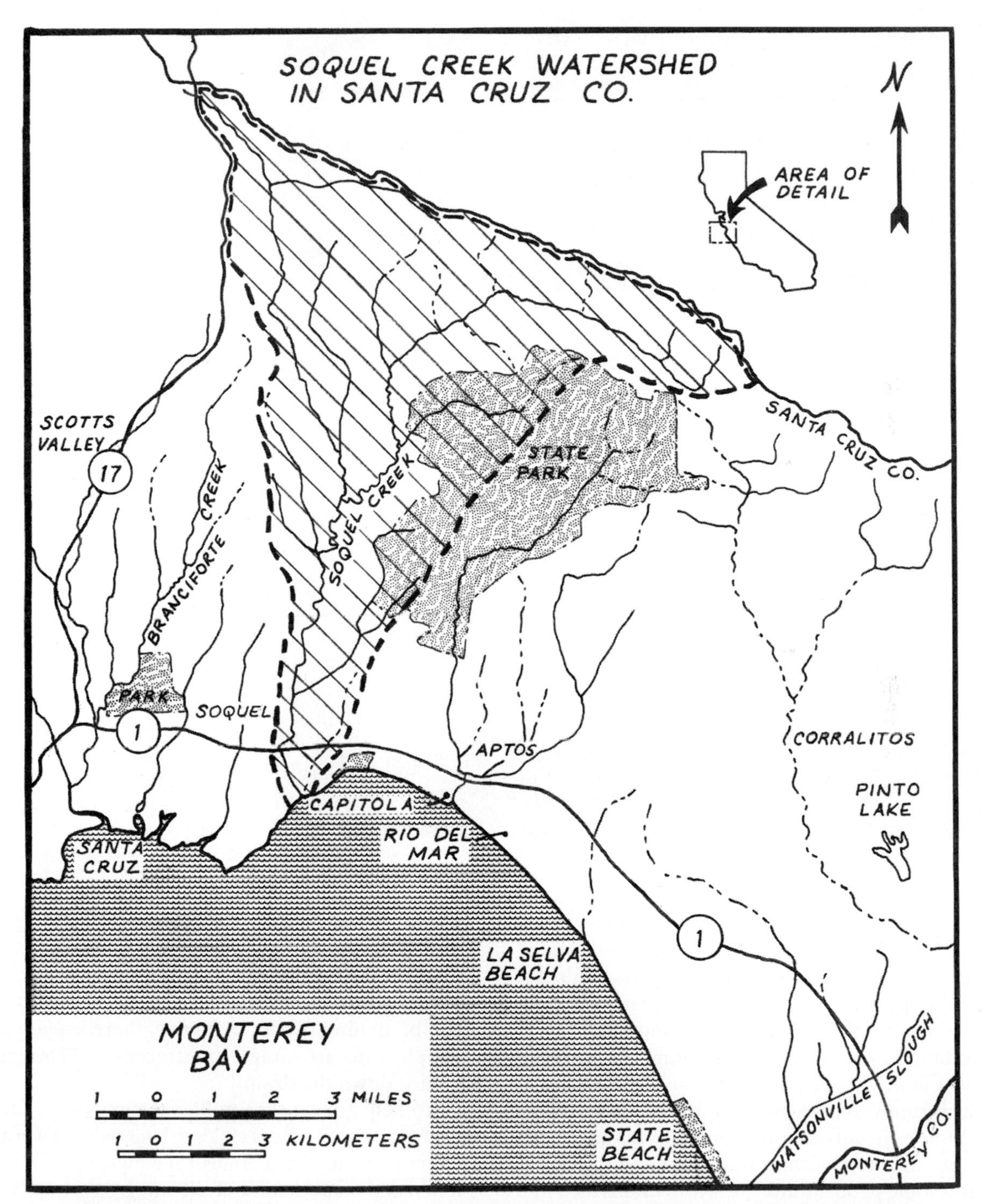

FIGURE 5-8 Soquel Creek watershed in Santa Cruz County, California. (*Source:* Map by Char Holforty)

FIGURE 5-9 Soquel Creek, California. (a) During the winter, Soquel Creek flows out to the Monterey Bay. Note the exposed outlet culvert with fish gate/shroud at the inlet to Soquel Creek; (b) Right before the summer tourist season, the city of Capitola uses a tractor to close the mouth of the creek to create an artificial lagoon for bathers and the Begonia Festival. Note that the conduit is now covered by the artificial sand bar. Also note the eutrophication that is occurring in the artificial lagoon. (*Source:* Photos by author)

have "pack it in/pack it out" management policies when it comes to trash in wilderness areas, a "dog in/excrement out" policy may have to be enacted for riparian habitats. Since public opposition can be expected, an intensive educational program will likely need to be mounted.

Controlling the other types of pollutants in the creek will be even more difficult. Contaminants have entered the creek in runoff from parking lots and other paved surfaces. Motor oil and other toxic substances have been dumped into the storm drains that surround the creek. Apparently, some people think the storm drains feed into a sewage system where their discarded toxic substances would be neutralized, but these storm drains go directly into the creek. Other locals may know that what they pour down the storm drain goes directly to the Creek, but they mistakenly assume the environmental impact will be minimal. And, of course, there are always some people who are simply indifferent—"Who cares, just pour it down the drain!"

To help educate the public, the city of Capitola (with the help of local school children) has painted warning signs directly on the storm drains. (See Figure 5-12). The installation and maintenance of silt and grease traps within the surrounding storm drains is also a management strategy,

FIGURE 5-10 The Begonia Festival—Capitola's last fling with summer. (a) Boats and floats covered with begonias parade down Soquel Creek in 1997. (b) Float about to pass under Stockton Bridge on way to artificial lagoon. (c) Same float reaches end of creek and turns around within artificial lagoon. (*Source:* Photos by author)

but is more difficult to enforce as one moves upstream away from Capitola. Unlike the city of Capitola, the county of Santa Cruz—which is responsible for the creek upstream—does not require grease traps on storm drains.

Algae control. The creation of the artificial lagoon also intensifies the formation of surface algae mats that naturally occur in the late summer. These algae mats, which are popularly referred to as "green scum," are a result of an **algal bloom**—the temporary, rapid growth of algae in a shallow freshwater body. Simply stated, all it takes for an algal bloom to occur is the presence of algae and other aquatic plants in a shallow body of freshwater (e.g., a lagoon), an increase in water temperatures in the summer heat, and an abundance of nutrients (e.g., usually nitrogen or phosphorus). Algal blooms along the Central California coast are a *natural* occurrence. However, in recent years, algal blooms have intensified due to human activities (e.g., the application of fertilizers to lawns, gardens, and flower beds around the creek).

In the past, the city of Capitola applied copper sulfate to control the algal blooms, which helped minimize the blooms and helped control the noxious odor that arose from the decomposing algae. However, copper sulfate is toxic to fish and is a known carcinogen. Consequently, in 1988, the California Department of Fish and Game banned the use of copper sulfate in Soquel Lagoon, prompting a switch to the use of Aquazine which inhibits photosynthesis but is less toxic to fish. In recent years, even the use of Aquazine has been phased out due to concerns with other creek fauna including the Tidewater goby. As a result, managers are using a safer, less intrusive mechanical procedure which involves skimming the surface of the water with long wood poles to remove floating algae. This technique was made practical through recently successful nutrient loading and algal growth reduction strategies such as the closing of streamside storm drains during summer, which reduces polluted run-off into the creek. Kelp and sea grass are also removed from the lagoon before summer closure, eliminating the nutrients that would be released upon its decomposition.

The city is also considering some more benign management strategies. One strategy is to make it easier for the native wild ducks to feed on what they already like to eat—the algae and pondweed. In theory, this can be done by removing the geese which compete with the ducks for space. A second management technique is to allow only the partial dismantling of the begonia floats in the lagoon. Traditionally, the floats were fully dismantled in the lagoon with no concern as to the amount of debris that was left in the lagoon; the sandbar would be artificially breached at the end of the festival so the debris could be washed out to sea. Today, however, the Department of Fish and Game no longer allows this breaching of the sandbar. Unless there is a danger of flooding, the sandbar must be allowed to breach naturally.

Fisheries management. In order to regularly create an artificial lagoon, while at the same time not disturb steelhead trout and other fish that use the creek for spawning grounds, the city of Capitola had to build a **flume**—a concrete, open culvert with fish gate/shroud that allows the passage of water and fish between the creek and the bay. (See Figure 5-13). So, despite the presence of a sandbar at the mouth of the creek, steelhead smolts can still migrate to the ocean through this flume. In 1992, Capitola was one of the few cities on the California coast, if not the only city, to use a flume on an urban creek to protect its fisheries.

Another fisheries management technique is to minimize the use of heavy equipment when cleaning the creek. In the 1950s and 1960s, the city used bulldozers to scour the creek clean of algae mats and kelp that had washed upstream. The California Department of Fish and Game and the California Coastal Commission now both recommend that the city clear the large mats of kelp by hand just prior to the building of the sandbar. They maintain that the use of heavy equipment negatively impacts the vegetation and wildlife food sources necessary to maintain the delicate balance of the ecosystem. These agencies also recommend that during drought years the city make an effort to capture steelhead that have been stranded in pools upstream and transport them to the lagoon where they can swim out to sea.

(a)

(b)

FIGURE 5-11 (a) Geese and other nonnative waterfowl on Soquel Creek—one of the causes of eutrophication of the creek; (b) "Do Not Feed The Birds" signs are posted along Soquel Creek to discourage bird feeding. Citizens caught feeding birds are even 'ticketed' (fined) by the local police. (*Source:* Photos by author)

FIGURE 5-12 Storm drains that feed into Soquel Creek are now painted with signs to warn and educate potential polluters. (*Source:* Photo by author)

Revegetation along the creek. Associated with Capitola's efforts at fisheries and wildlife protection are its plans for riparian habitat restoration. The city plans to remove much of the non-native vegetation along the creek banks, such as pampas grass *(Cortaderia selloana),* giant reed (*Arundo donax*), and green wattle (*Acacia decurrens*). Native plants would then be planted. The large eucalyptus grove, near the Rispin mansion, which serves as a habitat for monarch butterflies would remain, though this non-native tree in another locale would be removed. The implementation of this plan, of course, requires ample funding, at leat part of which the city hopes to get from the state Coastal Conservancy.

GENERAL POLICY RECOMMENDATIONS

The following general recommendations apply to controlling pollution along America's coasts and within her marine sanctuaries. Although several of these recommendations may not be economically or politically feasible for certain communities at this time, they can still serve as goals to strive towards. First, there are five underlying assumptions to what follows: (a) Cleaning up the coastal sea will require proper management of *the land;* (b) Pollution management will require an integrated approach which includes natural resource and land use decisions; (c) marine organisms, being in closer chemical contact with their environment than terrestrial organisms, are more vulnerable to pollution and may require closer scrutiny; (d) projects with unacceptable pollution potential should be discouraged; and (e) enforcement of existing pollution regulations may be as im-

(a)

FIGURE 5-13 Outlet culvert with fish gate/shroud at the mouth of Soquel Creek. The purpose of the culvert is to allow out-migration to the ocean for steelhead smolts when the creek is artificially closed. (a) During the summer months, only the tips of the conduit protrude from underneath the beach sand.

(b)

FIGURE 5-13 (Continued) (b) After winter storms, the conduit is fully exposed. (*Source:* Photos by author)

portant as writing new pieces of legislation. What now follow are some general recommendations as they apply to specific categories of pollution.

Point Source Pollution

1. *Sewage outfalls*
   - *Question the type of sewage system being proposed for a coastal site.* As shown earlier, replacing a primary treatment plant with a secondary treatment plant in a coastal area is not necessarily better. Scientists are beginning to question the general "upgrading" from primary to secondary treatment in certain locations.
   - *Discourage inland cities from looking at nearby coastal waters as sewage disposal sites.* Inland cities must be severely discouraged from this path of solving their sewage disposal problems. As long as communities can transport it elsewhere—the "out of site, out of mind" management strategy—no serious effort will be made at reducing and recycling sewage waste. If inland cities are allowed to dump their sewage effluent in coastal waters, then strict regulations and rigorous monitoring of that discharge must be religiously enforced.
   - *Encourage the reclamation of sewage water.* Sewage effluent is 99 percent water. When the 1 percent of pollutant is removed, it is possible to have water that is purer than the original source. For example, Los Angeles daily discharges 64 million liters (17 million gal) of processed sewage water. The treated water then filters through aquifers before it is pumped out of town wells. The water turns out to be of higher quality than its source, the Colorado River. Such *reclaimed water* is already being used in the Bethlehem Steel plant in Baltimore, Maryland to cool steel. San Francisco uses reclaimed water to irrigate its golf courses. San Bernadino, California uses treated sewage water to irrigate ornamental shrubs along highways, and San Antonio, Texas, and Fort Collins, Colorado even use treated sewage effluent to irrigate crops.
2. *Industrial wastewater*
   - *Require that dischargers regularly demonstrate that other alternatives are not available and/or that the best possible pretreatment systems are being used.* This recommendation could be carried out by the agency that issues permits under the National Pollution Discharge Elimination System (NPDES).
   - *Levy discharge fees according to the toxicity of the effluent.* Dischargers should have to pay the true cost of disposal. Only then will many companies get serious about reducing their output of effluent.
   - *Establish comprehensive monitoring programs.* Outfalls within bay waters could be monitored to see if dischargers are actually limiting output or reducing toxicity of effluent. NOAA's Status and Trends Program could possibly oversee the monitoring operation.
3. *Solid waste disposal*
   - *Continue the ban on coastal dumping of solid waste.* Although solid waste disposal was not discussed in this chapter, but rather in Chapter 6, *Ocean Dumping,* any set of recommendations dealing with minimizing coastal pollution must also address the solid waste disposal problem. As this country runs out of landfill space, there will be increasing social pressure to use coastal waters as dumping grounds. Ocean

dumping only reinforces the out of sight, out of mind management philosophy. The communities along the Monterey shore should continue their existing policy of not dumping garbage or sludge into bay waters. Other U.S. coastal communities should follow suit.

Nonpoint Source Pollution

1. *Harbors and marinas*
   - *Increase monitoring and enforcement of waste disposal at boatyards.* The state of Washington, in conjunction with new federal water pollution laws, is tightening its control over businesses that discharge industrial wastewater into harbors. Since 1992, boatyard owners must have a National Pollution Discharge Elimination System (NPDES) permit. The permit contains limits and monitoring requirements for the most common pollutants discharged from boatyards. The new limits are designed to prevent dangerous, environmentally damaging heavy metals, such as lead, copper, and zinc, from entering our coastal waters. NPDES permits for boatyards are available from the Washington State Department of Ecology. Boatyard owners, not surprisingly, were originally concerned that the cost of complying with the new environmental regulations would force them out of business. It has turned out, however, that most boatyards can treat their wastewater economically and effectively so that it meets the limits required for discharge to the sewer system. To help boatyard owners move in the right direction, the Municipality of Metropolitan Seattle (1991) produced a "how-to" manual, *Wastewater Treatment Guidelines for Boatyards,* which includes information on obtaining permits, designing economical collection and treatment systems, and other helpful resources.
   - *Establish regulations and enforcement policies regarding boat maintenance within slips.* Currently, there are very few restrictions on the methods or products used for boat maintenance while the boat sits in its harbor slip. Enforcing boatyards to clean up their act while allowing the individual boat owner to pollute at will makes little sense. Since most harbors are already patrolled hourly by harbor officials, the method of monitoring and enforcement is already set in place. It would be merely a matter of ticketing *serious* pollution offenders as harbor officials already ticket boats that are docked in the wrong slip.[1]
2. *Agricultural runoff*
   - *Increase soil and water conservation practices along coasts.* Working with the local district office of the U.S. Natural Resources Conservation Service (NRCS, formerly the Soil Conservation Service), identify coastal farmers that are not incorporating good soil and water conservation practices. Encourage such conservation farming techniques as terracing, minimum tillage, drip irrigation, dryland farming, etc., where and when appropriate. Representatives from the NRCS are experts in conservation farming techniques. Also, look into the feasibility of linking farm subsidy support to conservation compliance. For further details, see Hegyes and Francis, eds., 1997.
   - *Reduce use of pesticides on coastal farmlands.* The general public, and farmers themselves, are becoming increasingly concerned about the negative effects of using large and regular doses of herbicides, insecticides, and various other pesticides. **Integrated Pest Management (IPM),** an ecological farming method that combines cultivation practices, biological controls, and limited chemical usage in proper sequence and time, and various other "environmentally-friendly" management strategies should be encouraged.
3. *Urban runoff*
   - *Replace CSOs (Combined Sewer Overflow systems).* A CSO is a combined sewage/storm runoff system. This means that during heavy rains, raw sewage combines with urban storm runoff, bypasses the treatment plant, then dumps directly into coastal waters. Carrying out this recommendation will be extremely capital intensive, but some way of controlling these overflow discharges is necessary.
   - *Enforce water conservation strategies during months of normal rainfall, not just during drought years.* Practicing water conservation year round not only conserves water supplies, it also reduces the amount of urban wastewater that drains into creeks, streams, rivers, and eventually into coastal waters.
4. *Marine debris*
   - *Further restrict ocean dumping.* See Chapter 6, *Ocean Dumping,* for recommendations.
   - *Encourage increased recycling of plastics.* Facing increasing consumer outrage over wasteful overpackaging, not to mention the higher costs for the virgin resins used to manufacture plastics, the plastics industry is taking a new look at recycling. In 1989, the National Wildlife Federation reported that the Dupont Company and Waste Management, Inc. plan to jointly construct several recycling plants, each capable of processing some 40 million pounds of plastic milk jugs and soft drink bottles every year. The amount of plastic packaging that goes straight from the garbage can to the dump must be reduced. Some plastics have been recycled into paint brushes, industrial straps, and stuffing, but much more must be done.
   - *Encourage the development of photodegradable plastics.* Some manufacturers have already developed photodegradable plastics—plastics that de-

[1] Harbors *are* places where boats are to be maintained. Half the enjoyment of owning a small boat is weekend "tinkering" (i.e., maintenance). It is only the very gross offender that should be cited.

grade when exposed to ultraviolet light from the sun. These products (e.g., plastic six-pack frames, plastic bags) should dissolve quickly and with no breakdown products that are harmful to the environment. It does not take long for a seagull to suffocate with a six pack frame around its neck!

- *Promote recycling programs in all coastal communities.* See Chapter 6, *Ocean Dumping,* for further details.
- *Integrate environmental education into school curricula.* All schools that are located within the "lets go to the beach" range should incorporate an understanding of human impacts on coastal ecosystems within their existing science or social science courses. The impacts of marine debris on coastal wildlife should be stressed. Ways to reduce waste should also be discussed.

Physical and Hydrological Modifications

1. *Harbor dredging*
   - *Monitor dredge spoils for toxic wastes and regulate accordingly.* Approximately 1 of every 3 tons of dredge spoils in the United States is contaminated with both urban and industrial waste, as well as agricultural runoff. See Chapter 6, *Ocean Dumping,* for further details.
   - *Investigate ways to reuse or recycle dredge spoils.* Excess non-toxic dredge spoils not used for beach nourishment could possibly be "dewatered" and recycled to create new coast wetlands.
2. *Groundwater withdrawal*
   - *Encourage farmers to become more efficient water users.* Methods exist for conserving water on the farm, such as good soil management practices (e.g., terracing and contour tillage), efficient irrigation methods (e.g., use of computers and drip irrigation), good systems management (e.g., precise water application scheduling), control of losses in water transit (e.g., use of plastic or cement lining), increased use of reclaimed water (e.g., use of treated municipal sewage water on crops), new cropping patterns (e.g., dryland farming), and altering the crops grown (e.g., planting drought-resistant crops).
   - *Encourage city dwellers to become more efficient water users.* Methods also exist for conserving water in the city, such as replacing high-flow toilets with more efficient models (e.g., the Watsonville example discussed above). In Palo Alto, California, the city's 1992 toilet rebate program was so successful that the entire $39,000 fund set aside for rebates was used up in just three weeks, and 500 people were on a waiting list for the program to renew.

Other urban water conservation techniques include detecting and eliminating leaks (e.g., repairing leaking water mains), reusing reclaimed wastewater for park landscapes, requiring mandatory water restrictions (e.g., restrictions on lawn-watering during drought years), issuing higher water rates (e.g., an increase in water prices to encourage conservation), requiring water-conserving landscaping (such as requiring the use of drought-tolerant plants in park design), encouraging conservation education (use of the news media to promote the use of low-flow showerheads), and improving management techniques (e.g., a regional approach to water management).

CONCLUSION

Pollution management is certainly one of the coastal resource managers largest challenges. Accordingly, a solid understanding of the types and sources of coastal pollution will create greater opportunity for the development of creative solutions.

We have seen that coastal pollution comes in two forms, water pollution and marine debris, and there are point sources and nonpoint sources of coastal pollution. Point sources include petroleum development and transportation facilities, sewage outfalls, industrial wastewater discharges, and solid waste disposal. Nonpoint sources include boat maintenance chemicals in harbors and marinas, ballast water from international ships, agricultural run-off, logged forest areas, urban run-off, air pollution, and noise pollution. Coastal areas are also polluted by physical and hydrological modifications such as harbor dredging, groundwater withdrawal, and dam, canal, and levee construction.

It should now be clear that coastal pollution comes from many sources and in many forms, which cumulatively form a complex management problem. As a result, the solution must be integrative, capable of addressing each individual problem as part of a larger, holistic management strategy. The Integrated Coastal Management (ICM) concept being deployed in our National Marine Sanctuary Program is an example of this management methodology. In the Monterey Bay National Marine Sanctuary, interested parties have come together to create a watershed-wide water quality management plan. This plan seeks to address the cumulative nature of coastal pollution through interagency cooperation and a balanced assessment of the needs of individual resource users and the requirements for a healthy, fully functioning ecosystem (Galasso 1993). The challenges encountered with this ambitious project are highlighted in our case study of Soquel Creek Watershed, part of the greater Monterey bay area watershed. Coastal resource managers must balance the cultural demands upon this creek with the need for pollution control and fisheries management—a complex task, even for this small ecosystem.

Coastal pollution is a formidable problem, fortunately a wide range of potential solutions exist, and remain to be implemented and evaluated. Point source pollution problems can be improved by updating pollution control infrastructure, such as sewage treatment plants, and harnessing new technology, such as wastewater reclamation. Nonpoint source pollution may be reduced by increasing monitoring and enforcement of existing laws, and further encouragement of agricultural pollution control methods, such as a reduction in the amount of pesticides used, and better soil and water conservation practices. The opportunities for reducing coastal degradation are as varied as the types and sources of pollution. What is needed is understanding, resolve, creativity and the will to do it.

REFERENCES

ABA Consultants. 1989. *Elkhorn Slough Wetland Management Plan.* Capitola, California: ABA Consultants.

Advanced Research Projects Agency. 1995. *Final Environmental Impact Statement/Environmental Impact Report for the California Acoustic Thermometry of Ocean Climate Project and its Associated Marine Mammal Research Program.* Vol. I, II, Silver Springs, Maryland: National Oceanic and Atmospheric Administration.

Alley, Don W., and Associates. 1992. *Soquel Creek Lagoon Monitoring Report, 1990–91.* Brookdale, California: Don Alley & Associates.

Association of Monterey Bay Area Governments. 1987. *Annual Report.* Monterey, California: Association of Monterey Bay Area Governments.

Aubrey, David G. and Michael Stewart Connor. 1993. "Boston Harbor: Fallout Over the Outfall." *Oceanus.* Vol. 36, No. 1, pp. 61–70.

Basta, Daniel J., Blair T. Bower, Charles N. Ehler, Forest D. Arnold, Barton P. Chamvers and Daniel R. G. Farrow. 1985. *The National Coastal Pollutant Discharge Inventory.* Rockville, Maryland: National Oceanic and Atmospheric Administration (NOAA).

Bunte, L. S., Jr. et al. 1967. Basic Data and Operation Report for Water Year October 1, 1966, to September 30, 1967. Monterey County Flood Control and Water Conservation District, Hydrology Section.

California Native Plant Society (The), et al. 1992. *Soquel Creek: Streamside Care Guide.* Capitola, California: City of Capitola.

Carlton, James T. and Jonathan B. Geller. 1993. "Ecological Roulette: The Global Transport of Nonindigenous Marine Organisms." *Science.* Vol. 261, No. 5117, pp. 78–82.

Carter, R. W. G. 1988. *Coastal Environments.* New York: Academic Press.

Center for Marine Conservation. 1994. "1993 Beach Cleanup Results Set New Record." *Coastal Connection.* Spring. [No volume nor number].

Champ, M. A. and F. L. Lowenstein. 1987. "TBT: The Dilemma of High-Technology Antifouling Paints." *Oceanus.* Vol. 30, No. 3, pp. 69–77.

Clark, John. 1996. *Coastal Zone Management HANDBOOK.* Boca Raton, Florida: CRC Lewis Publishers.

Committee on Evaluation of the Safety of Fishery Products. 1991. *Seafood Safety.* Washington, D.C.: National Academy Press.

Cowell, Alan. 1991. "A Poisoned Season: Dead Dolphins, Abused Pups." *New York Times.* 4 September, Sec. A: 4.

Cox, George. 1993. *Conservation Biology.* Dubuque, Iowa: Wm C. Brown Publishers.

Crutchfield, Stephen. 1987. "Controlling Farm Pollution of Coastal Waters." In *Agricultural Outlook.* Washington, D.C.: U.S. Department of Agriculture, pp. 24–26.

Dixon, John A., Lee M. Talbot, and Guy J. M. Le Moigne. 1989. *Dams and the Environment: Considerations in World Bank Projects.* Washington, D.C.: The World Bank.

Environmental Defense Fund. 1988. *Polluted Coastal Waters: Role of Acid Rain.* New York: Environmental Defense Fund.

Environmental Protection Agency, EPA. 1991. *Nonroad Engine and Vehicle Emission Study.* Washington D.C.: Environmental Protection Agency.

Galasso, George A. 1993. "Monterey Bay National Marine Sanctuary Water Quality Management: The Use of Coordinating Mechanisms." *Coastal Management.* Vol. 21, No. 4, pp. 333–345.

Goldberg, E. D. 1986. "TBA: An Environmental Dilemma." *Environment.* Vol. 28, No. 8, pp. 17–20, 42–44.

Group of Experts on the Scientific Aspects of Marine Pollution. 1990. *The State of the Marine Environment.* Nairobi: United Nations Environment Programme.

Habitat Restoration Group (The). 1988. *Soquel Lagoon Management and Enhancement Plan.* Scotts Valley, California: John Stanley and Associates, Inc.

Harmstead, John. 1995. "Pollution is Whales' New Bane; Chemicals Take Hunters' Place, Researcher Says." *The Arizona Republic.* May 30, p CL 11.

Hegyes, Gabriel, and Charles A. Francis (eds.). 1997. *Future Horizons: Recent Literature in Sustainable Agriculture.* Lincoln, Nebraska: Center for Sustainable Agricultural Systems, University of Nebraska-Lincoln.

Heikoff, Joseph M. 1981. *Marine and Shoreland Resources Management.* Ann Arbor, Michigan: Ann Arbor Science.

Hodgson, Gregor and John A. Dixon. 1988. *Logging Versus Fisheries and Tourism in Palawan: An Environmental and Economic Analysis.* Honolulu, Hawaii: East-West Environment and Policy Institute.

Lee, Martin R. 1990. "Waste Disposal." In *CRS Issue Brief: Oceans and Coastal Management Issues.* Congressional Research Service. Washington D.C.: The Library of Congress.

Levy, Paul F. 1993. "Sewer Infrastructure: An Orphan of Our Times." *Oceanus.* Vol. 36, No. 1, pp. 53–60.

Linker, Lewis. 1996. Chesapeake Bay Program. Personal Communication. July.

Luecke, Daniel F., and Carlos De La Parra. 1994. "From Pollution to Park." *California Coast & Ocean.* Vol. 10, No. 1, pp. 7–19.

McDowell, Judith E. 1993. "How Marine Animals Respond to Toxic Chemicals in Coastal Ecosystems." *Oceanus.* Vol. 36, No. 2, pp. 56–61.

Mele, Audre. 1993. *Polluting for Pleasure.* New York: W. Norton.

Minerals Management Service. 1987. *Five Year Outer Continental Shelf Oil and Gas Leasing Program. Mid-1987 to Mid-1992: Final Environmental Impact Statement.* Washington, D.C.: Minerals Management Service, U.S. Department of the Interior.

Municipality of Metropolitan Seattle. 1991. *Boatyard Wastewater Treatment Guidelines.* Seattle, Washington: Metro.

National Oceanic and Atmospheric Administration (NOAA). 1990a. *Coastal Environmental Quality in the United States, 1990: Chemical Contamination in Sediment and Tissues.* Rockville, Maryland: National Oceanic and Atmospheric Administration.

National Oceanic and Atmospheric Administration (NOAA). 1990b. *Draft Environmental Impact Statement and Management Plan for the Proposed Monterey Bay National Marine Sanctuary.* Washington, D.C.: National Oceanic and Atmospheric Administration.

National Oceanic and Atmospheric Administration (NOAA). 1992. *Monterey Bay National Marine Sanctuary: Final Environmental Impact Statement/Management Plan.* Vol. I & II. Washington, D.C.: National Oceanic and Atmospheric Administration.

National Oceanic and Atmospheric Administration (NOAA). 1994. *Monterey Bay National Marine Sanctuary Water Quality Protection Program. Workshop Summary: Preliminary Identification of Issues and Strategies.* Silver Springs, Maryland: National Oceanic and Atmospheric Administration.

National Oceanic and Atmospheric Administration (NOAA). 1995a. *Monterey Bay National Marine Sanctuary Water Quality Protection Program Working Document: Comparison of the California Water Quality Assessment and the January 1994 Workshop Results,* April. Silver Springs, Maryland: National Oceanic and Atmospheric Administration.

National Oceanic and Atmospheric Administration (NOAA). 1995b. *Strategy for Stewardship: Florida Keys National Marine Sanctuary. Draft Management Plan/Environmental Impact Statement.* Vol. I, II, III. Silver Springs, Maryland: National Oceanic and Atmospheric Administration.

Nikolov, Kalin, Jose A. Revilla, Cesar Alvarez, and Albert Luceno. 1994. "A Design Methodology for Combined Sewer System Elements with Overflows in Coastal Zones." *Journal of Coastal Research.* Vol. 10, No. 3, pp. 531–538.

Paul, Elizabeth. 1998. *Testing the Waters VIII: Has Your Vacation Beach Cleaned Up Its Act?* New York: Natural Resources Defense Council.

Peierls, Benjamin L., Nina F. Caraco, Michael L. Pace, Jonathan J. Cole. 1991. "Human Influence on River Nitrogen." *Nature.* Vol. 350, No. 6317, pp. 386–387.

Shabecoff, Philip. 1988. "Pollution Is Blamed for Killing Whales in the St. Lawrence." *New York Times.* January 12, 1988, Sec. C: 1.

Shaw, S.B. 1972. "DDT Residues in Eight California Marine Fishes." *California Fish and Game.* Vol. 58, No. 1, pp. 22–26.

Sindermann, Carl J. 1996. *Ocean Pollution: Effects on Living Resources and Humans.* Boca Raton, Florida: CRC Press.

Sun Sentinel. 1995. "Now or Never" December 10, pp. 1A, 12A, 13A.

Trim, Alan H. 1987. "Acute Toxicity of Emulsifiable Concentrations of Three Insecticides Commonly Found in Nonpoint Source Runoff into Estuarine Waters to the Mummichog, *Fundulus Heteroclitus,*" *Bulletin of Environmental Contamination and Toxicology,* Vol. 38, No. 4, pp. 681–686.

World Resources Institute (The). 1992. *World Resources:* 1992-93. New York: Oxford University Press.

FURTHER READING

Capuzzo, Judith E. McDowell. 1990. "Effects of Wastes on the Ocean: The Coastal Example." *Oceanus.* Vol. 33, No. 2, Summer, pp. 39–44.

Davis, W. Jackson. 1990. "Global Aspects of Marine Pollution Policy—The Need for a New International Convention." *Marine Policy.* May, Vol. 14, No. 3, pp. 191–197.

Environmental Protection Agency, EPA. 1993a. *Created and Natural Wetlands for Controlling Nonpoint Source Pollution.* Boca Raton, Florida: Lewis Publishers.

Environmental Protection Agency, EPA. 1993b. *Guidance Specifying Management Measures for Sources of Nonpoint Pollution in Coastal Waters,* Washington, D.C.: Environment Protection Agency.

Environmental Protection Agency (EPA). 1993c. *Water Quality Protection Program for the Florida Keys National Marine Sanctuary.* Unpublished Draft. April 1993. Washington D.C.: Eivironmental Protection Agency.

Hershman, Marc J., ed. 1988. *Urban Ports and Harbor Management.* New York: Taylor & Francis.

Hinrichsen, Don. 1998. *Coastal Waters of the World: Trends, Threats, and Strategies.* Washington D.C.: Island Press.

Kennish. Michael J. 1997. *Practical Handbook of Estuarine and Marine Pollution.* Boca Raton, Florida: CRC Press.

Mooney, Kailen, Ashley McLain, Beth Hanson, and Sarah Chasis. 1992. *Testing the Waters: A National Perspective On Beach Closings.* New York: Natural Resources Defense Council.

Naiman, R.J., ed. 1992. *Watershed Management: Balancing Sustainability and Environmental Change.* New York: Springer-Verlag.

National Oceanic and Atmospheric Administration. 1990. *Coastal Environmental Quality in the United States.* Rockville, Maryland: National Oceanic and Atmospheric Administration.

National Research Council. 1993. *Managing Wastewater in Coastal Urban Areas.* Washington D.C.: National Academy Press.

Natural Resources Defense Council. 1992. *Ebb Tide for Pollution: Actions for Cleaning Up Coastal Waters.* Washington D.C.: The Natural Resources Defense Council.

Pait, Anthony S., Alice E. De Souza, and Daniel R.G. Farrow. 1992. *Agricultural Pesticides in Coastal Areas: A National Summary.* Rockville, Maryland: National Oceanic and Atmospheric Administration.

Young, Terry F. and Chelsea H. Congdon. 1994. *Plowing New Ground: Using Economic Incentives to Control Water Pollution from Agriculture.* Oakland, California: Environmental Defense Fund.

6 Ocean Dumping

Water may flow in a thousand channels, but it all returns to the sea.

An African proverb

It has often been said that the ocean is the "ultimate sink." In addition to receiving natural runoff, the world's open ocean and coastal areas receive agricultural and urban runoff (Chapter 5), accidental oil spills from offshore oil drilling platforms and tankers (Chapter 7), as well as contaminants from deliberate ocean dumping (e.g., dredged material, sewage sludge, and garbage). (See Figure 6-1).

This chapter will concentrate on the practice of **ocean dumping**—the direct disposal by dumping by barge or ship of waste materials at a particular site at sea. This form of pollution accounts for only 10 percent of the waste that enters the oceans each year (Guarascio 1985). However, regulating (or banning) this form of ocean pollution is important for two reasons: (a) depending on what they are, the wastes can be extremely harmful or toxic to living organisms; and (b) direct dumping is more easily regulated than indirect dumping (e.g., agricultural or urban runoff, polluted waterways) (Moore 1992). The problem of ocean dumping did not come to the attention of the American public until the summers of 1987 to 1989, when beaches in numerous New Jersey townships were closed as a result of medical debris and other garbage washing ashore. Americans clamored for Congress to do something about this threat to their coastlines (Kitsos and Bondareff 1990). Despite more strict regulations since the late 1980s, the problem of ocean dumping continues to plague America's shores.

Types of Wastes Dumped looks at the various sources of possible ocean and coastal contaminants, such as dredged materials, sewage sludge, solid wastes (garbage), industrial wastes, military wastes, radioactive wastes, and contaminants resulting from the use of ocean incineration vessels. Under each category, there will be a brief history of the practice, its effect on the environment, typical management strategies, and its practice around America's coastlines, including near national marine sanctuaries. *Case Study: Radioactive Waste Dumping at the Farallones* discusses the historic practice of dumping low-level radioactive materials just outside the Golden Gate Bridge of San Francisco, in an area that is now part of the Gulf of the Farallones National Marine Sanctuary—a sanctuary that is adjacent to two

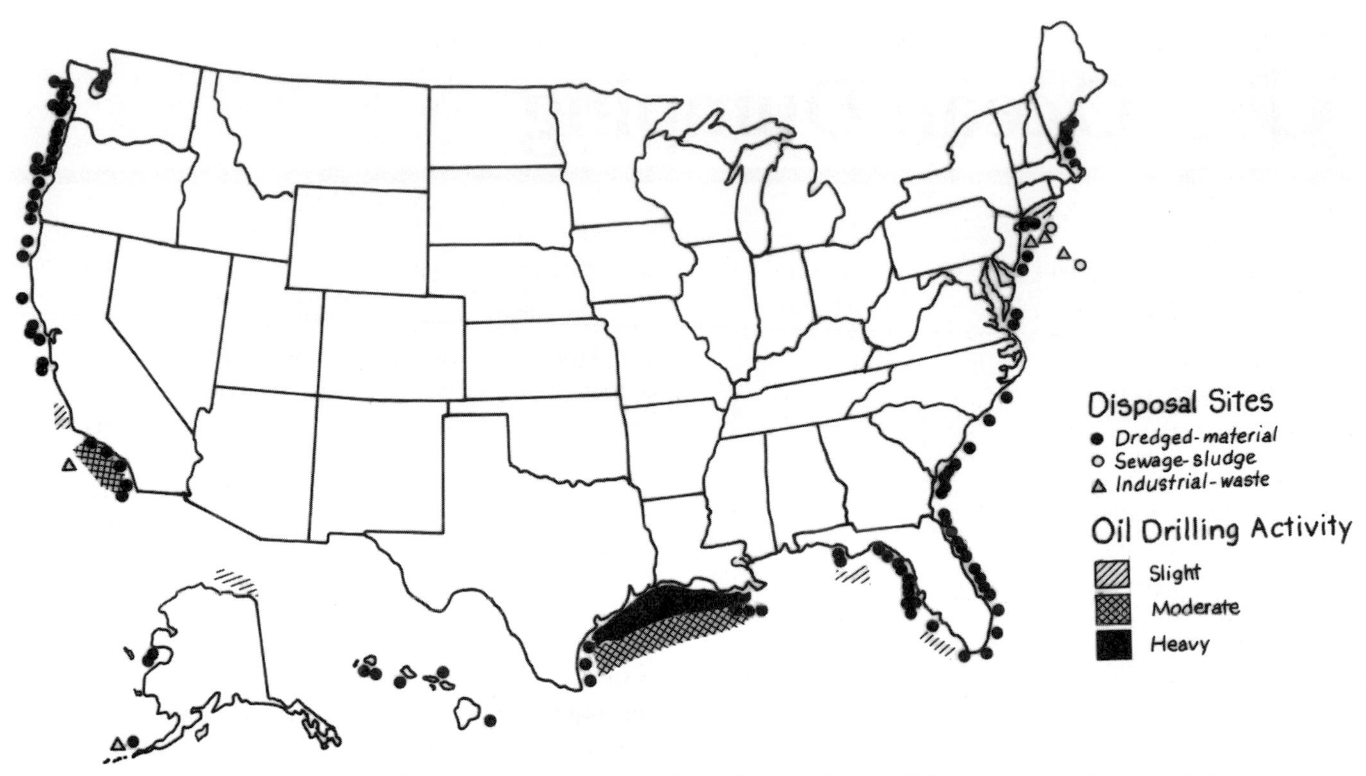

FIGURE 6-1 Major types and location of dumping sites along U.S. coastline in 1990, just prior to 1991 ban of all dumping except dredge-material. (*Source:* Illustration by Susan Giles: Adapted from Duedall 1990, p. 29)

other marine sanctuaries—Cordell Bank NMS and Monterey Bay NMS. *Future Ocean Disposal* provides an introduction and brief analysis of what is called "deep ocean dumping"—one of the most controversial proposals for future waste disposal in oceans. *General Policy Recommendations* provides coastal managers and others with guidelines appropriate for discussions related to ocean dumping, and finally, *Conclusion* discusses what some have referred to as the "new environmentalism," and its relevance to the subject of ocean dumping.

TYPES OF WASTES DUMPED

Ocean dumping occurs world-wide and is done by almost all countries. In 1972, representatives from over 80 countries convened to formulate a policy for the reduction of marine pollution caused by the dumping of wastes. This meeting—the *Convention on the Prevention of Marine Pollution by Dumping Wastes and Other Matter*—is generally referred to as the *London Dumping Convention,* or *LDC.* It was negotiated in the London-based United Nation's International Maritime Organization—the agency concerned with maritime safety and the prevention of pollution from ships. It came into force on August 30, 1975. The member nations, known as Contracting Parties, record dumping data and meet annually to discuss progress being made toward establishing national systems to control dumping from their respective shores.

The LDC is one of many International and National protocols, treaties, and laws that regulate ocean dumping, but the LDC was one of the first International protocols which brought the attention of the world to ocean pollution issue. Since the LDC, some of the other significant protocols and laws include the International **MARPOL** Protocol and its Annexes I, II, III, and V (the U.S. Congress passed the Marine Plastic Pollution Research and Control Act [MPPRCA] in order to bring Annex V into effect within U.S. waters); the Marine Protection, Research, and Sanctuaries Act (MPRSA); and the Clean Water Act.

Today, the United States dumps only one major type of waste off its coasts: dredged materials. In the recent past, however, this country has dumped (*and in some cases still does with special permits*) other categories of wastes, such as solid wastes (e.g., municipal garbage), industrial wastes (e.g., construction and demolition debris), military wastes (e.g., explosives and chemical munitions), and radioactive wastes. Although actually "burned" rather than "dumped" at sea, wastes burned at sea on ocean incinerating vessels also come under the general topic of ocean dumping. We will now look at all of these categories of ocean dumping in more detail, with an emphasis on understanding (a) the history of the practice,

(b) its effect on the environment, (c) the regulatory agencies, laws, and management strategies associated with the practice, and (d) whether or not it has occurred (or still occurs) within, or adjacent to marine sanctuaries.

Dredged Materials

History of the practice. Navigable channels, small boat harbors, and deep water ports for large ships must be regularly dredged to allow vessels to pass safely in and out of bays and estuaries without running aground. **Dredging** is often done to remove sediment from waterways to improve navigation. This practice generates 80 percent of the wastes that are dumped into our coastal waters (Owen and Chiras 1990). In the San Francisco Bay, for example, over 4,000 commercial ocean-going vessels pick up and deliver cargo every year. Many of these container ships require a draft (water depth) of at least 13.7 m (45 ft). That required ship draft may not sound like very much until you realize that two thirds of San Francisco Bay measures less than 5.5 m (18 ft) deep, and that 8 million cubic yards of new sediment pours into the bay (really an estuary) every year. Without regular dredging, deep draft ocean shipping would come to a halt in San Francisco Bay. In 1993, this type of shipping alone provided 79,000 jobs and approximately $11 billion in revenues to Bay Area counties.

In addition to keeping shipping channels, turning basins, and docking slips navigable, there are other reasons for dredging. Dredged materials can be used for construction aggregate in building delta levees, flood control, securing footings for bridges and piers, beach nourishment, enhancement of wetlands and wildlife habitats, offshore mound and island construction or other forms of land development, agriculture, and even mariculture. The more positive uses of dredged material we can find, of course, the fewer disposal sites would be needed. In 1987, 1.3 billion metric tons of sediment were dredged worldwide, with an estimated 35 percent due to U.S. activities (Office of Technology Assessment Task Force 1987). A significant number of dredge spoil dumping sites can be found in the Atlantic and Pacific Oceans, as well as the Gulf of Mexico (refer back to Fig. 6-1).

Effect on the environment. Hundreds of millions of cubic meters of dredged material are brought up annually from the world's harbors, and some of this dredged material is ecologically harmful. It is important to remember that there are two types of dredged material: (1) uncontaminated, or "clean" sediment, and (2) contaminated, or "unacceptable" sediment. *Uncontaminated* dredged material consists of inorganic (e.g., sand, silt, and clay, as well as rock and gravel) and organic matter that is a result of natural erosion and mineralization processes. This clean dredged material can be used in the broadest range of ways and types of locations, many of which are listed above. Environmental concern is limited to physical impacts, such as habitat modification in an aquatic environment.

Unfortunately, some dredged material is *contaminated* by pollutants from urban, industrial, and agricultural sources. In a typical industrialized harbor, *toxins in the sediments* could result from excess nutrients or harmful microbes from fertilizers on farmlands or effluent from waste treatment plants, oil and petrochemical by-products from urban runoff, heavy metals from gold or other mining operations, or PCBs and other organohalogens from coastal industries. According to Engler (1990), approximately 10 percent of the U.S. and global dredged material is contaminated. The EPA and the US Army Corps of Engineers use a variety of tests for contaminants, ranging from simple water leachate tests to multiorganism benthic bioassays. As might be suspected, the highest concentrations of contaminants in organisms and sediments do not occur in the central portions of bays and estuaries, but rather within the more narrow, confined, and heavily-used areas such as harbors and shipping channels.

In addition to toxins in the sediments, other environmental dredging concerns are turbidity, burial of the benthos, and food web contamination. Dredging causes *turbidity* (sediment-laden water), whether it be at the haul up or release stage. (See Figure 6-2). The disposal process creates the most turbidity. A plume of sediment-laden water can impede plant photosynthesis by blocking sunlight, as well as clog the gills, mouth organs, and respiratory surfaces of marine organisms. Most plumes of suspended sediment disappear within 20 minutes, but this depends on such factors as site conditions, currents, and type of disposal equipment used. Although turbidity can stress an ecosystem and its creatures, it is usually nonlethal. *Burial of the benthos* can be a concern when rich biological communities are disturbed or buried by dredging materials. Released sediments can smother worms, clams, crabs, and other life on a bay or estuary floor. Finally, *food web contamination* might result from dredging operations. Most organic contaminants and heavy metals remain connected or bound to particles of sediment. However, dredging operations redistribute these contaminants into the water column, possibly making these contaminants available to estuarine and marine organisms. Whereas many worms, mussels, clams and other bottom-dwelling organisms may tolerate or metabolize contaminants, organisms higher up the estuarine and marine food web seem to have less tolerance (e.g., reproductive problems or skin disease in certain fish). The fate of disposed sediments is still largely unknown and is the subject of much research and scientific debate.

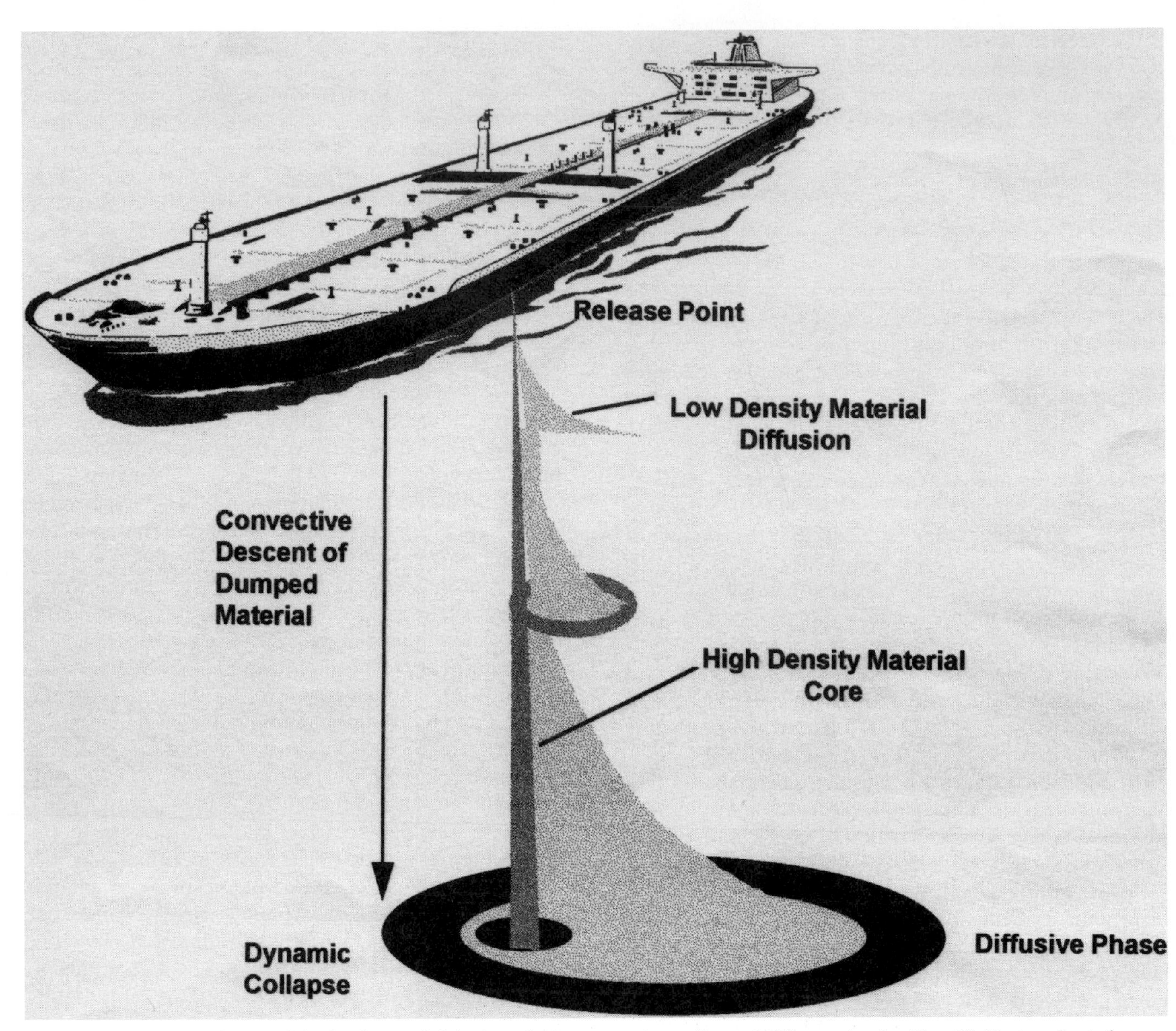

FIGURE 6-2 Phases of dredged material descent during open-water disposal. (Illustration by Char Holforty; adapted from San Francisco Estuary Project 1993)

Regulatory agencies, laws, & management strategies. The *U.S. Environmental Protection Agency (EPA)* establishes criteria or regulatory controls through the authority of the *Marine Protection Research and Sanctuaries Act (MPRSA) of 1972.* This act also authorizes the *U.S. Army Corps of Engineers* to issue permits, and to apply the EPA's controls. Dredging operations must also fit state *Coastal Management Plans (CMPs)* that are authorized under the Federal *Coastal Zone Management Act (CZMA).* On an international level, ocean placement of dredged material is regulated by the *London Dumping Convention (LDC).* Fortunately, the EPA's regulations and the LDC's approach are compatible. They both have procedures for assessing dredged material, such as analyzing its toxicity, reviewing placement site characteristics and placement methods, and considering placement site alternatives.

There are five basic management strategies for dredged materials: (1) local redistribution; (2) long distance transportation; (3) beach nourishment; (4) habitat development; and (5) agricultural land development. Perhaps the simplest and least expensive dredging management strategy is *local redistribution.* A good example of this strategy would be a suction dredge that clears out a navigation channel by sucking in the sediments then discharging the dredge spoils just a few hundred yards away. Sometimes, however, *long distance transportation* of dredged material is required, when dredge

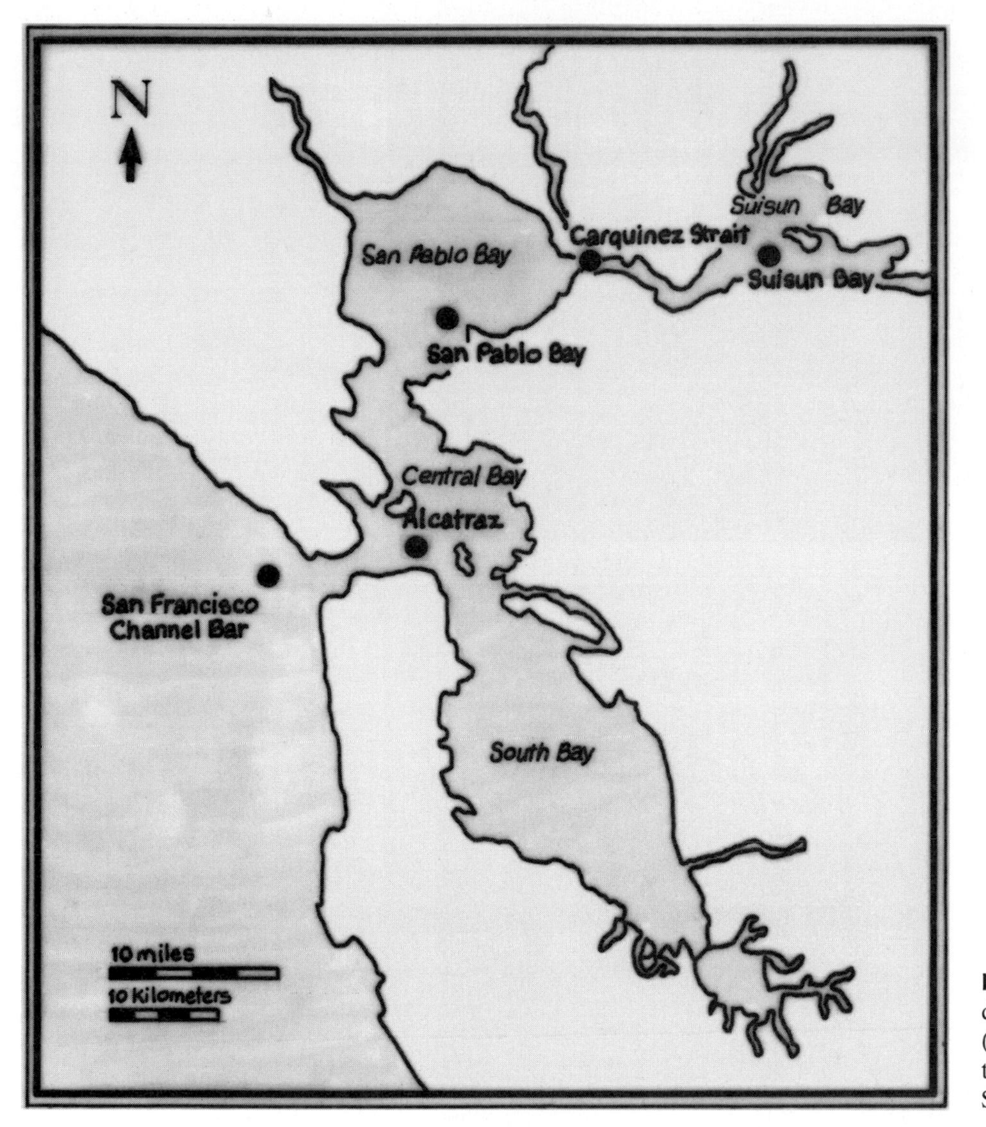

FIGURE 6-3 Active aquatic dredge disposal sites in the San Francisco Bay (estuary), California. (*Source:* Illustration by Susan Giles. Adapted from San Francisco Estuary Project 1993)

material is emptied directly onto barges which transport the material elsewhere. If, and only if, the dredge material is of the uncontaminated type (and if it is of compatible grain size), it can be used in positive ways, such as in *beach nourishment* (building or restoring a beach with periodic additions of dredged sand), and *habitat development,* such as creating artificial islands for wildlife.

Dredging in or near marine sanctuaries. Dredged material can be disposed of *adjacent* to marine sanctuaries as long as the disposal process conforms to federal law. A case in point is the current dredge dump site proposed outside the Golden Gate Bridge, San Francisco that could possibly affect two marine sanctuaries—the Monterey Bay National Marine Sanctuary and the Gulf of the Farallones National Marine Sanctuary. In the next 50 years, it is projected that 400 million cubic yards of dredged spoils will have to be deposited from San Francisco Bay's shipping channels, ports, and small marinas. In the past, dredgers dumped sediments throughout S.F. Bay and even in the Gulf of the Farallones.

Current disposal sites for dredged spoils in the S.F. Bay are restricted to sites at Alcatraz Island, the Carquinez Strait, Central San Pablo Bay, Suisun Bay, and San Francisco Channel Bar (See Figure 6-3). The Alcatraz site has been used the longest (since 1890), and has taken the most dredged material. This site may receive more than two loads per day (San Francisco Estuary Project 1993) (See Figure 6-4).

In 1982, the U.S. Corps of Engineers became concerned about "Alcatraz Mounding"—the buildup of dredge material from a previous site depth of 33.5 m (110 ft) to merely 8.5 m (28 ft). This, of course, posed a serious navigational hazard for large draft shipping. Despite having redredged the site, the mound reappeared within a

FIGURE 6-4 Dredge barge near Alcatraz Island, San Francisco Bay. (*Source:* Photos by author)

few years. Natural resource agencies and local fishermen were also beginning to complain that the Alcatraz disposal operation may be reducing the Central Bay fish catch. It was believed by some that the turbidity from dumping drives forage fish (e.g., anchovies, smelt) out of the bay, thereby sending the larger predator species away from mid-bay fishing grounds.

To help tackle the "Alcatraz Mounding" problem, as well as to facilitate long-term planning, several regulatory agencies in 1990 established the Long Term Management Strategy (LTMS)—a cooperative effort to launch a 25 year plan for dredging within the region. LTMS brings together state and federal agencies, port officials, fishing groups, environmentalists, and others into a planning group. This planning group has made progress on overcoming decades of regional stalemate on the issue of dredging. In 1996, the EPA, the U.S. Corps of Engineers, and NOAA, issued the LTMS' draft environmental impact statement which identified the most desirable ocean disposal site. Four sites were initially considered, but a fifth site 16 km (10 mi) from the boundary of the Gulf of the Farallones National Marine Sanctuary and about 32 km (20 mi) from the northern edge of the Monterey Bay Sanctuary seemed to offer the least impact on fisheries, adjacent sanctuaries and other marine resources. (See Figure 6-5).

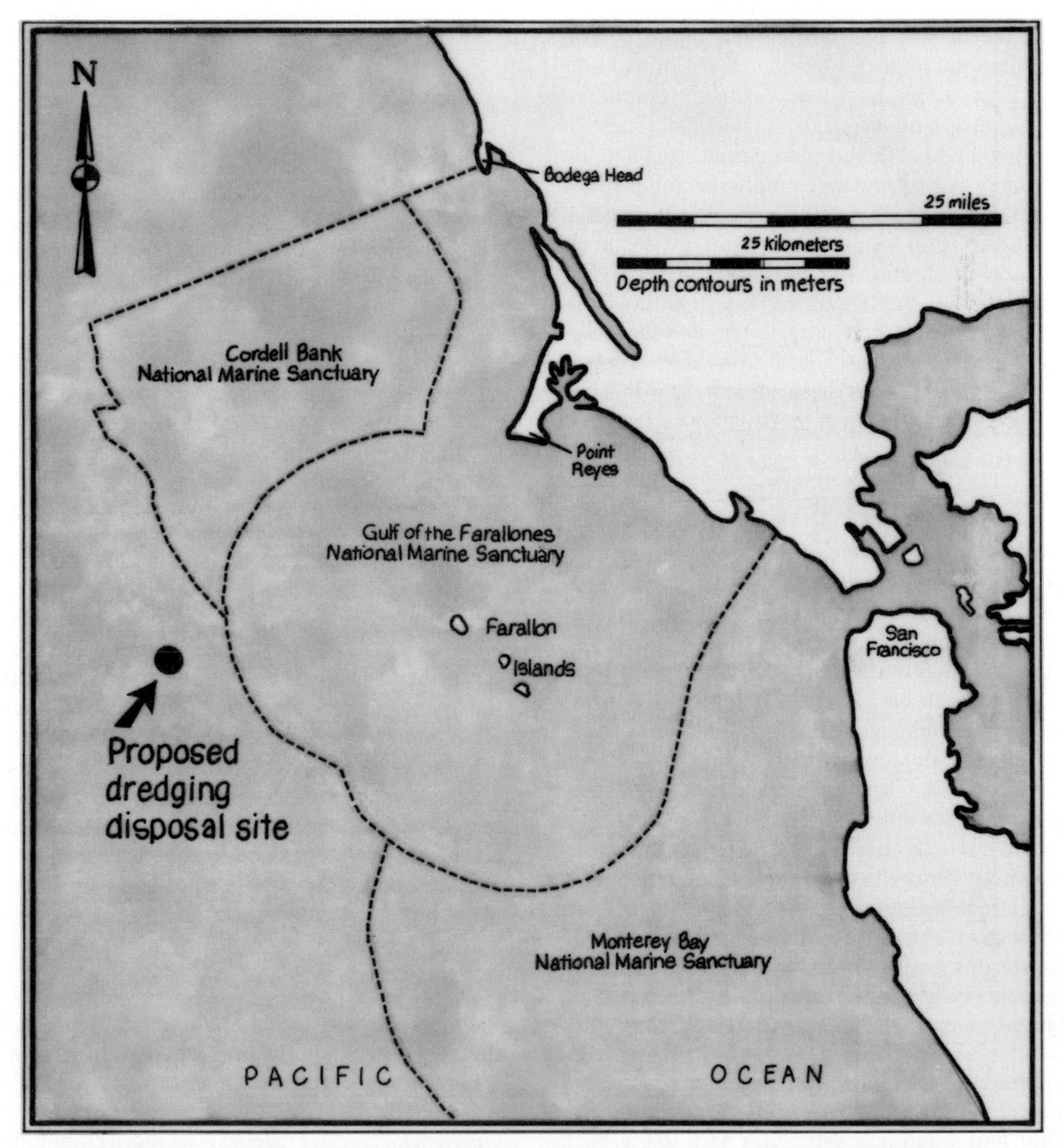

FIGURE 6-5 Proposed dredging disposal site near the Gulf of the Farallones National Marine Sanctuary. (*Source:* Map by Susan Giles)

Being 2.4 km (1.5 mi) beneath the ocean surface, this proposed dumping zone would be the nation's deepest dredge spoil dump site. The site is favored by many because of its depth and because old chemical weapons were dumped in the area from 1951 to 1954. In other words, the area has not been "pristine" for decades. Questions still remain, such as how will the material be transported, and how often? Will the impacts on marine life be monitored, and if so, who will pay for the monitoring? And, how reliable are computer models that attempt to predict sediment movement and suspension? Other questions remain as well, such as what will be the impact on northern fur seals and California gray whales, and what are the land disposal alternatives that may or may not be available?

These questions are to be addressed in the final edition of the Long Term Management Strategy. The LTMS' draft environmental impact statement also outlined four long-term disposal plans which variously combined aquatic disposal, upland disposal, and beneficial reuse of dredged spoils. The preferred plan emphasizes aquatic disposal initially, with increasing emphasis on beneficial reuse as sites become both available and feasibly usable. A finalized Long Term Management Strategy is expected to begin implementation late in 1998. Until the LTMS is completed, the S.F. Bay Regional Water Quality Control Board and the S.F. Bay Conservation and Development Commission (BCDC) are allowing dredged materials to be placed at existing in-Bay disposal sites (San Francisco Estuary Project 1993).

Sludge

History of the practice. One of the problems associated with primary, secondary, and tertiary sewage treatment plants is what to do with the slurry or mud-like settlings called sludge. **Sludge** is the combination of liquid and solid waste that remains after sewage treatment. Prior to the 1991 federal ban on all ocean dumping of sewage sludge, sludge was "managed" by disposing it in conventional landfills (42 percent), which could contaminate groundwater, by using it as fertilizer or as a soil conditioner for agricultural lands (25 percent), by incinerating it (21 percent), which could pollute the air with toxic chemicals, and by dumping it into the ocean (6 percent), which could affect the marine environment (Miller 1992). In the late 1980s, over 8 million wet tons of sewage sludge per year were dumped into U.S. coastal waters, making this the second largest amount of permitted material dumped into U.S. coastal waters (Moore 1992).

Effect on the coastal environment. The practice of dumping sludge into the **neritic** zone occurred on our Pacific, Gulf, and Atlantic coasts (refer back to Fig. 6-1). Depending on its source, sludge can be relatively benign or highly toxic. The toxic variety can contain disease-causing microorganisms and pathogens, pesticides, polychlorinated biphenyls (PCBs), and even heavy metals such as lead, copper, cadmium, chromium, and zinc. Sludge can have one or all of the following effects on marine ecosystems (Owen and Chiras 1990):

- *Reduced Dissolved Oxygen.* Because of the high biological oxygen demand (BOD) of much of the waste, the concentration of dissolved oxygen at an ocean dump site may be reduced—in some cases, to less than 2 ppm. This can cause populations of crustaceans to fall sharply, which in turn, would reduce the number of commercially valuable species of fish.
- *Chromosome damage to fish.* Highly toxic sludge has been noted to cause a number of harmful mutations in fish (e.g., young mackerel) resulting from chromosome damage.
- *Abnormal items ingested by fish.* Not all sewage waste is "filtered and refined" into a slurry-like consistency. In other words, some waste items (e.g., cigarette filters, hair, bandages) somehow get through the filter process and end up whole in the stomachs of bottom-dwelling fish like flounders that are caught near ocean dump sites.
- *Increased occurrence of disease within fish.* A greater occurrence of *Black gill disease* of fish has been found at some sludge ocean dumping grounds. Fish suffering from this disease have abnormally dark gill membranes and reduced respiratory function. *Fin rot disease,* a skin necrosis that infects summer flounder and bluefish, has also been associated with sludge dumping (Bulloch 1989).
- *Abnormal levels of toxic metals within fish.* Some fish have been noted to have extremely high levels of such toxic metals as chromium, lead, and nickel within their systems.

The former "Deep-Water Dumpsite 106" off the coast of New Jersey serves as a case in point. (See Figure 6-6). According to Dover et al., (1992), serious pollution of the open ocean had occurred since 1986 through disposal of municipal sewage sludge at 2500m (8202 ft) underwater at a site 198 km (123 mi) off the New Jersey coast. Through analysis of the ratios of certain isotopes, these scientists found that the sewage sludge had accumulated on the sea floor and entered the benthic food web. Dispersal and dilution of the sewage had proved inadequate to prevent the accumulation. Urchins and sea-cucumbers feeding on surface deposits provided the point of entry into the food web.

Regulatory agencies, laws, and management strategies. In the United States, the ocean dumping of sewage sludge is regulated by the *Environmental Protection Agency (EPA)* under the *Marine Protection, Research and Sanctuaries Act (MPRSA)*—more commonly known

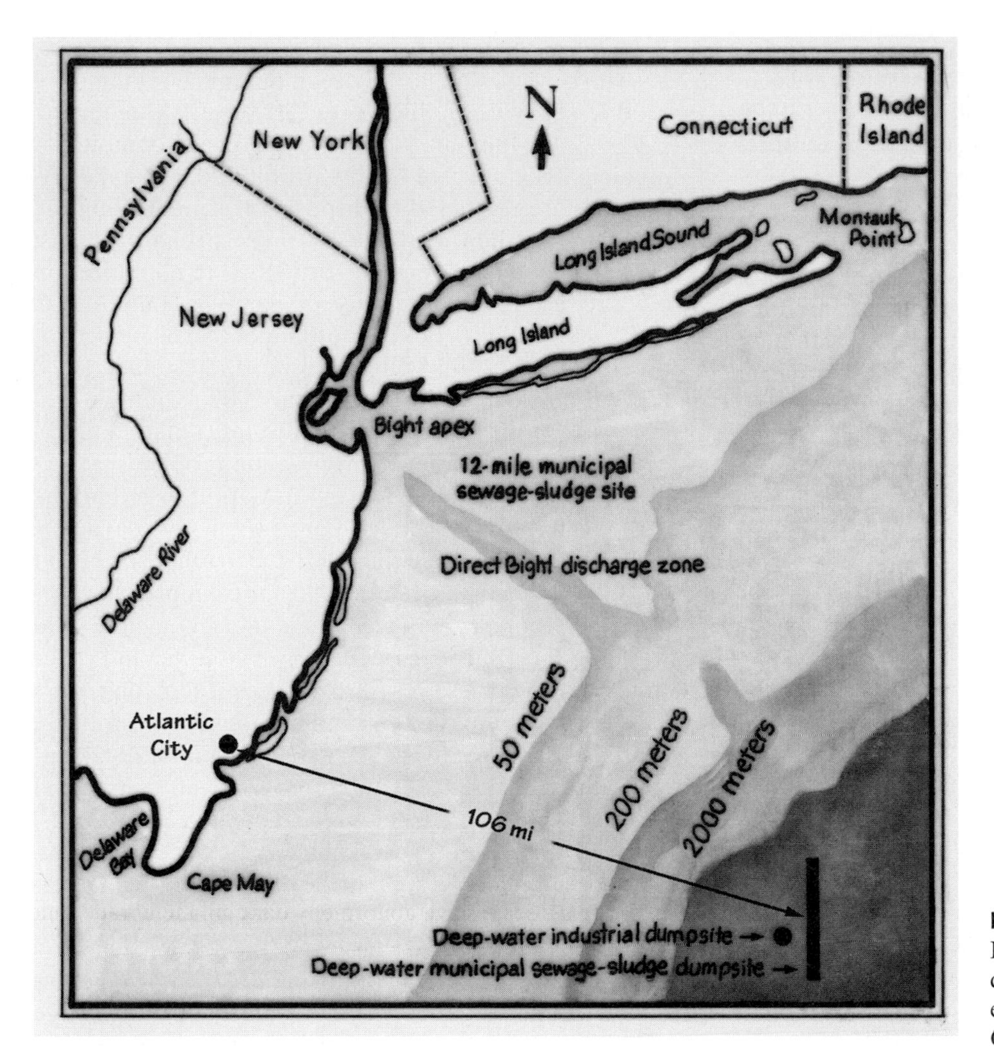

FIGURE 6-6 The New York Bight. Illustration of the 12-mile municipal dump site versus the newer, and deeper, 106-mile site. (Adapted by Susan Giles; from Stegeman 1990, p. 61)

as the *Ocean Dumping Act.* The Ocean Dumping Act, passed by Congress in 1988, mandated that all ocean dumping of sewage sludge and industrial waste end as of December 31, 1991. Whereas other U.S. cities were able to comply, New York City (the nation's biggest ocean dumper) had to request a postponement until June 1992. By that time, over 32 million wet tons of sewage sludge had been dumped at the "106" site (Refer back to Fig. 6-6). (From 1924 until EPA's 1984 decision to move the dump site to deeper waters 106 miles off the coast of Atlantic City, sewage sludge had been dumped at the New York Bight Apex, called the "12-mile site" [Kitsos and Bondareff 1990]). Although the effects of ocean dumping at the "106" site originally did not seem severe, later studies found that pollutants were penetrating into the deepest layers of the sea. For additional information on the dumping of sludge in the New York Bight area, refer to NOAA (1991) and Swanson (1993).

American cities, including New York, no longer dump sludge into the ocean, but their sludge disposal problem has yet to be resolved. City politicians maintain that it has become almost impossible to export, bury, or burn their city's sludge. Furthermore, most American citizens are not eager to live near a plant designed to process sludge, either by turning that waste into useful compost or by baking it into fertilizer pellets at tremendous heat. Yet, reusing "clean" sludge for agricultural or urban landscaping purposes seems to make the most environmental sense, since sludge contains high levels of nitrogen and phosphorus, and because its water-retention capacity makes it a good soil conditioner. When metals and other poisons are eliminated, it can be better than most manure.

For example, the flowers and lawns of Gracie Mansion in New York City have prospered with its use, as well as the shrubs and grasses of Bryant Park in midtown Manhattan. Compared to just 25 percent in the United States, 48 percent of the weight of sludge produced in Denmark is used to fertilize land (Miller 1992). If the sludge is left untreated, it still can be used on land not used for livestock or crops, such as surface-mined land, golf courses, forests, highway medians, and cemeteries. For additional insight into how East Coast states are converting to land application, composting, chemical stabilization, and pelletization of sludge, as well as to other conversion strategies, see Steuteville (1992).

Sludge dumping in or near sanctuaries. For the time being, sludge dumping is prohibited in all coastal waters—not just within or near marine sanctuaries. Although the federal Ocean Dumping Act (1991) bans sewage sludge dumping, sewage still finds its way into coastal waters. According to Moore (1992), Los Angeles, Boston, and other cities skirt the Marine Protection, Research, and Sanctuaries Act by either occasionally discharging a less concentrated version of sludge (not raw sewage, but not yet sludge) into the ocean, or by discharging sewage sludge into rivers and harbors. No doubt some of this sewage sludge and raw sewage dumping has affected a number of the sanctuaries within our national marine sanctuary program. Further research needs to be done in this area.

For a detailed account of Publicly Owned Treatment Works (POTWs) in the nation in 1988, and the estimated concentrations of pollutants of concern in the sewage sludge of those plants, see EPA (1992).

Solid Waste (Garbage)

History of the practice. For decades, barges and ships dumped solid waste at designated sites off the Pacific, Gulf, and Atlantic coasts. In some cases, the accumulation of waste materials became so great that its spatial distribution could be mapped. The New York Bight, for example, was one of the most heavily used dumping areas in the world. Every year, approximately 10 million tons of various types of waste were dumped into this area, covering 104 sq km (40 sq mi) of the ocean bottom (Miller 1988). The dumping of solid waste in New York Harbor and the Bight ceased in 1934 (Dieterichs 1994).

Effect on the environment. The effect on the marine ecosystem has been lower oxygen concentrations and the reduction or elimination of populations of small marine plankton (protozoans, crustaceans, and algae) and crabs. Plastic debris alone, for example, injures and kills marine mammals, sea birds, and sea turtles. Marine debris often causes injury among sea animals, which can lead to infection and death. Sea animals may become entangled, which affects their mobility and ability to feed, and somtimes causes them to drown. They may ingest debris and subsequently starve. Plastic debris also poses a threat to humans, such as when divers get tangled in discarded or lost plastic fishing lines and nets.

Regulatory agencies, laws, and management strategies. As with sewage sludge, the ocean dumping of municipal solid waste is also administered by the EPA under the 1991 Ocean Dumping Act. Since U.S. ocean dumping of municipal solid waste is now prohibited, coastal communities are restricted to disposing (or minimizing the problem) of their garbage by three primary means: (1) landfills, (2) incineration, or (3) waste minimization (i.e., precycling, recycling, reuse of materials, composting).

By the year 2000, the EPA estimates that 50 percent of the nation's landfills will be closed, and as many as 80 percent may be closed by the year 2009 (Kaufman and Franz 1993). Hazards (e.g., groundwater contamination), noise, odors, and other environmental problems associated with landfills are causing this nation to rethink its waste management strategy. For years, many heavily urbanized states exported their garbage to other states, and even to foreign countries. In 1990, for example, New Jersey exported more than 55 percent of its municipal solid waste [MSW] (Spencer 1990). However, fewer and fewer states and countries are willing to accept someone else's garbage, as exemplified by the infamous New York City "garbage barge" that was rejected by a number of countries in Central America and the Caribbean (Kitsos and Bondareff 1990).

Incineration is also not a long term management strategy for our municipal solid waste. **Incineration** can be defined as the controlled process by which solids, liquids, or gaseous combustible wastes are burned and changed into gases. **Waste-to-energy incinerators,** one type of incinerator, at first seem attractive since they burn garbage to produce heat and steam, which can then be used to produce electricity. In 1990, 128 waste-to-energy plants in the United States incinerated 16 percent of the nation's solid waste and produced enough electricity for 1.1 million homes (Kaufman and Franz 1993). During the same year, 29 new onshore incineration plants were under construction in the United States; another 64 proposed plants were held up in litigation because of a combination of factors (Spencer 1990).

Why the delay? Incineration has three major drawbacks: First, there is the threat to human and environmental health. Even with the best pollution control devices installed, incinerators may release into the atmosphere small amounts of acid gases (e.g., hydrogen chloride), metal particulates (e.g., mercury), and various toxic chemicals (e.g., dioxins). Consequently, few communities want an incineration plant in their backyard. Since incineration only reduces the volume of waste by 75 to 90 percent, there is still the environmental question of what to do with the fly ash. This fly ash must be landfilled in a safe place, since it may contain high concentrations of toxic heavy metals (e.g., lead and cadmium).

Second, there is the economic drawback. A large incineration plant fully equipped with state-of-the-art pollution control devices may cost as much as $400 million to build. Finally, incineration is wasteful in itself. Much of what it uses for fuel (e.g., paper) are unrealized resources that can be either recycled or reused.

As the public becomes aware of the myriad of environmental and economic problems associated with ocean dumping, landfills, and incinerators, municipal agencies are being pressured to look more seriously at *waste minimization* as a management strategy for dealing with the solid waste problem. **Waste minimization,** also known as **source reduction,** is an umbrella term that includes reducing the volume of material goods, eliminating unnecessary packaging, and decreasing the amount of toxic substances in products. In Germany, for example, a 1991 law now requires that manufacturers, wholesalers, and retailers *take back* packaging returned to them by their customers. The law places the burden of disposal of unnecessary packaging on the businesses that create the waste in the first place (Kaufman and Franz 1993). The concept of waste minimization includes **precycling** (i.e. consciously purchasing merchandise that has a minimal adverse effect on the environment) and **resource recovery** (i.e., extracting useful materials or energy out of solid wastes before ultimate disposal; methods include **recycling** and **composting**). Put waste minimization practices together with the use of landfills and incineration and you have a *combined* management strategy that is now known as an **integrated solid waste system.** Related to the notion of waste minimization is **green marketing**—the promotion of products that help or are benign to the environment.

Solid waste dumping in or near sanctuaries. It is no longer legal in the United States for municipalities to dump their garbage into the sea, regardless of whether it is in or near a marine sanctuary. However, garbage is still generated on and disposed of by ships during their ocean voyages, though commercial ships are facing more and more restrictions on such practices. For example, *MARPOL Annex V* of the *International Convention for the Prevention of Pollution from Ships* became effective on December 31, 1988. In brief, it changed the way shipboard wastes were disposed. The *Marine Plastics Pollution Research and Control Act (MPPRCA)* of 1987 implemented *MARPOL Annex V* in the United States. The Act prohibits the discharge of any plastics into the waters of the United States, and requires that every port, harbor, or marina provide and maintain adequate waste reception facilities (NOAA 1992; Hollin and Liffmann 1993). In other words, "ports of call" now receive wastes that were traditionally dumped at sea.

Industrial Wastes

History of the practice. Today's younger Americans would probably be shocked to learn that this nation used to regularly dump shiploads of highly toxic chemicals and other industrial wastes annually into our coastal waters. As late as the 1980s, for example, approximately 300,000 wet tons of industrial wastes were dumped every year at New York's 106-Mile Dump Site. Hydrochloric acid and liquid solutions of organic chemical wastes from industrial processing were dumped at the site (NOAA 1985). Additional industrial wastes at other dump sites have included scrap metal, fish by-products, coal ash, and flue-gas desulfurization sludges.

Effect on the environment. Many physical, biological, chemical, and benthic processes affect the distribution and fate of industrial (and other) waste in coastal waters. (See Figure 6-7). Unfortunately, scientists have only begun to study the ecological complexities of ocean dumping. We need to know more about the physical processes (e.g., how currents influence contaminant distribution), chemical processes (e.g., how chemical processes affect availability, persistence, and degradation of materials in waters and sediments), and long term biological effects (e.g., how fisheries and other animal populations are impacted over the long haul). As Capuzzo (1990) has nicely pointed out, the answers to these questions will not be found by any one academic discipline. It will take multidisciplinary/interdisciplinary teams of toxicologists, ecologists, oceanographers and other scientists to truly understand the causal relationship between pollution (especially industrial waste dumping) and coastal degradation.

In the meantime, there has been a growing concern over the impact of chemical contamination on fisheries and other coastal resources. Health advisories have warned the public against frequent consumption of fish caught off Southern California (early 1980s), of eating striped bass caught off New York and Rhode Island (1986), and of eating tomalley (the green liver of lobsters) from Quincy Bay, Massachusetts (1988) (Capuzzo 1990). Why the concern? Much of it was because of the fear of possible PCB contamination.

Regulatory agencies, laws, and management strategies. The ocean dumping of industrial wastes is regulated in the U.S. by the Environmental Protection Agency under the authority of two federal laws—the *Marine Protection, Research, and Sanctuaries Act (MPRSA, the so-called "Ocean Dumping Act")* and the *Resource Conservation and Recovery Act (RCRA).* After reexaming the weaknesses of the original 1972 Ocean Dumping Act,

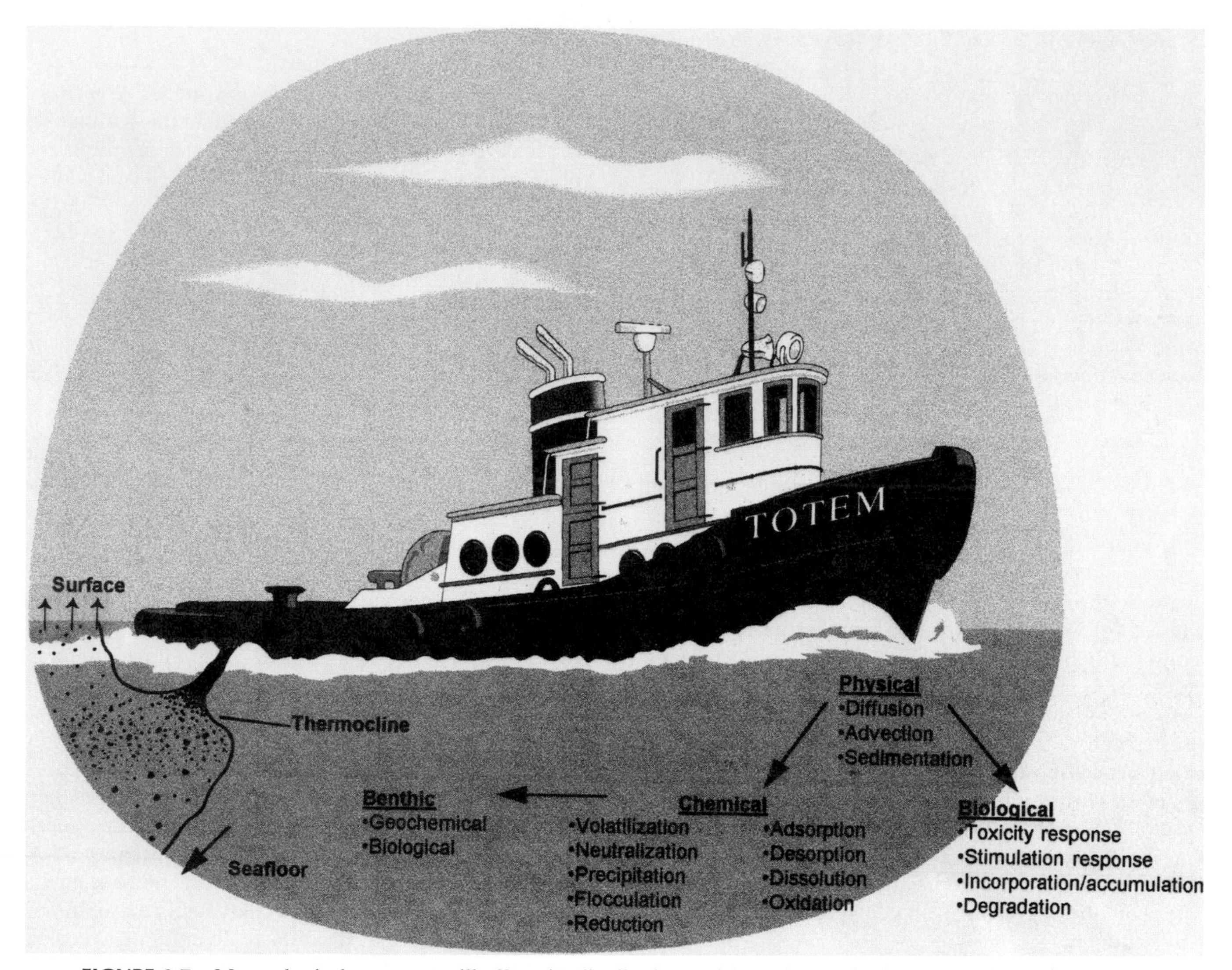

FIGURE 6-7 Many physical processes will affect the distribution and fate of waste in the water column. (*Source:* Illustration by Char Holforty; adapted from Duedall 1990, p. 31)

Congress came up with the *Ocean Dumping Ban Act of 1988* which made it very clear that all ocean dumping of industrial waste (and sewage sludge waste), whether or not it unreasonably degraded the marine environment, would cease after December 31, 1991 (Kitsos and Bondareff 1990).

On the international front, European and other international communities have also strengthened their laws regarding ocean dumping of industrial wastes, through the *London Dumping Convention* (*LDC* [Bakalian 1984]).

Although coastal disposal of industrial waste has been steadily declining because of national and international laws and enforcement, as the landfill alternative becomes harder to come by, pressure will no doubt mount to again dump these wastes into the open sea (See Section III "Future Ocean Disposal" in this chapter). To counter this future threat, greater efforts need to be made at waste minimization as a management strategy for industrial wastes. In the United States, for example, the Minnesota Mining and Manufacturing (3M) Company has won a lot of praise from conservationists for its "3P" program (Pollution Prevention Pays). By such waste minimization strategies as process modification, product reformulation, equipment redesign, reduction of packaging, and recovery of wastes for reuse, the 3M Company has been able to conserve resources, reduce its waste stream, and reduce its costs of production. The company maintains that by achieving a 50 percent reduction in potential hazardous waste generation over the past 10 years—the environmental benefits—it has saved over $300 million (Kaufman and Franz 1993).

Some farmers are also experimenting with innovative ways to make their farms more economically viable while

FIGURE 6-8 From industrial waste to resource. (Left) A northern California farming operation that uses industrial wastes as a soil additive; (Right) Piles of industrial wastes (e.g., crushed wallboard, paper pulp effluent, sawdust and wood chips) used as soil additives. (*Source:* Photos by author)

helping industries reduce their waste management problems. For example, Veale Tract Farms in Contra Costa County, California is experimenting with those resources that can be used as *soil amendments* (any substance such as lime, used to alter the properties of a soil), as *fertilizers,* or as *recycled industrial materials.* Since 1975, they have been recycling wallboard, sheet rock (gypsum aids in the leaching of toxic salts), paper pulp effluent, sawdust, and wood chips (valued for their organic matter), lime sludge (used for pH control and soil structure), and many other materials. (See Figure 6-8).

Although it is a time-consuming process to search out usable by-products and to clear their use with various governmental agencies, the outlook for such recovery operations is promising. As the costs of disposing unwanted materials rise, or as regulations against such dumping become more severe, it becomes more and more economical (and wise) for industrialists and farmers to give serious attention to resource recovery.

Industrial waste dumping in or near sanctuaries. Fortunately, industrial waste dumping is no longer permitted off the coasts of the United States. The U.S. Environmental Protection Agency (EPA) "undesignated" its last industrial waste dumping site in 1990 (Moore 1992). Consequently, industrial waste dumping is no longer permitted in or near U.S. marine sanctuaries.

Military Wastes

History of the practice. The U.S. military has historically dumped two types of wastes in the world's oceans: (a) food garbage and (b) outdated war materials (e.g., outdated ships, weapons, and ammunition). The public would probably be surprised (if not appalled) to learn how much garbage the military dumps at sea. For example, a 1990 Coast Guard study indicated that the U.S. Navy dumps 63,356 tons of garbage a year into U.S. waters, more than all commercial passenger ships combined.

In 1993, an "ecological sailor" from Santa Cruz, California helped focus public attention on the U.S. Navy's disposal practices. Aaron Ahearn (America's first environmental conscientious objector?) abandoned the aircraft carrier *USS Abraham Lincoln* after refusing to dump tons of trash a day into the ocean. His job included heaving up to 200 plastic bags a day into the ocean, each filled with plastic and food garbage. Ahearn also maintained that sailors routinely dumped solvents, corrosives, and large debris—such as desks and broken computers—into the sea. Because of his background as a surfer and participation in beach cleanups as a child, Ahearn decided he could no longer in good conscience continue the dumping. After being denied a change of assignments, he decided to abandon ship. Immediately, this 20-year-old surfer from Santa Cruz became a celebrity to environmental groups that have been fighting for years to eliminate loopholes in U.S. dumping laws.

Navy rules and federal loopholes. Government vessels, including those of the U.S. Navy and U.S. Coast Guard, are exempt from the 1988 federal law (*Marine Plastics Pollution Research and Control Act,* also known as *MARPOL*) that bans cruise ships and other boats from discharging plastics in U.S. water. Cruise ships, for example, can be fined up to $500,000 per offense for the dumping of plastics to the ocean, and MARPOL gives the U.S. Coast Guard authority to prosecute dumpers within

322 km (200 mi) of the U.S. coast. In the "Aaron Ahearn case," most of what this surfer-sailor saw go over the *USS Abraham Lincoln* was probably *legal.* This 333 m (1,092-ft) aircraft carrier carries a crew of approximately 5,500 sailors—more people than many U.S. towns. According to Navy estimates, each sailor at sea generates an average of 1.4 kg (3.1 lbs) of garbage a day. Because ships have limited storage space and sanitary facilities, Congress allowed this loophole in the dumping law. Consequently, Naval ships can legally dump plastics, food, and other nontoxic trash overboard more than 80 km (50 mi) offshore. Greenpeace, the Center for Marine Conservation, the Surfrider Foundation, and other environmental groups interested in ocean and coastal protection see such federal loopholes as a double standard.

To help eliminate the controversy, the Navy intends to install new plastics compactors by 1998, as well as become more aggressive with their existing recycling programs. Navy regulations require all toxins and medical waste to be brought ashore. This does not mean, however, that all sailors follow the law. According to Ahearn, for example, sailors aboard the *USS Abraham Lincoln* routinely chose to dump paint and other toxic chemicals directly overboard as opposed to using the special hazardous waste containers that were supposed to be used. Commanders on the *Lincoln* eventually investigated Ahearn's charges, and announced in August 1993 that his claims of toxic dumping were without merit and that his allegations that sailors dumped broken desks, chairs and computers could not be corroborated. Regardless of what the real truth may be, ocean and coastal advocacy groups maintain that the "Ahearn Incident" alerted the public to loopholes in the ocean dumping regulations, and, consequently, helped spark the Navy to write more strict rules regarding their ocean dumping practices.

The Navy has made some progress in its garbage processing research program. It has developed new technology that can be transferred to the world's private fleets, including a plastic waste processor that shreds, compresses, and heats plastic into solid disks that can be stored onboard ships for later disposal on land or even for recycling. Coastal advocates maintain that it must do more to reduce the amount of plastics taken aboard its vessels *in the first place,* and to devote more attention to *recycling.*

The military has also dumped defense equipment that was no longer needed or wanted. In Spring 1993, for example, an underwater exploration scientific team discovered—and videotaped—discarded military waste objects on the ocean floor in Cambridge Bay in the Canadian Arctic. For years during the cold war, the U.S. air force dumped tons of highly toxic polychlorinated biphenyl-(PCB) filled capacitors and other forms of contaminated equipment (Bochove 1993). Later that same year, Russia admitted dumping 900 tons of liquid nuclear waste from its aging fleet of nuclear-powered submarines into the Sea of Japan. This violation of the 1983 international moratorium on ocean dumping sparked widespread protest in Japan and the United States, resulting in a promise from Russia to temporarily refrain from future dumping. However, Russia also made it clear that it would resume the practice unless richer countries helped to process the waste for underground burial.

On February 21, 1994, an international ban on the dumping of radioactive waste at sea went into effect, but as stated above, Russia would not comply. The ban follows an agreement reached in November by the *London Dumping Convention*—the international organization concerned with the prevention of marine pollution by dumping wastes.

Despite the fact that the U.S. and Britain have signed an international agreement that bans nuclear dumping at sea, these two countries plus Russia are still considering the seabed as a future graveyard for old submarines and other discarded military equipment. In 1994, for example, the U.S. Navy was studying the options for disposing of 100 nuclear reactors from decommissioned vessels, including the possibility of sending them to the bottom of the ocean. International arms limitation agreements further complicate the issue. One of the reasons that Russia has yet to sign the dumping ban is that international arms limitation agreements require that it now scrap more than a hundred nuclear submarines, and they see their only alternative is to dump (scuttle) these submarines in the deep ocean (MacKenzie 1994).

Military waste dumping in or near marine sanctuaries. Military waste dumping has long been practiced in or near current date marine sanctuaries. Even in the relatively pristine Monterey Bay National Marine Sanctuary, there have been three areas used for the disposal of military explosives and wastes. (See Figure 6-9).

One military dumping site, about 64 km (40 mi) west of San Gregorio Beach, is the approximate location where the Navy scuttled the USS Independence. This small aircraft carrier was used as a target ship during the atomic bomb tests near Bikini Atoll, Micronesia in 1947. Considered too "hot" (radioactive) to be stored at Hunters Point (pier) in South San Francisco, it was purposely sunk in 1951 approximately halfway between what is now the Gulf of the Farallones NMS and the Monterey Bay NMS. A few miles south, about 48 km (30 mi) west of Pigeon Point Light House, is an approximately 583 sq km (225 sq mi) area that had been used for military explosive and waste dumping. Both of these two northern sites are no longer used by the military for dumping. The third site, the dunes and adjacent ocean waters off Fort Ord, is more of a "sacrifice area" than an official dumping site. This area

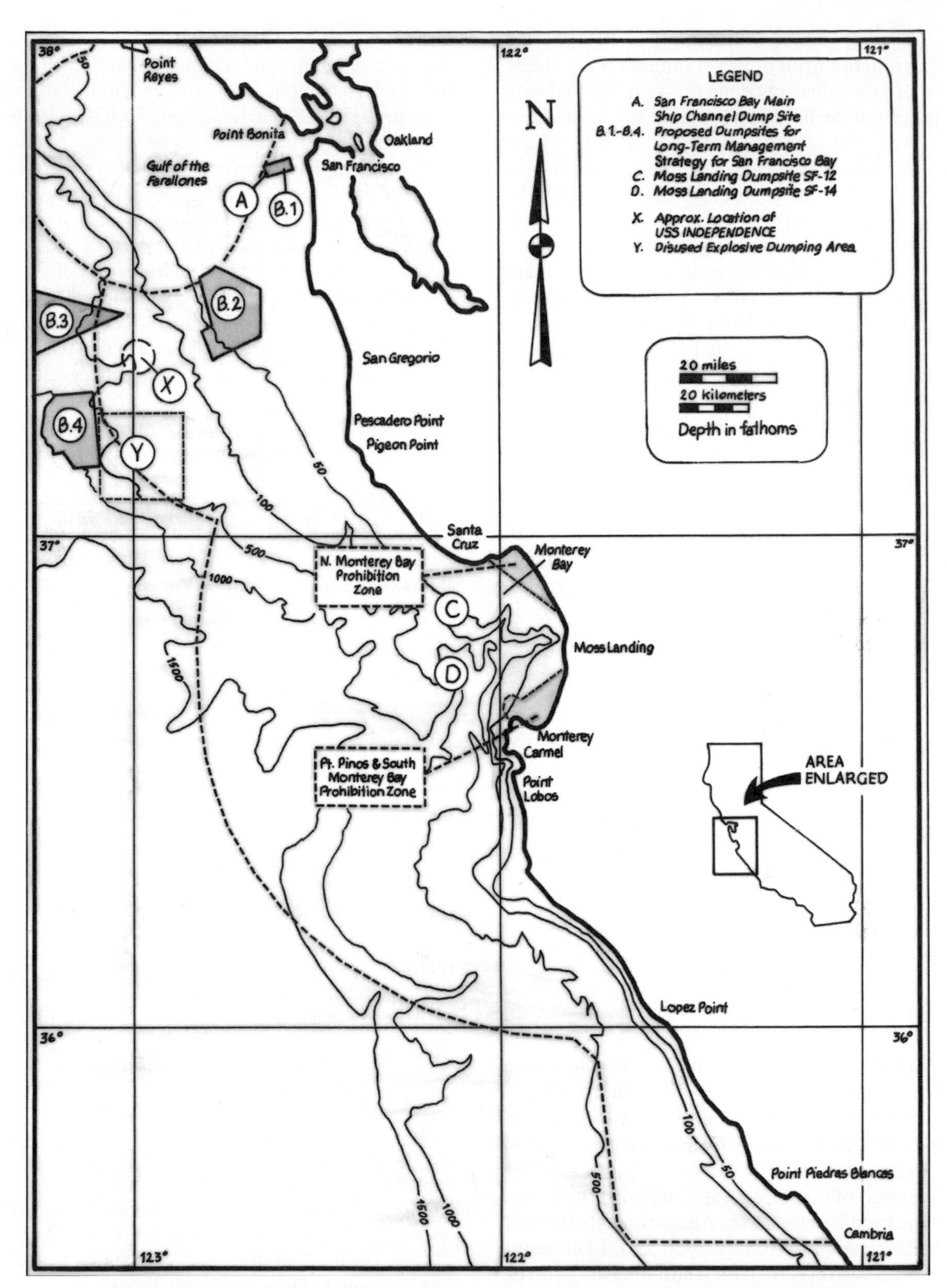

FIGURE 6-9 Ocean discharge and dump sites in the area of the Gulf of the Farallones NMS and the Monterey Bay NMS. (*Source:* Map by Susan Giles; adapted from Fig. 23 of NOAA 1992, p. II-80)

contains the spent rounds of ammunition fired by the army during practice drills at target ranges on the dunes. The dunes and the adjacent ocean floor are littered with the steel jackets of the fired bullets. In the ocean, the steel jackets of the bullets erode leaving behind a lead core (NOAA 1992). Whether these remaining lead cores have a significant negative environmental effect has yet to be determined. Now that Fort Ord is fully decommissioned, this area is also no longer in use by the military.

Some of these "abandoned ordinance devices" are reappearing in our national marine sanctuaries. In February 1995, for example, a World War II-vintage naval mine was snared by a fishing boat in the Farallone Islands National Marine Sanctuary. The mine came up with a net full of fish as the fishing boat Irene's Way was trawling for Dover sole and bottom fish. After bringing it aboard, the crew lashed the rusty device to the deck to keep it from rolling around and headed back to port in Monterey. Unable to remove the detonating trigger, the Navy eventually exploded the mine in 46 m (150 ft) of water in the Monterey Bay National Marine Sanctuary in a site off Fort Ord where no known marine mammals were believed present. The site where the bomb was detonated was first approved by the California Environmental Protection Agency. This was not the first incident of this nature. World War II era ordinance devices had previously been found in or adjacent to what is now the Farallone Islands NMS. Mines had been placed in the area to protect the Golden Gate (the entrance to San Francisco) against enemy vessels.

The placement of ammunition dumping is not found only within the MBNMS. Within the Olympic Coast NMS, several sites were once used for military dumping of ammunition. Military shipwrecks from World Wars I and II have been recorded within the Florida Keys NMS. As they slowly corrode they may someday become an environmental contamination problem for the sanctuary.

Nonmilitary Radioactive Wastes

History of the practice. America's coastal waters have also seen the dumping of nonmilitary radioactive wastes. After World War II, this country began producing a lot of "low-level" radioactive wastes from mining, industrial, research, and medical activities. Between 1946 and 1970, most (97 percent) of this waste was dumped at three sites off major metropolitan areas: the Massachusetts Bay site outside of Boston; the Atlantic 2800 and 3800 meter dumpsites off Newark, New Jersey; and the Farallone Islands dumpsite approximately 80 km (50 mi) west of San Francisco (EPA 1984).

Effect on the environment. Scientists simply do not know the effects of at-sea radioactive waste dumping. There are profound gaps in their knowledge about the complexities of how (or whether) radionuclides are transferred from the deep sea to marine organisms back to humans (Curtis 1986). To complicate matters, radionuclides enter the seas through other means as well, such as (a) fallout from past atmospheric nuclear weapons testing, (b) fallout from nuclear plant disasters (e.g., 1987 Chernobyl accident in the Soviet Union), (c) discharges by on-shore nuclear power plants, (d) discharges by nuclear fuel reprocessing plants, and (e) discharges by the world's ever-expanding fleet of nuclear-powered submarines and other warships. We do know, however, that only low levels of radioactive contaminants need be present to affect marine organisms. Marine biologists, for example, report that 0.2 microcuries of radioactivity will seriously disrupt fish egg development (Owen 1985). When it comes to the cumulative synergistic effects of radioactive waste dumping along with all these other activities, scientists are in a quandary.

Regulatory agencies, laws, and management strategies. The dumping of high-level radioactive waste has always been illegal in U.S. coastal waters. However, until recently, it has been legal to dump low-level waste with the proper permits from the Environmental Protection Agency. The EPA, in turn, is restricted under the guidelines of the MPRSA and London Dumping Convention. For example, the EPA can issue a dumping permit for low-level radioactive materials if Congress passes a joint resolution conferring such authority within 90 days of EPA's request. Since 1985, there have been no new applications for a waste dumping permit. However, as land-based storage sites become more difficult to establish, there may be a new push for at-sea and/or seabed disposal of low (and high) level radioactive waste. See "Future Ocean Disposal" in this chapter for further details.

Radioactive waste dumping in or near marine sanctuaries. As mentioned above, the legal dumping of radioactive wastes in U.S. coastal waters, including in or near sanctuaries, has not been practiced since 1985. However, past practices continue to haunt two marine sanctuaries—the Stellwagen Bank NMS and the Gulf of the Farallones NMS. Some potential exists for contamination of bottom sediments and organisms due to expected residual radioactivity and possible leakage (EPA 1993). For information on the historical practices of radioactive waste dumping and current environmental concerns related to Stellwagen Bank NMS, see Raulinaitis (1994) and NOAA (1993a, 1993b). The story of the Farallone Islands experience is provided in a case study below.

Ocean Incinerated Wastes

History of the practice. The burning or incineration of wastes, so-called "purification by fire," *on land* dates back to biblical times. In 1991, the U.S. had 136 waste-to-energy land-based plants burning 16 percent of the trash generated in this country, with 100 additional facilities at various stages of completion (Marine Policy Center 1993).

Far more controversial has been the concept of burning hazardous wastes on "incineration vessels"—incinerators mounted on ocean-going vessels. One drawback is that only liquid hazardous wastes can be incinerated at sea, such as PCBs and other organohalogens (Duedall 1990). In 1986, this country produced approximately 250 million metric tons of hazardous wastes; but only 8 percent of this waste was of the type and consistency that could be incinerated at sea (Office of Technology Assessment 1986).

Effect on the environment. The transfer of hazardous wastes from the source to the dock, or from the dock to the incinerator ship, is one source of spills and leaks that can have a negative impact on coastal waters. Even the EPA admits that 70 percent of ocean incineration spills are likely to occur in highly populated coastal areas (EPA 1985a). The incineration process itself is a potential threat to marine life, though the EPA maintains that the potential threat is minimal (Millemann 1986). Incinerator vessels release (a) tons of materials (e.g., hydrochloric acid) for every hour of sea burning, (b) unburned materials that escaped the incineration process, (c) and new synthesized products that are more lethal than the original waste stream (EPA 1985b).

Regulatory agencies, laws, and management strategies. The U.S. Environmental Protection Agency had a major program in the early 1980s to assess the viability of ocean incineration, promulgate regulations, designate sites, and evaluate permit applications. But after public outrage was exhibited at an EPA-sponsored public hearing on ocean incineration in Brownsville, Texas in 1983, the agency decided to temporarily halt the development of ocean incineration in this country. After a long battle on many fronts, the incineration program was terminated in 1988. All site designation activities were halted and the only designated site, the Gulf of Mexico Site, was de-designated after the Ocean Dumping Ban Act (ODBA) was passed by Congress.

On an international level, ocean incineration is also regulated by the London Dumping Convention. Contracting parties (LDC countries) adopted a resolution prohibiting the ocean incineration of noxious liquid wastes. According to the EPA, none of the LDC countries are currently incinerating wastes at sea (Redford 1993).

Ocean incineration in or near sanctuaries. In U.S. coastal waters, ocean incineration has only occurred at the Gulf of Mexico Site—approximately 177 km (110 mi) off the coast of Texas. However, this site is only 100 km (62 mi) south of the Flower Garden Banks National Marine Sanctuary. It is not known whether the practice of vessel incinerated wastes have negatively impacted this sanctuary. In 1995, vessel incineration was no longer occurring near the sanctuary.

On the Atlantic coast, there was a proposal in 1993 calling for the construction of vessels and facilities to allow the offshore incineration of trash from metropolitan Boston. The incineration vessels would operate within the area of the then proposed Stellwagen Bank NMS. Now that sanctuary status has been approved for Stellwagen Bank, regulations are in place that ban incineration within the sanctuary.

CASE STUDY: RADIOACTIVE WASTE DUMPING AT THE FARALLONES

It may be hard to believe, but one of the nation's marine sanctuaries—The Gulf of the Farallones National Marine Sanctuary—encompasses what was once the nation's West Coast Low-Level Radioactive Waste Dumping Site. As we will see, much controversy still surrounds the past dumping of radioactive waste at this site. Scientists and local citizens continue to debate the immediate and long-term impact on marine organisms and whether the waste is slowly moving toward the densely populated San Francisco metropolitan area. Much of the information for this case study comes from NOAA (1990) and Tetra Tech, Inc., (1992a,b,c).

History of the Practice—Where, When, and How Much?

The Farallon Islands Radioactive Waste Dump (FIRWD) is located 10 to 23 km (6–14 mi) south and west of the Farallon Islands off San Francisco, California. (See Figure 6-10). The practice began at the end of World War II (1946) and continued, interestingly enough, up to the beginning of the modern environmental movement (1970). Between those years, over 47,750 55-gallon (208-liter) steel drums, concrete boxes and other containers were dumped in this area. Four private companies carried out the actual dumping, under the auspices of the Atomic Energy Commission. A team of University of California scientists estimate that 14,500 curies of low-level radioactive wastes (transuranic radionuclides) were dumped during this period (Noshkin et al., 1978).

There were three designated dumping sites at the FIRWD: Sites A, B, and C (refer back to Fig. 6-10). The oldest site, Site A, was used only briefly in 1946. At this 90 m (295 ft) deep site, which is only 35 km (22 mi) from the city

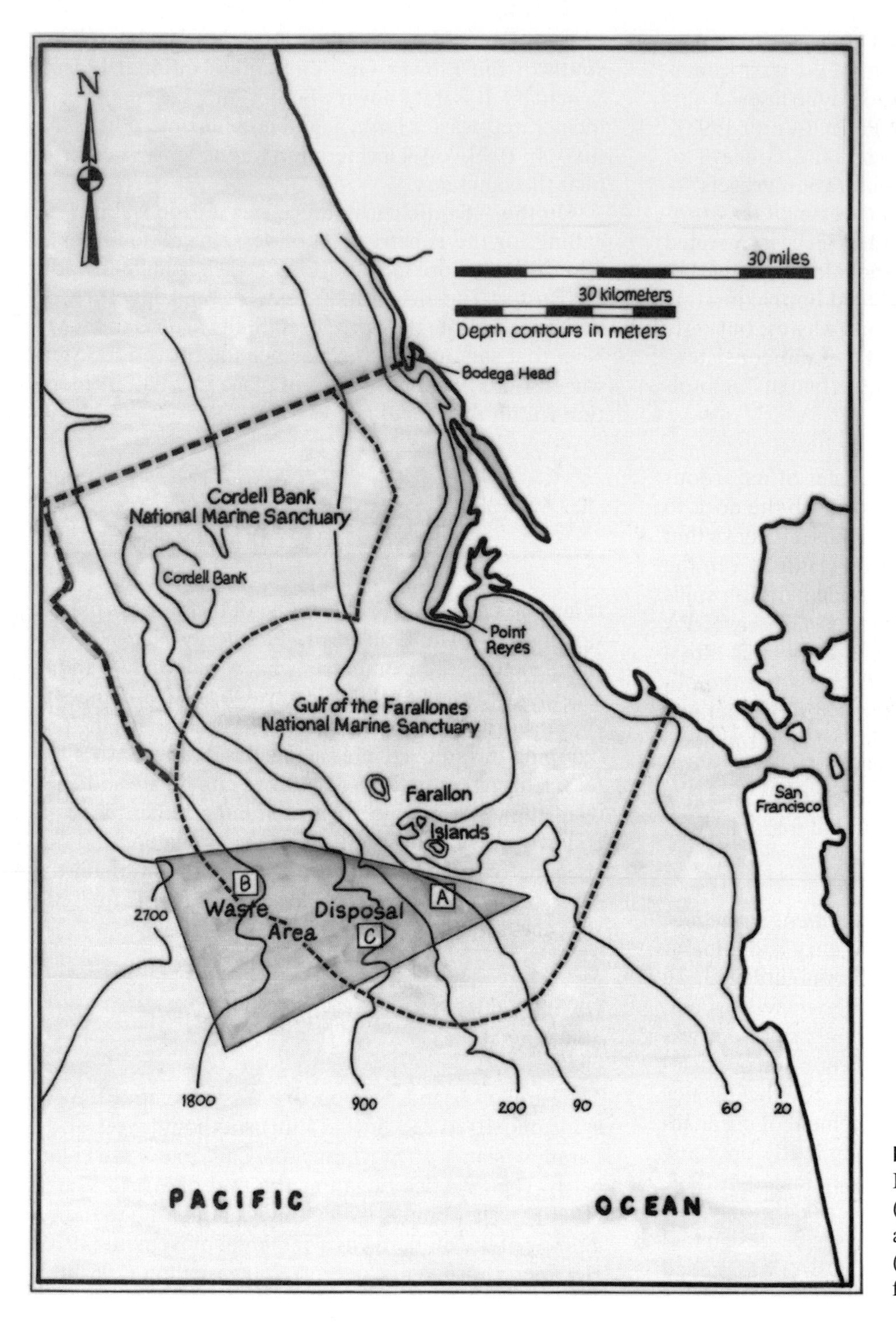

FIGURE 6-10 Map of the Farallon Islands Radioactive Waste Dump (FIRWD) within the Gulf of the Farallones National Marine Sanctuary. (*Source:* Map by Susan Giles; adapted from NOAA 1990)

of San Francisco, an estimated 150 containers were dumped. Site B, a 1,800 m (5906 ft) deep site, was used between the years 1946 and 1951 and 1954 and 1970. Here, an estimated 44,000 containers were dumped. Site C, in 900 m (2953 ft) of water, approximately 3,600 containers were discarded during the interim years of 1951 and 1954 (Noshkin et al. 1978; Diridoni 1988). Because of bad weather or malfunctioning equipment, an unknown number of containers were dumped within the region, but not necessarily at these three specific sites (NOAA 1990).

Although it is generally referred to as a former radioactive waste dump site, this area of the Farallon Islands also received several other types of products, such as aerospace and other industrial wastes (e.g., phenols, cyanides, mercury, beryllium), harbor and municipal wastes (e.g., dredge spoils, garbage), and military wastes (e.g., explosives).

Pathways for Contaminant Transport

These contaminants can possibly be transported by either ocean currents or bioturbation by benthic organisms (Crabbs 1983; Reish 1983; Suchanek 1987). The California Current is the dominant oceanic surface current near the area. During July to October, it flows southward from Alaska to Mexico. This general movement, however, is affected by several other oceanographic phenomena, such as the northward flowing Davidson Counter Current (October to March), upwelling (March or April to July), local gyres and eddies, and tidal exchanges with San Francisco Bay (estuary) (James Dobbin Associates 1987). So which way might the contaminants move?

The most recent studies of bottom currents in the vicinity of the FIRWD arrive at different conclusions. One study found that bottom currents were predominantly northward with an average flow rate of 1.33 cm (0.5 in) per second (Crabbs 1983) while another study found the bottom current to be eastward at speeds of at least 0.58 cm (0.2 in) per second (Suchanek 1987). Furthermore, contaminants can also be transported by **bioturbation**—the burrowing activity of benthic organisms.

Chemical Contaminants Present at FIRWD

Until 1992, there had only been four major studies at the FIRWD. Plutonium 239 and 240 (239 + 240Pu: half-life of 24,360 years) and Cesium 137 (137Cs: half life of 30 years) were the two most commonly found radionuclides at the FIRWD (NOAA 1990). The *sediments* sampled contained 137c and 239 + 240Pu, but the researchers differed on the significance of the levels found, depending on which comparison values were used for background levels. Despite their differences in interpretation of the data, however, the studies all concluded that the sediments at the site were contaminated beyond what would be expected from background sources. The researchers also agreed that the *water* within the vicinity of the FIRWD was found to be similar to the amounts from nuclear fallout debris delivered to the open ocean. But in terms of the *biota* sampled for contamination, the scientists once again reached different conclusions. The study by PneumoDynamics (1961), for example, reported that none of the biological samples collected showed contamination above background levels, whereas the Schell and Sugai (1980) study detected measurable levels of 239 + 240Pu and 137Cs in fish and invertebrates collected. Using Schell and Sugai's data, Davis (1980) calculated that 239 + 240Pu levels in muscle and liver tissues were as high as 8,500 and 5,071 times background levels respectively (NOAA 1990). For the actual technical data of these studies, see Dyer (1976), Noshkin et al. (1978), PneumoDynamics Corp. (1961), and Schell & Sugai (1980).

Environmental Impact on Habitats and Species

The biotic community in the Farallones region is diverse and abundant. The Farallones was selected and designated as a national marine sanctuary because of its rich populations of marine mammals (e.g., cetaceans, pinnipeds), commercial fish (e.g., Pacific Herring, Rockfish, Sablefish, Chinook Salmon, Dover Sole), marine flora (especially kelp), eelgrass beds and salt marsh vegetation, and benthic fauna. Furthermore, the Farallon islands support the largest concentrations of breeding sea birds in the continental United States. According to James Dobbin Associates (1987), twelve of the sixteen species of sea birds known to breed along the U.S. Pacific flyway have colonies on the Farallon Islands and feed in the sanctuary. Because of the extraordinary diversity and abundance of wildlife, in 1989 the Gulf of the Farallones National Marine Sanctuary was designated as part of the Central California Coast Biosphere Reserve[1], which is a part of the Marine Biosphere Program of the United Nations Educational Scientific and Cultural Organization (UNESCO).

The Farallon Islands Radioactive Waste Dump (FIRWD) covers a substantial portion of the 3,252 sq. km (1256 sq. mi) Gulf of the Farallones National Marine Sanctuary (GFNMS). Yet even today, we do not know the degree of risk this former dumping ground has on the GFNMS, on adjacent sanctuaries (Cordell Banks and Monterey Bay NMS), or the San Francisco Bay Area. The limited scientific information on this subject precludes any real conclusions being reached. What is obvious is that more site-specific information is needed, as well as studies that evaluate the toxicity of radionuclides, particularly 239 + 240Pu and 137Cs. Some new studies are beginning to appear, such as the draft report prepared for the Hazardous Materials Response Branch of NOAA (Tetra Tech 1992a, b, c). The report was based on data from previously published studies, the biota present, the levels of radioactive contamination observed, and "estimates" of the amount of radionuclides disposed and "presumed" location of disposal. Although the report concludes that "Risk to resources in the vicinity of the GFNMS due to past disposal of low-level radioactive wastes is well below the level of concern (Tetra Tech 1992a, p. ii), the *admitted uncertainties* about (a) the amounts of waste disposed (e.g., the number of containers), (b) the actual contents of the containers, and (c) the

[1] In 1996, the name was changed to the Golden Gate Biosphere Reserve.

exact disposal location of the containers (Tetra Tech 1992c, p. iii) indicates that many more studies are needed before we can rest assured that "all is well" in the GFNMS.

If nothing else, this brief case study should make one thing absolutely clear: The United States practiced radioactive waste dumping with absolutely *no idea* of the immediate or long-term consequences of its action. It is also especially disturbing to think that this country is now intensifying its efforts to "manage" its nuclear waste disposal problem by shipping it to less developed countries (e.g., Belau [Palau] Islands, Micronesia) that are *even less* equipped to handle and monitor the problem. This is merely the latest form of the "out of sight, out of mind" management strategy of the 1940s.

FUTURE OCEAN DISPOSAL

Closing landfills and the dangers associated with land disposal of chemicals have stimulated renewed interest in ocean dumping—despite current U.S. bans on the practice. There is a particular interest among some scientists and political leaders in what is called *deep ocean disposal.* **Deep ocean disposal** refers to a waste management scheme whereby various forms of wastes (e.g., sludge, industrial chemicals, radioactive wastes) are dumped into non-coastal waters—waters that are 914 m (3,000 ft) deep or more. The deep, dark, relatively barren "deep ocean" accounts for 98 percent of the ocean, whereas "coastal waters"—the seashores, harbors, and habitats for fish—only account for 2 percent of the ocean. When Congress passed legislation that prohibited the dumping of sewage and industrial wastes into the ocean after December 31, 1991, it ignored several prestigious scientific organizations such as the National Research Council of the National Academy of Sciences and the National Advisory Committee on Oceans and Atmospheres that argued for keeping our options open. These groups argued that this country should not foreclose deep ocean dumping because, in some cases, it is preferable to any of the alternatives.

The Deep Ocean Disposal Debate

There is a whole string of hotly debated arguments for and against deep ocean disposal as a waste management strategy. Some MIT and Woods Hole Oceanographic scientists are pitted against conservation groups and environmentally concerned politicians over issues ranging from the ocean's ability to dilute waste products, to our technological know-how regarding ocean placement and monitoring of wastes, to the notion of "sanctity of the seas". (See Table 6-1).

Charles Osterberg—a retired professor of oceanography at Oregon State University and author of the well known text, *Chemical Oceanography*, maintains that the best way to protect the coastal ocean (where humanity lives) is to dump our waste in the deep ocean (where humanity does not live). According to Osterberg, Mother Nature will merely thwart the intent of Congress by flushing wastes, legally confined to the land and air we try to live on, illegally to the coastal ocean. In other words, despite our well intended ocean dumping laws, gravity will prevail over legislation. Percolating rains, changes in groundwater levels, land erosion, earthquakes, and various other natural processes will see that landfills leak, and gravity in turn will assure that this landfill leachate reaches creeks, aquifers, streams, rivers, estuaries, and eventually our coastal shores.

Even well known ecologist Sir James Lovelock, author of the Gaia hypothesis, argues in his book *The Ages of Gaia,* that deep ocean waste disposal would merely put "Gaia" (the Earth's naturally self-correcting system) to work. Lovelock argues that pollution (waste) from industries is necessary, and that legislating against such waste (e.g., bans on ocean dumping) make as much sense as legislating against the emission of dung from cattle. On the other hand, Elliott A. Norse, a marine ecologist and chief scientist at the Center for Marine Conservation in Washington, seriously questions deep ocean dumping as a wise waste management strategy. According to Dr. Norse, if there are unforeseen problems, it would be impossible to retrieve the waste.

From Sludge to Radioactive Waste

Environmental groups thought they had won the battle over ocean dumping. However, if some marine scientists at the Woods Hole Oceanographic Institution (WHOI) get their way, everything from sludge to radioactive waste may eventually receive "deep ocean placement." Although only at the research stage now, some members of the environmental community are concerned that these research projects may become a springboard for commercial ocean dumping.

Sludge. According to a controversial research proposal developed at a January, 1991, workshop convened by staff of the Woods Hole Oceanographic Institution, the United States produces over 300 million tons of sewage sludge annually, with most of it applied to the land, put in landfills, or burned in incinerators. One million tons of this sewage sludge could be transported by specially designed "sludge ships" to one of two deep sites in the Atlantic about 483 km (300 mi) off shore, roughly halfway between the East Coast and Bermuda. There, the sludge could be

TABLE 6-1
Summary Of Arguments For And Against Deep Ocean Disposal

Arguments For	*Arguments Against*
	Ocean Dilution Capacity
Adequate: Oceans can *dilute, disperse,* and *degrade* large amounts of wastes. They are more *resilient* than scientists earlier believed.	*Inadequate:* Our throwaway lifestyle will eventually overwhelm the ocean's dilution and renewal capacity.
	Scientifc Knowledge
Sufficient Knowledge: An abyssal landfill will be completely isolated from the biological productive surface waters.	*Insufficient Knowledge:* We know less about the deep ocean than we do about outer space; for example, the basic biochemistry, physiology, and population biology of deep sea creatures are less known than those of coastal species.
	Technology
Available Technology: Technology exists for placing surface wastes in exact locations adjacent to venting systems.	*Failed Technologies:* Just because a technology exists does not necessarily mean it should be used. Many "technological promises" have turned out to be technological failures.
Environmental Hysteria: The Ocean Dumping Ban was propelled by "environmental hysteria"	*Environmental Concern:* Scientists, themselves, concede that ocean dumping could disturb the ecosystem.
Deep Ocean Disposal is Best Choice: Landfills on the coast pollute onshore waters with runoff, degrading beaches and fishing grounds.	*Waste Reduction is Best Choice:* To advocate deep ocean dumping will delay urgently needed waste reduction and other forms of pollution prevention.
Ineffective: Ocean dumping law does not protect the coastal ocean. Waste, when burned or dumped on land, eventually ends up in the ocean.	*Conservation laws help:* Past ocean dumping bans have helped conserve our oceans and coastal waters.
Reduced visibility: Deep ocean disposal is less visible than landfills and or incineration plants.	*Problem:* Lack of visibility equals "Out of sight/Out of Mind" mentality regarding waste management
Fewer Regulations: Ocean dumping is less regulated and less politically difficult than land disposal.	*More regulations:* More regulations will be needed if deep ocean disposal ever gets approved.
Precious land space must be preserved: Deep ocean dumping will reduce the pressure for new landfills and incineration plants which require much land space.	*We must run out of landfill space:* People won't take waste reduction seriously until we run out of landfill space
	Philosophy
Deep Ocean is a "Desert": Most of the ocean floor is lifeless, motionless, and geological stable, except around vents. The ocean bottom is not "delicate" or "fragile," or "sacred."	*Oceans are Sacred.* The seas are sacred places. They should not be used as a human garbage can. To advocate deep ocean dumping would further degrade this vital part of the earth's life support system.

lowered through 4877 m (16,000 ft) of water in giant, leak-proof buckets (each larger than an average house) and deposited on the ocean bottom, where scientists maintain it would be decomposed by bacteria. (See Figure 6-11).

A remote-controlled submersible vehicle, a so-called "autonomous benthic explorer," could be built to monitor the dump site to help assure environmentalists and regulators that the practice was safe. This proposal has proved to be one of the most politically sensitive ideas put forward by the Woods Hole Institution since the early 1970s, when its scientists proposed disposing of high-level radioactive wastes deep in the ocean bottom. After spending $100 million of research funds from the Department of Energy, this 1970s radioactive waste proposal was scrubbed when then President Ronald Reagan decided to use a barren mountain in Nevada as the site for the nation's repository for nuclear waste.

The 1991 Woods Hole Oceanographic Institute's proposal got immediate rebuttal from the nation's environmental groups, including the Environmental Defense Fund, Friends of the Earth, the Coast Alliance, Greenpeace, the Natural Resources Defense Council, American Oceans Campaign, the American Littoral Society, Clean Ocean Action, and the U.S. Public Research Interest Group. Their rebuttal was summarized by Sarah L. Clark, Staff Scientist, Environmental Defense Fund, and Boyce Thorne-Miller, Senior Scientist, Friends of the Earth (Clark and Thorne-Miller 1991). They argued several

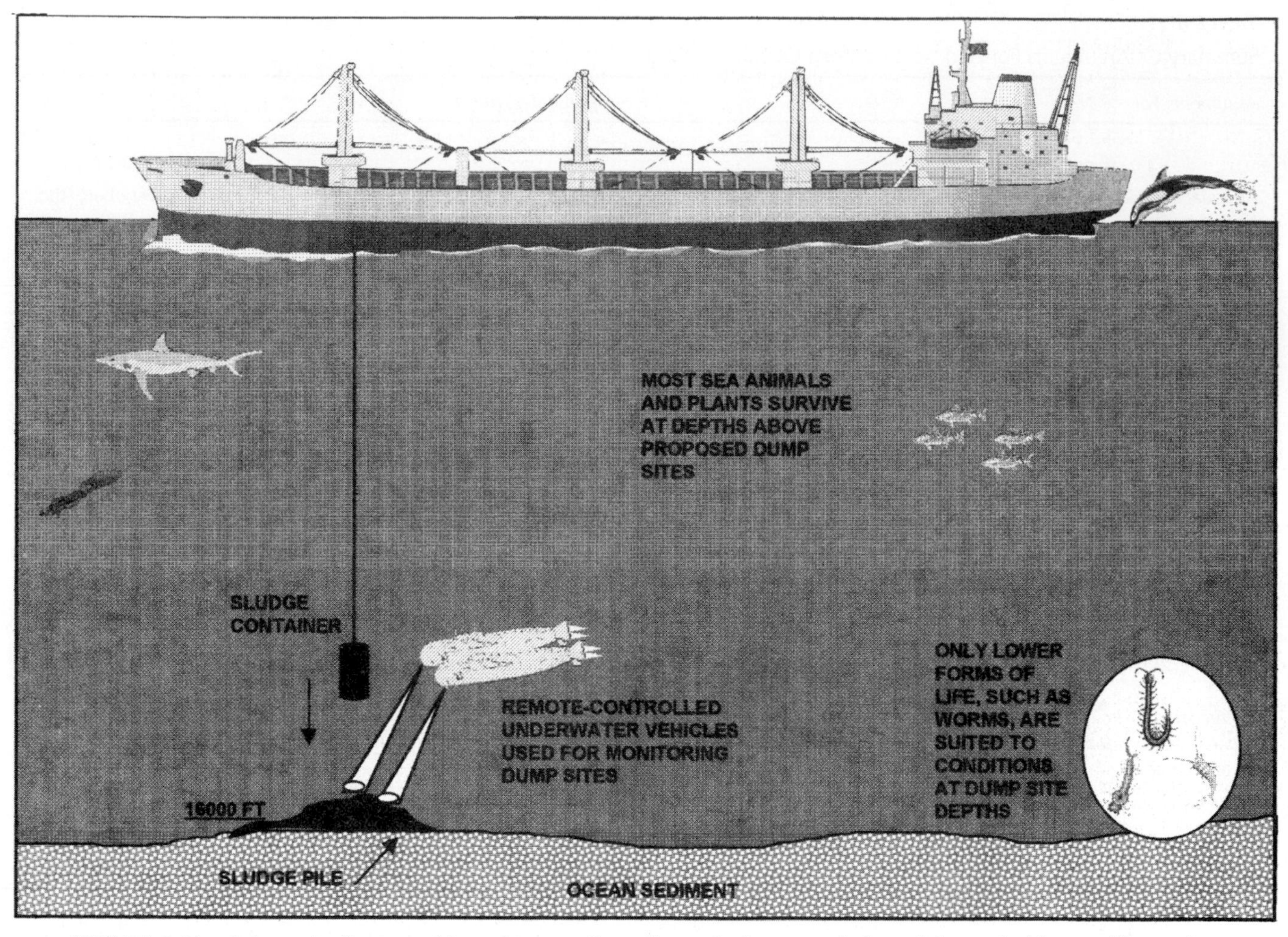

FIGURE 6-11 Schematic diagram of how deep sea floor disposal of sewage sludge might work. (*Source:* Illustration by Char Holforty)

points, including (a) the U.S. does not produce 300 million tons of wet sludge per year (as alleged by the Woods Hole Oceanographic Institute's proposal), that (b) safe land-based sludge management options are readily available; and that (c) sewage sludge should not necessarily be considered a "waste." For these and numerous other reasons, they concluded that the Woods Hole Oceanographic Institute's experiment was not an appropriate research direction to pursue. Research at WHOI is continuing on deep ocean disposal, but there is yet no serious effort to formally pursue this option.

Industrial wastes. In the past, industrial waste dumping included such materials as sulfuric acid, hydrochloric acid, and liquid solutions of organic wastes from chemical manufacturing. In the future, there may be great pressure to once again open up the seas for industrial waste dumping, especially for the dumping of flue gas desulfurization sludges and coal ash by industrial boilers and coal-burning electric utilities. One of the strategies for stretching our existing oil resources is to place greater reliance on our nation's coal supply. Unfortunately, burning coal produces great quantities of fly ash, bottom ash, and flue-gas desulfurization sludge. In fact, 25–40 percent of all coal burned results in ash (Bulloch 1989). In order to trap the pollutants that help cause acid rain, industries are required to use "scrubbers" or some other form of pollution control device. A **scrubber,** for example, is a technological device that removes particulates and sulfur dioxide from air pollution in smokestacks. As more and more industries "capture their waste product," there will be increased pressure on our government to permit dumping of this voluminous waste into the sea.

How hazardous would this material be to the marine environment? We simply do not know. We do know that unconsolidated ash is very fine and has the potential to smother sea-bottom life. Consequently, experiments have already been done on converting ash into solid blocks and creating rubble reefs for marine life. In early experiments, the ash block reefs have been colonized quickly by fish and lobster (Bulloch 1989). However, little is known about the long term effects of such a practice, which, unfortunately, is also the case with many common practices.

Radioactive wastes. Pressure is also mounting for the resumption of dumping both low-level (e.g., clothing and other forms of medical waste) and high-level radioactive waste (e.g., outdated radioactive reactors; "hot waste" from plutonium manufacture). For example, the U.S. Navy, eager to dispose of the radioactive reactors of outdated nuclear submarines, wants to sink the vessels in deep water. The U.S. Department of Energy (DOE) hopes to use the oceans as a dumping ground for thousands of tons of contaminated soil from abandoned atomic weapons facilities. And the nuclear power industry, already under attack for disposing of low-level wastes onshore, hopes to return to the disposal of waste materials in the ocean.

Advocates of "ocean-bed repositories" for nuclear waste generally argue three points: (1) *Numerous sites*—the ocean is vast, thereby providing numerous appropriate sites for disposal; (2) *Away from human populations*—ocean-bed repositories would be away from humans, and consequently safer than land-based repositories; (3) *Resource potential*—limited land space would be freed-up for its maximum resource potential, such as for agriculture, recreational activities, or for some other human use. These same advocates point out that there are various methods for developing ocean-bed repositories, including the simple "freefall" method (dropping it overboard), to winch controlled, drilling, and trenching. (See Figure 6-12).

Those that oppose open dumping of low and high-level nuclear waste cite the instability of plate boundaries due to earthquakes and volcanism, and the fact that deep-bed water circulation might eventually surface the

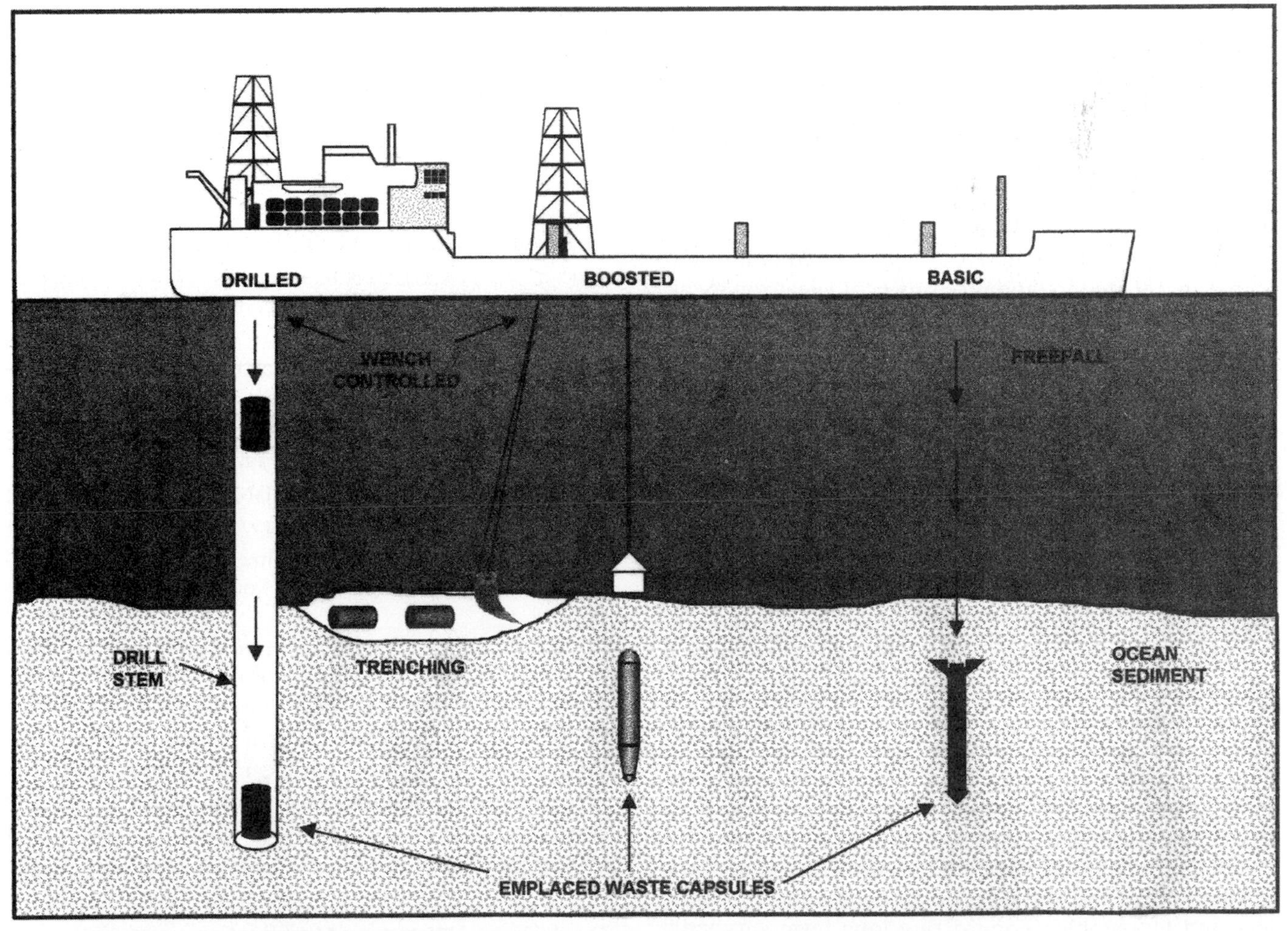

FIGURE 6-12 Proposed ocean-bed repositories for radioactive waste. Various methods for developing ocean-bed repositories. Material may simply be dropped to the ocean floor in a free-fall, or its fall may be controlled by a winch attached to a transport ship. Drilling and trenching are two other proposed methods. (*Source:* Illustration by Char Holforty; adapted from Kaufman and Franz 1993, p. 386)

radionuclides. Although mid-ocean water may take thousands of years to surface, the long life expectancy of some of these radioactive wastes are even longer—tens or hundreds of thousands of years or more. Whether it is a plan to implant radioactive wastes beneath the sea in **subduction** zones—the border where one crustal plate goes under the edge of an adjoining plate, or whether the proposal is to bury the wastes in the ocean's abyssal red clay away from plate boundaries and allow continental drift to eventually carry the wastes to the nearest plate boundary, then downward, or whether the proposal is to fire torpedo-shaped canisters filled with nuclear waste to 20 to 30 m beneath the ocean floor (Skerrett 1992), one thing is certain: *No one knows what the long-term environmental impacts will be of such experiments in radioactive waste dumping.*

GENERAL POLICY RECOMMENDATIONS

The world's oceans are capable of breaking down or absorbing certain wastes. They are less capable of resisting other contaminants, however, including radioactive wastes, persistent and nondegradable synthetic organic compounds, metals that are toxic to marine organisms, and toxic chemicals that are biologically magnified. The ban on dumping of such toxic materials must be either continued or the practice of dumping limited until we gain a better understanding of its ecological ramifications. Some progress is being made on restricting dumping in the world's oceans, such as the global moratorium on radioactive waste dumping at sea that has been in place since 1983, the regional (North Sea) decisions to end sewage-sludge and industrial dumping at sea, and the proposed global phase-out of ocean incineration of toxic wastes (Curtis 1990).

However, as decision-makers are enticed and lobbied to try new disposal options (e.g., deep ocean disposal), the threat to the oceans and coastal areas will worsen. Through the process of normal water circulation and upwelling within the oceans, deep ocean floor nutrients *and* "discarded materials" may be transported to the ocean surface and consequently to the coastal shores or into the intestines of marine animals. According to Clifton E. Curtis, Director of the Oceanic Society, "putting toxic waste in the ocean, whatever the disposal method, is akin to a business owner myopically concentrating on increasing profits in the next quarter, while his company's infrastructure and state-of-the art capabilities become less and less stable, jeopardizing long-term survival" (Curtis 1990, p. 20).

Over the last 25 years, the United States has made significant progress in restricting ocean dumping from its own shores. The *Marine Protection, Research, and Sanctuaries Act (MPRSA)* and *Clean Water Act (CWA)* have been powerful weapons against unrestricted ocean dumping (Moore 1992). Congress needs to continue to periodically strengthen these acts (and others) to meet new challenges. In the Summer of 1994, for example, a coalition of ten or more coastal environmental groups (e.g., American Littoral Society [NJ], Coastal Advocates [CA], The Coast Alliance [DC]), and fishing groups (e.g., the United Fishermen's Association [NY], Jersey Coast Anglers Association [NJ], and The Pacific Coast Federation of Fishermen's Associations [CA]) filed suit against a recent U.S. EPA decision to drop one of the prescribed tests that determine the advisability of open ocean dumping of contaminated sediments. EPA adopted an "interim rule" eliminating what is called the "suspended solid phase bioaccumulation test" which these groups maintain is critical in determining the environmental impact of silty material suspended in the water column during dumping.

The Office of the London Dumping Convention, International Maritime Organization, in London provides some general recommendations for reducing environmental pressures on the ocean and our coastal areas. They are as follows (Duedall 1990, p. 38):

- Wherever possible recycle and reuse waste products.
- Treat wastes that cannot be recycled or reused at the source to the extent feasible.
- Use sea disposal, whether by outfall or by dumping, only for those materials that are compatible with the marine environment.
- Use locations for sea disposal of wastes that will not interfere with other uses of the sea.
- Use waste disposal practices at sea that minimize local impacts at the point of disposal.
- Carefully evaluate the potential environmental impacts of new developments and seek to mitigate adverse impacts.
- Manage the use of the resources of the sea so as to prevent depletion of resources worldwide.

The above LDC recommendations are very broad in scope and, in many cases, are almost too general to really serve much purpose. More specific recommendations can be made for each of the types of dumping materials—dredge materials, sewage sludge, municipal solid wastes, industrial wastes, military wastes, radioactive wastes, and ocean incinerated wastes. What follows are just a few examples:

Dredging and Waterway Modification

- *Fate of sediments.* Conduct studies to evaluate the sediment dynamics of the area (e.g., accretion and erosion processes in ocean trench, marsh, or mudflat areas).

- *Contaminant bioavailability.* Develop and set sediment quality objectives (e.g., conduct laboratory and field bioaccumulation investigations to determine suspended sediment effects on sensitive life stages and the food chain).
- *Regional strategy.* Develop a dredge project needs assessment (e.g., a prioritization plan; a reuse/nonaquatic disposal opportunities and constraints).
- *Dredge material disposal.* Determine dredge material disposal options (e.g., traditional and alternative disposal options, cost estimates).
- *Future threats and benefits.* Identify areas subject to future flooding and erosion, and determine causes (e.g., acquire diked historic baylands identified as buffer areas against coastal flooding and sea level rise).

Sewage Sludge

- *Sludge disposal.* Encourage the disposal of sludge by application to agricultural lands. Sludge contains high levels of nitrogen and phosphorus, and its water-retention capacity makes it a good soil conditioner. Care must be taken, however, not to apply sludge that contains heavy metals or organic chemicals, particularly PCBs. Whether or not sludge contains these troublesome compounds is determined by the types of industrial plants that feed wastewater to the sewage treatment plant.
- *Sludge conversion.* Encourage the conversion of sludge into a useful resource. For example, some treatment plants are already using anaerobic bacteria to digest sludge, thereby producing methane gas for sale as a commercial fuel source. Since sludge even has nutritive value, scientists are exploring the possibilities of sludge as a poultry or cattle feed supplement.
- *Ban on sludge dumping.* Continue the U.S. ban on all ocean dumping of sewage sludge. Encourage the global community to do the same.

Municipal Solid Wastes

There are basically only two ways to deal with the growing mountains of solid waste that we produce: *waste management* and *waste prevention.* To date, the national priority has been on waste management (e.g., burying waste in landfills, dumping wastes into the ocean, incinerating wastes at sea, and, most recently, shipping our wastes overseas). This country, and the world, needs to shift to *waste prevention.* There are two priorities:

- *First priority.* First, we need to work toward primary pollution and waste prevention, such as reducing unnecessary packaging, using less material per product (e.g., lighter cars), making products that are more durable, can be easily recycled, or reused. Industrial processes should be changed to eliminate as many harmful chemicals as possible. This is perhaps the most fundamental "anti-waste" technique, since it is founded on the premise that Americans overconsume resources. With only 5 percent of the world's population, which inhabits only 6 percent of the global land area, the United States gobbles up 23 percent of the world's nonfuel mineral resources (Moran et al. 1986). It is no wonder that environmentalists call for conservation strategies that minimize the consumption of non-fuel mineral resources.
- *Second priority.* Second, we need to work toward secondary pollution and waste prevention, such as finding creative ways to repair, recycle, or reuse existing products. Recycling, for example, increases the *residence time* of minerals (i.e., the time a mineral remains in use), while at the same time decreases the use of virgin resources, the amount of land disruption, the degree of air pollution, and the need for new landfill sites. See Klee (1991) for further details.

Industrial Wastes

- *Product durability and miniaturization.* Products that are smaller, last longer, and consequently utilize fewer natural resources, also produce less waste. Industry is already beginning to adapt automobiles, calculators, computers, refrigerators, ovens, and other "American necessities" to the pending mineral-scarce world. Inventions that lead us toward greater product durability (e.g., solid-state components) and miniaturization (e.g., microchips) should be encouraged.
- *Institutional and industrial procedures.* Tax breaks and depletion allowances that encourage primary production at the expense of recycling need to be modified or eliminated. Industries need to simplify, and, where possible, to standardize the specifications for certain materials, for example, standardizing the composition of the alloys in automobile parts. This would make recycling easier and less expensive, since it would eliminate the problems presented by a multiplicity of specifications.

Radioactive Wastes

- *Continue the ban on radioactive waste dumping.* Until two decades ago, the United States dumped radioactive waste at sea in steel drums, and there are organizations and agencies that would like to resume that practice. Remember, only low levels of radioactive contaminants need be present to affect marine organisms. Marine biologists, for example, report that 0.2 microcuries of radioactivity will seriously disrupt fish egg development (Owen 1985). Although the London Dumping Convention currently provides an indefinite moratorium on all radioactive-waste dumping at sea, pressures from military and civilian sources are underway to reverse that decision. Fortunately, public resistance to any ocean disposal of radioactive waste is growing.

- *Increase monitoring of known sites previously used for radioactive dumping.* In 1985 and 1986, for example, EPA's Office of Radiation Programs, using the navy's manned deep submersible, the *DSRV Avalon,* surveyed the ocean bottom and water column in the region of the two Farallon Islands low-level radioactive waste disposal sites. These and other known radioactive dumping sites need to be carefully monitored.

Incineration

- Use incineration only as a "land-based backstop technology." The United States should continue its ban on the use of incineration vessels. The global community should be encouraged to work toward the same goal.

CONCLUSION

Our oceans have become the repository for every type of imaginable waste, including very dangerous chemical and radioactive products. Fortunately, national and international laws are working to stop the disposal of these destructive wastes in the marine environment.

It is clear that waste disposal is a complex matter with no simple solution. Indeed, there is disagreement among waste management experts on whether ocean disposal is safe and/or necessary. In March, 1993, the *New York Times* documented the emergence of a "new environmentalism"—one that puts greater emphasis on economic tradeoffs and cost-benefit analysis. It even cited ocean dumping as a case of the premature abandonment of potentially cost-effective environmental solutions. It is particularly interesting, therefore, when the Marine Policy Center of the Woods Hole Oceanographic Institution concluded in July 1993 that their "new environmentalist" extended economic research tended to support the claims of the "old environmentalists," that (a) waste management can be dealt with without resorting to ocean dumping; that (b) waste reduction and beneficial use strategies are promising; and that (c) there is no great urgency in developing deep ocean options for the future. With all the uncertainties associated with ocean disposal, particularly the deep ocean option, it would behoove us to follow the path of waste reduction improvements in association with traditional land disposal technologies.

Future multiple and sustainable use of our ocean resources will require continued careful consideration of the types of materials disposed and the methods used. Promising, new, environmentally benign technologies, such as Ocean Thermal Energy Conversion (OTEC), are dependent upon clean (non polluted) coastal water. OTEC uses the ocean's natural temperature gradient to drive an electrical generator. This source of renewable energy may help reduce human dependence upon fossil fuels. As will be discussed in the next chapter, reduction of oil consumption is integrally tied to a clean coastal environment.

REFERENCES

Bakalian, Allan. 1984. "Regulation and Control of United States Ocean Dumping: A Decade of Progress, an Appraisal for the Future." In *Harvard Environmental Law Review,* Vol.8, No. 1, pp. 193–256.

Bochove, Danielle. 1993. "The 'Conspiracy of Ignorance.'" *World Press Review.* October, Vol.40, No. 10, p. 43.

Bulloch, David K. 1989. *The Wasted Ocean.* New York: Lyons & Burford Publishers.

Capuzzo, Judith E. McDowell. 1990. "Effects of Wastes on the Ocean: The Coastal Example." *Oceanus.* Vol. 33, No. 2, Summer: pp. 39–44.

Clark, Sarah L. and Boyce Thorne-Miller. 1991. *Comments of the Environmental Defense Fund, Friends of the Earth, the Coast Alliance, Greenpeace, the Natural Resources Defense Council, American Oceans Campaign, the American Littoral Society, Clean Ocean Action, and the U.S. Public Research Interest Group on the Report of a Workshop to Determine the Scientific Research Required to Assess the Potential of the Abyssal Ocean as an Option for the Future Waste Management—January 10, 1991.* Mimeograph.

Crabbs, D. E. 1983. *Analysis of Ocean Current Meter Records Obtained from a 1975 Deployment Off the Farallones Islands, California. EPA 520/1-83-019.* Washington, D.C.: Office of Radiation Programs, U.S. Environmental Protection Agency.

Curtis, Clifton E. 1986. "Radioactive Wastes: Reflections on International Policy Developments Under the London Dumping Convention." *Sixth International Ocean Disposal Symposium.* Pacific Grove, California, April 21–25, 1986, pp. 213–214.

Curtis, Clifton E. 1990. "Protecting the Oceans." *Oceanus.* Vol. 33, No. 2, Summer, pp. 19–22.

Davis, W. Jackson. 1980. "Radioactive Dumpsites in U.S. Coastal Waters: An Analysis of Documents Released by the USEPA Pertaining to the 1977 Surveys of Radioactive Dumpsites at the Farallon Islands and in the Atlantic Ocean." Issued by The Honorable Quentin L. Kopp, Supervisor, City of San Francisco, September 15, 1980.

Dieterichs, Bob. 1994. Environmental Protection Agency. Telephone Interview, August 10, 1994.

Diridoni, J. 1988. *Reassessment of the Farallones Islands Nuclear Waste Dump Site* (Pacific Ocean-Boeing). Memo. February 4, 1988.

Dover, Cindy Lee Van, J.F. Grassle, Brian Fry, Robert H. Garritt, Victoria R. Starczak. 1992. "Stable Isotope Evidence for Entry of Sewage-Derived Organic Material into a Deep-Sea Food Web." *Nature.* Vol. 360, No. 6400, pp. 153–155.

Duedall, Iver W. 1990. "A Brief History of Ocean Disposal." *Oceanus,* Vol. 33, No. 2, Summer, pp. 29–38.

Dyer, R. S. 1976. *Environmental Surveys of Two Deepsea Radioactive Disposal Sites Using Submersibles. International Symposium on the Management of Radioactive Wastes from the Nuclear Fuel Cycle.* Vienna: International Atomic Energy Agency.

Engler, Robert M. 1990. "Managing Dredged Materials," *Oceanus.* Vol. 33, No. 2, Summer, pp. 63–71.

Environmental Protection Agency (EPA). 1984. *"Report to Congress January 1981–December, 1983, on Administration of the Marine Protection, Research and Sanctuaries Act of 1972, as Amended (P.L. 92-532) and Implementing the International London Dumping Convention."* Washington, D.C.: Office of Water Regulations and Standards, Environmental Protection Agency.

Environmental Protection Agency (EPA). 1985a. *Incineration-at-Sea Research Strategy.* February 19, 1985, Washington D.C.: Office of Water Regulations and Standards, U.S. Environmental Protection Agency.

Environmental Protection Agency (EPA). 1985b. *Report on the Incineration of Liquid Hazardous Wastes by the Environmental Effects, Transport and Fate Committee, Science Advisory Board.* Washington, D.C.: Environmental Protection Agency. April, p. 17.

Environmental Protection Agency (EPA). 1992. *Statistical Support Documentation for the 40 CFR, Part 503. Final Standards for the Use or Disposal of Sewage Sludge. Vol. I & II.* Washington, D.C.: Environmental Protection Agency.

Environmental Protection Agency (EPA). 1993. *Environmental Impact Statement for Designation of a Deep Water Ocean Dredged Material Disposal Site off San Francisco, California.* San Francisco, California: Environmental Protection Agency.

Guarascio, John A. 1985. *The Regulation of Ocean Dumping After City of New York v. Environmental Protection Agency,* 12 B.C. Envtl. Aff. L. Rev. 701, 703. Citing U.S. EPA, *A Guide to Regulations and Guidance for the Utilization and Disposal of Municipal Sludge,* p. 34 (1980).

Hollin, Dewayne, and Mike Liffmann. 1993. "Use of MARPOL ANNEX V Reception Facilities and Disposal Systems at Selected Gulf of Mexico Ports, Private Terminals and Recreational Boating Facilities." In *Coastal Zone '93: Proceedings of the Eighth Symposium on Coastal and Ocean Management,* edited by Orville T. Magoon, W. Stanley Wilson, Hugh Converse, and L. Thomas Tobin, pp. 1966–1980. New York: American Society of Civil Engineers.

James Dobbin Associates Incorporated. 1987. *Gulf of the Farallones National Marine Sanctuary Management Plan.* Washington, D. C.: U.S. Department of Commerce, National Oceanic and Atmospheric Administration, Marine and Estuarine Management Division.

Kaufman, Donald G., and Cecilia M. Franz. 1993. *Biosphere 2000: Protecting Our Global Environment.* New York: HarperCollins College Publishers.

Kitsos, Thomas R. and Joan M. Bondareff. 1990. "Congress and Waste Disposal at Sea." *Oceanus.* Vol. 33, No. 2, Summer, pp. 23–28.

Klee, Gary A. 1991. *Conservation of Natural Resources.* Englewood Cliffs, New Jersey: Prentice Hall.

MacKenzie, Deborah. 1994. "Doubts Lurk in Graveyard for Nuclear Subs." *New Scientist.* March. Vol. 141, No. 1916, pp. 4–5.

Marine Policy Center. 1993. *Optimal Strategies for Waste Management: Economic Perspectives on the Ocean Option.* Woods Hole, Massachusetts: Marine Policy Center, Woods Hole Oceanographic Institution.

Millemann, Beth. 1986. *And Two if By Sea: Fighting the Attack on America's Coasts: A Citizens Guide to the Coastal Zone Management Act and Other Coastal Laws.* Washington D.C.: Coast Alliance, Inc.

Miller, G. Tyler. 1988. *Living in the Environment.* 6th ed. Belmont, California: Wadsworth.

Miller, G. Tyler. 1992. *Living in the Environment.* 7th ed. Belmont, California: Wadsworth.

Moore, Steven J. 1992. "Trouble in the High Seas: A New Era in the Regulation of U.S. Ocean Dumping." *Environmental Law,* Vol. 22, No. 3, Spring, pp. 913–951.

Moran, Joseph M., Michael D. Morgan, and James H. Wiersma. 1986. *Introduction to Environmental Science.* 2nd ed. New York: W. H. Freeman.

National Oceanic and Atmospheric Administration (NOAA). 1985. *National Marine Pollution Program: Federal Plan for Ocean Pollution Research, Development & Monitoring, Fiscal Years 1985–1989.* Washington, D.C.: National Oceanic and Atmospheric Administration. September 1985.

National Oceanic and Atmospheric Administration (NOAA). 1990. *Preliminary Natural Resource Survey: Farallon Islands Radioactive Waste Dumps, Farallon Islands, California.* CAD 981159585, Site ID: 09, March 29. Rockville, Maryland: National Oceanic and Atmospheric Administration.

National Oceanic and Atmospheric Administration (NOAA). 1991. *Response of the Habitat and Biota of the Inner New York Bight to Abatement of Sewage Sludge Dumping: Third Annual Progress Report—1989.* Rockville, Maryland: National Oceanic and Atmospheric Administration.

National Oceanic and Atmospheric Administration (NOAA). 1992. *Monterey Bay National Marine Sanctuary: Final Environmental Impact Statement/Management Plan. Vol. I & II.* Washington, D.C.: National Oceanic and Atmospheric Administration.

National Oceanic and Atmospheric Administration (NOAA). 1993a. *Stellwagen Bank National Marine Sanctuary. Final Environmental Impact Statement/Management Plan.* Silver Spring, Maryland: National Oceanic and Atmospheric Administration.

National Oceanic and Atmospheric Administration (NOAA). 1993b. *Stellwagen Bank National Marine Sanctuary Final Environmental Impact Statement/Management Plan.* Vol. II: Appendices. Silver Spring, Maryland: National Oceanic and Atmospheric Administration.

Noshkin, V. E., K. M. Wong, T. A. Jokela, R. J. Eagle and J. L. Brunk. 1978. *Radionuclides in the Marine Environment Near the Farallon Islands.* UCRL-52381. Livermore, California: University of California.

Office of Technology Assessment Task Force. 1986. *Ocean Incineration: Its Role in Managing Hazardous Waste.* Washington, D.C.: Government Printing Office.

Office of Technology Assessment Task Force. 1987. *Wastes in Marine Environments, OTA-O-334.* Washington, D.C.: Government Printing Office.

O'Hara, Kathryn J., Suzanne Iudicello, and Jil Zilligen, eds. 1994. *A Citizen's Guide to Plastics in the Ocean: More Than a Litter Problem.* Washington, D.C.: Center for Marine Conservation.

Owen, Oliver S. 1985. *Natural Resource Conservation: An Ecological Approach.* 4th ed. New York: Macmillan.

Owen, Oliver S., and Daniel D. Chiras. 1990. *Natural Resource Conservation: An Ecological Approach.* 5th ed. New York: Macmillan.

PneumoDynamics Corp. 1961. *Survey of Radioactive Waste Disposal Sites, Technical Report. ASD 4634-F.* El Segundo, California: PneumoDynamics Corps., Advanced Systems Development Division.

Raulinaitis, John. 1994. "FDA Goes Fishing for Toxic Waste." *FDA Consumer.* March. Vol. 28, No. 2, pp. 18–25.

Redford, David. 1993. Chief, Ocean Dumping and Marine Debris Section, U.S. Environmental Protection Agency, Personal Correspondence, September 9.

Reish, D.R. 1983. *Survey of the Marine Benthic Fauna Collected from the United States Radioactive Waste Sites off the Farallon Islands, California. EPA 520/1-83-006.* Washington, D.C.: U.S. Environmental Protection Agency, Office of Radiation Programs.

San Francisco Estuary Project. 1993. *Comprehensive Conservation and Management Plan.* San Francisco: San Francisco Estuary Project.

Schell, W. R. and S. Sugai. 1980. "Radionuclides at the U.S. Radioactive Waste Disposal Site Near the Farallon Islands." *Health Physics,* No. 39: pp. 475–496.

Skerrett, P. J. 1992. "Nuclear Burial at Sea." *Technology Review,* Feb–March, Vol. 95, No. 2, pp. 22–23.

Spencer, Derek W. 1990. "The Ocean and Waste Management." *Oceanus.* Vol. 33, No. 2, Summer, pp. 5–12.

Stegeman, John. 1990. "Detecting the Biological Effects of Deep Sea Waste Disposal." *Oceanus,* Vol. 33, No. 2, Summer, pp. 54–62.

Steuteville, Robert. 1992. "East Coast States Find New Sludge Routes." *BioCycle.* Vol. 33, No. 5, May. pp. 66–68.

Suchanek, T. H. 1987. *Potential Bioaccumulation of Long-Lived Radionuclides by Marine Organisms in the Vicinity of Farallones Islands Nuclear Waste Dump Site. Symposium on Current Research Topics in the Marine Environment.* San Francisco: Gulf of the Farallon National Marine Sanctuary, March 21, 1987.

Suchanek, T. H. and M. C. Lagunas-Solar. 1987. *Bioaccumulation of Long-Lived Radionuclides by Marine Organisms from the Farallon Islands Nuclear Waste Dump Site: First Quarter Report (Jan/March 1987).* Submitted to the State of California Department of Health Sciences. March 31, 1987.

Swanson, R. Lawrence. 1993. "The Incongruity of Policies Regulating New York City's Sewage Sludge: Lessons for Coastal Management." *Coastal Management.* Vol. 21, No. 4, pp. 299–312.

Tetra Tech, Inc. 1992a. *Baseline Ecological Assessment of Disposal Activities in the Gulf of the Farallones: Ecological Risk Assessment (Draft).* December. Lafayette, California: Tetra Tech, Inc.

Tetra Tech, Inc. 1992b. *Baseline Ecological Assessment of Disposal Activities in the Gulf of the Farallones: FARADD User's Guide.* November. Lafayette, California.: Tetra Tech, Inc.

Tetra Tech, Inc. 1992c. *Baseline Ecological Assessment of Disposal Activities in the Gulf of the Farallones: Information Identification, Evaluation, and Analysis (Draft).* Lafayette, California: Tetra Tech, Inc.

FURTHER READING

Brown, Lester R., Christopher Flavin, and Sandra Postel. 1991. "Reusing and Recycling Materials." In *Saving the Planet: How to Shape an Environmentally Sustainable Global Economy.* New York: Norton. p. 17.

Carless, Jennifer. 1992. *Taking Out the Trash: A No-Nonsense Guide to Recycling.* Washington, D.C.: Island Press.

Davis, W. J. and J. M. VanDyke. 1990. "Dumping of Decommissioned Nuclear Submarines at Sea—A Technical and Legal Analysis." *Marine Policy.* Vol. 14, No. 6, (Nov.) pp. 467–476.

Davis, W. Jackson. 1993. "Contamination of Coastal Versus Open Ocean Surface Waters—A Brief Meta-Analysis." *Marine Pollution Bulletin.* Vol. 26, No. 3 (March), pp. 128–134.

Denison, Richard A., and John Ruston. 1990. *Recycling and Incineration: Evaluating the Choices.* Washington, D.C.: Island Press.

Gershey, Edward L., et al. 1990. *Low-Level Radioactive Waste: From Cradle to Grave.* New York: Van Nostrand Reinhold.

Lamb, John and Peter Gizewski. 1993. "The Policy Makers' Challenge: Radioactive Dumping in the Arctic Ocean." *Oceanus.* Vol. 36, No. 3, pp. 89–91.

Leskov, Sergei. 1993. "Lies and Incompetence." *Bulletin of the Atomic Scientists.* Vol. 49, No. 5 (June), pp. 13, 55.

McNair, E. Clark, Jr., and Russell K. Tillman. 1993. "The Dredging Research Program of U.S. Army Corps of Engineers." In *Coastal Zone '93: Proceedings of the Eighth Symposium on Coastal and Ocean Management,* edited by Orville T. Magoon, W. Stanley Wilson, Hugh Converse, and L. Thomas Tobin, pp. 1505–1516. New York: American Society of Civil Engineers.

Monastersky, Richard. 1993. "Hazard from Soviet Nuclear Dumps Assessed." *Science News,* May. Vol. 143, No. 20, pp. 310–311.

Murphy, Patricia, and Jody Zaitlin. 1993. "Upland Reuse of Dredged Material As an Alternative to Aquatic Disposal: The Port of Oakland's Berth 30 Terminal Redevelopment, A Case Study." In *Coastal Zone '93: Proceedings of the Eighth Symposium on Coastal and Ocean Management,* edited by Orville T. Magoon, W. Stanley Wilson, Hugh Converse, and L. Thomas Tobin, pp. 3110–3118. New York: American Society of Civil Engineers.

Serpas, Ricardo W., and L. Phil Pittman. 1993. "Beneficial Utilization and Wetland Creation From Dredge Spoil." In *Coastal Zone '93: Proceedings of the Eighth Symposium on Coastal and Ocean Management,* edited by Orville T. Magoon, W. Stanley Wilson, Hugh Converse, and L. Thomas Tobin, pp. 3131–3143. New York: American Society of Civil Engineers.

Varmer, Ole, and Amy J. Santin. 1993. "Ocean Management Under the Marine Protection, Research and Sanctuaries Act: Sanctuaries, Dumping and Development." In *Coastal Zone '93: Proceedings of the Eighth Symposium on Coastal and Ocean Management,* edited by Orville T. Magoon, W. Stanley Wilson, Hugh Converse, and L. Thomas Tobin, pp. 1990-2003. New York: American Society of Civil Engineers.

7 Offshore Oil Development & Transport

It may be said that America's coastal dwellers love only one thing more than their beaches, and that is their high energy intensive way of life. The epitome of this lifestyle, of course, is Southern California with its flashy cars and monumental freeways that dominate the landscape. For years, the "dune buggy with a surfboard in the back" was an acceptable symbol of "the good life" in this region of America. More recently, however, images of oil-slicked birds, black gooey sand, and brown foam waters have reminded us of the connection between our energy intensive lifestyle and our coastlines. It is on our coasts that we often locate conventional and nuclear power plants; it is off our coasts that we have "offshore" oil drilling and the movement of supertankers full of oil from distant locations, and, it is also off our coasts that a few experiments are being done with alternative forms of harnessing coastal energy, such as using tidal barrages and wave energy devices.

There is no getting around the fact that energy production has had an uneasy relationship with the coast and dealing with this problem—especially the potential for oil spills—is one of the more important issues facing coastal resource managers (Leslie 1993). The first part of this chapter, *Sources and Effects of Oil Pollution,* opens with a discussion of the origins and overall ecological impacts of oil pollution in coastal waters. *Offshore Oil Development: Components and Impacts* goes into a more detailed discussion of the stages and associated environmental impacts of offshore oil and gas development. *Oil Transportation Options* reviews the advantages and disadvantages of transporting oil by pipeline versus tanker. *Cleanup Strategies* begins by reviewing the various cleanup techniques, as well as their limitations, then provides some examples of typical oil spill response programs—including a case study of the 1990 Huntington Beach oil tanker spill, from the cause of the accident to clean-up efforts, and to lingering effects. *General Policy Recommendations* provides students, coastal managers, and others with appropriate guidelines for coastal energy conservation and development, with an emphasis on offshore oil and gas development and oil tankers. *Conclusion* provides some final thoughts about future directions of energy development and related impacts to the coastal environment.

Although there may be a day (though unlikely) when "coastal energy issues" refer primarily to the siting of tidal and wave generating machines or nuclear power plants on our coastlines, for the present and near future, the U.S. and other nations will continue to rely heavily on offshore development and transport of fossil fuels, especially oil. Consequently, it would behoove coastal resource managers, students of coastal environments, and the general citizenry to at least be familiar with the sources and effects of oil pollution in coastal waters, the stages and associated environmental impacts of offshore drilling and transport, the cleanup strategies available when oil spills do occur (and they will), as well as some general policy guidelines which they can follow when addressing the offshore energy development issue.

SOURCES AND EFFECTS OF OIL POLLUTION

Sources

Coastal oil pollution comes from a number of sources—some natural and some from humans. Oil seeping out of the ocean floor has produced approximately half of the oil in the world's oceans (Owen and Chiras 1990). This natural process has traditionally been of little environmental concern to scientists, since the oil is leaked slowly over a wide area. (Recent studies completed at the University of California, Santa Barbara, indicate that we may have reason to be concerned after all.) The other half of the oil pollution in the ocean comes from various human activities. Human-caused coastal oil pollution is more ecologically (and economically) damaging since it often concentrates a lot of oil pollutants in a small area in a short period of time. *River runoff* is the source of over 40 percent of ocean oil pollution from humans. (See Figure 7-1). Individuals, businesses, industries, cities, and armies accidentally or deliberately dump waste oil on the land that eventually reaches rivers that flow out to sea. Although they receive most of the publicity, *tanker accidents* account for only 20 percent of human-caused coastal oil pollution, and *offshore oil and gas production* account for even less—just 5 percent (Owen and Chiras 1990).

Environmental terrorism. The worst oil spill in world history occurred because of *environmental terrorism*—perhaps the most unnecessary reason for humans to contaminate coastal waters. In February 1991, approximately 525 million liters (139 million gallons) of oil spilled during the Persian Gulf war (Miller 1992). It is believed that

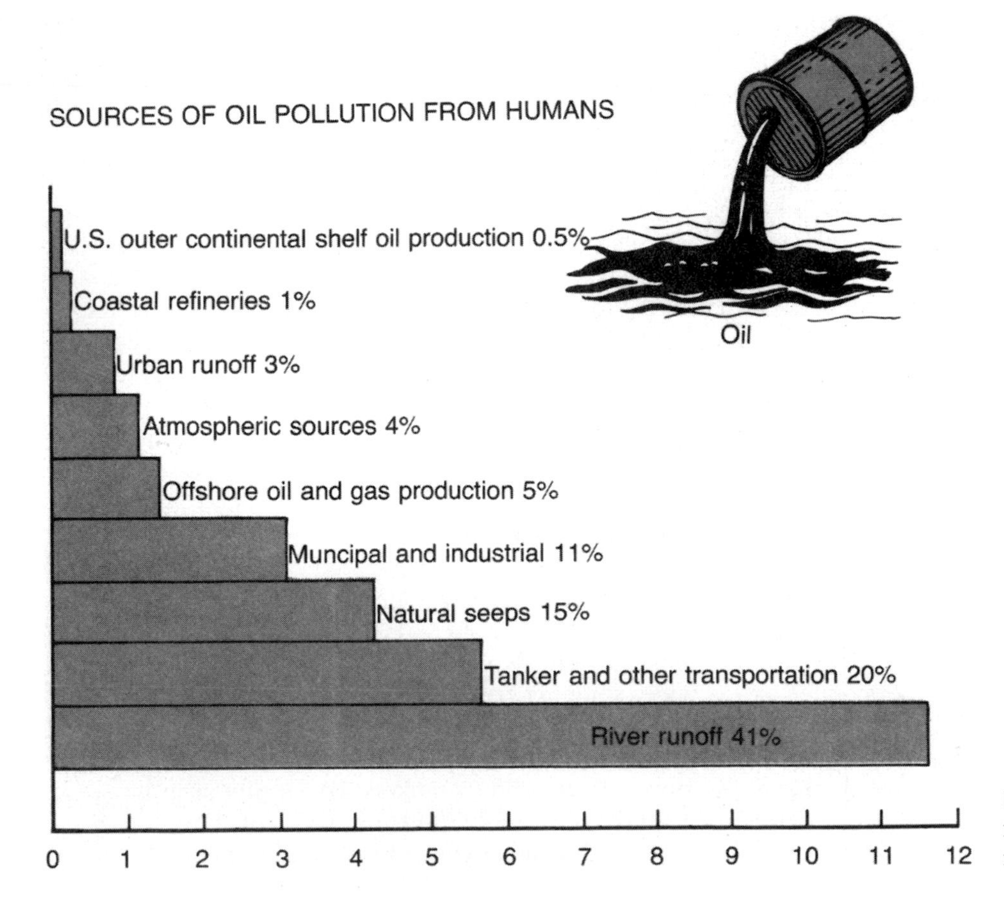

FIGURE 7-1 Sources of oil pollution from humans. (*Source:* Owen and Chiras 1990, p. 233)

Iraq's Saddam Hussein ordered that oil be released from Kuwait's Sea Island oil storage terminal and from several fully loaded Kuwaiti oil tankers. Some of the oil spill, however, may have been caused by bombing raids carried out by the American led coalition of military forces. Regardless, the volume of oil released was 13 times the volume of oil released in 1989, when the *Exxon Valdez* supertanker hit submerged rocks on a reef in Prince William Sound near Valdez, Alaska (Miller 1992).

Oil tanker accidents. Although the Valdez tragedy was a dramatic event and the worst tanker oil spill in U.S. history (See Figure 7-2), one must remember that oil spills are a *daily event.* Unlike the widely publicized disasters like the Valdez accident, or the February 7, 1990 spill off Huntington Beach, California (see Case Study), most spills are small amounts in highway accidents, from pipeline ruptures, or in tanker transfers. In the 12 month period after the 1989 Valdez accident, America experienced 10,000 of these smaller, less publicized spills—an average of *27 spills per day* (Wilderness Society 1991). All of this, of course, adds up to a tremendous impact on our coastal ecosystems.

Pipeline leaks. Some coastal oil spills have only been a day- or week-long "event" (e.g., offshore oil platform blowout; tanker collision; pipeline rupture), but others have gone on for over a decade. In March 1994, for example, Unocal Corporation was found guilty of causing what had quietly become the largest oil spill in California history. For *15 years,* employees of the Unocal Guadalupe oil field—32 km (20 mi) south of San Luis Obispo on the Central California coast—failed to report frequent leaks and pipeline breaks at their site. As a result, over 32 million liters (8.5 million gallons) of petroleum thinner contaminated the coastal groundwater and ocean with a clear, diesel-like diluent (a diluting substance)—at times covering beaches along the Central California coast with an oil sheen. Residents of the Central California coast had long complained that some sort of oil fluid from the Guadalupe field was contaminating beaches and harming dozens of sea lions, seals, and other marine animals. But, it was not until 1990 that Unocal first reported the spills. It took two raids of Unocal offices by the state's recently created Office of Oil Spill Prevention and Response team to capture the 80 boxes of records that eventually incriminated the oil company.

Although accidental events may capture a media blitz and immediate public outcry, it is the quieter oil spill that may actually do more damage. For example, the infamous Santa Barbara oil spill of 1969—the offshore oil well blowout that spilled 15.9 million liters (4.2 million gallons) on the ocean and onto the beaches—not only received immediate public reaction, but it has been credited by many scholars of environmental history for being *the launching pad for the modern American environmental movement.* It is noteworthy that the less visible (and thus less dramatic) Guadalupe oil field contamination that spilled twice as much oil over a 15 year period barely got any media attention at all, yet it is the largest oil spill in California's history.

Effects

It is not easy to predict the effects of oil pollution on an ocean ecosystem. Why? Because several factors come into play: Was the spilled oil *crude petroleum*—oil as it

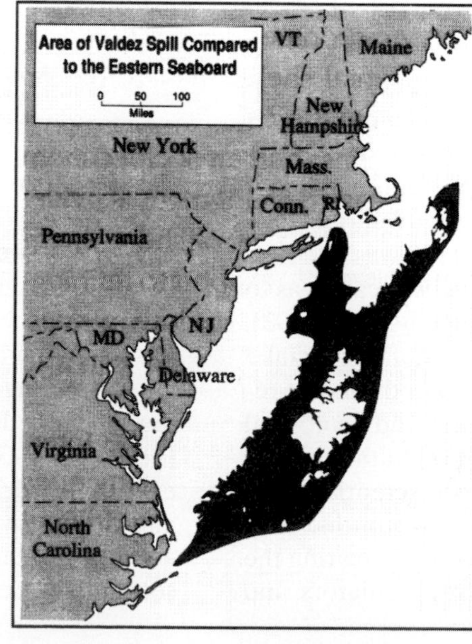

FIGURE 7-2 Area of *Exxon Valdez* spill compared to the California coast and the Eastern Seaboard. (*Source:* Granted with permission from *Coastal Alert,* Dwight Holing, © 1990. Published by Island Press, Washington, D.C. and Covelo, CA.)

comes from the ground, or was it *refined petroleum*—such as fuel oil or gasoline that had been obtained through distillation and chemical processing of crude petroleum? How much oil was spilled and how close was it to organisms? Was it spilled in an enclosed bay or in the open ocean? In what season of the year did it occur? What about the water temperature, direction of currents, wind velocity? All of these and more are factors that determine how an oil spill will affect marine life.

In general, however, oil spills can have the following effects: They can (1) *concentrate chlorinated hydrocarbons* (e.g., increase the rate of absorption and concentration of pesticides, such as DDT, toxaphene, and dieldrin from surrounding waters); (2) *reduce photosynthetic rates and food production* (e.g., an opaque oil slick blocks sunlight, which in turn reduces the photosynthetic activity of marine algae); (3) *contaminate food chains* (e.g., by increasing the concentration of benzopyrene, a known carcinogen); (4) *upset marine behavior patterns* (e.g., oil chemicals confuse the normal "chemical communication" between marine mammals, resulting in deranged feeding, mating, and migration patterns); and (5) *kill marine animals* (e.g., the 1989 *Exxon Valdez* oil tanker spill is known to have killed 22 whales, 5,500 sea otters, 580,000 birds, and an unknown number of fish [Wilderness Society 1991]).

Long-term effects on natural ecosystems. The "lingering effects" of oil spills on marine ecosystems have many scientists concerned. For example, in 1969, the barge *Florida* ran aground off Buzzards Bay, Massachusetts. Scientists found that five years after the spill, marine populations had not returned to normal. In the late 1970s, fiddler crabs, as well as other marine organisms in the area, were found to be less abundant and smaller than before the spill (Sanders et al. 1980; and Teal and Howarth 1984). Field researchers from Woods Hole Oceanographic Institution found oil that has remained for at least 20 years in the most heavily oiled and protected intertidal sediments at Wild Harbor and Buzzards Bay, and the researchers also found traces of oil in some animals, such as the Wild Harbor Marsh fiddler crab that burrows into the mud and feeds at the mud surface. Although 99 percent of the oil spilled in 1969 was believed to be gone, there is still enough oil in a few local areas to kill animals that burrow into those sediments (Teal 1993).

The barge *Florida*'s oil spill illustrates that it may take two to three decades before a contaminated site can recover, and that the long-term biological and ecological effects may never be known. Ironically, when the spill occurred, the oil industry ridiculed those scientists who claimed the oil would persist beyond six months. They were accused of "crying wolf" and needlessly scaring the American public. Fortunately, scientists, regulators, and the American public are beginning to question business and industry representatives who maintain a spill or other possible negative circumstance "has no long-term impact." Studies are now beginning to cause debate on the long-term impacts of other spills too, such as the *Exxon Valdez* oil spill (*Sport Fisheries Institute Bulletin* 1993; *Alaska's Marine Resources* 1992).

For general studies on oil spills and their effects, see Cotton (1992); and Kearney (1991a, b, c). The most serious overall impacts of marine oil pollution are on seabirds, and Ohlendorf et al. (1978) is an excellent source for details. Sea otters are also highly affected by oil pollution, as reported by Geraci and Williams (1990). Unfortunately, the most heavily oiled birds and sea otters die, even after painstaking efforts have been made to clean and rehabilitate them.

Long-term effects on human cultures and societies. Oil spills can disrupt human cultures and livelihoods, as well as natural ecosystems. In May 1994, five years after the crude oil from the *Exxon Valdez* blackened Alaskan waters, 10,000 fishermen, property owners, and Alaskan natives took the giant Exxon Corporation to court, seeking to prove that the pollution damaged their culture and livelihood. This was the first civil trial resulting from the March 24, 1989 accident that tainted an estimated 2,414 km (1,500 mi) of coastline. Scientists for the fishermen, property owners, and Alaskan natives argued that oil trapped under mussel beds would slowly leak and cause long-term chronic pollution, thus affecting even the long-term viability of the local economy and way of life. In September 1994, a jury in Anchorage slapped Exxon with a $5.3 billion judgment to pay fishermen, native Alaskans and others for their losses.

OFFSHORE OIL DEVELOPMENT: COMPONENTS AND IMPACTS

Most Americans think of "offshore drilling" as only that—an ugly drill platform "offshore." The oil drilling platform, however, is merely one of many offshore *and* onshore components that are required in offshore oil and gas development. What follows is a brief description of the offshore and onshore components required in oil and gas development on the **outer continental shelf (OCS)**—the submerged lands lying seaward of the state tide lines. (See Figure 7-3). OCS development components include exploratory drilling rigs, platforms, pipelines, separation and treatment facilities, supply bases, marine terminals, and storage tanks. Potential environmental impacts associated with each component will also be discussed.

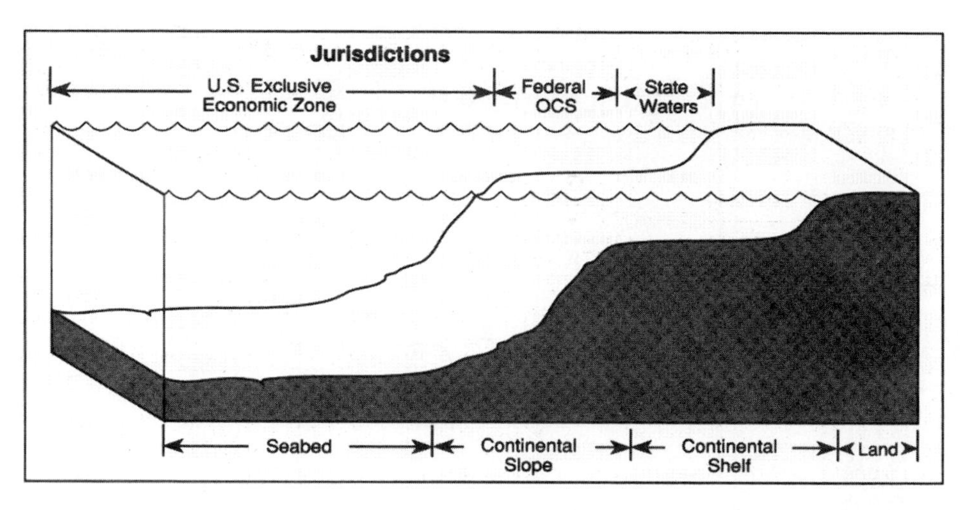

FIGURE 7-3 Federal control over the outer continental shelf (OCS) begins three miles from shore. (*Source:* Granted with permission from *Coastal Alert,* Dwight Holing, © 1990. Published by Island Press, Washington, D.C. and Covelo, CA.)

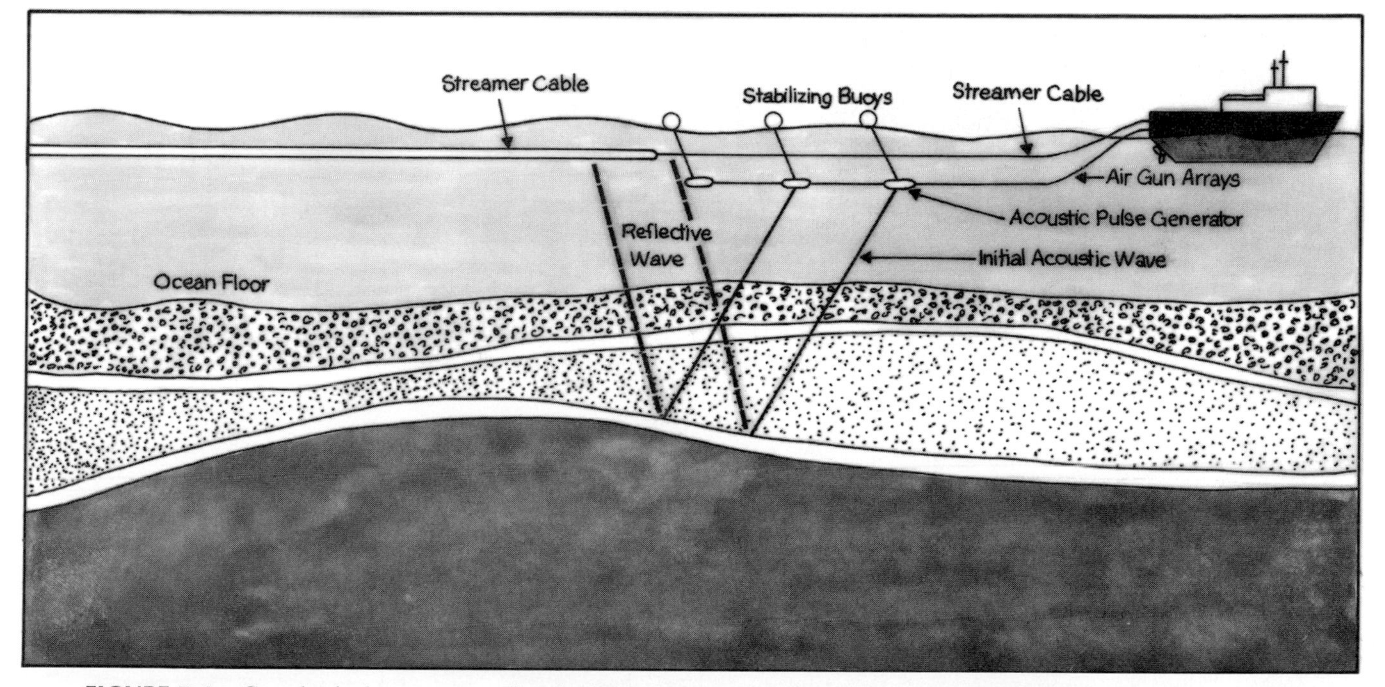

FIGURE 7-4 Geophysical survey vessel and seismic survey equipment. (*Source:* Drawing by Susan Giles, adapted from ERC Environmental and Energy Services Co., 1989 (Fig. 3-1) on p. 3-3)

Exploration Activities

Geophysical surveying. Prior to a **lease sale** (the public opening of a specified area for development), an interested oil company will perform a *geophysical survey* to determine if the underlying geologic formations are the types that can accumulate oil and gas. The most common technique is a seismic reflection from a geophysical survey vessel. (See Figure 7-4). Compared to the other stages of OCS oil and gas development, the environmental impact is minimal. However, some scientists are beginning to fear that "noise pollution" has a more serious effect on marine fish and mammals than originally suspected. This has already been discussed in Chapter 5, *Coastal Pollution.*

Exploratory drilling. If the seismic survey reveals a potential petroleum-rich site, the oil company will then bid for a lease to the tract. Once the company obtains a lease and all necessary permits, the company may begin the second phase of exploration—*exploratory drilling.* There are three major types of drilling rigs used in offshore exploration: (1) jackup rigs, (2) semisubmersible drilling rigs, and (3) drillships. (See Figure 7-5).

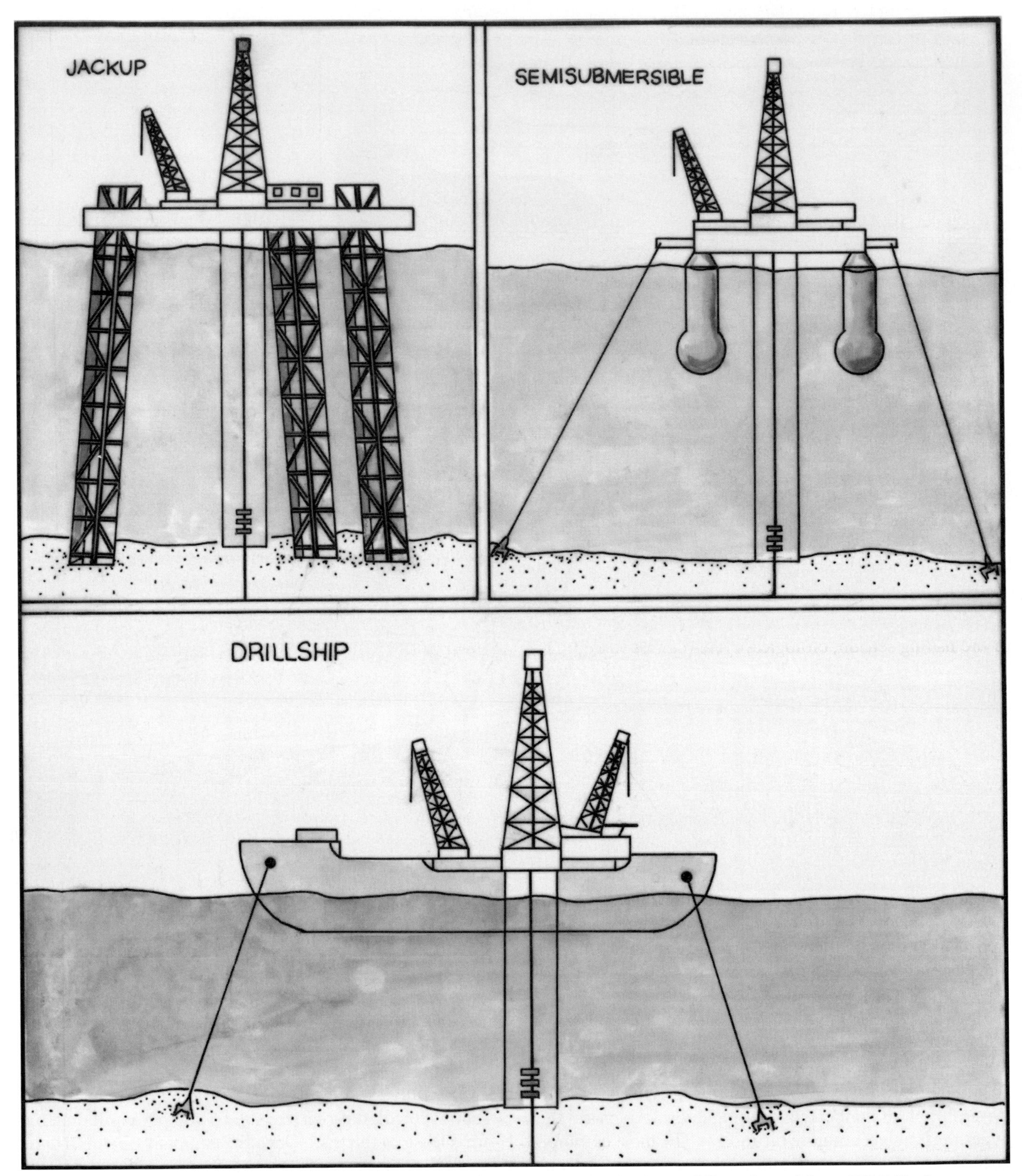

FIGURE 7-5 Examples of exploratory drilling rigs. (*Source:* Drawing by Susan Giles, adapted from ERC Environmental and Energy Services Co. 1989 [Fig. 3-20], on p. 3-4)

A **jackup rig** is a floating barge which supports a *rig*—the equipment used for drilling an oil or gas well. At the drill site, legs from the barge are lowered to the sea bottom, which ultimately "jackup" the barge so that it becomes a platform. One disadvantage of the jackup rig is that it is restricted to waters less than 114m (375 ft). A **semisubmersible** drilling rig is mounted on a barge-like hull with buoyant pontoons. Once at the drill site, the pontoons are flooded, thereby sinking the hull below the waterline. To keep the floating platform from moving during the drilling operation, the derrick and its equipment are anchored to the sea bottom. Semisubmersible drilling rigs are designed to operate in depths of 91 to 457 m (300 to 1,500 ft). The **drillship** is a ship with a drilling rig on its decks. Like the semisubmersible drilling rig, the drillship must also be anchored to the sea bottom prior to drilling. If not anchored, thrusters on the ship would keep it in position during drilling. Drillships can operate in waters as deep as 457 m (1,500 ft).

Impacts from exploratory drilling are essentially the same as those experienced by fixed platforms (see below). The main difference, however, is that exploratory drilling is a *temporary* and relatively *short term* activity—usually only 60–90 days. Typical exploration programs only drill four wells per lease tract. If not appropriately timed, however, exploratory drilling can take a toll. For instance, if it is scheduled in major fishing grounds during the height of the fishing season, biological cycles can be disrupted, and severe economic hardships can occur to the commercial fishing industry.

Offshore Development and Production

If the exploratory program finds quantities of oil and gas that are economically recoverable, the development and production program begins. This involves five major phases: (1) platform construction, (2) platform installation and hookup, (3) development drilling, (4) oil and gas production, and (5) waste disposal and air emissions from platforms.

Platform construction. *Platform construction* occurs in fabrication and assembly yards that have the necessary space requirements and easy access to the open ocean. The most common **platform** is the fixed-leg platform—an enormous multi-layered, steel structure that will appear from a distance as something that is assembled from a child's erector set. What is seen above the water surface, however, is just the *deck* of the platform. The deck is supported by what is called a **jacket**—large-diameter pipes welded together with braces to form a multi-legged stool-like structure that is secured to the ocean floor with pilings.

Platform installation and hookup. The second phase, *platform installation and hookup,* begins with the towing of the platform jacket to the offshore site (lease tract). Once the platform jacket is up-ended and attached to the ocean floor with piles, deck sections twice the size of a football field are assembled atop the jacket. This process can take several weeks to several months. Many additional months are needed for "hookup"—the installation of electrical wiring, instruments, and operating equipment.

There are a number of environmental impacts at this phase. For example, the large anchoring devices used to secure the jacket to the seafloor can cause the loss of hard-bottom **benthos**—the rocky outcrops on the ocean floor which support marine organisms, sometimes potentially unique biological species. Once the jacket is anchored to the sea floor, and the above surface components (e.g., **derrick**) are fully erected, additional environmental impacts occur. Some impacts are glaringly obvious (e.g., visual intrusion on the seascape), and others are less obvious (e.g., the disruption of seasonal whale migrating patterns; the displacement of commercial fishing grounds; and the creation of a major navigational hazard for freighters, tankers, and even small boats). Additional impacts occur, however, with the beginning of the next phase of offshore oil and gas development.

Development drilling. The third phase, *development drilling,* begins when the platform is operational. Using a crane attached to the derrick, a thirty-foot long tubular steel pipe with a "drill bit" on the end is lowered into the water. **Drill mud,** sometimes just called mud, is used to lubricate and cool the drill bit. Drill mud is a mixture of special chemicals, clay, water, and barite. In addition to acting as a lubricant and coolant, drill mud is essential for raising the drill cuttings to the surface, coating the bore hole wall, and containing the underground pressure. Since drilling muds have been known to contain organic and inorganic chemicals, biocides, and even heavy metals, disposal of this waste product is quite controversial. The routine discharge of the toxic drilling muds can reduce the biological productivity of nearby ocean waters. Over the 20–30 year lifespan of a drilling platform, these muds accumulate on the ocean floor and may affect fish and benthic invertebrates (bottom dwelling organisms). Scientists disagree, however, as to the severity of these impacts.

Oil and gas production. *Oil and gas production,* the fourth phase, refers to the "lifespan" of the platform in terms of oil and gas production. Typically, this period lasts 15 to 20 years, yet a 30-year lifespan is not uncommon. When the liquid emerges from the platform's numerous wells, it is usually a mixture of oil, gas, water, and sediments. Consequently, the product must first undergo separation before the product is shipped to refineries.

Oil spillage from a well blowout has the greatest potential for ecological and economic disruption. A **blowout** can be defined as an uncontrolled flow of oil, gas, or other fluid from a platform well into the surrounding environment. It can happen whenever pressure within the rock strata becomes greater than the overburden of a column of drilling fluid. Perhaps the most famous case of a blowout in this country was the 1969 Santa Barbara blowout. It is historically important not only because it coated 241 km (150 mi) of beaches with thick oil but also because it ignited a national environmental movement, as mentioned earlier. The spill smothered intertidal and subtidal biota, killed thousands of seabirds and marine mammals, fouled beaches, and generally brought economic havoc to the tourist and recreational industries in the area.

Waste disposal and air emissions from platforms. Unfortunately, one cannot have offshore development and production without the fifth phase—*waste disposal and air emissions from platforms.* In addition to the continual outflow of drilling muds, the platform over its lifespan also produces a number of other waste products. Additional *liquid wastes* are created from the accumulation of water from the oil formations, the use of water for cooling machinery, the hydrostatic testing of pipelines, deck drainage, and the brine from desalinization units. Whereas non toxic wastes are discharged directly into the ocean, the more toxic wastes should to be stored in containers on the platforms and then shipped to treatment or disposal centers on shore. There are also *air quality impacts* from the diesel-powered machinery on the platform and from "flaring" (release of gas during system malfunctions). Whenever there are generators and other machinery, helicopters, crew boats, and so on, there are *noise impacts.* As mentioned earlier, noise pollution may have serious effects on marine mammals, as well as on humans.

Subsea and Onshore Pipelines

Offshore oil drilling requires miles of subsea and onshore pipelines to transport oil and gas from the platform to its various destinations. Most oil pipelines are approximately 24 inches in diameter capable of moving 250,000 barrels of oil per day (Holing 1990). Pipeline impacts vary according to whether they are subsea or onshore.

Subsea pipelines. Subsea pipelines can be installed by either the "pipelay" or "pipepull" technique. With the *pipelay method,* workers on a "Pipeline Lay Barge" weld the lengths of pipe together then "lay" the connected pipe onto the sea floor. A second barge, the "Pipe Bury Barge," follows close behind and "buries" the pipe with sand. (See Figure 7-6). With the alternative "*pipepull method,*" the required lengths of pipe are welded together onshore and then "pulled into place" by offshore tugs. This latter method requires a temporary onshore pipeline staging area of 1 to 3 hectares (3 to 8 acres) for welding and launching the pipeline.

Seabed pipelines can have adverse impacts. For example, pipelay barge anchoring activities can result in

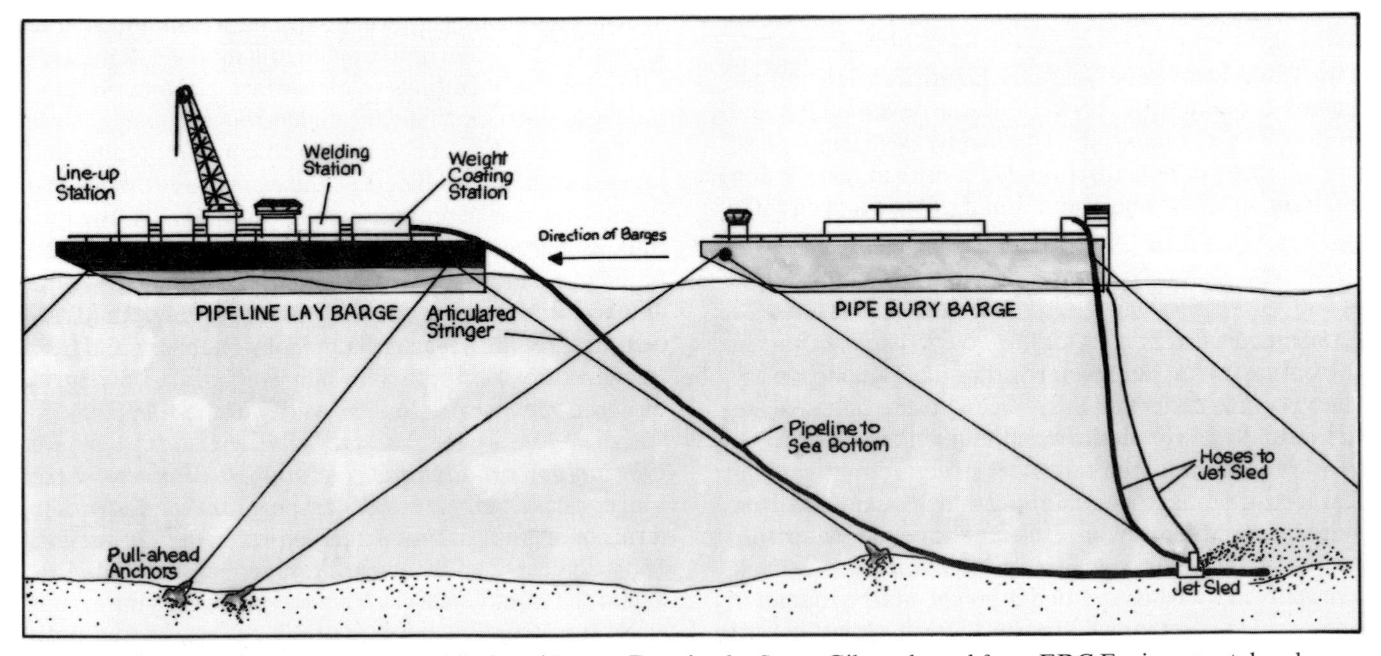

FIGURE 7-6 Subsea pipeline installation. (*Source:* Drawing by Susan Giles, adapted from ERC Environmental and Energy Services Co., 1989 [Fig. 3-6] on p. 3-19)

loss of hard bottom benthos. Seabed pipelines require a 1500 ft Right-of-Way (ROW). Many of these "underwater freeways" are plowed right through kelp forests that once served as habitat for dependent fish and invertebrate populations. Since blasting is sometimes used to clear these swaths, there is an additional potential for disruption and mortality of marine mammals and seabirds. If seabed pipelines are not adequately buried, or become exposed over time, pipelines in commercial fishing areas can snag fishing nets and other gear. Of course, there is always the possibility that an operating seabed (or onshore) pipeline could break and spill oil—fouling recreational beaches, coastal wetlands, and bays (Wilderness Society 1991). As with platform spills, pipeline oil spills reduce both tourist dollars as well as lifeforms (e.g., seabirds and sea mammals mortality).

Onshore pipelines. The construction of permanent onshore pipelines requires an even larger staging area. As much as 6 to 18 hectares (15 to 45 acres) may be needed for the storage of materials and equipment. (See Figure 7-7). In addition to this temporary staging area, an 18 to 30 m (60 to 100 ft) wide construction right of way (ROW) or corridor is needed for laying the pipe. The trench, itself, is generally only 1.5 m (5 ft) wide and 1.2 m (4 ft) deep. After the pipeline is laid in the trench and backfilled, a permanent 9 to 15 m (30 to 50 ft) wide ROW is mandatory for operation and safety purposes.

A number of environmental impacts are created with onshore pipeline construction, operation, and maintenance. To create staging areas, for example, the land is completely stripped of all vegetation, then graded. In some cases, this has left permanent scars that are highly visible across the coastal landform. Temporary impacts include increased tractor and truck traffic and its associated noise and air pollutants. Depending upon the route of the onshore pipeline corridor, the pipeline ROW can adversely impact the stability and erosional characteristics of slopes, archaeological resources, surface water resources, wildlife, and wildlife habitats (wetlands, lagoons, riparian communities). Many of these can become long-term impacts since trucks and other vehicles must regularly use the pipeline ROW for maintenance. Pipeline breakage, as previously mentioned, can have detrimental impacts upon land and aquatic species, soils, surface and groundwater supplies, and runoff from the land to the ocean.

Oil and Gas Processing

As mentioned earlier, a mixture of oil, gas, water, and suspended mineral impurities flows out of an offshore well stream. Although *initial* "separation" occurs directly on the platform, further **processing** (e.g., dewatering) must occur before oil and gas can be sent to refineries. This processing stage can be performed either on the platform itself, at an offshore storage and treatment vessel (OS&T), at a separate onshore facility, or at a combination of both offshore and onshore processing sites. A number of factors go into the decision of where and how to process oil: the distance between the platform and the shore, the availability of suitable onshore sites, the characteristics of the well stream, and the economic advantages and political feasibility of each strategy.

Onshore processing facilities. Most oil and gas is processed at onshore facilities. Although oil and gas can be processed at separate facilities, the two are usually processed at the same onshore site. For example, the Chevron Company has a 25-hectare (62-acre) oil and gas

FIGURE 7-7 Onshore pipeline construction activities and environmental impact. (*Source:* Photo by John Storrer from *Coastal Alert* © 1990 by Dwight Holing. Published by Island Press, Washington, D.C. and Covelo, CA.)

FIGURE 7-8 Dramatic alteration of onshore environments accompanies offshore drilling, such as the Chevron Processing Plant. (*Source:* Photo by Tyler Johnson from *Coastal Alert* © 1990 by Dwight Holing. Published by Island Press, Washington, D.C. and Covelo, CA.)

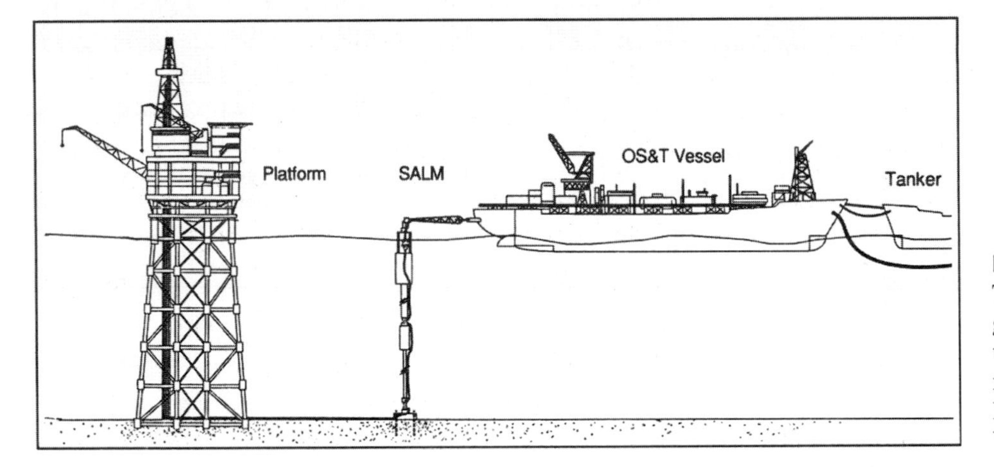

FIGURE 7-9 (a) Offshore Storage and Treatment (OS&T) Vessel and A Single Anchor Leg Mooring (SALM) System. (*Source:* Granted with permission from *Coastal Alert* © 1990 by Dwight Holing. Published by Island Press, Washington, D.C. and Covelo, CA.)

processing plant on the Santa Barbara coast in Southern California. (See Figure 7-8). According to the Minerals Management Service (1985), this plant has the ability to process 250,000 barrels per day (**BPD**) of oil and 120 million standard cubic feet per day (**MSCFD**) of gas. In addition to the clearing of land (sometimes up to 57 hectares, or 140 acres), onshore processing facilities create other environmental impacts. Interestingly enough, oil processing plants that essentially "dewater" the oil must also use large amounts of water in their operations. A 200,000 barrel-a-day facility can use between 112 to 425 acre feet of water per year (Holing 1990). By comparison, the average single family residence uses about 1/2 an acre foot of water per year. Obviously, constructing onshore processing facilities along a coast, especially in a drought-prone area like Santa Barbara, can place a tremendous drain on the community's water resources. Onshore processing plants must also discharge large amounts of waste water. In most cases this waste water is flushed into the sea; however, this is usually in accordance with regulatory stipulations, which include the requirement for treatment.

Offshore Storage and Treatment (OS&T). Sometimes oil and gas processing is carried out on converted tankers called *offshore storage and treatment vessels,* or simply *OS&Ts.* An **OS&T** is basically a converted oil tanker that is permanently moored relatively close to a platform by a "single anchor leg mooring" (SALM). (See Figure 7-9). Crude oil is transferred from the platform via subsea pipeline to the OS&T for processing. Once the crude oil is processed, the oil stream can be transferred directly to tankers or transported via subsea pipelines to onshore storage facilities. Should weather conditions (or some other reason) cause the OS&T to suddenly break away from the SALM, self-closing safety valves exist between the universal joints of the SALM. Offshore storage and treatment vessels can typically process between 40,000 to 200,000 BPD of oil.

The impacts of OS&Ts are minimal compared to onshore processing facilities and their related pipeline systems. One problem, however, is with air emissions. OS&Ts often operate in federal waters beyond a state's coastal three-mile jurisdiction. Consequently, state and local jurisdictions cannot exercise regulatory controls. This would not be a problem if it were not for the fact that federal air pollution standards are often lower than that of states. Coastal cities have a hard enough time meeting their state's more stringent air pollution standards without pollutants blowing to shore from offshore storage and treatment vessels that are allowed to operate under less severe restrictions.

Support Facilities

Offshore oil and gas exploration, development, and production requires onshore industrialization—everything from supply and crew bases, to roads and heliports, and to electrical power facilities, fresh water wells, and waste disposal operations.

Supply and crew bases. A *supply base* is a staging area where heavy equipment and supplies are temporarily stored and then transferred by boat to offshore platforms. The typical supply base requires a relatively flat terrain of 20 to 32 hectares (50–80 acres) that is as close to the offshore oil platform as possible. Since hundreds of workers may be onsite at one time, food service, temporary housing, office space, and warehouse/storage facilities are all necessary components of the supply base. Port Hueneme in Southern California is an example of a supply base; it supports the offshore platforms in the Santa Barbara Channel and north of Point Conception. The construction and operation of a supply base can have numerous environmental impacts, including the destruction of acres of aquatic and terrestrial biotic resources, impacts to surface and ground water resources, degradation of recreational resources (e.g., beaches), and the loss of archaeological resources and coastal aesthetics.

Some of the same support activities can occur on a *crew base*—a smaller onshore facility. Because they are smaller, and typically operate from existing piers, crew bases generally have less of an impact than supply bases. Carpinteria and Ellwood piers on Santa Barbara's southern coast are examples of crew bases.

Roads and heliports. Supply bases also need increased road access and capacity, as well as heliports. If a supply base is located in a relatively isolated area, new roads are often constructed or widened to accommodate large trucks, cranes, and other needed vehicles that transport the heavier equipment to the shore. Lighter supplies and personnel are regularly flown via helicopters to the offshore platform. All this activity adds up to increased traffic, as well as noise and air pollution.

Electrical power, fresh water, waste disposal. Whenever there are hundreds of workers together for any industrial activity, provisions must be made for electrical power, fresh water to drink, and industrial and domestic waste disposal. *Electricity* is needed to power the supply base, the processing plants, and the offshore platforms. (Some platforms provide their own electricity, but that requires regularly shipping fuel from onshore bases). *Potable water* is needed for the obvious reasons (hygiene, cooking, sanitary needs) and less obvious (preparation of some drilling muds, emission control of diesel-powered turbines). *Waste disposal* is part of any support facility. Waste generated by offshore oil and gas activities that cannot be legally disposed of in the ocean must be disposed elsewhere. For example, toxic drilling muds and other cuttings must be shipped to shore and then transported by rail or truck to an appropriate landfill disposal site.

Refineries and Petrochemical Plants

In addition to initial "separation" at the offshore platform and "processing" at an offshore storage and treatment vessel (OS&T) or at an onshore processing plant, crude oil must go through one or two additional steps before it can become marketable. *Refineries* are facilities for heating crude oil so that it separates into its chemical components. After these separate components are distilled, they become the daily products that we have all become accustomed to using, such as gasoline, fuel oil, kerosene, propane, and a variety of lubricants. *Petrochemical plants* can take the "refining" process one step further. Using feedstocks produced by either petroleum refineries or natural gas processing plants, petrochemical plants produce such items as fertilizers, pesticides, industrial chemicals, explosives, synthetic fibers, paints, and even medicines.

In terms of potential environmental impact, both refineries and petrochemical plants require large expanses of land. For example, a refinery that can produce 250,000 barrels per day requires from 405 to 607 hectares (1000 to 1500 acres) of flat, industrially zoned land. Only about 81 hectares (200 acres) is actually needed for the processing unit; the rest of the land is needed for storage and as a buffer zone against adjacent properties. Petrochemical plants, in addition, produce an almost limitless variety of gaseous pollutants. Accidental release of pollutants occurs even under stringent safety regulations. And, when industrial accidents do occur, they can cause untold damage to the environment and human life. For example, at Bhopal in India in 1985, an estimated 2,300 people were killed and another 500,000 injured when a petrochemical plant accidentally released toxic gases into the atmosphere.

Shutdown and Abandonment

Once the hydrocarbon reserves are depleted, production stops, and the disassembling and removal process begins. Platform equipment, deck sections, and the jacket are disassembled and transported to the support facility for reuse or disposal. The pilings are cut below the mudline, thereby leaving no obstructions on the seafloor. Divers seal the actual wells with concrete. The seabed pipelines are purged with seawater, filled with an inert liquid, such as a barite-water mixture, and then "abandoned" (left in place). The onshore facilities are usually taken over by some other industry. This author is not aware of any onshore facility that has been totally abandoned and has reverted to its natural state. Current law requires that offshore oil drilling facilities must be fully dismantled and out of the area one year after officially declaring shutdown.

OIL TRANSPORTATION OPTIONS

Once the oil and gas have been processed (either onshore or offshore), they must then be transported. Oil needs to be transported to a refinery; gas needs to be transported to a utility distribution system. Since the transportation of oil has the greatest potential for environmental disaster, it will be the focus of this section. There are only two options for the transportation of oil and gas: pipelines or marine tankers. These two methods and their associated impacts are discussed below.

The Pipeline Option

As with most issues that can have a profound impact on land and water, various opinions exist on how best to carry out a particular activity. In terms of transporting oil up and down the U. S. West Coast, for example, local communities and their associated coastal commissions generally prefer pipelines over the use of marine tankers. They maintain that tankers have a greater impact on air quality and increase the probability of oil spills. The oil industry generally prefers to use tankers, arguing that using marine terminals and tankers allows them greater operational flexibility. Rather than being permanently " locked in" to one destination, tankers allow them to go to different destinations as the demand for crude oil warrants. (Of course, the United States must rely on marine tankers for importing oil from distant foreign countries. For this purpose, " the pipeline option" is really not an option.)

Despite these typically held positions, there are a number of questions that are always considered before onshore oil transportation is proposed and approved. For example, are there any existing policies by the affected local governments that encourage pipeline transportation? Is there an existing crude oil pipeline network nearby that could be connected into? Are there any existing marine terminals in the area, and could they handle the required amount of OCS crude? What are the possible refinery destinations? And, what are the possible environmental impacts and risks associated with pipelines versus tankers to those destinations? All of these questions, and more, must be answered before a decision is made regarding which oil transportation option is to be used.

Size, capacity, and ownership. Onshore pipelines for crude oil and natural gas range in size and capacity depending on a variety of factors (e.g., projected production volumes; quality of oil and gas), but generally the pipe diameter ranges from 20 to 91 cm (8 to 36 in) and has a capacity of between 35,000 and 350,000 barrels per day. Most pipelines are privately owned common carriers, regulated by the Federal Energy Regulatory Commission. A number of U.S. counties have "consolidation policies" regarding the construction of new pipelines. By encouraging common carrier or multiple-use of new pipelines, the amount of new pipeline construction and associated environmental impacts is reduced.

Pipeline leaks. One good reason to have as few pipelines as possible is the obvious fact that they can leak, bringing about much environmental damage. For example, on January 1, 1990 an underwater pipeline connecting Exxon plants in Linden, New Jersey, and Bayonne, New Jersey leaked 956,949 liters (252,800 gal) of heating oil into the Arthur Kill—an elongated body of water that runs between New Jersey and Staten Island, New York (Wilderness Society 1991). Among other things, the spill drastically reduced breeding of the wading birds that had been making a dramatic comeback in the area. Furthermore, the Manomet Bird Observatory of Massachusetts, that has studied the spill's impacts, also noticed delayed nesting activity among snowy egrets, glossy ibises, and much higher mortality rates for young birds.

The Marine Terminal and Tanker Option

Oil tankers (ships) and marine terminals (oil loading/unloading and storage facilities) go hand and hand; one cannot exist without the other. Yet, it is the "oil spill from a tanker accident" that one reads about in the newspaper or sees on television. It is therefore necessary to spend a little time here to discuss exactly what a marine terminal is and identify some possible environmental impacts.

Components of a marine terminal. There are four major components of a marine terminal: (l) storage tanks, (2) subsea oil and gas pipelines, (3) a power transmission system, and (4) anchoring buoys. *Storage tanks* are needed to accommodate any volume of oil from 200 to 600,000 barrels per day. Such a "tank farm" requires a large area of flat land with stable soil. For example, a marine terminal with a throughput of 150,000 to 300,000 barrels per day would require approximately 6.5 hectares (16 acres) for a tank farm with a storage capacity of 2.7 million barrels. Since storage tanks might leak, buffer zones must also be designed around the facility. Typically, marine terminals must be able to store oil for at least 7 to 10 days. *Subsea oil and gas pipelines* are needed to transfer the crude from the SALM (Single Anchor Leg Mooring) to the onshore storage tanks. A *power transmission system* is also needed to supply energy to the terminal and tanker for pumping the crude through the pipes. And finally, *anchoring buoys* and other devices are needed to stabilize the tankers and terminal components.

Impacts, tankers, and oil spills. The construction of an onshore tank farm has the same impacts as identified earlier for the construction of a processing or treatment facility, such as impacts to landforms, archaeological sites, land use resources, terrestrial flora and fauna, and visual aesthetics. The laying of subsea oil and gas pipelines, the creation of a power transmission system, and the use of anchoring buoys has basically the same types of impacts as the construction of a platform or the laying of subsea pipelines for transmission of oil to processing facilities. In both cases, there is a substantial impact to benthic organisms, marine and intertidal biota, as well as fishing and recreational resources.

Crude oil tankers, themselves, vary in size according to the volume of oil to be transported. On the West Coast of the U.S., tankers generally range from 17,000 dead weight tons (DWT) to 150,000 DWT. A 50,000 DWT tanker, for example, has the capacity to carry 365,000 barrels of oil and an average pumping rate of 30,000 barrels an hour.

Studies of worldwide tanker accidents have provided some interesting points. For example, researchers from the National Oceanic and Atmospheric Administration (NOAA) in collaboration with the American Petroleum Institute (API) have found that (a) smaller vessels seem to exhibit higher casualty rates than larger ones (i.e., the smaller the vessel the greater the accident rate); (b) vessels operating under certain flags of registry have much higher casualty rates than others in the world fleet (e.g., Greece and Liberia had the highest accident frequencies; the United States and the United Kingdom rated in the middle range; France, Norway, and Italy had the lowest accident rates); and most surprisingly, (c) older vessels appear to be no more accident prone than newer vessels (e.g., vessels less than 20 years old had higher accident rates than vessels older than 20 years [Meade, LaPointe, and Anderson 1983]).

According to some reports, oil tanker spills appear to be decreasing in U.S. waters. Golob found that only 208,197 liters (55,000 gallons) of oil were spilled in U.S. waters in 1991—the least since 1978 (*Golob's Oil Pollution Bulletin* 1992). He maintains that the decrease is due to tougher environmental laws passed after the 1989 *Exxon Valdez* spill. The Valdez spill shocked the nation. Americans turned on their nightly television news program and witnessed the killing of thousands of sea otters and hundreds of thousands of birds, the degradation of over 1770 km (1,100 mi) of pristine Alaskan coastline, and the futile effort of 11,000 workers at an expense of $2.5 billion trying to clean oil-slicked rocks. It took this disaster to shock Congress into passing the *Oil Pollution Act of 1990.* In a nutshell, what this complicated act did was (a) raise liability limits for spills (i.e., make tanker owners more liable for their actions), (b) establish a $1 billion spill fund, and (c) mandate double hulls for tankers calling at U.S. ports. Golob believes the recent decrease in oil spills is not a statistical oddity, but rather the direct result of the passage of this act. For additional information on tankers and coastal zone management, see Trench (1993).

Oil spills and marine sanctuaries. Not all spills are reported, however. For example, in January 1993, globs of oil appeared off the shores of Central California. With over 50 Alaskan oil tankers passing outside the Monterey Bay National Marine Sanctuary on their way to refineries and with many returning up the coast with refined products, an offshore tanker spill was highly suspect. Although the MBNMS does not regulate vessel traffic, tankers are not allowed to discharge their tanks within approximately 74 km (46 mi) off the coast. How can a state Department of Fish and Game be prepared for phantom spills at sea? If a phantom vessel was caught purging its bilge within the boundaries of the sanctuary, it can be charged with a federal regulation violation and fines can be imposed.

Until Congress agrees on a tanker traffic safeguards amendment for NOAA and the U.S. Coast Guard, the California Department of Fish and Game has implemented its own set of proposed regulations, the so-called Oil Spill Prevention and Response (OSPR) plan. Numerous environmental groups, however, have challenged the regulations because the OSPR designation of the

Moss Landing Marine Terminal as a "Facility/Transfer Area" *only* meant that Moss Landing harbor would *not* need to have resident emergency equipment. If a major oil spill did occur, the local environmental groups maintained that it would take at least 24 hours before clean-up equipment would arrive at the site. The Sierra Club, the Center for Marine Conservation, the Friends of the Sea Otter, and Save Our Shores assert that the Department of Fish and Game needs to re-write the OSPR plan to include better and larger distribution of emergency equipment at strategic coastal locations, as well as the availability of trained local personnel to deploy it.

CLEANUP STRATEGIES

Regardless of how careful oil companies are in the development, processing, and transportation of their product, oil spills are going to occur. Consequently, it is necessary that we all have an understanding of what we can and cannot do in terms of spill cleanup. We will now briefly discuss cleanup techniques and their limitations, the lingering effects of oil spills, and some oil spill response programs.

Cleanup Techniques and Limitations

Cleanup technologies are presently both rudimentary and dependent upon the resources available at or near the site of the spill. The currently available technologies are:

- *Inflatable booms.* One of the most common methods of controlling or impeding the spread of a floating spill is by encircling it with a mechanical barrier known as an inflatable boom. As can be imagined, the technique is rather ineffective (or impossible) with large spills, in high seas, or in ice-congested water.
- *Skimmers.* Skimmers are small ships that have pumps aboard to gather in the oil-water mixture, as well as machines that can separate the oil from the water and store the oil in storage tanks. As with inflatable booms, skimmers cannot work in rough seas or in ice-congested water.
- *At-sea absorbents.* Cleanup crews have used a variety of absorbent materials to clean oil spills at sea, including pillows containing chicken feathers. As one can imagine, this technique is highly labor intensive and ineffective in rough seas or with large spills. Plus, imagine the secondary mess of having to dispose of hundreds of thousands of oil-drenched pillows or other absorbent materials.
- *Chemical dispersants.* Rather than "contain it" (inflatable booms), "pump it up and transport it" (skimmers), or "absorb it and dispose of it" (sea absorbents), cleanup crews have also used more controversial techniques, such as using chemical dispersants. Specially equipped airplanes spray chemical dispersants over the spill; the chemicals are designed to cause the oil to either disperse, dissolve, or sink to the ocean floor. One problem is that a chemical dispersant relocates the oil rather than eliminates the problem. More seriously, the technique can be highly toxic to the environment. Many biologists contend that more marine life is killed by the chemical dispersants than by the oil itself. The process has only been used in limited field experiments, such as when Exxon tried the technique on a few beaches after the Valdez spill. In these experiments, Exxon did find that the dispersant helped accelerate the dispersal of oil.
- *Ignition.* "Burning it off" is another technique that has been tried. In this case, special equipped helicopters use lasers to ignite the oil spill. On the positive side, ignition is more effective and less expensive than using chemical dispersants, and it is about the only technique that effectively works in ice-congested waters. On the negative side, however, ignition can create an on-land fire danger (if used too close to the shore), can create severe air pollution problems, and can create falling ashes that can be toxic to fish and other marine life.
- *Genetic engineering.* Even more controversial than chemical dispersants is the development of "superbugs" that can "gobble up" the oil. The idea is to develop special bacterial strains that are more efficient "degraders" than natural bacterial strains already in the ocean. This form of development, sometimes called "**bioremediation,**" needs to be very carefully investigated before widespread use. For further information on the pros and cons of genetic engineering in general, see Suzuki and Knudtson (1989).
- *Chemical engineering.* Chemists are also trying to engineer quick and relatively cheap ways to clean up huge oil and chemical spills. Dr. Adam Heller of the University of Texas at Austin, for example, is working on a mechanical agent, "microscopic glass beads." At a 1992 national meeting of the American Chemical Society, Heller demonstrated the process. Dr. Heller first poured a layer of crude oil over a glass dish filled with water. He then poured a few teaspoons of his specially designed hollow glass beads (about the thickness of human hair) that had been coated with titanium dioxide (a common pigment used in white paint). The oil from the "spill" almost instantly combined with the floating glass bubbles, congealing into floating clumps and leaving the rest of the water surface completely free of oil. In sunlight, the titanium dioxide coated beads can oxidize the hydrocarbons from oil spills that adhere to them, rendering the remaining oil components soluble in water. To put it another way, this method essentially creates "wet combustion" in which the titanium dioxide beads act as a match. According to Heller, the remaining beads would have no more effect on the environment than beach sand. Extensive field testing will be needed before any real conclusions can be made about this technique.
- *On-land absorbents/detergents/high-pressure hoses.* Once the oil reaches the shore, clean-up crews are limited to only a few techniques, all of which are labor intensive.

Hay is an "on-land absorbent" that is often used to help clean beach sand. *High-pressure hoses* are used to force the heavier crude from boulders and rocks. *Detergents, buckets, and rags* are used on rocky shores, jetties, and groins. A more controversial method is to spread phosphorus and nitrogen fertilizers on the oil-soaked shoreline to accelerate the growth of natural bacteria that break down the oil. This method has other environmental impacts asociated with its use.

Lingering Effects

"Woefully inadequate" is the phrase most often used by oil spill cleanup specialists when asked about the state of oil spill technology. For example, after years of cleanup efforts and over $2 billion spent, the oil-impacted beaches—from Prince William Sound, Alaska to Kodiak Island—are not as they were before the March 24, 1989 *Exxon Valdez* spill. Millions of gallons of oil and tar remain, much of it hidden several feet below the rocks, while the oil that remains on the surface is turning to asphalt. Of the 257,000-barrels from the Valdez oil slick, only 32,500 barrels were recovered by Exxon, and 77,100 barrels evaporated. As a result, more than 147,000 barrels remain in the environment (Holing 1990). This led the National Wildlife Federation, as well as many other environmental organizations, to conclude that oil companies (with their present technologies) will never be able to thoroughly clean up sites degraded from oil spills.

The Valdez spill made it very clear that oil companies had spent very little research and development money on oil spill cleanup technology up to that time. Until the Valdez spill, oil companies basically operated on the philosophy of satisfying the public's need for fuel and their stockholder's profit margin at the expense of America's "environmentally secure future." The Valdez spill brought the Exxon company so much bad publicity that after it repaired the tanker, the company re-located and re-named the ship. The "*Exxon Valdez*" is now named "*Exxon Mediterranean*" and transports oil from the Middle East to Europe.

Oil Spill Response Programs

Several coastal states have "Oil Spill Response Programs." These programs generally provide guidance documents for industry planners regarding safety requirements, waste disposal, wildlife rehabilitation, and other spill procedures.

The California Program. In California, for example, all facilities designated as marine facilities under the *California Oil Spill Prevention and Response Act of 1990* must have on file an *Oil Spill Contingency Plan,* including amount, types, and location of all spill response equipment. Amended versions of this act require that the equipment be appropriate for a specified area, taking into account geography, tides, water depths, and other environmental conditions. These plans are administered by *California's Office of Oil Spill Prevention and Response (OSPR).* When 15,142 liters (4,000 gal) of diesel fuel surfaced from a sunken fishing boat in October 1992 off Humboldt Bay, OSPR field response staff were on the scene and helped prevent detrimental impacts on the environment.

In addition to "official" state oil spill response teams, other organizations are also concerned (and involved) with oil spill containment and prevention. In California, for example, the *Fishermen's Oil Response Team (FORT)* in Ventura County responded to an oil spill at Avila Beach in 1992. One hundred fifty barrels of oil spilled at the Avila Beach UNOCAL pipeline facility. FORT is made up of commercial fishermen who have received training on cleanup procedures. This organization was joined by 67 members of the *California Conservation Corps,* some of whom had previous oil spill response training. As with many oil spill cleanup operations, the *United States Coast Guard* also played a major role. Many of the deep draft skimmers used in the Avila Beach spill, however, proved ineffective because of the extensive kelp forests and shallow, rocky coastline. To help rescue wildlife impacted by the spill, the *International Bird Rescue Research Center* and *Pacific Wildlife Cares* were called. Many coastal states are now establishing "mobile veterinary laboratories" as part of their oil spill response program. This allows response teams and trained volunteers to provide immediate care for oiled birds and mammals. As part of some mobile units, trailers functioned as on-site holding facilities which can transport injured animals to permanent facilities, such as the Marine Mammal Center at Fort Cronkite, California.

The Monterey Bay Program. Shaken by the *Exxon Valdez* oil spill of 1989, many coastal communities such as those along the Monterey Bay National Marine Sanctuary have been developing a land-based response plan for ocean spills of hazardous materials. Monterey Bay's plan began in 1989 with an ad-hoc committee headed by a Watsonville Fire Chief and a Santa Cruz County Hazardous Materials Advisory Commission Chair. Over the years, other individuals and agencies became involved, such as the Santa Cruz and Monterey County Planning Departments, the Association of Monterey Bay Area Governments (AMBAG), the California Coastal Commission, the U.S. Coast Guard, the California Department of Fish and Game, emergency response agencies, utility companies, animal rescue groups, and local environmental conservation groups such as Save Our Shores. After the initial draft was approved by the Boards of Supervisors of both Santa Cruz and Monterey Counties in March 1992, the California's Office of Oil Spill Prevention and Response donated $100,000 to continue work on the plan. The major objective of the task force was *to clarify the responsibilities and hierarchy of the agencies involved.* The *Exxon Valdez* oil spill of 1989 well illustrated the need for clear lines of responsibility. In 1997, California's Office of Oil Spill Prevention and Response completed a $5 million oiled wildlife veterinary care and research facility at the University of California at Santa Cruz Long Marine Laboratory in Santa Cruz. Its purpose is to assist sea otters and other marine animals that have been impacted by an oil spill, and to carry out related research.

A second major purpose of the task force was to write *a field handbook*—a quick, easy to use, reference tool for appropriate agencies and industries. On the task force, the principle agencies involved in drafting the plan included the Monterey and Santa Cruz County Planning Departments. The plan, known as the *Santa Cruz/Monterey Counties Oil Spill Contingency Plan,* was completed and approved by the California Department of Fish and Game's Office of Oil Spill Prevention and Response (OSPR) in March 1994 (Planning Department 1994). The Plan provides for a coordinated oil spill response and cleanup effort between local, state and federal agencies, to assure the best achievable protection of the counties' coastal resources. Whereas local response plans cannot prevent an offshore disaster from happening, these plans (once established) can enable local agencies to have a "first response strategy" in place for that inevitable offshore chemical, oil, or hazardous materials spill.

The Channel Islands Program. The Channel Islands National Marine Sanctuary lies off the coast of Southern California. (See Figure 7-10). This sanctuary could be contaminated by oil or other hazardous materials by one of three major ways: a blowout from one of the Channel's oil platforms; a tanker collision with a platform or another ship; or a tanker grounding on the Channel Islands.

Should a major spill occur in state waters off the coast of Santa Barbara, the following events are planned to occur (Sutkus 1994):

1. *Office of Emergency Services (OES) is notified.* Once a spill is discovered, the Office of Emergency Services in Sacramento (the state capital) is notified;
2. *OES notifies Oil Spill Prevention and Response Office (OSPR);*
3. *OSPR opens Command Center.* OSPR opens a command center in Sacramento and dispatches a team of marine experts and liaison staff to the accident location;
4. *A Spill Site Command Center is Established.* The OSPR liaison staff set up a command station at accident site;
5. *The State Interagency Oil Spill Committee (SIOSC) notifies Other State Agencies.* SIOSC consists of 16 state agencies covering coastal protection, water quality, toxic waste, transportation, safety, and many others.
6. *A Liaison Officer is Established.* At the spill site command center, a state liaison officer is designated. This liaison officer coordinates the activities between local, state, and state capital personnel; and
7. *Federal Agencies are Integrated into Response Effort.* The U.S. Environmental Protection Agency (EPA), the U.S. Fish and Wildlife Service (FWS), and other relevant federal agencies are integrated into the proper element of the response activity. Federal coordination is facilitated by the FOSC (Federal On-Scene Coordinator) who works for NOAA.

With proper command structures, responsibilities, and strict coordination, an accidental oil spill at the Channel Islands National Marine Sanctuary (or any other site, for that matter), has a better chance of being cleaned up quickly with minimal impact to the environment.

The Florida Keys Program. The Florida Keys National Marine Sanctuary (FKNMS) extends approximately 354 km (220 mi) southwest from the southern tip of the Florida peninsula. The Florida Keys are vulnerable to oil spills in two major ways: a tanker to tanker collision; or a tanker grounding on the coral reefs. The strategy of the FKNMS Management Plan for hazardous spills is intended to reduce the likelihood that a spill of oil or other hazardous materials will have a significant impact on Sanctuary resources. Responsible parties for the plan implementation are the U.S. Coast Guard (USCG) and the Florida Department of Environmental Protection (FDEP). NOAA, Monroe County, and the Florida Department of Community Affairs (FDCA) will assist.

The FKNMS Management Plan is broken into three distinct categories: Hazardous Materials (HAZMAT) Response, Spill reporting, and Hazardous Materials (HAZMAT) Handling. The following is a condensed outline of strategies to be implemented for each category (NOAA 1996, pp. 229–231):

HAZMAT RESPONSE:

1. *Develop and periodically revise the Sanctuary Spill Contingency Plan. The Plan should include:*
   - Strategic placement of crews and equipment throughout the Keys.

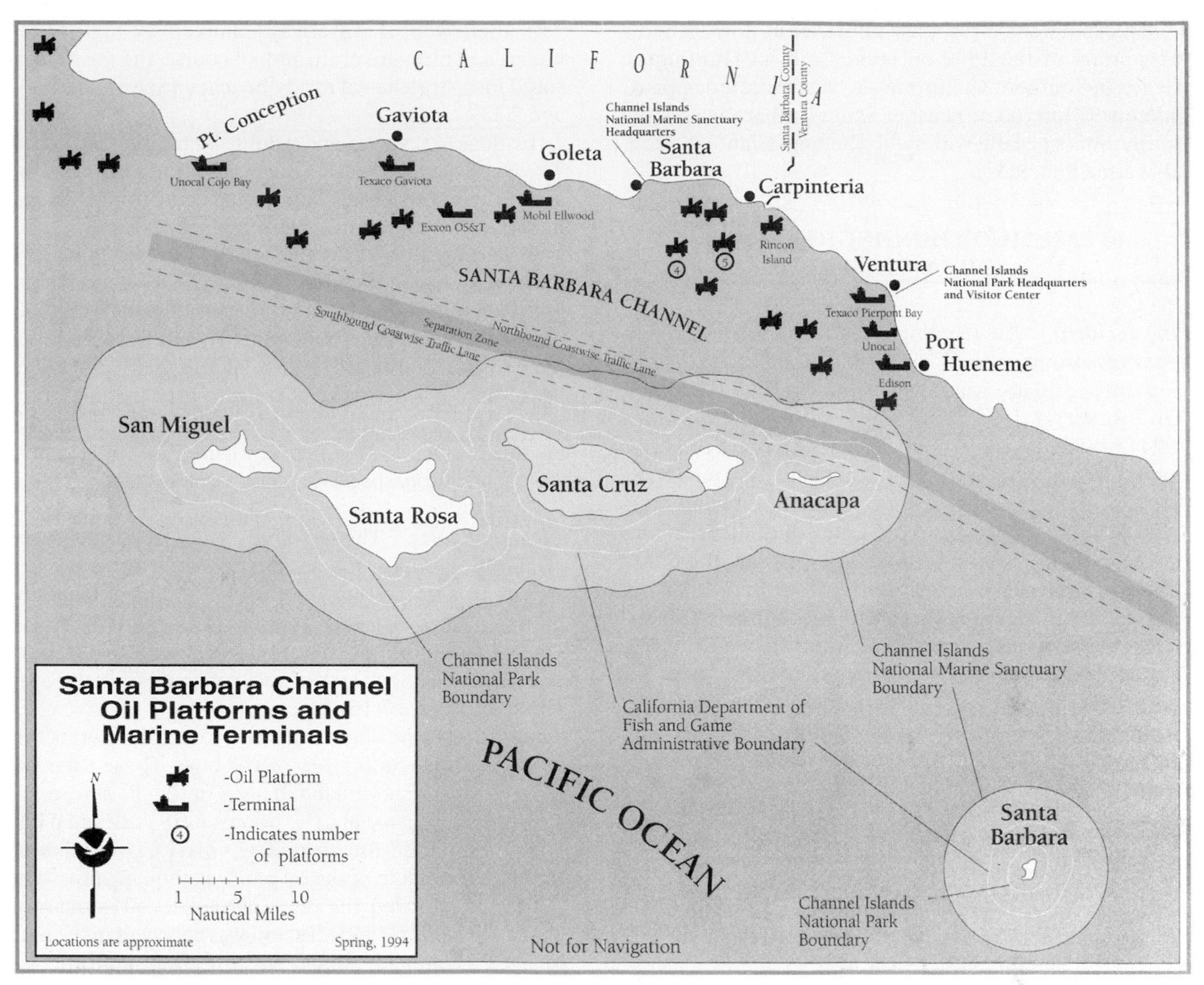

FIGURE 7-10 Santa Barbara Channel Oil Platforms and Marine Terminals. Map by John Dutton. (*Source:* Sutkus 1994, pp. 8–9).

- Coverage of spills of a size not responded to be USCG.
- Training and education of local response teams.

2. *Improve coordination and cooperation among the Federal, State, and local agencies reporting to spills.*
   - All marine HAZMAT responses will be coordinated from the Marine Safety Office in Miami.

Spill Reporting

1. *Establish spill reporting system*
   - All spills documented by various agencies, regardless of their size, shall be reported to Sanctuary managers.
   - The responsible reporting agency shall be the USCG. The FDEP shall assist in reporting all land-based spills that might affect Sanctuary waters.
   - The National Response Center is to be notified of all spills.
2. *Establish and maintain Sanctuary spills database.*
   - Establish and maintain a geo-referenced database for the Sanctuary that tracks all spill information regarding the Sanctuary.
   - The responsible agency shall be NOAA with assistance from the USCG and FDEP.

HAZMAT Handling

1. *Conduct HAZMAT assessment/inventory.*
   - Assessment and inventory of all hazardous materials handling and use in the Florida Keys. This assessment shall include facilities, types and quantities of materials, and transport/movement of materials. Information will be referenced in the FDEP/EPA/Monroe County GIS database.
   - The responsible agency will be the FDEP. The EPA and the FDCA will assist.

As an example of a response effort, let us now turn to a case study of the 1990 oil tanker spill at Huntington Beach in Southern California—a beach located approximately 97 km (60 mi) east of Santa Barbara Island, the southernmost island within the Channel Islands National Marine Sanctuary.

CASE STUDY: HUNTINGTON BEACH OIL TANKER SPILL

The accident. The Huntington Beach tanker accident is a perfect example of how "inevitable" a spill can be if large amounts of oil are transferred from one site to another. On February 7, 1990, about 3 p.m., the *American Trader,* a 244 m (800 ft) tanker, arrived off the coast of Huntington Beach, Southern California. Shortly after 4 p.m., the tanker began to gush an estimated 1,514,160 liters (400,000 gal) of Alaskan crude into the Pacific. It was the worst spill in Southern California since the Santa Barbara offshore oil platform blowout in 1969. What caused this accident? Severe weather conditions? No, the seas were calm and the visibility was excellent. The typical collision at sea? No, there were no other ships for miles, and certainly no hidden icebergs. In a sense, the *American Trader* merely collided with itself.

Specifically, the *American Trader* was attempting to tie up to what is called a "spread mooring"—an offshore terminal used by tankers to load and unload oil at sea. Rather than pull into ports, tankers are held in position by buoys and are hooked up to undersea pipelines which lead to refineries on shore. The *American Trader* was in relatively shallow water, and although the seas were calm, there were enough swells to rock the ship from port (left) to starboard (right). It is believed that the vessel drifted over the port anchor, wrapping the chain under its bow (front of ship). As the crew raised the port anchor, it may have struck the starboard side of the hull, leaving a 0.9 m (3 ft) gash and another hole. Some veteran pilots suspect that the tanker may have struck the anchor as it rested on the sea floor in shallow water. Regardless, anyone who operates even a small 6 m (20 ft) motor or sail boat knows how tricky it is to lay or retrieve an anchor at sea. Think of how difficult it must be for ship's crewmen to keep an anchor from swinging and striking their own ship.

Threatened coast. For days, the oil slick continued to spread, moving onto and off the shore with shifting winds and tides. (See Figure 7-11). Over 32 km (20 mi) of coastline was at risk, from Long Beach Harbor in the north to Newport Beach in the south. Much was threatened, including Palos Verdes Peninsula with its rocky and picturesque coastline; Long Beach harbor with its extensive commercial development; Alamitos and Anaheim Bays, with their web of waterways flanked by high-priced homes and pleasure craft, and, of course, the much-treasured long stretches of sandy beaches. (See Figures 7-12 and 7-13).

Ecological reserves and wildlife were also threatened. The Newport and Bolsa Chica ecological reserves were areas where the greatest harm could come from oil contamination. Both reserves act as temporary homes for thousands of migratory birds and marine life. Newport Bay's upper reserve, for example, is home to 165 species of migratory and resident birds, 60 species of fish, 20 amphibians and reptiles, and 10 mammals. There was concern that the birds could bring oil residue back to their nests, killing eggs. There was also special concern over the fate of the surface-diving birds, such as grebes, terns, and pelicans, that are most vulnerable to oil. (See Figure 7-14). Fortunately, several migratory species had not yet arrived in the region.

Cleanup efforts. The *Offshore clean-up activities* relied primarily on *skimmer boats* and *booms.* A boom was first spread around the *American Trader* to help confine the oil that gushed from the tanker. (See Figure 7-15). For days, as many as 17 skimmer boats continued the laborious, round-the-clock task of sucking up oil from the open seas. (See Figure 7-16).

Each boat was rigged with two 15 m (50 ft) arms that extended from both sides of the boat. These arms deployed the booms which had a 0.8 m (2.5 ft) skirt made of rubberized materials. The booms were positioned into U-shapes behind the arms. The vessel moved directly through the middle of the oil patch at about 2 knots. With both arms extended, the vessel can collect oil from a path about 44 m (145 ft) wide. The oil was collected in the apex of each U-shaped boom. The oil/water mixture was pumped to a holding tank on board the vessel.

The mixture was then separated (using gravity), and the seawater was returned to the ocean. Each vessel could store only about 1,200 barrels of recovered oil. Although this may all sound rather efficient, it really was not. After the first two days of this activity, the skimmer boats had collected only a small number of the lost barrels of Alaskan crude. The process was slowed by the spill's transformation from a concentrated slick (the easiest spill to collect) into a more distilled, light sheen which is practically impossible to collect.

In addition to the booms used alongside the skimmer boats, *lead-weighted booms*—absorbent barriers of a synthetic material—were strategically placed to protect key coastal areas from the slick. For example, booms were placed at the mouth of the Santa Ana River, which hosts a least tern refuge, and at the entrance to the Newport harbor.

Onshore cleanup activities used available tools and techniques, but were rather inefficient. *Absorbent paper*

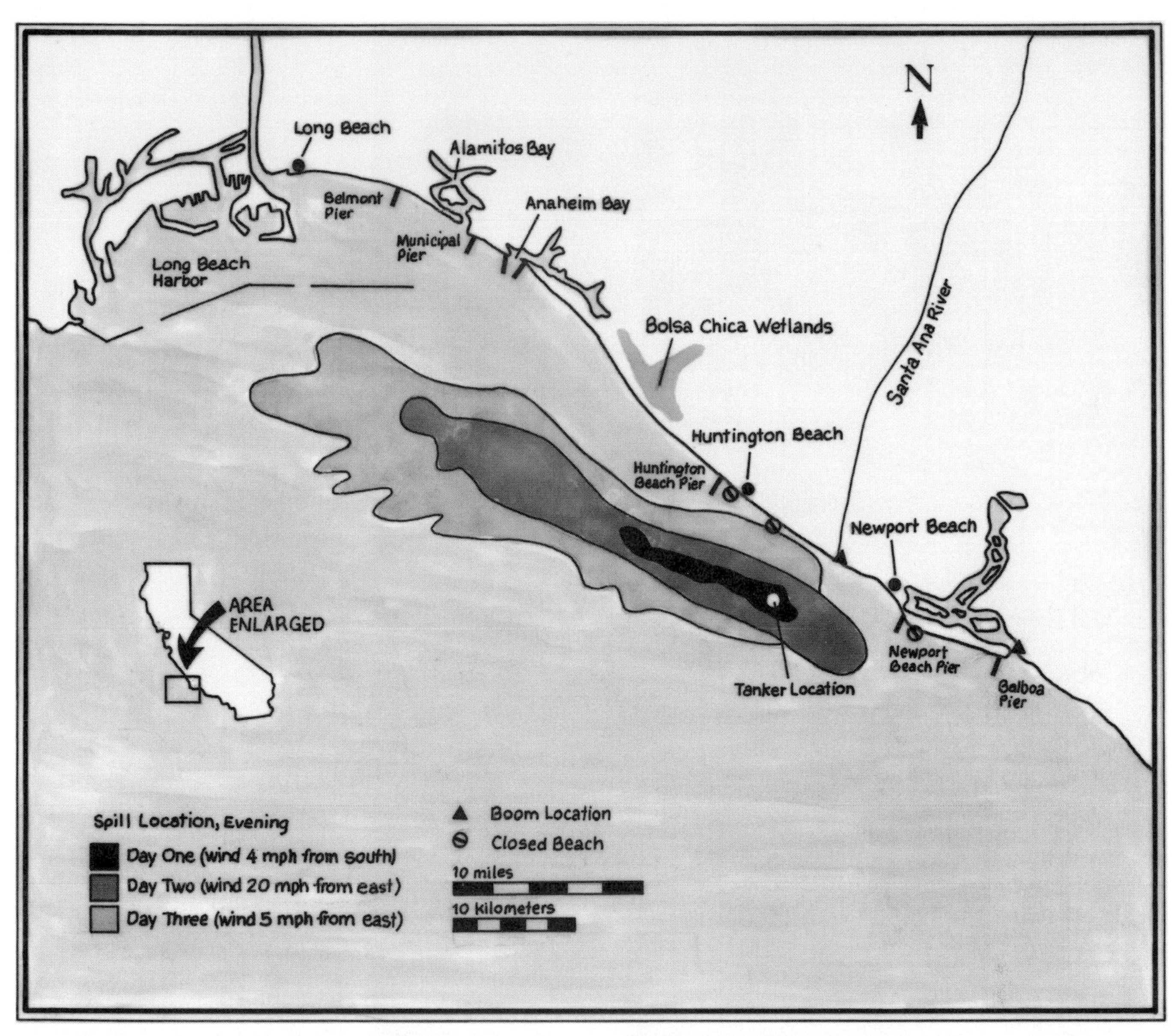

FIGURE 7-11 The Southern California coast at risk. (*Source:* Map by Susan Giles, adapted from *Los Angeles Times*, February 12, 1990, Section T5)

was distributed to paid workers and volunteers so that they could mop up the oily scum as it rolled on shore, or to scrub down the rocks that make up the many groins along Newport Beach; other workers were given hand-held *sieves* to sift gray scum from beach sand. (See Figure 7-17).

Oiled birds were brought to the Huntington Beach Treatment Center. Since the animals were in shock, they were first put in a warm, dark container to calm them down and to prevent them from cleaning themselves and swallowing oil. Once the birds were rested, they were tube-fed a high-protein liquid diet of "Bird Gatorade," which helps clean out ingested oil. A mixture of one-part dish soap and three-parts warm water was used to wash the birds' feathers. Working in pairs—one person to hold the bird down, the other to wash the bird—it took 15 washes to remove the oil from one bird. A cotton swab was used to clean the bird's bill.

Lingering effects. One common denominator to such oil debacles as the 42 million liter (11.1-million gal) spill by *Exxon Valdez* in Alaska, the 12 million liter (3.2-million gal) Santa Barbara spill in 1969, and the 1,514,160 liter (400,000 gal) spill at Huntington Beach in 1990 is that there is no good barometer for judging just how quick or how successful a cleanup operation will be. Most of the

(a)

(b)

FIGURE 7-12 (a) Alamitos Bay, Long Beach, was one of several bays, wetlands, rivers, and beaches threatened by the Huntington Beach spill. (b) Alamitos Bay contains hundreds of high-priced homes on the waterfront, all with their private pleasure craft. (*Source:* Photos by author)

visible oil is now gone from the surf and shore from the *American Trader* spill, scrubbed away by time and a $35 million cleanup effort. But there is a lingering effect on the environment and the local economy.

Although officials of BP America, which owned the oil that spilled from the *American Trader,* maintain that their consultants have found no lasting environmental damage, state officials disagree. The state sued BP America in order to gain restitution for the harm to wildlife and beaches. Approximately 1,000 birds were killed and 24 km (15 mi) of beaches had to be closed for several weeks. According to the California Department of Fish and Game, marine life has yet to return to normal. There was also an economic impact to this high tourist traffic area. According to many local businessmen, the oil spill was the start of a financial disaster.

The *American Trader* incident, however, did spark some positive change. Due to public outcry, California legislators passed the *Lempert-Keene-Seastrand Oil Spill Prevention and Response Act of 1990.* Under the law, oil companies are responsible for writing plans and providing equipment to clean up the "worst case" spill their tankers could generate. Furthermore, when the *American Trader* spill occurred in 1990, there were only two employees in the California Department of Fish & Game to handle oil spills; by May 1995, there were 140. State lawmakers had created the *Office of Oil Spill Prevention and Response* which is funded by a 4-cent-a-barrel tax on oil. One of the first tasks of this new office was to present a detailed report to the Legislature, outlining California's state of readiness. In May 1995, a draft of the report noted that California had not experienced a major oil spill in the last five years. Furthermore, it cited such advances as the following:

- *Tugboat escorts.* Tugboats are now required for all oil tankers and barges sailing through San Francisco Bay.
- *Ships remain 80 km (50 mi) offshore.* In 1992, the ten largest oil companies that transport crude oil from Alaska to refineries in San Francisco and Los Angeles signed a voluntary agreement to sail 50 miles offshore.
- *Creation of a nationwide cleanup group.* Since 1991, the oil industry has spent $800 million to create a nationwide cleanup group, the *Marine Spill Response Corporation.* With gear and ships stockpiled in Richmond, Eureka, San Diego and Point Hueneme, near Ventura, the response corporation has increased the amount of cleanup equipment in California by one-third.
- *Equipment contracts.* More than 300 contractors have signed state agreements to provide everything from helicopters to catering, vacuum trucks, and portable toilets within hours of a spill.
- *Large-scale cleanup drills.* New laws also require oil companies and state officials to conduct large-scale cleanup drills. Each year there are about 20 around the United States; three are in California.

One of the big differences since the 1990 *American Trader* incident is the price of cleaning up a spill. Initially, spills were assumed to be part of the cost of doing business. Since oil companies are now more liable for those spills, ship captains and others are sometimes fired when they happen. Because of the cost of cleanup, the oil industry is moving toward "zero tolerance" for errors.

The National Academy of Sciences has documented the aftermath of various oil spills around the world

(a)

(b)

FIGURE 7-13 Beach closure at Newport Beach due to Huntington Beach spill. (a) Upper photo shows piles of plastic bags containing oil soaked pads used for beach cleanup. They are eventually picked up and trucked off to an appropriate landfill. (b) Lower photo shows "Danger Beach Closed" sign in lower left. Clean up crews can be seen close to the shore. (*Source:* Photos by author)

Western grebe
(Aechmophorus occidentalis)

Because it spends most of its time on the surface of coastal waters, this wintering bird is among the most vulnerable to the early effects of a spill.

California least tern
(Sterna antillarum browni)

This endangered gull-like bird nests on the sand in wetlands such as Bolsa Chica.

Brown pelican
(Pelecanus occidentalis)

This endangered bird flies along the surface and dives for fish.

Common murre
(Uria aalge)

Like the grebe, this sea bird bobs on the water's surface, diving occasionally for fish.

Light-footed clapper rail
(Rallus longirostris levipes)

More than half of the state's population of this sedentary marsh bird lives in the dense cordgrass stands of Upper Newport Bay. Any damage to this nesting site would be devastating to the bird's future prospects.

FIGURE 7-14 (Above) Birders standing in Bolsa Chica Ecological Reserve viewing Huntington Beach oil spill on other side of highway. (*Source:* Photo by author). (Below) Surface-diving birds most vulnerable to oil in the area of the Huntington Beach spill. (*Source:* Adapted from *Los Angeles Times*, compiled by Rick Vaiderknyff, illustrations by Russ Arasmith)

FIGURE 7-15 A boom was spread around the *American Trader* as one of the methods used to confine the oil that gushed from the tanker. (*Source:* Drawing by Heather Theurer)

(National Academy of Sciences 1991). After the massive 1969 Santa Barbara spill, for example, oil was found embedded in deep layers of sand a decade later. If the *American Trader* spill has an aftermath similar to spills elsewhere around the world, the ocean and shore of Southern California will conceal oil (and its ill effects) for decades.

GENERAL POLICY RECOMMENDATIONS

What follows are a number of general policy, technique, and design recommendations. Recommendations specific to (a) offshore oil and gas development and (b) oil tanker traffic will first be discussed, followed by a (c) set of overall policy strategies that relate to both development and transportation of the product.

Offshore Oil and Gas Development

- *Prohibit oil drilling in ecologically sensitive areas.* In 1990, President Bush decided to cut back oil and gas drilling areas off the coasts. (See Figure 7-18). His decision was a significant departure from the offshore leasing policies of the Reagan administration, which tried to make the entire 0.6 million hectares (1.4-million acres) outer continental shelf available to industry. Although the Bush decision was a step in the right direction, it only offered *temporary* protection that was open to political whim.
- *Develop cleaner ways to drill.* Prodded by strong public sentiment, stringent state laws, and aggressive enforcement, some oil companies are striving for cleaner ways to drill for offshore oil. In 1991, for example, the drilling platform Glomar Highland IV off Dauphin Island, Alabama was taking extraordinary steps to protect Mobile Bay—an estuary renowned for its rich catches of shrimp, oysters and fish. Alabama state officials and various environmental groups maintain that Alabama has established a model of ecological safety that should be the nationwide standard for drilling operations close to shore. This "cleaner way" of doing business costs Mobil Oil approximately $150,000 a week. Although this may seem expensive, Mobil considers it a worthwhile expense, since it allows the company access to a natural gas reserve beneath Mobile Bay which is estimated to be worth at least $650 million at 1991 prices. Mobil did not, however, abide by Alabama's rules without pressure. In the early 1980s, investigators discovered that Mobil was pouring untreated sludge and toxic drilling mud into the bay. The company had to pay a $2-million penalty to the state and was forced to pay at least $500,000 to dredge up large quantities of chemical-soaked bay bottom.
- *Extend federal jurisdiction boundary to create larger buffer zone.* Federal jurisdiction currently extends only 4.8 km (3 mi) from shore in California. One policy or legislation option is to extend the federal jurisdiction boundary for states further offshore. In 1993, Representative Anna Eshoo, Congressperson for southern San Mateo and Northern Santa Clara counties in California, for example, introduced in the House of Representatives the *California Ocean Protection Act of 1993 (OPA, H.R. 2583).* If eventually approved, new offshore oil and gas development could be banned within 322 km (200 mi) offshore along the entire 1352 km (840 mi) California coast under federal legislation. The bill would also ban other activities within this extended boundary, including deep-sea mining, at-sea incineration of toxic wastes, and harmful ocean dumping. It would exclude, however, some vessel waste: dredged materials that meet federal regulations, discharges authorized by a National Pollutant Discharge Elimination System permit, and disposals carried out by federal agencies under the Marine Protection, Research and Sanctuaries Act of 1972. The idea of a 322 km (200 mi) wide protected zone has growing support in Congress. The OPA has the support of a number of local representatives and Senator Barbara Boxer has introduced a companion bill in the Senate—The *California Ocean Protection Zone* (S.727). Environmentalists see the passage of these two bills as the first step to real protection for California's coastline and the state's $27 billion fishing and tourism industries. The next step, they maintain, is coastal protection for all the coastal states in the nation.

  The oil industry, however, is less enthusiastic. It maintains that the conservation strategy of extending boundaries would serve as just another "nail in the domestic oil industry's coffin," driving the country to further dependence on foreign oil, and the increased use of tankers which are vulnerable to accidents at sea.

Oil Tankers

- *Promote better tanker design.* Engineers, regulators, and environmentalists have suggested several ways by which tankers could be better designed. After the 1989 wreck

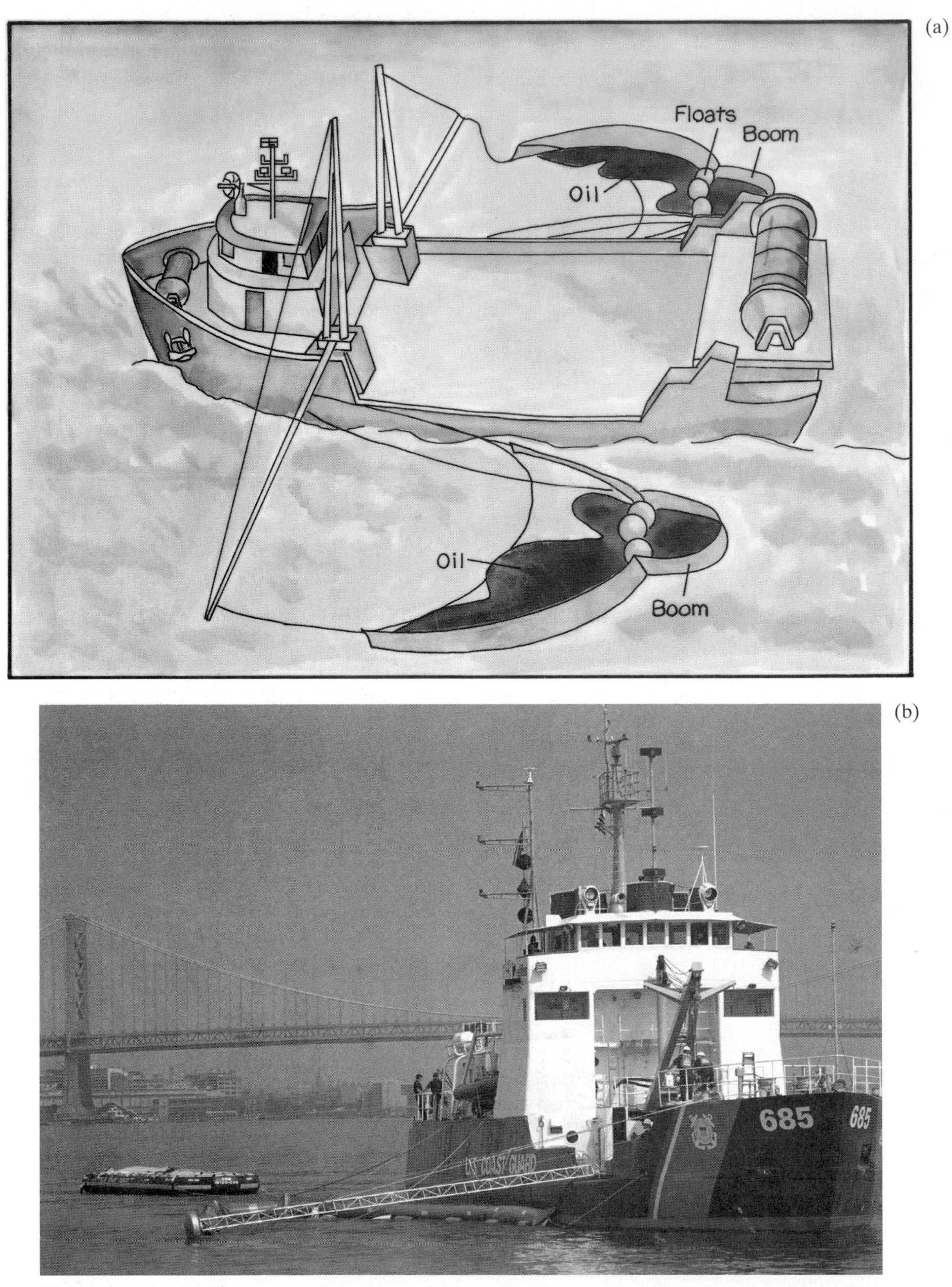

FIGURE 7-16 (a) Diagram of how a skimmer ship works. (*Source:* Diagram by Susan Giles) (b) The Coast Guard cutter, Redwood, demonstrates the Vessel of Opportunity Skimming System, VOSS, Tuesday, Sept. 16, 1997, on the Delaware River near the port of Philadephia during an oil spill cleanup drill. The drill involved more than 200 people from federal offices around the country. Oil companies based in Norway, Cyprus, and the United States also participated. (*Source:* AP Photo/David Maialetti)

(a)

(b)

FIGURE 7-17 (a) Professional clean-up crews working on Newport Beach (*Source:* Photo by author); (b) A man shovels lumps of crude oil and sand into plastic bags on Coppet Hall beach in southwest Wales Wednesday February 28, 1996. The oil is from the tanker Sea Empress which went aground February 15, 1996 further up the Welsh coast causing massive pollution in the area. Official estimates put this spill at 20 million gallons—one of the 10 biggest tanker spills ever. The Exxon Valdez spilled 11 million gallons in Alaska in 1989, creating environmental havoc. (*Source:* AP Photo/ Max Nash)

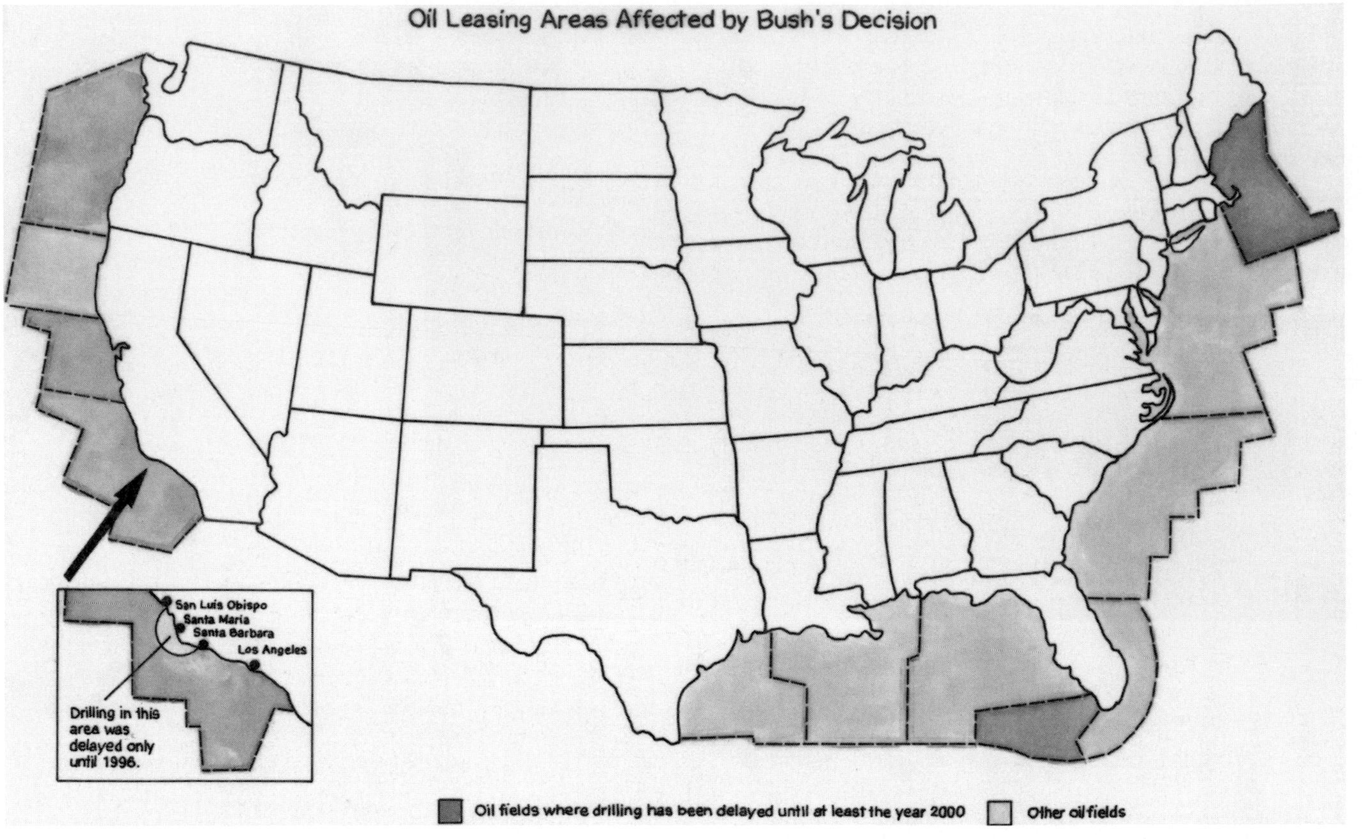

FIGURE 7-18 Oil leasing areas affected by former President Bush's decision. (*Source:* Map by Susan Giles, adapted from *The New York Times*, June 27, 1990, p. A13)

of the *Exxon Valdez* which had a single hull, there was much discussion about requiring all new supertankers to have *double hulls* (*bottoms*) to provide an added layer of protection in a grounding. In fact, the Oil Pollution Act of 1990 now requires that all *new* U.S. tankers have double hulls to lessen the chances of severe leaks. Unfortunately, U.S. oil companies are getting around this law by operating tankers under the flags of other countries. This same act also requires that *existing* U.S. tankers have double hulls by 1998. However, the law allows oil companies a grace period of 25 years to accomplish this goal. The need for double-hull construction remains one of the most debated safety measures. Oil companies, for example, argue that if the outer layer of a double hull is pierced, water rushing into the space between the hulls would make the ship settle lower, impaling itself more firmly. Some engineers maintain that explosive gases can gather in the space and corrosion can proceed uninspected. To help resolve this debate, in 1992, Richard Golob, publisher of *Golob's Oil Pollution Bulletin* of Cambridge, Massachusetts, began an analysis of accidents involving tankers with double hulls to see if the double-hull construction actually lessens spills. As of this writing, the study was still in progress.

Tankers could also be required to have *redundancy in engineering systems* (*a backup for a ship losing power*) and *towing bridles*. For example, on January 6, 1993, the *Braer*—a Liberian-registered tanker carrying 95 million liters (25 million gallons) of crude oil— ran aground off Great Britain's Shetland Islands, which are located about 161 km (100 mi) off the north coast of Scotland. The accident resulted because the ship lost power amid 70 mph winds, with waves of 5.8 to 6.7 m (19 to 22 ft). Regulators argue that this accident could have been prevented if the ship had a backup system aboard so that it could have maintained power. It is interesting that many recreational boaters and commercial fishermen with their small crafts of only 4.9 to 11m (16–35 ft) carry a small outboard motor as a spare in case their primary power system fails (as well as using it for trawling purposes). Yet, 550,819 deadweight ton supertankers are designed without a backup power system.

- *Load tankers less than full.* Since the above methods would require years to accomplish, a temporary quick-fix is using a strategy called *partial loading* which is based on the principle of hydrostatic balancing. The idea is to load only enough oil to make the pressure on the inside of the hull equal to the pressure from seawater on the outside (oil weighs 86 percent as much as salt water). Oil will only gush from a torn ship as long as the pressure of the oil against the inside of the ship is greater than the water pressure against the outside. The Valdez accident is proof that this concept is valid. After about 15 percent of the cargo was lost, the Valdez stopped hemorrhaging. A less quick fix approach to achieving hydrostatic balancing is by building ships with ballast tanks all the way down each side of the ship, giving the vessel an elaborate "bumper system". (See Figure 7-19). The estimated cost for this *bumper system* would be an additional 15 to 20 percent, similar to a double hull or partial loading.

- *Establish vessel control measures in environmentally sensitive areas.* In environmentally sensitive areas such as in the waters off the Florida Keys and in the waters off the Central California Coast, special vessel control measures could be established. One "vessel control measure" often mentioned is to map out "Tanker-free Zones" or "Areas to be Avoided"—portions of the world's oceans and bays where tankers would not be allowed. A step in this direction became effective on June 1, 1992, when ten of the largest U.S. oil companies agreed to *voluntarily* stay outside of certain zones. Specifically, they agreed that their tankers traveling from Alaska to California would remain at least 50 nautical miles from the mainland. These ten companies represent almost 85 percent of all tankers delivering crude oil to California. In theory, moving oil tanker routes farther to sea should reduce shipping traffic congestion and diminish the likelihood of a collision at sea.

Environmentalists, however, would like to see *mandatory* tanker-free zones established. For example, Save our Shores would like to see the Monterey Bay National Marine Sanctuary designated as an "Area to Be Avoided" (ATBA) under international maritime law. Currently, the Central California coast line between Point Año Nuevo and Point Arguello has no formalized vessel traffic separation system, yet the region has a high degree of ecological sensitivity and hazardous seas. Ships that stray into these areas that are off-limits to maritime traffic could be severely penalized. There is some evidence, for example, that the tanker *Braer* that lost power and grounded off Great Britain's Shetland Islands in 1993 drifted for 2 1/2 hours into an exclusionary zone before notifying authorities. The Center for Marine Conservation has recommended an even larger ATBA—some 66 nautical miles off the coast of Central California—that would encompass the three national marine sanctuaries in Northern and Central California. (See Figure 7-20). In the event of a spill, less oil would hit beaches and cleanup crews would have more time to respond.

Likewise, the Center for Marine Conservation recommended an "Area To Be Avoided" for the Olympic Coast National Marine Sanctuary. In Spring 1995, the International Maritime Organization approved a 48 km (30 mi) buffer zone for tankers when transiting the northern half of Washington State. (See Figure 7-21). This buffer, advocated by the Center and developed by the U.S. Coast Guard and NOAA, ensures that oil tankers and ships carrying hazardous cargo will remain far enough offshore to minimize the risk that a spill will contaminate the relatively new Olympic Coast National Marine Sanctuary.

A second "vessel control measure" in highly environmentally sensitive areas could be to adopt a *vessel traffic separation scheme.* For waters off the Central California Coast, the Center for Marine Conservation has recommended establishing a 8 km (5 mi) wide fairway as a traffic separation scheme. In 1992, there were virtually no regulations regarding the approximately 1,000 commercial vessels, including oil tankers, which pass *through* the Monterey Bay National Marine Sanctuary. As of this writing, NOAA and the U.S. Coast Guard had yet to establish safeguards for tanker traffic, including a Vessel Traffic Separation Scheme.

A third "vessel control measure" would be to have continuous *vessel traffic service* in highly sensitive areas. This would mean having emergency offshore rescue vessels (ERVs) permanently stationed at strategic locations. When not being used for emergency reasons, these fire-fighting-equipped tugboats could provide an "escort service" for tankers in waters of sensitive areas, or in and out of harbors and enclosed sounds and bays. An ERV could perhaps have come to the rescue of the *Sea Life Pacific* in 1984, when it lost power 19 km (12 mi) off Point Lobos, California. The tanker drifted 32 km (20 mi) before its anchors stopped the ship just 1.6 km (1 mile) from Point Sur, in the heart of the Monterey Bay sanctuary.

- *Seek improved monitoring control of vessel traffic.* One recommendation is to *extend coastal radar coverage.* Currently on the Central California Coast, for example, the Coast Guard can only monitor ships out to about 48 km (30 mi) from the Golden Gate Bridge with its Vessel Traffic Service, based on Yerba Buena Island in San Francisco Bay. Conservationists, including the well known Cousteaus (Jacques and his son, Jean-Michel), have long maintained that tankers must be monitored as we currently monitor airline traffic. They have called for a maritime equivalent of the federal "air traffic control system." All tankers and the Coast Guard could have state-of-the-art radar equipment so that if a tanker strays off course (e.g., into a "tanker free zone"), an alarm would automatically sound on the tanker and at the Coast Guard stations.

  In addition to extending coastal radar coverage, an aerial surveillance program could be established. An ongoing aerial surveillance program, for example, could monitor vessel movements, their compliance with routing measures, and their compliance with pollution prevention regulations.

- *Stronger international maritime law.* Only international maritime law can regulate all tankers, regardless of flag. The International Maritime Organization, a branch of the United Nations, needs to adopt stricter safety measures.

Thinking big

Maximum tanker sizes at the end of World War II and today. The supertankers are harder to maneuver, and likely to spill more oil when damaged.

1945 16,500 deadweight tons

1989 550,819 deadweight tons

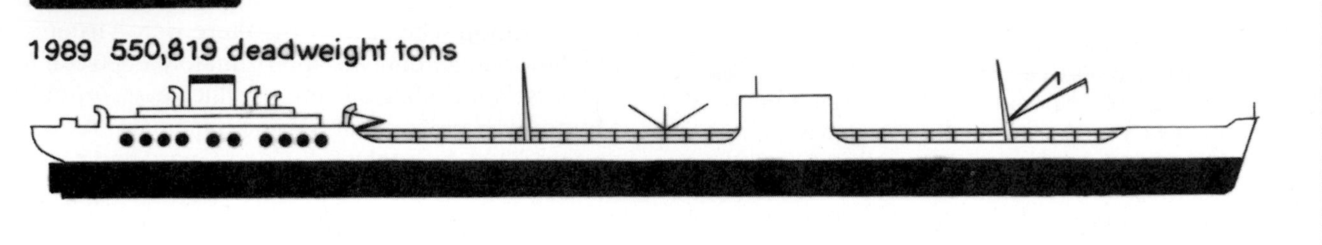

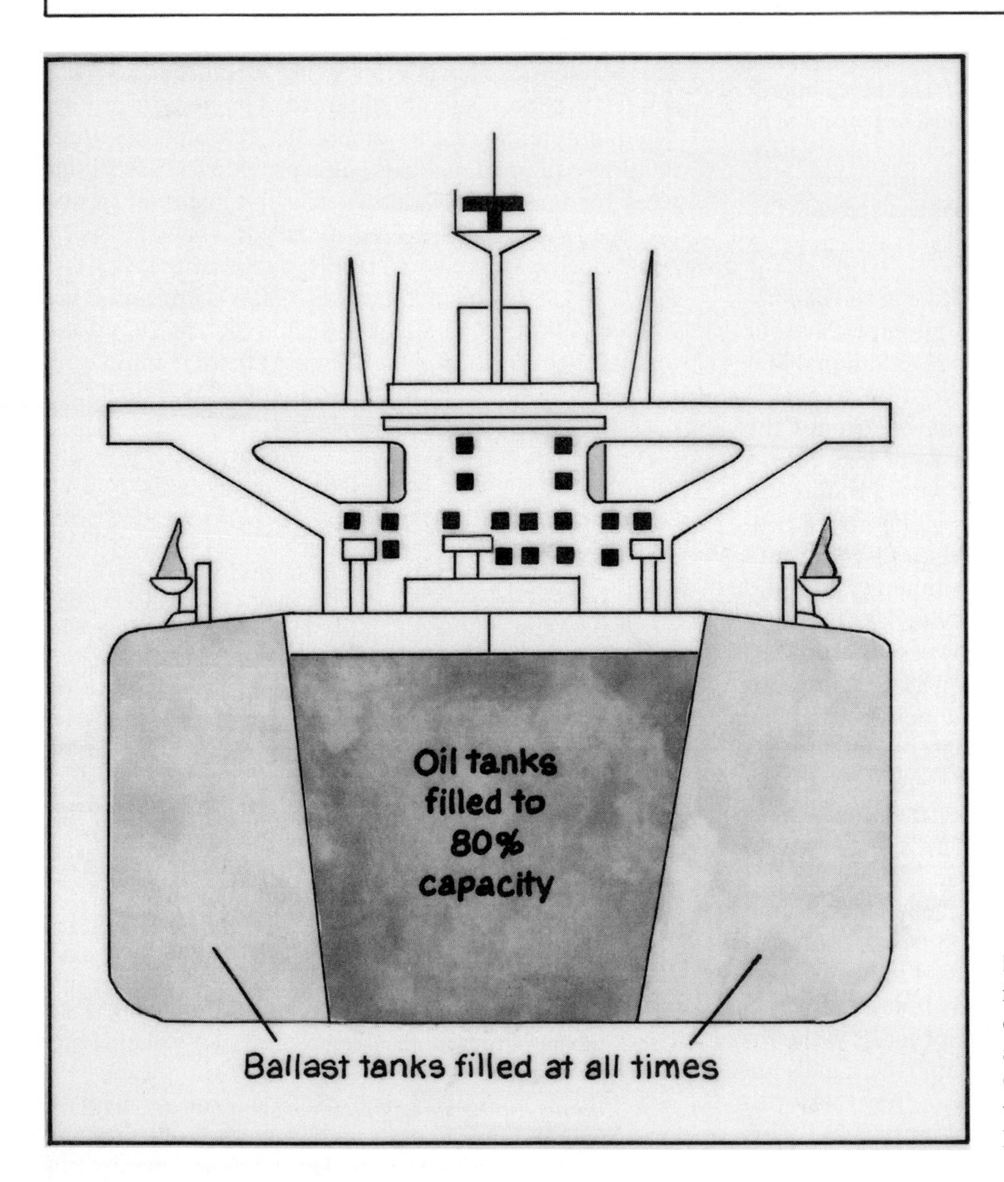

FIGURE 7-19 (a) The ever increasing size of oil tankers; (b) The strategy of "partial loading" or "hydrostatic balancing." (*Source:* Drawings by Susan Giles, adapted from information provided by the American Petroleum Institute)

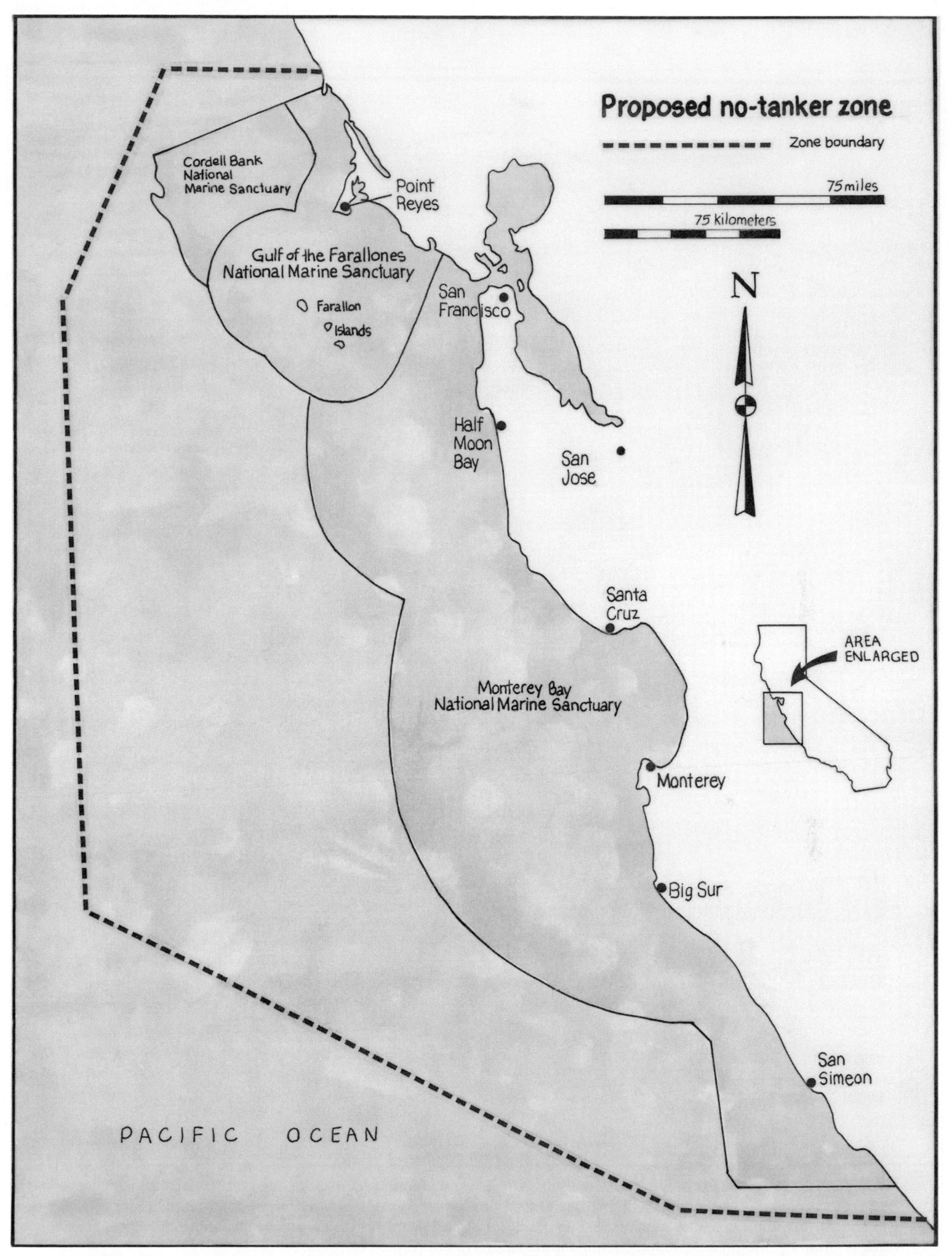

FIGURE 7-20 A "no tanker zone" proposed by the Center for Marine Conservation to include three national marine sanctuaries: Cordell Bank, Gulf of the Farallones, and Monterey. (*Source:* Map by Susan Giles)

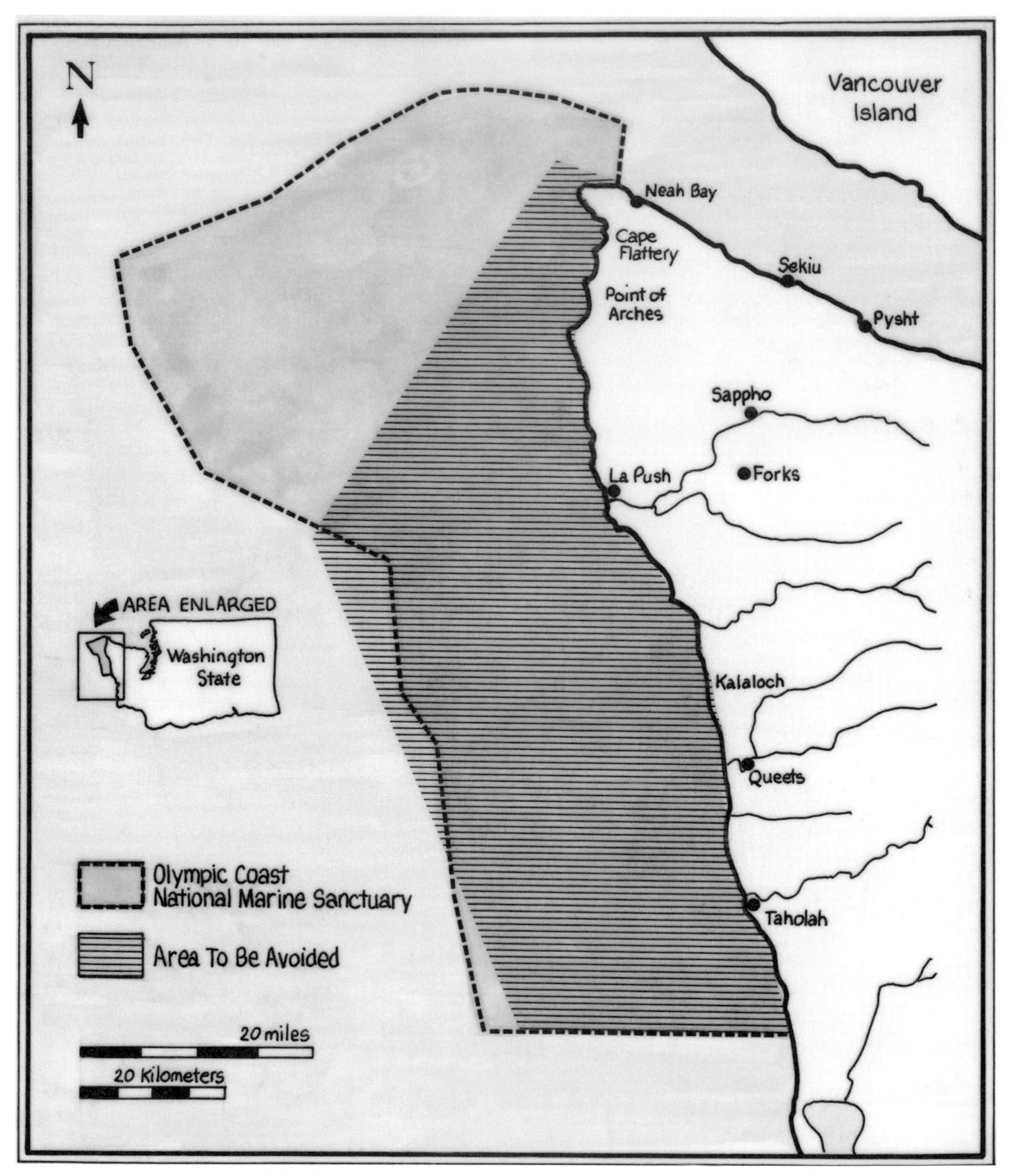

FIGURE 7-21 The U.S. Coast Guard's proposed "Area to be Avoided" (ATBA) for vessels carrying hazardous materials off the Olympic Coast of Washington State. The proposal was recently approved by the International Maritime Organization, making the ATBA an "international advisory" for hazardous cargo vessels to remain outside this area. (*Source:* Map by Susan Giles, adapted from *Marine Conservation News*, Vol. 7, No. 1, Spring 1995)

For example, the Center for Marine Conservation has suggested that there should be an international requirement that all large commercial vessels be equipped with *transponders for improved vessel identification* purposes. (A transponder is a radio or radar receiver-transmitter activated for transmission by reception of a predetermined signal.) Since the shipping industry is generally opposed to stricter rules, conservationists will likely have to rally all the "earth citizens" who are horrified by the ecological defilement and destruction of livelihoods that results from tanker accidents.

- *Convert to oil transport by pipeline.* For over a decade there has been a controversy brewing on the Southern California coast. State officials have tried to get the Chevron Corporation to build a pipeline to carry oil from Santa Barbara to Los Angeles, arguing that it would be safer than using tankers. Finally, a "carrot-and-stick approach" was used to get the giant corporation to agree to build the pipeline. The California Coastal Commission prohibited the movement of oil in tankers from Santa Barbara to Los Angeles after January 1, 1996.
- *Collect information on "near misses."* The Center for Marine Conservation recommends that a system for reporting near misses be developed (e.g., collisions, groundings, loss of power, etc.) so that the data can be analyzed for lessons to be learned regarding casualty avoidance.

Overall Policy Strategies

- *Encourage better business management.* According to Karlene H. Roberts, professor of business administration at Haas School of Business, University of California at Berkeley, the troubles of the oil tanker industry show that industrial disasters have less to do with acts of God than failures of management. According to Roberts, executives need to upgrade their command and control system to a communication and culture system. In *"command and control,"* the old hierarchical structure of the maritime industry (e.g., the captain controls all power and decision making) needs to be replaced by a sharing of authority by front-line personnel, as mechanics on aircraft carrier flight decks can shutdown flight operations if they perceive a danger. Such a system might have prevented the *Exxon Valdez* incident, since no one on the bridge was sufficiently trained to see the "big picture," as the captain might have, had he been present. *"Communication,"* likewise, can be improved upon. Too often, shippers hire seamen with questionable skills, and worst of all, no common language. The 1993 oil tanker *Braer* accident resulted because the bridge and engine-room crew literally did not speak the same language. And, since any business has its own *"culture,"*—a set of beliefs and values that determine human behavior—building a safe and reliable culture can do much to prevent accidents in the oil industry. The oil industry, and other high risk businesses, must reexamine their values (e.g., what happens when they subcontract their shipping to businesses that have a lower set of safety values). According to Roberts, catastrophes are all too often rooted in top management's neglect of the all important *"Cs"—command and control, communication, and culture.*
- *Establish stronger regulations.* Better regulations regarding the construction and operation of oil tankers, offshore oil rigs, and oil refineries need to be established. Regulations regarding safety and disposal procedures also need to be strengthened. For example, oil companies need to routinely test employees for alcohol and drug abuse, and convicted "drunk drivers" must be banned from future tanker commands and from management positions on offshore platforms. If you recall, the *Exxon Valdez* tanker disaster off the Alaskan coast is believed to be the result of a tanker captain who had a history of alcohol abuse. The oil companies will argue that new policies will not necessarily safeguard completely against human error and the unpredictability of nature. This, of course, is true, just as the "traffic cop" cannot prevent all motorists from speeding or driving under the influence of alcohol. Regardless, we know that things would be worse if we did not have highway patrol officers to keep us somewhat under control, and likewise, we are now finding out that the oil industry is in need of "tanker patrol officers" as well.
- *Increase criminal prosecution.* The financial liability of oil companies for cleaning up oil spills needs to be increased. The *Oil Pollution Act of 1990* was a first step in this direction. The act raises the liability for owners of an average-size supertanker operating in U.S. waters to around $100 million from about $14 million under the United Nations' International Maritime Organization—the nominal traffic cop of the seas. In addition, it authorizes unlimited liability if the spill results from a violation of federal law—operation of a ship while intoxicated, for example. According to many experts, the Oil Pollution Act has unquestionably improved the quality of tankers coming to the United States. Currently, federal agencies determine whether oil companies have to pay lesser charges for the "lost use value," or pay full restitution. This arbitrary decision should not be left to federal agencies, according to the Wilderness Society (1991). Rather, it maintains that Congress should require companies to pay the total cost of full restoration of damaged natural resources, including the cost of cleanup and scientific assessments.
- *Increase security measures.* Although one rarely hears about it, pirates are once again becoming the scourge of commercial shipping. According to the International Maritime Bureau, a London-based organization that has a "piracy-monitoring center," between 1991 and 1993, ship captains reported nearly 300 acts of piracy, more than twice as many as were recorded over the previous 10 years. Today's pirates use high-speed boats and are much more heavily armed, often using machine-gun fire and even rocket launchers to stop ships. It is amazingly

easy to hijack a petroleum tanker or LNG (Liquified Natural Gas) carrier. Although most of the strikes have been in the heavily trafficked sea lanes of Southeast Asia, Americans cannot be complacent. Terrorism (in association with piracy of cargo ships and other large commercial vessels) is a growing phenomenon. Imagine the environmental (not to mention the personal) disaster if something went wrong.

- *Increase funds for research and assessment.* Since the culprits of many spills cannot be identified, a way is needed to "fingerprint" oil. Scientists at the U.S. Environmental Protection Agency are currently experimenting with using a gas-liquid chromatograph to see if they can give oil samples distinctive fingerprints. If oil can be fingerprinted, investigators trying to locate the source of an oil slick can trace its origin back to a passing tanker, a particular pipeline, or a natural oil seep. As shown earlier, much research also needs to be done on cleanup technology. The U.S. has "smart bombs" and space travel, but this country is still at the "mop and paper towel" level of technology when it comes to cleaning up oil spills.
- *Reverse energy course.* Oil continues to dominate U.S. energy policy. As long as "oil is king," our coastlines will suffer the consequence. Oil, however, has not always dominated the energy scene. Any study of the history of energy use in this country will show that our reliance on oil has been a 20th century post-World War II phenomena. And, no doubt, we will be relying on a variety of other forms of energy in the future. In the 21st century, this country (and the world) will need to have an energy strategy that stresses *diversity of energy sources*—a mixture of traditional nonrenewable energy sources (e.g., oil, natural gas, coal) and renewable energy (e.g., solar, wind, geothermal, tidal), coupled with improved energy conservation. Although there are signs that the U.S. is beginning to move in this direction, this trend needs to be accelerated.

Raising the gasoline tax will help move the country in the right direction. For years, environmentalists and many energy experts have called for it, despite the anticipated wrath of American drivers. Americans consume nearly one quarter of the oil produced in the world, yet we only have four percent of the known world's oil reserves. We can either remain "oil junkies," or pursue an alternative path that would reduce U.S. reliance on foreign oil. As a disincentive to use gasoline, increased gas taxes would not only reduce dependence on foreign oil, it would also help solve many of our other environmental problems, such as highway congestion and air pollution. Gas taxes are a market-based way to promote thrifty cars and mass transit that could work more efficiently than fuel economy standards or regional traffic reduction mandates. Higher gas taxes will not put the American economy at a competitive disadvantage. In Europe and Japan, the gas tax is in "dollars" per gallon, not pennies or dimes as in the United States. Although a gas tax would hit the poor disproportionately, this effect could be offset by other means (e.g., lessening income taxes for the poor). Bottom line: As long as energy is cheap, Americans are not going to get serious about investing in energy efficiency or alternative fuels. It is time to "dethrone" oil and accept the notion that energy conservation along with renewable energies are the only viable alternatives if we want clean coastlines, not to mention a healthy, sustainable planet.

The simple act of planting trees in cities and suburbs, for example, also helps reverse our dependence on fossil fuels. Planting trees can reap millions of dollars in long-term savings by making it less costly to heat and cool buildings, by reducing and absorbing air pollutants, and by reducing glare and noise. Additional benefits of planting trees include increased property values, improved aesthetics, and improved sense of community. According to the U.S. Forest Service, planting 95,000 trees in metropolitan Chicago, for example, would result in a net benefit of $38 million over 30 years. So convinced by these studies that "trees in many cases can do the job as well as fossil fuels," the Sacramento Municipal Utility District in Northern California said "no" to a new power plant in 1992. Instead, the utility company opted to fund the planting of half a million trees, thus beautifying the city, restoring the urban ecosystem, fostering community involvement, and reaping mega-public relations benefits, all while meeting the objectives of lowering peak summer energy demands, and lessening use of fossil fuels. *The tree is one of nature's most highly sophisticated, clean, and wondrous technologies.*

For a general review of America's energy situation, and a discussion of the pros and cons of various nonrenewable and renewable energy sources, consult one of several introductory textbooks in environmental studies that cover the subject, such as Klee (1991), Miller (1992), and Owen and Chiras (1990). For greater historical depth, consult Melosi (1987), Blackburn (1987), and the World Commission on Environment and Development (1987). The annual reports of the U.S. Energy Information Administration, such as its publication Energy Facts, are a good source for keeping up on the ever-changing figures on U.S. energy production and use.

CONCLUSION

Offshore oil development and transport is certainly one of the largest threats to the coastal environment and is an overwhelmingly important issue for the coastal resource manager to consider. The energy-intensive lifestyle of modern society guarantees a reliance upon fossil fuels, especially oil, in the short term. Therefore, the goal for the coastal resource manager will be to understand the threats this resource development imposes and to try to account for this in management decisions.

Offshore oil development can directly impact the marine environment through localized destruction of the sea floor and through platform-based oil releases. Meanwhile, terrestrial support facilities create habitat loss and the potential for contamination of the coastal environment. Finally, transportation of the resource may be perilous. Unfortunately, the capacity to control and mitigate the impacts of offshore oil development through technology are currently limited. When a facility is built, little can be done to make up for the loss of habitat, and when a spill occurs, clean-up efforts are often not very effective. Therefore, offshore oil development endeavors are high risk, and frequently exact a high price in environmental quality.

In the short-term, coastal resource managers can avail themselves of policies which will allow them greater control over the impact from offshore oil development. However, in the long run, the path is clearly different. Oil is a limited resource and will become increasingly scarce as population, and subsequently consumption, rise. As this occurs, there will be more demand for offshore oil development in increasingly remote, pristine, and often hostile environments, which may subsequently intensify the risk factor. Therefore, in order to preserve the health of the coastal environment, resource managers must focus their efforts on promoting energy efficiency and changing from nonrenewable fossil fuels, to renewable energy sources. Decreased reliance upon centralized, nonrenewable energy sources, may ultimately lead to lower risk of pollution and coastal habitat destruction. As a result, larger expanses of open space may be preserved, which is the subject of our next chapter.

REFERENCES

Alaska's Marine Resources. 1992. "After the Exxon Valdez Oil Spill." *Alaska's Marine Resources.* Newsletter of the Marine Advisory Program, Alaska Sea Grant College Program, School of Fisheries and Ocean Sciences, University of Alaska Fairbanks, pp. 1–15.

Blackburn, John O. 1987. *The Renewable Energy Alternative.* Durham, North Carolina: Duke University Press.

Cotton, Ulysses. 1992. *Oil Spills 1971–90: Statistical Report.* Herndon, Virginia: U.S. Minerals Management Service.

ERC Environmental and Energy Services Company, Inc. 1989. *Offshore Oil Development: Issues and Impacts for the Central California Coast.* Central Coast OCS Regional Studies Program.

Geraci, J. R., and T. D. Williams. 1990. "Physiologic and Toxic Effects on Sea Otters." In *Sea Mammals and Oil: Confronting the Risks,* edited by J. R. Geraci and D. J. St. Aubin, p. 221. San Diego, California: Academic Press, Inc.

Golob's Oil Pollution Bulletin. 1992. "Tanker Spills in United States Reach Lowest Level in 14 Years." Vol. IV, No. 16: p. 1.

Holing, Dwight. 1990. *Coastal Alert.* Washington D.C.: Island Press.

Kearney, Inc., A. T. 1991a. *Estimating the Environmental Costs of OCS Oil and Gas Development and Marine Oil Spills: A General Purpose Model: Overview and Summary.* Alexandria, Virginia: A.T. Kearney, Inc.

Kearney, Inc., A. T. 1991b. *Estimating the Environmental Costs of OCS Oil and Gas Development and Marine Oil Spills: A General Purpose Model. Volume I: Economic Analysis of Environmental Costs.* Alexandria, Virginia: A.T. Kearney, Inc.

Kearney, Inc., A. T. 1991c. *Estimating the Environmental Costs of OCS Oil and Gas Development and Marine Oil Spills: A General Purpose Model. Volume II—Model Methodology, Documentation and Sample Outputs.* Alexandria, Virginia: A.T. Kearney, Inc.

Klee, Gary A. 1991. *Conservation of Natural Resources.* Englewood Cliffs, New Jersey: Prentice Hall.

Leslie, Gay I. 1993. "Domestic Energy Policy and Coastal Zone Management." In *Coastal Zone 1993. Vol. 1. Proceedings of the Eighth Symposium on Coastal and Ocean Management,* edited by Orville T. Magoon, W. Stanley Wilson, Hugh Converse, and L. Thomas Tobin, pp. 2–16. New York: American Society of Engineers.

Meade, Norman, Thomas LaPointe, and Robert Anderson. 1983. *Multivariate Analysis of Worldwide Tanker Casualties.* Rockville, Maryland: National Oceanic and Atmospheric Administration.

Melosi, Martin. 1987. "Energy and Environment in the United States: The Era of Fossil Fuels." *Environmental Review,* Fall, pp. 167–188.

Miller, G. Tyler. 1992. *Living in the Environment.* Belmont, California: Wadsworth.

Minerals Management Service. 1985. *Facilities Related to Outer Continental Shelf Oil and Gas Development Offshore California: A Factbook,* prepared by Centaur Associates, OCS Study MMS #85-0053, October.

National Academy of Sciences. 1991. *Tanker Spills: Prevention by Design.* Washington D.C.: National Academy Press.

National Oceanic and Atmospheric Administration (NOAA). 1996. Florida Keys National Marine Sanctuary: Management Plan. Rockville Maryland: National Oceanic and Atmospheric Administration.

Ohlendorf, H. M., R. W. Risebrough, and K. Vermeer. 1978. *Exposure of Marine Birds to Environmental Pollutants: Wildlife Research Report 9.* Washington, D. C.: U. S. Department of the Interior, Fish and Wildlife Service.

Owen, Oliver S., and Daniel D. Chiras. 1990. *Natural Resource Conservation.* 5th ed. New York: Macmillan.

Planning Department. 1994. *Santa Cruz/Monterey Counties Oil Spill Contingency Plan.* Santa Cruz, CA: Planning Department, County of Santa Cruz.

Sanders, H. L., J. F. Grassle, G. R. Hampson, L. S. Morse, S. Garner-Price, and C. C. Jones. 1980. "Anatomy of an Oil Spill: Long-term Effects from the Grounding of the Barge *Florida* off West Falmouth, Massachusetts." *Journal of Marine Research Vol. 38: No. 2.* pp. 265–380.

Sport Fisheries Institute Bulletin. 1993. "The Lasting Effects of the *Exxon Valdez* Oil Spill." *SFI Bulletin,* No. 442, March, pp. 4–5.

Sutkus, Adam. 1994. "OSPR: California's Answer to How to Organize an Effective Oil Spill Response." *Alolkoy: The Publication of the Channel Islands National Marine Sanctuary* Vol. 7, No. 1: Spring, pp. 8–10.

Suzuki, David, and Peter Knudtson. 1989. *Genetics: The Clash Between the New Genetics and Human Values.* Cambridge, Massachusetts.: Harvard University Press.

Teal, John M. 1993. "A Local Oil Spill Revisited." *Oceanus* Vol. 36, No. 2: Summer, pp. 65–73.

Teal, John M., and R. W. Howarth. 1984. "Oil Spill Studies: A Review of Ecological Effects." *Environmental Management* Vol. 8, No. 1: pp. 27–44.

Trench, Cheryl J. 1993. "Petroleum Imports and Coastal Zone Management." In *Coastal*

Zone 1993, Vol. 1: Proceedings of the Eighth Symposium on Coastal and Ocean Management, edited by Orville T. Magoon, W. Stanley Wilson, Hugh Converse, and L. Thomas Tobin, pp. 43–55. New York: American Society of Engineers.

Wilderness Society (The). 1991. *Oil Spills: Just a Cost of Doing Business.* Washington, D.C.: The Wilderness Society.

World Commission on Environment and Development. 1987. *Energy 2000: A Global Strategy for Sustainable Development.* Atlantic Highlands, New Jersey: Zed Books.

FURTHER READING

Cairns, W. J. and P. M. Rogers, eds. 1981. *Onshore Impacts of Offshore Oil.* London, Applied Science.

Clark, John R. 1996. *Coastal Zone Management Handbook.* New York: CRC Lewis Publishers.

Flavin, Christopher, and Nicholas Lenssen. 1990. *Beyond the Petroleum Age: Designing a Solar Economy.* Washington, D. C.: Worldwatch Institute.

Gever, John, et al. 1991. *Beyond Oil.* 3rd ed. Niwot, Colorado: University of Colorado Press.

Hinrichsen, Don. 1998. *Coastal Waters of the World: Trends, Threats, and Strategies.* Washington D.C.: Island Press.

Minerals Management Service. 1987. *Five Year Outer Continental Shelf Oil and Gas Leasing Program. Mid-1987 to Mid-1992: Final Environmental Impact Statement.* Washington, D. C.: Minerals Management Service, U.S. Department of the Interior.

Tester, Jefferson, ed. 1991. *Energy and the Environment in the 21st Century.* Cambridge, Massachusetts.: MIT Press.

World Resources Institute (The). 1996. *World Resources: 1996–97.* New York: Oxford University Press.

8 Open Space Preservation And Management

Many Americans travel to the shore to find open space—a place on the earth where they can breathe fresh air, jog along the shore, bask in the sun, throw a Frisbee, or just "get away" from the traffic, noise, and hectic pace of urban living. **Open Space** generally connotes large expanses of land with few buildings, few posted regulations, and few people and cars. Although the term "open space" usually relates to an undeveloped green or wooded area *within* an urban area, such as Central Park in New York or Golden Gate Park in San Francisco, the term also applies to areas free of congestion *adjacent* to urban areas, such as coastlines. Beaches, dunes, tide pools, wetlands (e.g., estuaries, lagoons, lakes) and even rural agricultural communities along the coast fall under the category of open space. In other words, open space is more than just official Wilderness Areas (e.g., Ventana Wilderness, California) and city, county, state, or federal "parks."

One of the major problems in coastal resource management is maintaining (or restoring) this sense of open space. This chapter on open space preservation and management is divided into five parts. The first, *Population Pressures and the Coast,* summarizes 50 years of population change (and projected change) for various regions along America's coasts, from 1960 to 2010. In so doing, it will identify the "hot spots" of growth on our shores. *Environmental Concerns and Management Strategies* discusses the major environmental concerns and management strategies associated with open space preservation and management. It looks at such concerns as (a) the disappearance of coastal agricultural lands, (b) the loss of coastal wildlife habitat, (c) the decline in coastal recreational resources, (d) the loss of coastal village and small town character, and (e) the decline in respect for coastal property, as illustrated by graffiti and other forms of vandalism. *Case Study: Elkhorn Slough* provides an interesting case study in management of a major estuary on the shores of the Monterey Bay National Marine Sanctuary. The following part provides some general policy recommendations regarding open space management, while the *Conclusion* offers some final thoughts about individual beliefs, decisions, and actions versus the role of government agencies regarding open space management.

POPULATION PRESSURES AND THE COAST

In 1990, the Strategic Assessment Branch of the National Oceanic and Atmospheric Administration (NOAA) published the first comprehensive look at population trends on America's coasts: *50 Years of Population Change Along the Nation's Coasts, 1960–2010* (Culliton et al. 1990). Culliton, a geographer, and his interdisciplinary team of researchers used the best available data to analyze recent and projected trends regarding size, distribution, and density of coastal populations. They mapped the data according to county, state, and region. Coastal resource managers and other decision makers have much to learn from their analysis of coastal population trends. What follows is merely some of the highlights of that groundbreaking report.

You may be surprised to learn that 30 of America's 50 states have coastlines, and that nearly *one-half (50 percent), of all Americans live in* ***coastal areas.*** (This does not mean that 50 percent of all Americans live in a beach house on the shore. It means that they live in one of the 451 "coastal counties," or in one of the 1,569 inland "noncoastal counties" that can have a significant impact on the environmental quality of the coast.) Today's population of 110 million U. S. coastal dwellers is projected to reach 127 million people by the year 2010—an increase of almost 15 percent. These Americans will have to deal with the problems of coastal resource management. Evidence is mounting that many coastal environmental problems (e.g., disappearance of prime agricultural lands, loss of wetlands, declining recreational opportunities) are the result of poor planning and development. According to Culliton and his colleagues, general coastal development patterns seem to indicate that the present "site-by-site, permit-by-permit approach" to land use planning may be hazardous to the coast, and this system of decision making may cause intensified environmental problems when it must cope with the growth and development pressures of the future.

Most of America's 110 million coastal dwellers live in either the Northeast (35%) or the Pacific coast (26%), followed by the Great Lakes (17%), the Gulf of Mexico (13%) and the Southeast (8%). In terms of population change, Culliton and his associates discovered and mapped some interesting trends. During the study period (1960–2010), the 1960s was the decade of maximum coastal population growth, with California, Florida, and New York accounting for 58 percent of the increase. Although coastal growth slowed during the 1970s, it rebounded again in the 1980s, with California, Florida and Texas accounting for 73 percent of the growth in coastal areas.

Another indicator of environmental stress is *population by shoreline data,* which is derived by dividing a state's coastal population by its tidal shoreline mileage. On a national average, coastal areas had 1,177 persons per shoreline mile in 1988, with a projected ratio of 1,358 by the year 2010. Because of their relatively small shorelines, certain coastal states had particularly high population-to-shoreline ratios, such as Illinois (91,740), Pennsylvania (30,871), and Indiana (15,951). In addition to a national overview, Culliton and his associates did a regional analysis of coastal population, emphasizing population trends, population density, population by shoreline mile, and hot spots of growth.

Northeast Region—35 Percent of Coastal Population

The Northeast was found to be the most populated of the five regions, with one-third (35 percent) of the Nation's coastal population. (See Figure 8-1).

- *Population trends:* The coastal population here is projected to increase by 30 percent between 1960 and 2010. Since the coastal areas in this region are already relatively degraded, this anticipated growth is likely to cause severe coastal environmental problems.
- *Population density:* Projected environmental problems are intensified by the fact that this region has six of the Nation's seven leading states in coastal county population density. Overall, the region's coastal population has a density of 750 persons per square mile.
- *Population by shoreline mile:* In 1988, the state of Pennsylvania topped the region's list with 30,871 people per shoreline mile, while the District of Columbia ranked second highest with 15,049 people per shoreline mile. Both are expected to continue their lead into the year 2010. Pennsylvania receives the highest value because its major population center (Philadelphia) covers a large portion of the state's coastal zone. In other states within the region (e.g., New York), overall population per shoreline mile is less because the state's shoreline reaches less populated portions of the state.
- *Hot spots of growth:* The suburbs next to the large cities, such as Suffolk County, NY, are expected to grow fastest. This is a result of the so-called "suburban sprawl" phenomena that has taken place across the United States since World War II. Another pattern also emerged—the "sprawl" of an aging population. Counties with large retirement and resort areas grow the fastest, such as Virginia Beach, VA and Worcester, MD.

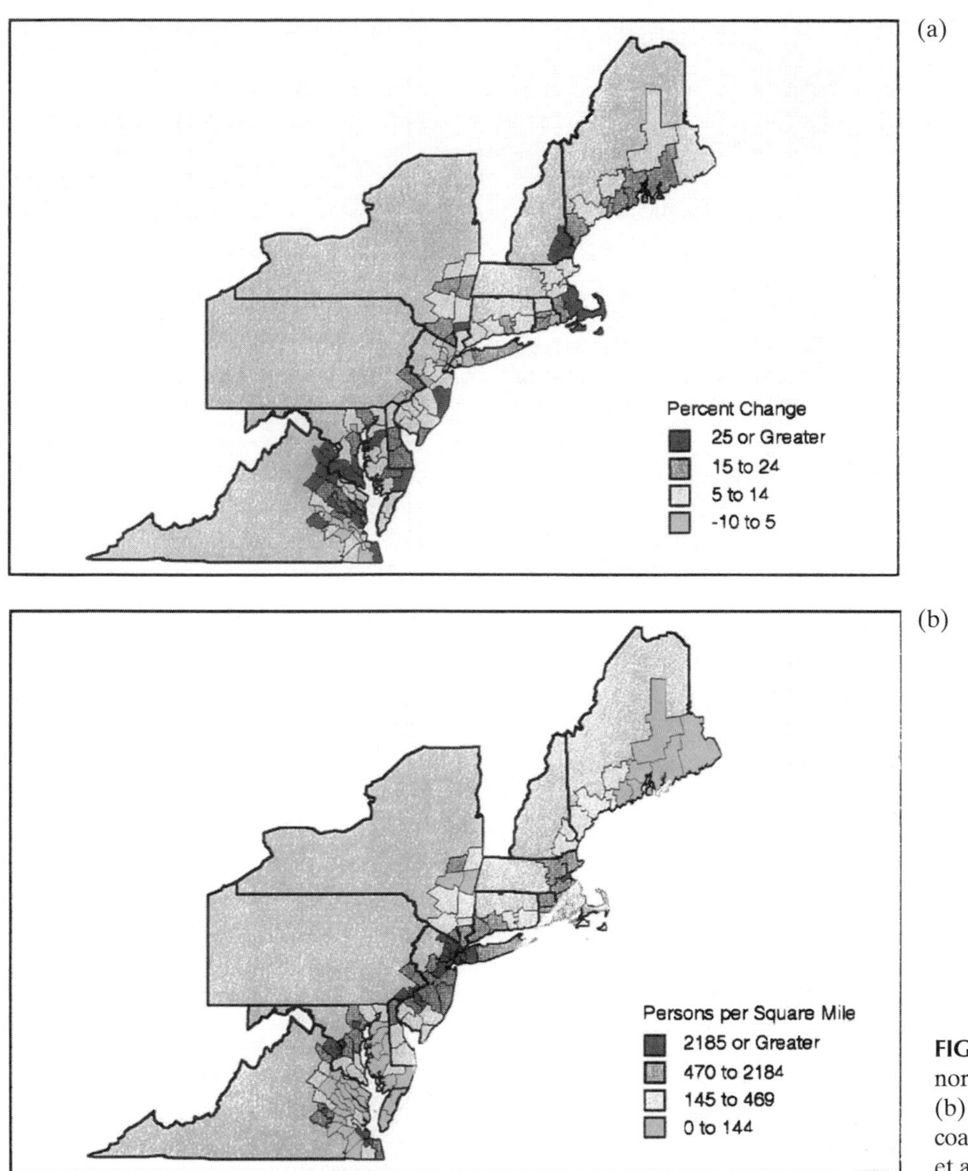

FIGURE 8-1 (a) Population change in northeast coastal counties, 1988–2010; (b) Population density in northeast coastal counties, 2010. (*Source:* Culliton et al 1990, p. 9)

Pacific Region—26 Percent of Coastal Population

The second most populated U.S. coastal region was the Pacific region. (See Figure 8-2). In 1988, over 26 percent of the U.S. coastal population lived within this region, with over three-quarters of them living in California.

- *Population trends:* The coastal population in this region is projected to double between 1960 and 2010. By 2010, Culliton and his associates project this region to increase by over 6 million persons—the largest of any coastal region. Alaska is expected to lead the coastal population growth rate (40%), followed by California with over 20 percent growth. Oregon is projected to receive the smallest growth rate.
- *Population density:* Of the five regions, the Pacific is the least densely populated coastal region, with only 36 persons per square mile—well below the national average. Obviously, population density is highly variable in this region—from the virtual absence of people on certain coastlines of Alaska to the "masses of humanity" on the beaches of Southern California. And, for Californians, those masses of humanity are projected to double—from 342 persons per square mile in 1960 to an estimated 718 persons per square mile in 2010.
- *Population by shoreline mile:* The Pacific has the lowest population-to-shoreline ratios of all the five regions. Within the Pacific region, California had the highest population to shoreline ratio, and it is projected to *double* by the year 2010.
- *Hot spots of growth:* Not surprisingly, Los Angeles, Orange, and San Diego counties are projected to increase the most. The counties projected to have the fastest percent increase between 1988 and 2010 are the

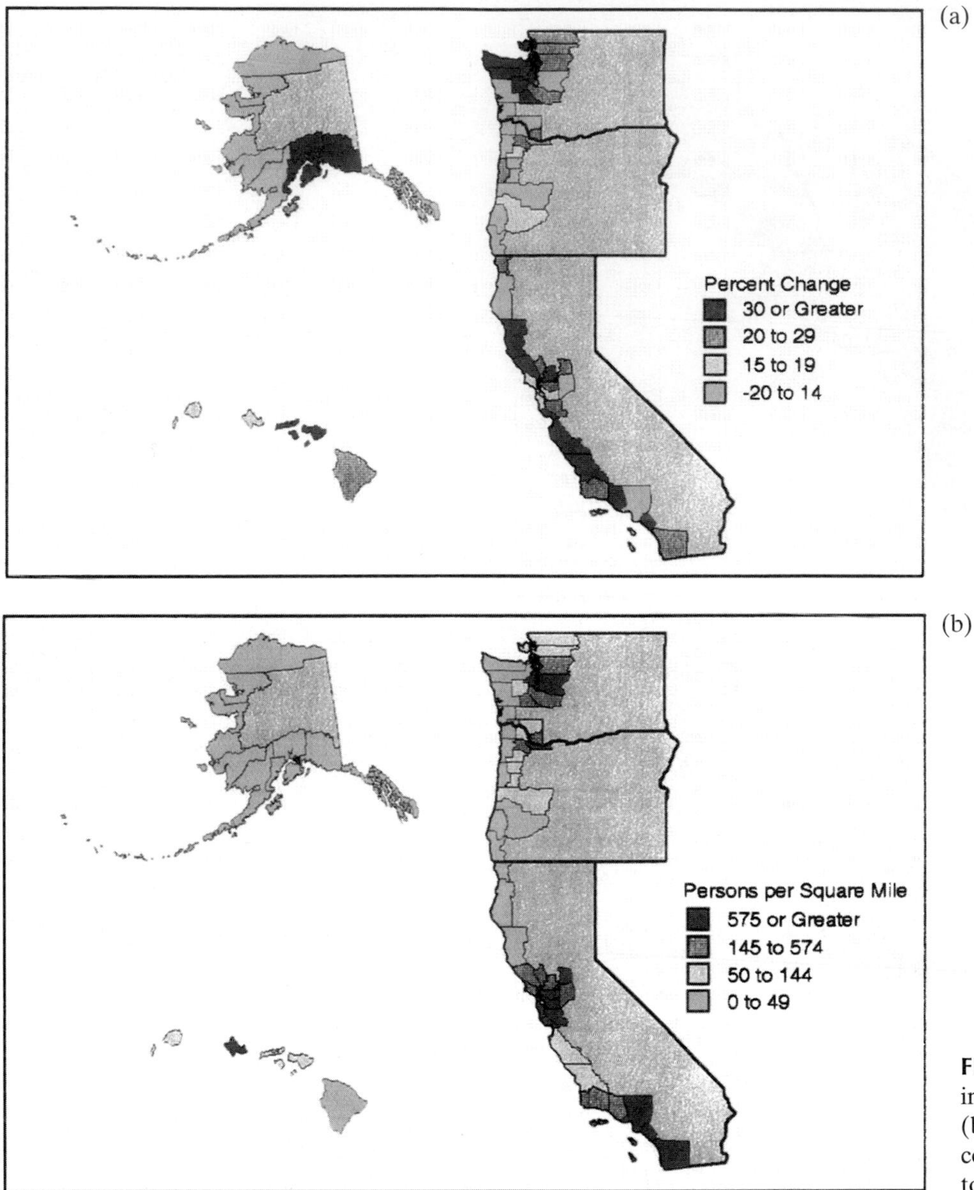

FIGURE 8-2 (a) Population change in pacific coastal counties, 1988–2010; (b) Population density in pacific coastal counties, 2010. (*Source:* Culliton et al 1990, p. 24)

Kenai Peninsula (65%) and Matanuska-Susitna (57%) areas in Alaska, followed by San Juan, Washington (51%), and Santa Cruz, California (50%) adjacent to the Monterey Bay National Marine Sanctuary. By the year 2010, Santa Cruz is expected to have a coastal population of approximately 760 people per square mile, which is substantially lower than the projected populations for the counties of San Francisco, CA (18,913), Orange, CA (3,710), and Los Angeles, CA (2,422).

Great Lakes Region—17 Percent of Coastal Population

Culliton and his associates found the Great Lakes region to be the third most populous coastal region. (See Figure 8-3).

- *Population trends:* Compared to the Northeast, only a small proportion (17 percent) of the population of the Great Lakes region is in coastal counties. With a projected coastal population growth of only eight percent between 1960 and 2010, this region's share of the U.S. coastal population is expected to drop.
- *Population density:* Population densities are highest along the southern shores of Lakes Michigan, Huron, and Erie, in the major metropolitan areas of the region.
- *Population by shoreline mile:* Illinois has 91,000 persons per shoreline mile, making it the *highest in the Nation.* Indiana comes in second within this region, with 15,951 persons per shoreline mile. Both Illinois and Indiana are expected to continue their lead in this region into the year 2010.

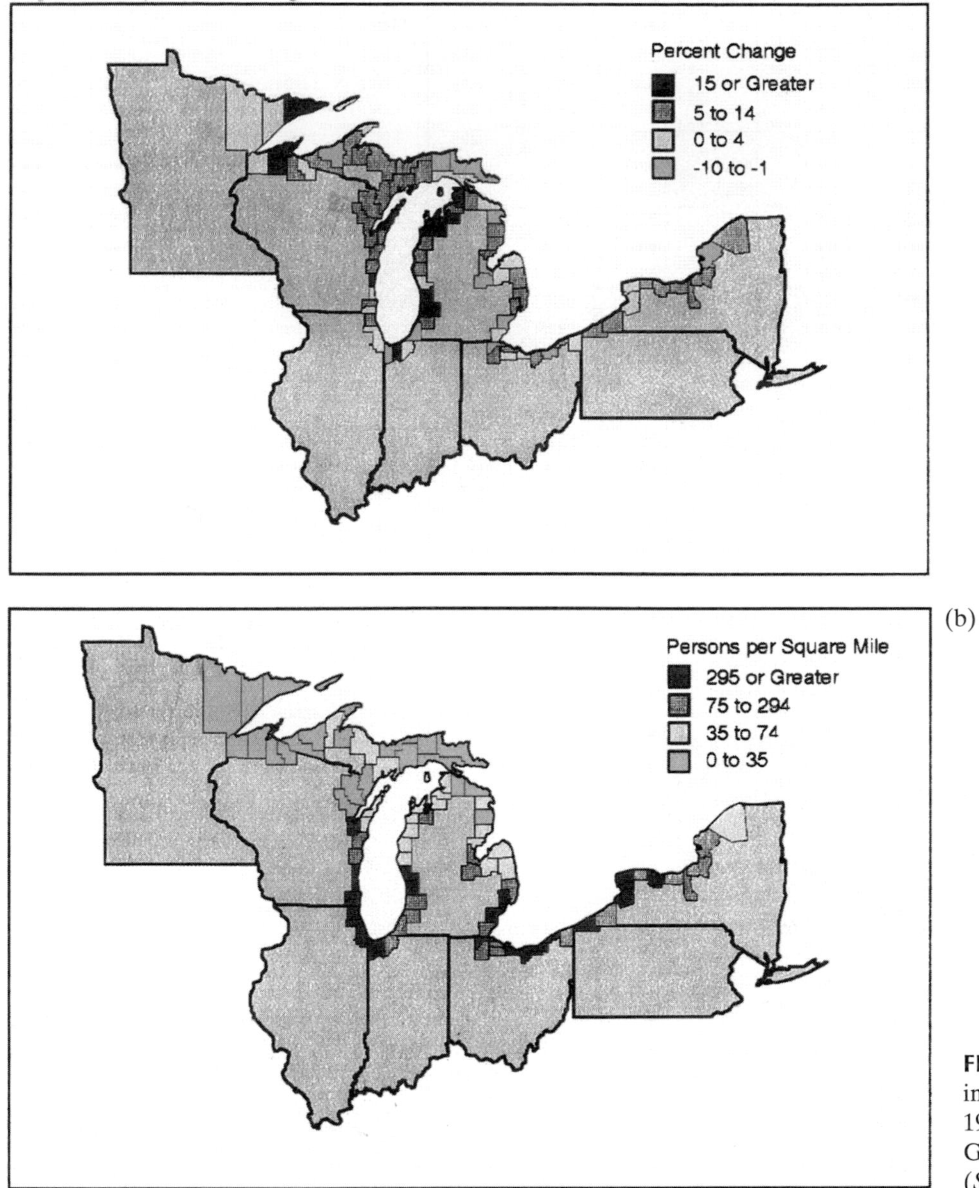

FIGURE 8-3 (a) Population change in Great Lakes coastal counties, 1988–2010; (b) Population density in Great Lakes coastal counties, 2010. (*Source:* Culliton et al 1990, p. 13)

- *Hot spots of growth:* The shores of Lakes Michigan and Superior are projected to have the fastest growth rate, with Porter County, Indiana, topping the list.

Gulf of Mexico Region—13 Percent of Coastal Population

With only 13 percent of the total U.S. coastal population, the Gulf of Mexico placed fourth in rank. In 1988, 14 million people lived within coastal counties in this region, with most of them living in Houston, New Orleans, Tampa, and St. Petersburg. (See Figure 8-4).

- *Population trends:* Although this region has a relatively small percent of the total U.S. coastal population, Culliton and his associates project its population to increase by 144 percent between 1960 and 2010—the second highest of the five regions. One reason for this projected growth rate is the recent tourist exposure that formerly "unknown areas" are now receiving. A case in point is Gulf Shores, Alabama. Until the 1990s, few people north of Mobile had heard of this town on the Gulf of Mexico, much less thought of going to vacation on its white beaches. But as people spread the word of a Florida-like beach resort at bargain prices, tourists began arriving in droves. Today, approximately four million visitors a year descend on Alabama's 51 km (32 mi) of Gulf Coast, vacationing in communities like Gulf Shores. The concept of "a Florida beach at half the price" has many of today's Gulf Shores' tourists dreaming of the day they can become permanent residents. Culliton and his associates expect Western Florida to

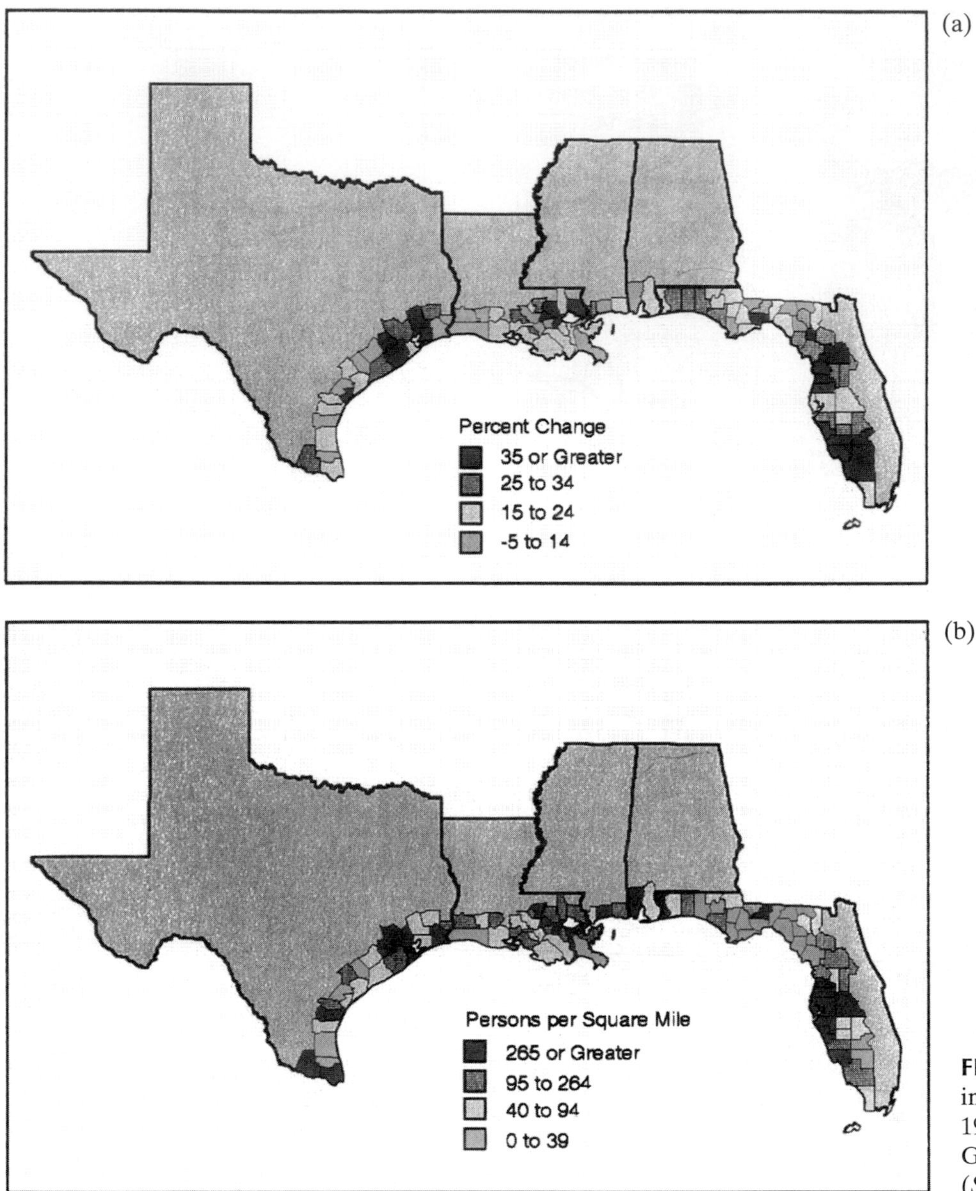

FIGURE 8-4 (a) Population change in Gulf of Mexico coastal counties, 1988–2010; (b) Population density in Gulf of Mexico coastal counties, 2010. (*Source:* Culliton 1990, p. 20)

continue to be the most rapidly growing area within the Gulf of Mexico Region.

- *Population density:* Western Florida is also projected to have the most rapid growth of population density, increasing to 195 persons per square mile in 2010.
- *Population by shoreline mile:* Population by shoreline mile is anticipated to double between 1960 and 2010. Texas is likely to take the regional lead with 1,956 people per shoreline mile, followed by Florida with 1,411 people per shoreline mile.
- *Hot spots of growth:* Counties projected to have the largest increases in population within this region are located primarily in Florida, from the Tampa area to the Florida Keys. Although coastal populations are increasing rapidly within the Gulf Region, coastal counties in the Northeast, the Great Lakes, and the Pacific regions are projected to be more densely settled in the near future.

Southeast Region—8 Percent of Coastal Population

Of the five regions, the Southeast region ranked last, with only 8 percent of the total U.S. coastal population. In 1988, nine million people lived within the coastal counties of the region, with four out of every five located in Eastern Florida. Major population centers within the Southeast region include Miami, Savannah, Jacksonville, and Charleston. (See Figure 8-5).

- *Population trends:* Although this region currently has the lowest total coastal population, Culliton and his associates project that coastal population in this region will in-

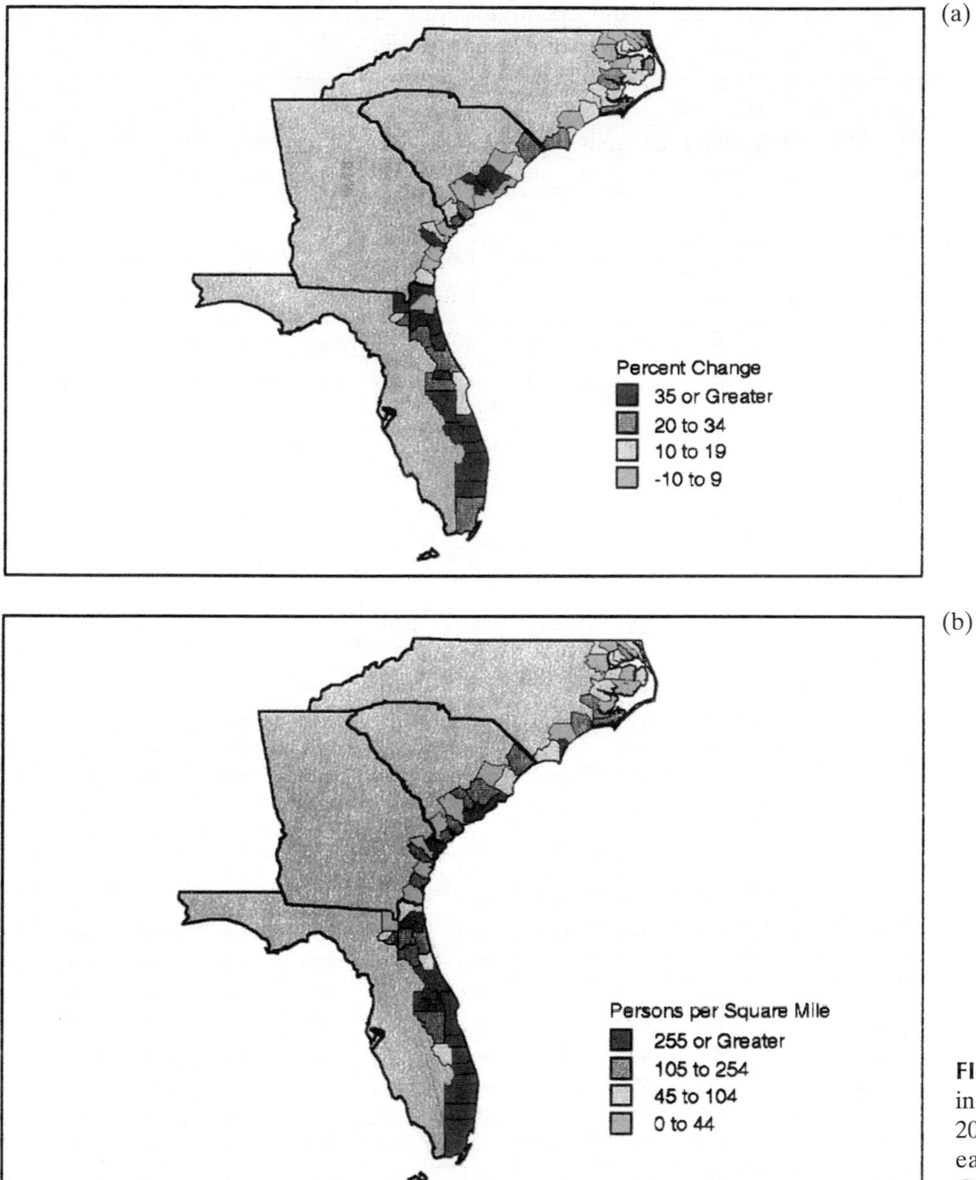

FIGURE 8-5 (a) Population change in Southeast coastal counties, 1988–2010; (b) Population density in Southeast coastal counties, 2010. (*Source:* Culliton 1990, p. 17)

crease by 181 percent—*the highest increase of the five regions.* Eastern Florida is expected to dominate population trends in this region.

- *Population density:* In 1988, the average population density for the region was 237 persons per square mile—well above the U.S. average of 70 persons per square mile.
- *Population by shoreline mile:* Between 1960 and 2010, the population-to-shoreline mile ratio is projected to almost triple.
- *Hot spots of growth:* Culliton and his associates maintain that Eastern Florida is likely to dominate in all three categories in this region—counties expected to increase by the most people; the counties projected to increase at the fastest rate; and the counties projected to have the highest population density. Osceola, (72%), Martin (61%) and Flagler (59%) counties in Florida are projected to experience the greatest percent of population change between 1988 and 2010.

Summary

Without a doubt, America's coastal areas are feeling the pressures of increasing population growth. Although one can always find fault with certain population projections, the underlying message should be clear—a "coastal population bomb" is building on America's shorelines. Furthermore, it is a population body that is becoming more *diverse* (and thus complex) by the minute—(a) diverse as to *form of competition* for coastal space and (b) diverse as to *ethnic background,* which affects perception and use of coastal space.

The major competitors for coastal space (in addition to housing for new residents) will be tourism and recreation (e.g., hotels, restaurants, boutique shops), environmental protection (e.g., habitat for wildlife), transportation (e.g., expanded harbor facilities for increasing transoceanic world trade), energy development (e.g., tidal power and ocean thermal energy conversion), and coastal mining (especially sand and gravel). Military activities and their associated space needs will play a lesser role in the near future (e.g., Fort Ord, California and other defense base conversions) [Goldberg 1993]. Although coastal space has a finite limit, the demands and competition for that space is clearly growing and becoming more intense.

Coastal resource managers will also have to respond to a second form of growing diversity—the diversity of color. As Thomas W. Gwyn (1992), past Chairman of the California Coastal Commission and a member of the Board of the State Coastal Conservancy, has pointed out, debates over how to manage America's coasts and oceans must now include the points of view of the emerging "minorities" (e.g., the Blacks, Asians, Hispanics, and Indians that will soon constitute the majority of the State of California). Both the environmental and the minority movements must *broaden their horizons* to include each other's concerns, or both will suffer. The environmental movement (in this case those who have a special interest in protecting the coast and ocean) will have to find a way to capture the allegiance of greater numbers of people of color. At the same time, according to former Coastal Commissioner Gwyn, the minority movements will have to make an effort to understand why coastal advocates and other environmentalists often seem to place more value on protecting "birds and trees" rather than "people and jobs."

As America's coastal population grows, so will the urgency to preserve and properly manage its remaining open space—its rural countryside, wetlands, dunelands, beaches, and marine sanctuaries. Planning for this population growth will probably be the biggest environmental challenge facing present (and future) coastal planners and marine sanctuary managers.

ENVIRONMENTAL CONCERNS AND MANAGEMENT STRATEGIES

Coastal resource managers must contend with five major problems affecting the amount and quality of open space: (1) the disappearance of coastal agricultural land; (2) the loss of coastal wildlife habitat; (3) the decline in coastal recreational resources, including public access; (4) the loss of coastal village or small town character; and (5) the loss of respect for coastal property. We will now take a closer look at these very different, but closely related, issues as they relate to open space preservation and management.

Disappearance of Coastal Agricultural Land

Coastal farmlands are being lost. Agricultural specialists, however, disagree as to how many acres are being lost, and at what rate the loss is occurring. For example, according to a recent study conducted for the American Farmland Trust, 155,810 ha (385,000 acres) of coastal California farmland was lost from 1978 to 1987 (Thompson et al. 1994). The authors ranked the coastal strip from San Diego to Sonoma as third on their list of the *Top 12 Most-Endangered Farm Regions in the United States.* California's Central Valley was at the top of the list, with South Florida ranked as second. Other farm regions on the group's Top 12 list were the Mid-Atlantic-Chesapeake area, North Carolina's Piedmont area, the Puget Sound Basin, the Chicago-Milwaukee-Madison metropolitan area, Oregon's Willamette Valley, Minnesota's Twin Cities, Western Michigan, the Shenandoah and Cumberland Valleys, and the Hudson River and Champlain Valleys. Although these 12 regions account for only 5 percent of the U.S. farm land, they include two-thirds of the nation's fruit, half the vegetable, and a fourth of all dairy producing areas.

Other organizations, such as the California Department of Conservation's Farmland Mapping and Monitoring Program, as well as the California Coastal Commission, take strong issue with the above report. Jack Leibster, a representative of the California Coastal Commission, maintains that the figure of 155, 810 ha (385,000 acres) of California coastal farmland lost is not only misleading, but wrong. According to Leibster, only 157,833 ha (390,000 acres) of agricultural land exists in the Coastal Zone (as defined by the Coastal Act of '73), and therefore within the jurisdiction of the California Coastal Commission. Of that land, there is about 16,188 ha (40,000 acres) of prime land (cropland) and about 141,645 ha (350,000 acres) of non-prime, or grazing land. Some of that marginal land has indeed been lost, but lost to things like overgrazing, not to development (personal communication, August 8, 1994).

No doubt the discrepancy in figures is a result of differences in opinion as to exactly what constitutes "coastal agriculture." Until federal agencies (e.g., U.S. Census of Agriculture), state agencies (California Department of Conservation; California Coastal Commission), and national farm preservation groups (e.g., American Farmland Trust) have an agreed-upon definition of "coastal agriculture," as well as an agreed-upon method of measuring its rate of change, it will be difficult to know how much coastal farmland is being lost, and at what rate.

Reasons for the loss. There is less disagreement, however, when it comes to the causes of this loss of coastal farmland. Urbanization, bad economics, high land prices, and conflicting governmental policies are generally the reasons for the loss of coastal agricultural land. One thief is *urbanization.* As cities grow, they devour the land around them, especially the flat land that is easiest and least expensive to develop. Urban sprawl, however, also "leap-frogs" across coastal rural communities, affecting even farmlands that have not themselves been converted. The farmers who survive must somehow cope with the hostility of suburbanites, trespassing and vandalism, depletion of groundwater, and increased costs of tillage.

A dramatic illustration of "mega-sprawl" in the San Francisco Bay Region was provided at the April 1994 meetings of the Association of American Geographers in San Francisco. Leonard Gaydos, a geographer and leader of the U.S. Geological Survey project, which is based at National Aeronautics and Space Administration's Ames Research Center in Mountain View, California, unveiled a powerful new video that portrays the Bay Area cities and suburbs in the act of sprawling. One hundred and forty years of urban development was squeezed into a 30 second computer animation, beginning with six red dots on a map marking Gold Rush-era settlements, and ending with today's megalopolis. The animation drew gasps from the audience. Even William Acevedo, the United States Geological Survey geologist who did the computer animation, said he was startled by the transformation in the region, even though he grew up in the Bay Area watching its orchards give way to electronics plants and tract housing. (See Figure 8-6).

Any remaining "checkerboard" farmlands (so-called because they are interspersed with pockets of urbanization) then become highly unprofitable to farm because of *bad economics.* Fresh fruits and vegetables once sold to local communities must be transported at great expense to market centers; farmers must travel greater distances to obtain seed and equipment; and farm labor grows scarcer and more costly as workers are attracted to higher-paying city jobs. *High land prices* also drive farmers off their land. As the price of land soars, farmers are faced with higher property taxes and legal and economic headaches. Unless they are protected by tax deferments and other relief, they are driven out of business. Finally, *conflicting governmental policies* add to the problem by fostering programs that encourage land conversion. For example, the U.S. Environmental Protection Agency—an agency dedicated to the conservation of resources—finances sewerage facilities in rural areas that encourage growth that would have otherwise been confined to the cities.

Need for farmland preservation. Coastal farmlands need to be cherished and preserved for their value as open space, their economic resource, and as our cultural heritage. As *open space,* farmlands give a coastal area scenic and aesthetic character by providing a visual and topographical background to the coastline. Rural coastal landscapes rest the eye, refresh the spirit, and provide weary urbanites a place to hike, bike, or jog. Humans are not the only organism to "seasonally" use coastal rural landscapes. Various forms of wildlife (e.g., birds of prey and small rodents) also use coastal farmlands as seasonal wildlife habitat. And, if properly managed, the higher parts of floodplains that transverse rural landscapes can be used for cropland and range, rather than for housing. Building housing on floodplains increases the risk for loss of life and property by flooding.

Coastal farms provide more than open space. Farming is an important component of the nation's *economic base.* It employs hundreds of thousands of people and produces billions of dollars in revenue. California's coastal agriculture, for example, since its earliest beginnings in 1812 as Russian subsistence farming at Fort Ross in Sonoma County, has expanded over regions of farm lands, producing a diverse variety of specialty crops. The moderating effect of a maritime climate creates the special conditions needed for certain specialty crops, such as artichokes, brussels sprouts, broccoli, strawberries, flowers, and greenhouse products. On the marine terraces and alluvial soils between Pillar Point and Año Nuevo (San Mateo County, CA), the coastal crop and grazing lands extend virtually to the water's edge. Furthermore, some of California's best grazing lands are located along the coast. Numerous acres just north of Santa Barbara in Southern California provide excellent grazing lands for cattle and sheep. In Monterey County, adjacent to Monterey Bay in Central California, the coastal river valleys provide deep alluvial and floodplain soils that are highly suited for irrigated crops such as vegetables, avocados, and citrus fruits. Because of the wide variety of vegetables grown there, Monterey County was once known as "The Salad Bowl of the World." Semi-agricultural land uses (e.g., dairies and feedlots) can also be found in the area. Amazingly, the coastal lands of Monterey County produce 90 percent of U.S. artichokes, 80 percent of its lettuce, 60 percent of its broccoli, and 50 percent of its cauliflower, not to mention a large percentage of other fruits and vegetables for America's dinner tables (Monterey County Farm Bureau 1988).

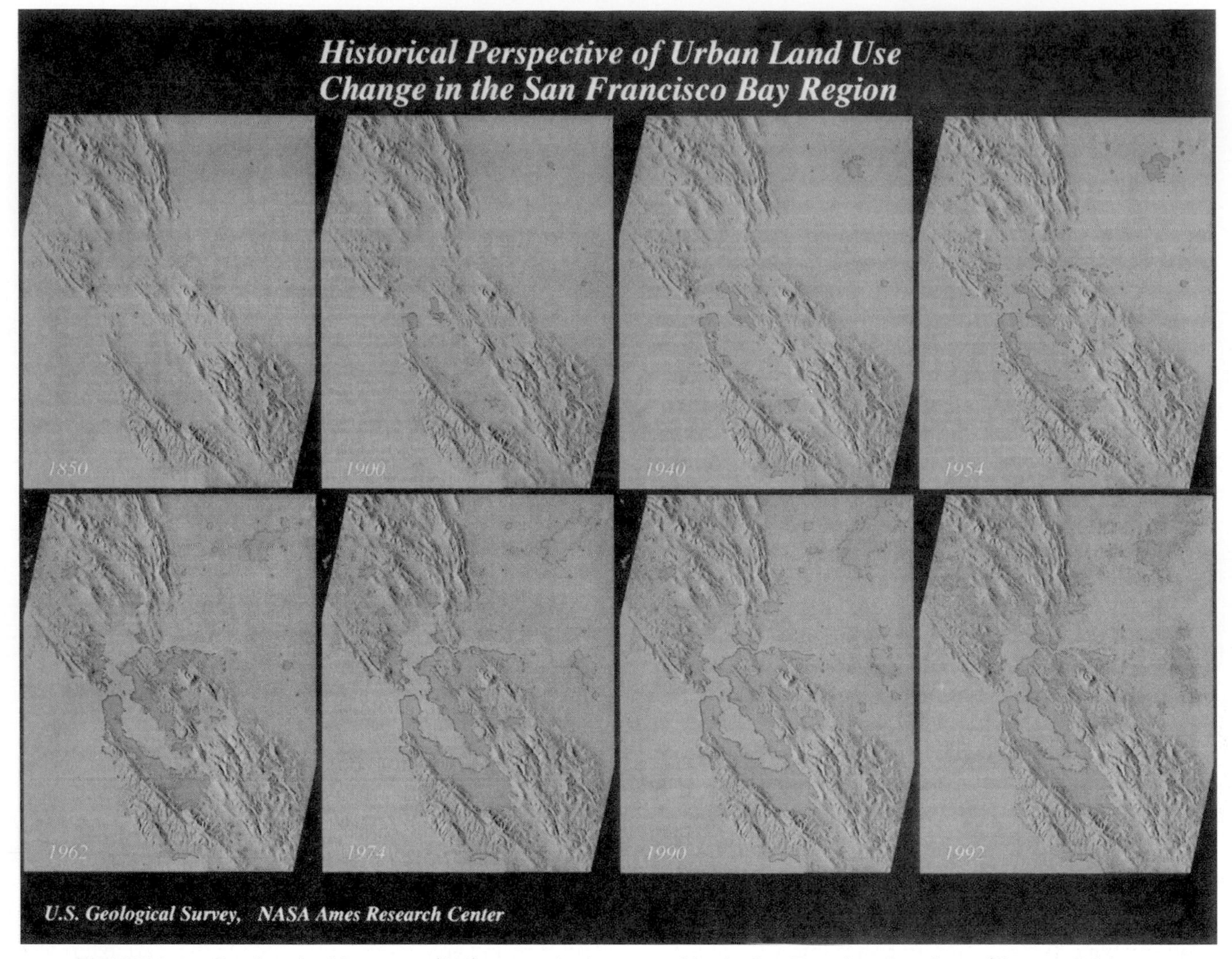

FIGURE 8-6 One hundred forty-two (142) years of urban sprawl in the San Francisco Bay Area. (*Source: California Agriculture,* Vol. 49, No. 6, Nov.–Dec. 1995, 13)

Finally, there is an often overlooked reason for protecting coastal farmland—it represents a valuable part of our *cultural heritage.* Country living nourishes the stomach and the spirit of millions of Americans, strengthening their sense of what makes life worthwhile. Wendell Berry, author of *The Unsettling of America: Culture and Agriculture* (1978), and others have explored the relationships between spiritual, environmental, and economic productivity of farmland culture in America. Coastal farming and ranching were major contributors to the growth of early settlements, and the coastal waters were often important in the transportation of agricultural products. In Central California, for example, the sternwheeler, *Vaquero,* made regular trips up and down Elkhorn Slough to transport potatoes and grain from Watsonville to Moss Landing harbor, then sailed on to San Francisco. Preserving coastal farmlands helps preserve the rich "farmland-sea" cultural heritage that is deeply ingrained in Americans and American history.

Loss of Coastal Wildlife Habitat

When humans modify, restrict, or eliminate natural habitat for wildlife, they are also unconsciously reducing the amount of open space for people. The term **wildlife** embraces all native life forms (except humans) that are not domesticated, including plants as well as animals. Wildlife connotes *wildness*—that is, *freedom,* not captivity or human manipulation; *independence,* not helplessness; and *renewability* through natural reproduction. When thinking of wildlife resources, however, most individuals have a more restrictive definition of wildlife—one that places an emphasis on animals, particularly land animals, such as birds, deer, and elk, rather than on fish in fresh water or fish and marine mammals in the ocean. Since fish and marine mammals in the ocean also have the same characteristics of wildness, they will also be included in our definition of wildlife. **Habitat** refers to the natural environment of a particular plant or animal, and it includes all the essentials of life for that organism, such as food, water, and cover. In coastal areas, predominant wildlife habitats are the various forms of wetlands (e.g., estuaries, coastal marshes, lagoons), forests, and grasslands that come right down to the sea, as well as tide pools and nearshore waters.

Many studies have been done on the loss of coastal wetlands. For example, Gosselink and Baumann (1980) estimated a loss of wetlands along U.S. coasts of 0.2 percent per year between 1922 and 1954, increasing to 0.5 percent between 1954 and 1974. Although data since the mid-1970s are poor, wetland conversion has probably dramatically decreased due to (a) various protection programs that were enacted in 1977 (Gosselink and Maltby 1990), and (b) heightened public understanding of the value of wetlands as a result of environmental education. In otherwords, the destruction of coastal wetlands continues, but at a lesser pace than in the past. For more details regarding wetlands as a threatened landscape, see Williams, ed., 1990; and Mitsch and Gosselink 1993.

Reasons for the loss. Several factors have fueled the disappearance of coastal wildlife habitat, especially timber harvesting, grassland conversion, inland waterway conversion, and general urbanization. *Timber harvesting* in America's coastal forests has disrupted much habitat that is indispensable to wildlife (Bandy and Tabler 1974). In September 1992, for example, scientists with the U.S. Fish and Wildlife Service in Portland, Oregon put the marbled murrelet (a robin-sized seabird) on the federal government's list of "threatened" species. British Columbia already lists the bird as a threatened species, and the state of California lists it as an "endangered" species. After a four-year study, the Portland scientists found that the murrelet was threatened in the Pacific Northwest primarily because of extensive logging of its habitat—most of the murrelets live in the Northwest's "old-growth" forests within 80 km (50 mi) of the Pacific Coast. As the coastal forests are harvested and converted to young growth, the habitats are made less favorable for some forms of wildlife, and more favorable for others. The same urban pressures that are causing conversion of farmlands are also causing conversion of forest lands to subdivisions and vacation homes, reducing the open space habitats for wildlife *and* humans.

Grassland conversion is also often detrimental to wildlife and wildlife habitats (Buttery and Shields 1975). The ranchers on the Hawaiian Islands are notorious for converting natural diversified coastal grasslands to artificial monocultures (e.g., pangola grass) for meat production. The fences that are erected to "keep in" the cattle also "keep out" the public, thereby reducing recreational opportunities in open space.

Inland waterway conversion is another cause of loss of wildlife habitat and open space. Dredging to deepen channels for navigation and filling to form solid land for construction has accounted for much of this loss. In Southern Louisiana, many wetlands have been lost to an intricate system of levees (earth and concrete mounds), floodwalls, steel floodgates, and pumps. There are 211 km (131 mi) of levees in New Orleans alone—all to protect this city located 1.8 m (6 ft) below sea level. In 1992, when Hurricane Andrew headed toward Louisiana, officials closed all 111 steel floodgates that make the levees solid against floodwaters.

Urbanization, in general, has taken a heavy toll on wildlife habitats, and consequently, open space. Urbanization means people, and people mean industrial parks, shopping centers, housing developments, and sanitary landfills—all of which need space.

The filling of San Francisco Bay is a classic example of urbanization. Since the 1850s, the Bay has been filled to provide space for various urban-related uses: harbor facilities, salt evaporation ponds, sanitary landfills, airports, industrial parks, and housing developments. In addition, the bay was indirectly filled by the heavy loads of silt sent down by the hydraulic mining for gold in the Sierra Nevada Mountains between 1850 and 1884. Since 1850, the total bay system (water areas and marshlands) has shrunk in size by approximately *30 percent.* (See Figure 8-7). If it were not for the creative ideas and determination of local conservation leaders, the filling of San Francisco Bay would not have been halted.

In the 1990s, urbanization continues to press into coastal wildlife habitats. In 1992, for example, a Huntington Beach developer proposed to build 4,884 homes in and around the Bolsa Chica Wetlands in Southern California. (See Figure 8-8). An environmental impact report for the project found that local automobile traffic would increase, the houses planned would be built on a seismic fault zone, the view along the Pacific Coast Highway would be marred, and some bird habitats would be displaced. The report was not all negative, however, it noted that the developer planned to dedicate 314 ha (775 acres) of "degraded wetlands" for restoration. Although leaving the land undeveloped was deemed an environmentally superior alternative, the Environmental Impact Report (EIR) maintained the project would expedite the costly process of restoration at a time when other sources of funding were running dry. Over 50 federal, state, and city agencies and groups have voiced an opinion about the project, including local

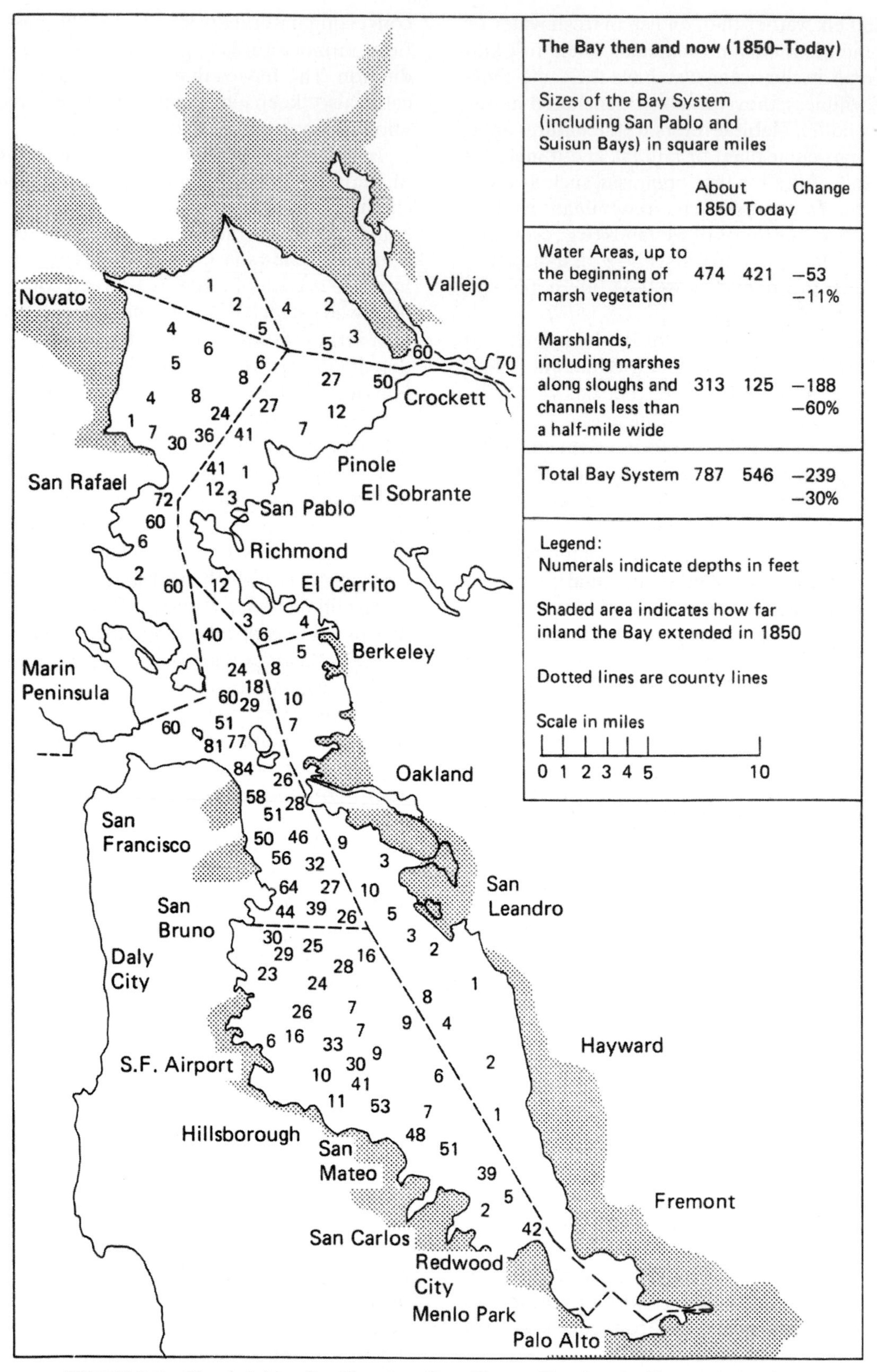

The Bay then and now (1850–Today)

Sizes of the Bay System (including San Pablo and Suisun Bays) in square miles

	About 1850	Today	Change
Water Areas, up to the beginning of marsh vegetation	474	421	−53 −11%
Marshlands, including marshes along sloughs and channels less than a half-mile wide	313	125	−188 −60%
Total Bay System	787	546	−239 −30%

FIGURE 8-7 The shrinking San Francisco Bay. (*Source:* Reprinted with permission from the San Francisco Examiner. Copyright 1989 *The San Francisco Examiner*)

FIGURE 8-8 Coastal wetlands such as Bolsa Chica Ecological Reserve provide numerous recreational opportunities, such as walking, hiking, and birding. On the other side of Highway 1, cars and RVs (Recreational Vehicles) line Huntington Beach. (*Source:* Photo by author)

conservation groups such as Amigos de Bolsa Chica. Despite the highly complex political, environmental, and developmental issues, projects such as this are generally permitted eventually, and thus—bit by bit, acre by acre—coastal wildlife habitats are lost to urbanization. For details regarding the Bolsa Chica Project, see Orange County Environmental Management Agency (1994).

Need for wildlife habitat preservation. It has been argued above that preserving wildlife habitat for wildlife also preserves open space for humans. Whereas humans may need estuaries, lagoons, marshes, and various other forms of coastal open space for their psychological well being (as well as for the economic riches they bring), it should be obvious that wild animals need abundant and quality habitat for an even more critical reason—*daily survival.* The habitat of a wild animal provides five major components: (1) cover (shelter); (2) food (nutrition and nourishment); (3) water; (4) home range (area habitually traveled); and (5) territory (area habitually defended).

It should go without saying that life forms other than our own deserve our respect and protection. The best way to protect wild animals is to protect their habitat. One of the most ambitious (and controversial) attempts to rescue and protect wildlife habitat is occurring on the Florida coast. In May 1994, Florida Governor Lawton Chiles signed an unprecedented piece of legislation—the *Everglades Forever Act.* This act, brokered by Interior Secretary Bruce Babbitt, effectively ended a mammoth six-year lawsuit brought by the federal government which charged the state of Florida of failing to protect America's greatest marshlands from the polluted waters of South Florida's vast sugarcane fields.

The Everglades region does not get enough water, or water at the right time. When it does get water, it is often polluted by the phosphorus fertilizers used by sugar and vegetable farms to the north. (See Figure 8-9). Phosphorus sets off an aquatic chain reaction by accelerating fecundity among bacteria and algae, encouraging the growth of cattails that choke the marshes. (See Figure 8-10). Thousands of acres have been taken over by cattails in the past decade, particularly in the Loxahatchee refuge and water conservation areas to the north of Everglades National Park.

The heart of the *Everglades Forever Act* is a restoration plan for 16,188 ha (40,000 acres) of newly created filtering marshes, which are being designed to "absorb" fertilizers from the farm runoff that is choking the Everglades. (See Figure 8-11). The plan also calls for the sugar industry to begin immediately cleaning up polluted water by using less fertilizer and reusing the water. Even the most optimistic schedule does not have the construction of the filtering marshes completed before the year 2003—one of the reasons that local environmentalists are highly critical of the project. Furthermore, critics charge that the *Everglades Forever Act* is a giveaway to the sugar industry that will delay for decades meaningful cleanup of the ecologically imperiled Everglades National Park and Loxahatchee National Wildlife Refuge. Only time will tell whether this unprecedented action to save wildlife habitat was a serious attempt at ecological rescue, or whether it spelled "a death sentence for the Everglades," as some local environmental advocacy groups contend.

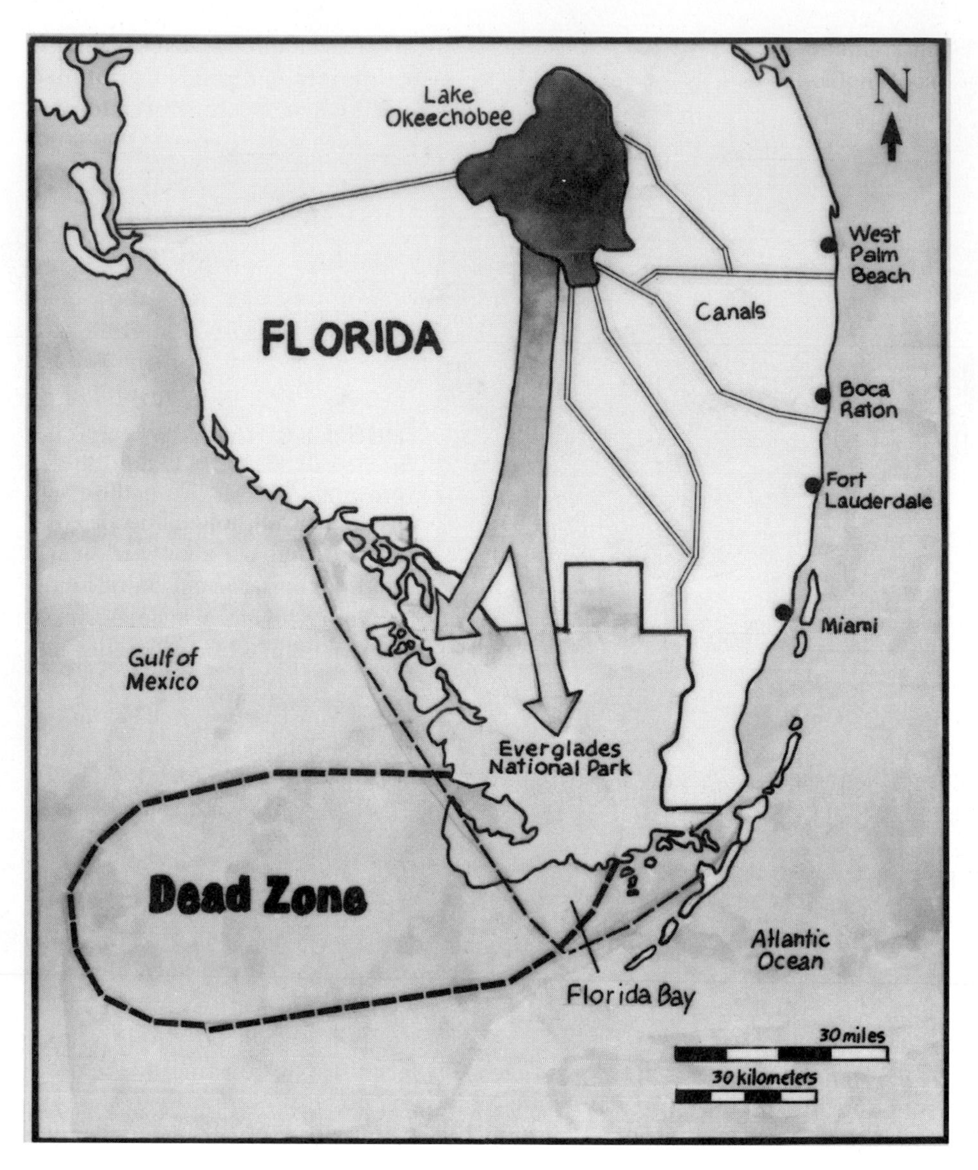

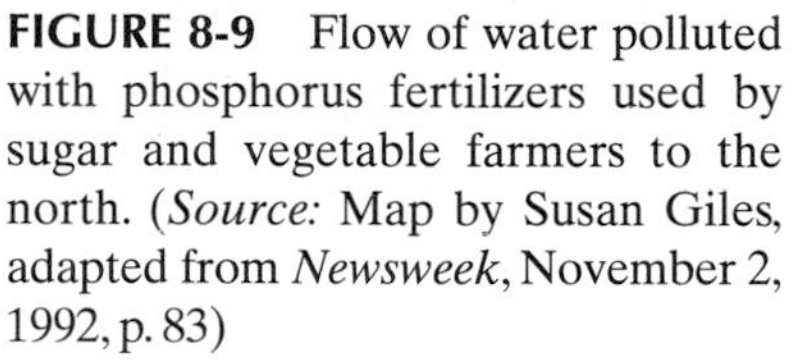

FIGURE 8-9 Flow of water polluted with phosphorus fertilizers used by sugar and vegetable farmers to the north. (*Source:* Map by Susan Giles, adapted from *Newsweek*, November 2, 1992, p. 83)

FIGURE 8-10 Some areas of the Florida Everglades state-managed conservation areas are being clogged by dense stands of cattails grown lush on fertilizer runoff from upstream farms (see "state-managed conservation areas" on map in Fig. 8-11 for specific location). (*Source:* Drawing by Heather Theurer)

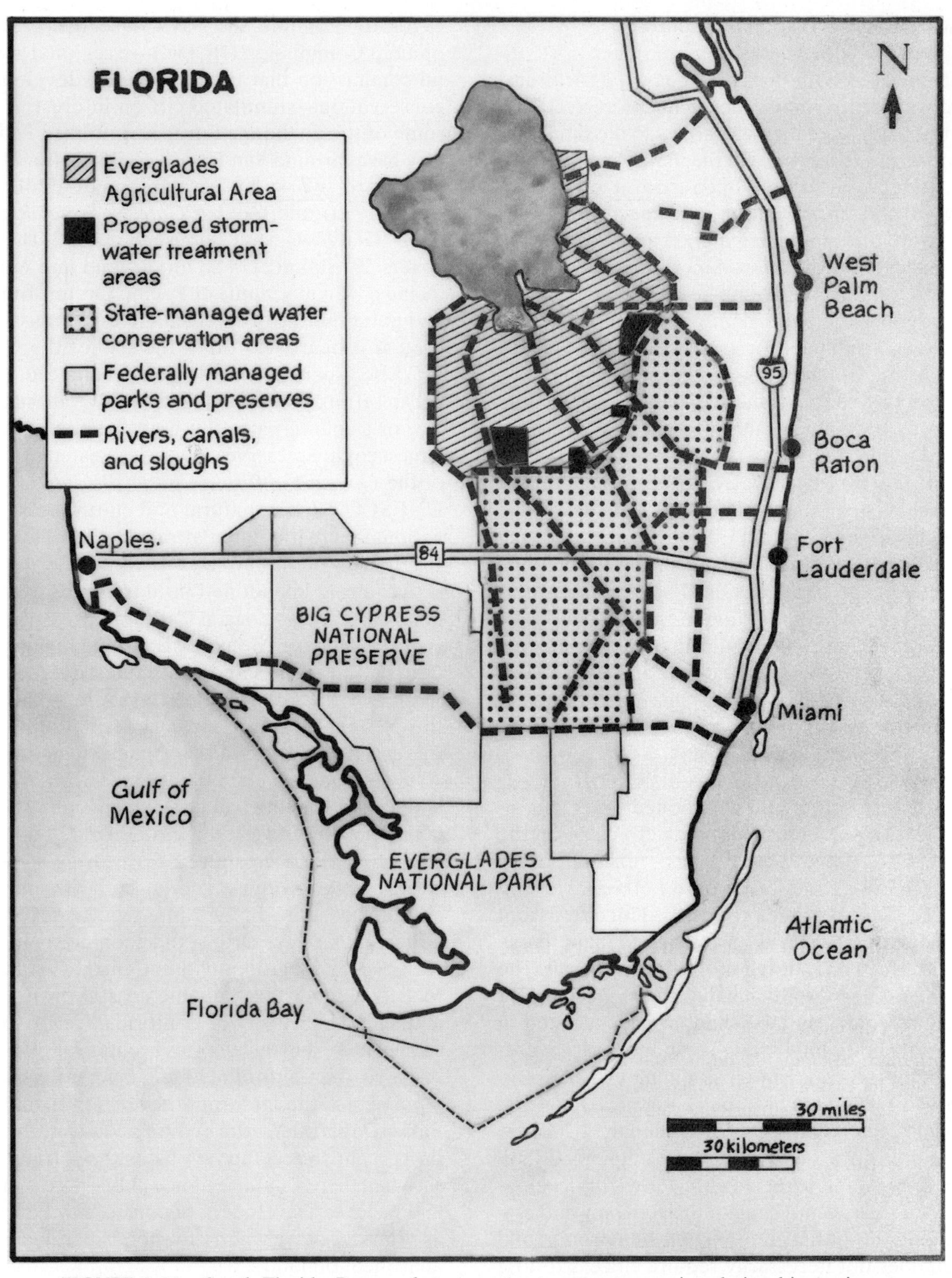

FIGURE 8-11 South Florida. Proposed storm water treatment areas in relationship to rivers, canals, and sloughs. (*Source:* Map by Susan Giles, adapted from *The New York Times,* Law Section, Friday, May 1, 1992)

Declining Coastal Recreational Resources

Closely associated with declining coastal agricultural lands and wildlife habitats is the decline of recreational opportunities. It is recreation and access to the shore that is generally first and foremost in the minds of Americans when they think about coastal open space; it is not maintaining good soil or protecting wild species. With increased leisure time and a heightened awareness of the environment due to education, Americans are demanding access to marshes, bluffs, and headlands for walking, nature study, and the enjoyment of coastal landscapes and shorelines. Swimmers, surfers, skin and scuba divers, and general "sun worshippers" increasingly need access to clean beaches; sailors, fishermen, and other boaters need marinas and launch ramps. Rapid coastal urbanization and other competing uses (e.g., heavy industry and power generation) have reduced the amount of shoreline that is open to the public.

There are various ways that U.S. states are now trying to increase or at least "hold on" to existing recreational access. Some of the strategies used are linking waterfront parks (e.g., Michigan), amending tideland permits (e.g., Massachusetts), acquiring access easements (e.g., California, Connecticut, Rhode Island, New York), encouraging early acquisition programs (e.g., Massachusetts), and converting abandoned railroad lines to shoreline recreation trails (e.g., California).

In the 1970s, for example, Detroit's waterfront was lined with deteriorating and abandoned ports and industrial facilities as well as empty lots. To clean up the waterfront and improve recreation and shoreline access, the city began in 1978 implementing a *Linked Riverfront Parks Master Plan*—a strategy that called for a riverfront park of linear, narrow parks connecting to major parks along a 16 km (10 mi) stretch of the Detroit River. The $82,000 invested by the Michigan Coastal Program for the waterfront plan, by 1988, had already resulted in $37 million of additional federal, state, and local government funds for the park as well as stimulating $210 million of private housing, office, and commercial retail and recreational developments, not to mention the 1,200 new jobs in the area (U.S. Department of Commerce 1988). The point here is that protecting, restoring, or developing new open space can stimulate the economy and create new jobs. To put it another way, environmental protection does not necessarily require citizens to lose their jobs. To the contrary, a clean environment is a *requirement* for most businesses to flourish.

Central Californians have used the techniques of acquiring both under-utilized or abandoned military bases and railroad lines to increase recreational opportunities and shoreline access for the general public. Back in the 1970s, for example, the Bay Conservation and Development Commission (BCDC)—a regional governmental commission that tries to balance development and conservation—stimulated citizen interest in acquiring some of the abandoned or under utilized Federal military bases around San Francisco Bay. After negotiation and hard work by Congressman Philip Burton (1926–1983), the *Golden Gate National Recreational Area (GGNRA)* was established in 1972. The GGNRA covers 295 sq km (114 sq mi) of land and water, which includes 45 km (28 mi) of Pacific Ocean, Tomales Bay, and San Francisco Bay coastline. In terms of area covered, it is nearly 2.5 times the size of the City of San Francisco, making it the largest area in the world in an urban setting that is managed by the National Park Service or a counterpart. Representing one of the nation's largest coastal preserves, it was designated in 1989 as part of the *Central California Coastal Biosphere Reserve* by UNESCO. Rich in natural and cultural resources, this national recreation area attracted 17 million visitors in 1991, making it the most heavily visited U.S. national park.

Acquiring abandoned military lands is also on the minds of Central Coastal Californians—specifically the Fort Ord lands on Monterey Bay. (See Figure 8-12). Fort Ord, on sand dunes and rocky hills covered with grass, scrub and oak, is located on the shore of the Monterey Bay National Marine Sanctuary, the nation's largest marine sanctuary. Fort Ord was a major staging area for U.S. Army units deployed to the Pacific during World War II, and a basic training and deployment center for soldiers in the Vietnam and Middle Eastern Wars. For years, there was no public access along Fort Ord's 9.7 km (6 mi) stretch of sandy dunes and beach between Highway 1 and the coastline. In 1992, the federal government announced that Fort Ord would soon close. Immediately various interest groups within Central California began to salivate over this valuable coastal property. For example, the University of California System expressed a strong interest in developing a center for marine science research, the California State University System sought a portion of the land for a new campus, and the State Parks Department worked on acquiring the shoreline property to protect the view along the corridor and make sure that there would be ample public access to the dunes and beaches. Developers, of course, saw Fort Ord lands as prime coastal property for hotels, condominiums, and apartment buildings.

On July 8, 1994, a Land Conveyance Ceremony on the Parade Grounds of Fort Ord completed the first transfer of Fort Ord lands to civilian use. The federal government gave 526 ha (1,300 acres) of Fort Ord lands to the state of California—land and buildings valued at more than $1 billion—to become California State University/

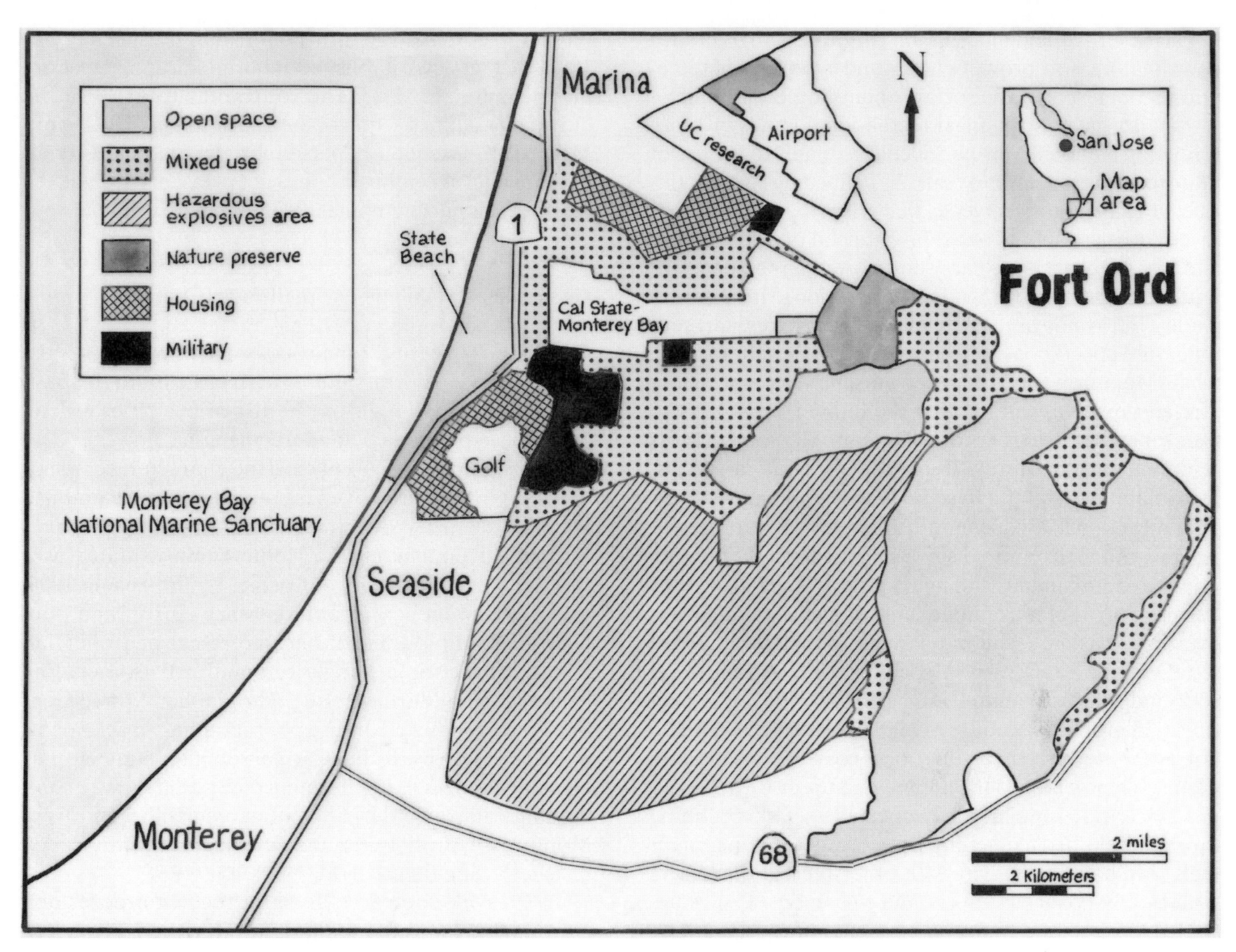

FIGURE 8-12 Fort Ord, California. Map illustrates former Fort Ord lands and future plans for the area. (*Source:* Map by Susan Giles, adapted from *San Jose Mercury News,* Friday, April 25,1997, p. 4B)

Monterey Bay (CSUMB). The Monterey Bay region boasts more than a dozen scientific research institutions, with which the CSU system envisions alliances. Classes began in the fall, 1995, and CSUMB had its first graduating class in May 1997.

Lands were also transferred to the University of California system for its planned science and technology research center. The whole notion is to replace the military in the region's economy with education—"from swords to plowshares." The California Department of Parks and Recreation also acquired major areas of dune habitat for recreation and restoration purposes. Of Fort Ord's 28,000 acres, approximately three-quarters have been allocated for open space. In May 1997, local concern over water supply and traffic has temporarily delayed the approval of the master plan for redeveloping the sprawling former Army base.

Also along the shores of the Monterey Bay National Marine Sanctuary, efforts have been underway to take an abandoned Southern Pacific Railroad corridor in Monterey County and convert it to a shoreline recreation trail. The project began when developers proposed that the abandoned corridor be broken into individual oceanfront properties. Finding this proposal unacceptable, the California Coastal Commission used the federal consistency provisions of the Coastal Zone Management Act (CZMA) to enable its purchase as a recreation trail. A 6.4 km (4 mi) regional recreation trail now runs from the City of Seaside down to Lover's Point in Pacific Grove. The biker or hiker can now travel completely separated from vehicular traffic through sand dunes, a Monarch Butterfly forest, along Monterey's waterfront harbor, the Presidio shoreline, Steinbeck's Cannery Row, the world famous Monterey Bay Aquarium and onto Lover's Point in Pacific Grove, and much more. This trail, with its branch coordinators, now provides hours and even days of recreational open space. The social benefits are enormous.

One of the controversial recreational issues in Monterey Bay National Marine Sanctuary relates to the use of **motorized personal water craft**. As defined by NOAA, this refers to any motorized vessel that is less than 4.6 m (15 ft) in length, is capable of exceeding a speed of fifteen knots, and has the capacity to carry not more than the operator and one other person while in operation. The term includes, but is not limited to, jet skis, wet bikes, surf jets, miniature speed boats, air boats, and hovercraft. In 1992, when Monterey Bay first received sanctuary status, motorized personal water craft were limited by NOAA to four zones off the harbors of Pillar Point, Santa Cruz, Moss Landing, and Monterey. These zones encompassed approximately 41 sq km (16 sq mi) in all. The management plan intentionally excluded thrill craft from surf breaks, beach swim areas, and sensitive kelp beds, thus affording safety to both humans and marine life. In the fall of 1993, despite overwhelming public support, this regulation was overturned as the result of a lawsuit brought by the Personal Watercraft Industry Association (PWIA). A district court judge in Washington, DC ruled that NOAA had not followed proper procedure in regulating only jet-skis and not other motorized vessels. Local coastal conservation groups, such as Save our Shores, were upset with the ruling, since they regularly get reports of jet skis running through kelp beds, speeding near or over marine mammals, and having conflicts with other recreational users such as sailors, surfers, divers, and swimmers. NOAA appealed the case, and won, but the agency must now decide on how to best manage thrill craft within the sanctuary. In 1997, the agency established its original four "offshore playpens for jet skis." The four areas total 39 sq km (15 sq mi), a tiny fraction of the 13,757 sq km (5,312 sq mi) MBNMS. Consequently, personal water craft industry officials and jet ski enthusiasts are not pleased. Time will tell whether this current management strategy will hold.

Loss of Coastal Village or Small Town Character

One of the concerns of many coastal communities is the loss of their village or small town character. In the past, this has not been high on the list of priorities among open space planners. Times are changing, however, and preserving that **sense of place** (the character, personality, or regional distinctiveness of a place) is becoming a greater issue.

A case in point along the Monterey Bay National Marine Sanctuary is the city of Carmel-by-the-Sea. In 1904, under the influence of author Mary Austin and poet George Sterling, Carmel became a sort of Bohemian colony, attracting other writers and artists, including painter Xavier Martinez and photographer Arnold Genthe. World renowned nature photographer Ansel Adams also lived close by. Today, Carmel retains a number of its unusual features (e.g., charming older homes and quaint boutique shops, winding and hilly streets with wind-swept Monterey cypress trees, and fine-grained white sandy beaches). (See Figure 8-13).

In Carmel, one can still sense the sea breezes and glowing fireplaces, the antique furniture and weathered

FIGURE 8-13 Quaint cottages by the sea help maintain "sense of place," Carmel, California. (*Source:* Photo by author)

hardcover books, the ancient oriental rugs and faded oil paintings, and, of course, the overall landscape beauty and social wealth of the community. Geographers and urban planners would say that Carmel has sense of place, sense of community, and is designed at a **human scale**—a comfortable balance between humans and nature. To the old-timers of Carmel, however, the small-town character and tranquillity of the area began to break down in the mid-1980s when local governing officials began to take on a more prodevelopment stance. In their opinion, buildings began to be erected that were out of scale—too big, too obtrusive, not fitting a small town character.

Carmel is but one coastal community that is trying to hang on to its small town or village character. Along the shores of the Monterey Bay National Marine Sanctuary, the towns of Big Sur, Monterey, Pacific Grove, Pescadero, Capitola, and Half Moon Bay are facing the same struggle. Immediately after the 1989 Loma Prieta Earthquake nearly flattened downtown Santa Cruz, the forces of "visual convergence" were at work. **Visual convergence** is a term used by cultural geographers to sum up all those forces in society that tend to "homogenize" a landscape—to make all towns and cities look alike. Similar technologies, similar materials, and similar architectural designs have a tendency to lead to similar looking strip development (the "franchised landscape"), similar shopping malls, and similar looking high-rises.

In summer 1994, when the people of Santa Cruz turned down a proposal by a well-known hamburger chain to build one of their easily recognizable franchise restaurants in downtown Santa Cruz, the citizens of this coastal town were resisting this latest move toward visual convergence. Whether it is Santa Cruz resisting unwanted development pressures and restoring Victorian cottages (Hyman 1993), or the citizens of Rockport, Massachusetts, preserving or restoring their weathered saltbox cottages and fishing shanties; or the citizens of small coastal towns in Maine fighting to preserve their New England townscape, they are all battling against "the forces of visual convergence" to maintain a sense of place. Analyzing strategies to conserve sense of place is a worthwhile and fascinating goal for any student of coastal resource management.

Declining Respect for Coastal Property

One of the reasons for walking barefoot through the surf, hiking on a shoreline bluff, observing birds in a coastal marsh, or even strolling down a Monterey cypress-lined street in Carmel, is to witness and experience the wonders of nature and human-designed landscapes without them being marred by some of the vulgarities of human civilization. In other words, the "open space adventure" should be an experience in beauty (both nature and human created) and relaxation—a period in which one can break the "nine to five," the daily grind (and grime?), of normal urban living. The symbols of the city—traffic, congestion, air and noise pollution, litter, and most recently, graffiti and vandalism—are invading our coastal towns, parks, and shorelines.

In 1992, the graffiti problem became so pervasive in the coastal community of Santa Cruz on the Monterey Bay that a task force was created—an umbrella organization that included law enforcement officials, chamber of commerce representatives, and citizen volunteers. The task force encouraged local citizens to keep their eyes peeled for graffiti painters, to testify in court cases against the vandals, and to volunteer to paint over graffiti in the community. When vandals were caught, the local judges required that they clean up or pay for removal of their handiwork and do volunteer community service. Some coastal communities, like Capitola on the Monterey shore, have a city ordinance against graffiti vandalism.

Other cities, like Gilroy in Santa Clara County, California, have gone so far as to post *bounties* on graffiti vandals. In 1992, for example, Gilroy Mayor Don Gage posted a $400.00 bounty to be paid to anyone who helped police identify and convict culprits who defaced public and private property with graffiti. The problem is so pervasive in Gilroy and other communities that people convicted of such things as traffic offenses must serve their term by being assigned to a graffiti cleanup detail. Since the graffiti painters are more often than not members of urban gangs, any real solution to the problem must deal with the bigger, more socially difficult issue—the role of gangs, the role of gang territorial markings, and so on.

Some surfers on the shores of the Monterey Bay National Marine Sanctuary are marking out their "surfing gang territories" with spray paint on the cliffs, boulders, gunnite, and riprap. Although graffiti does not have the seriousness of "Love Canal" or "Chernobyl," it is nevertheless a very serious *environmental problem* that must be addressed. We first saw the graffiti and other forms of "transported urban blight" in our national parks, and now the problem has spread to our coastlines. Graffiti is *not* an art form, as some defenders have claimed. Nor is it something that we should ignore or become "numb" to. Fortunately, Capitola, Santa Cruz, and many other coastal communities on Monterey Bay (and across the nation) are beginning to take the problem seriously.

Let's now turn for a closer look at Elkhorn Slough—one of the richest estuarine habitats on the West Coast. Although not an example of an area with a serious vandalism problem, Elkhorn Slough provides an interesting

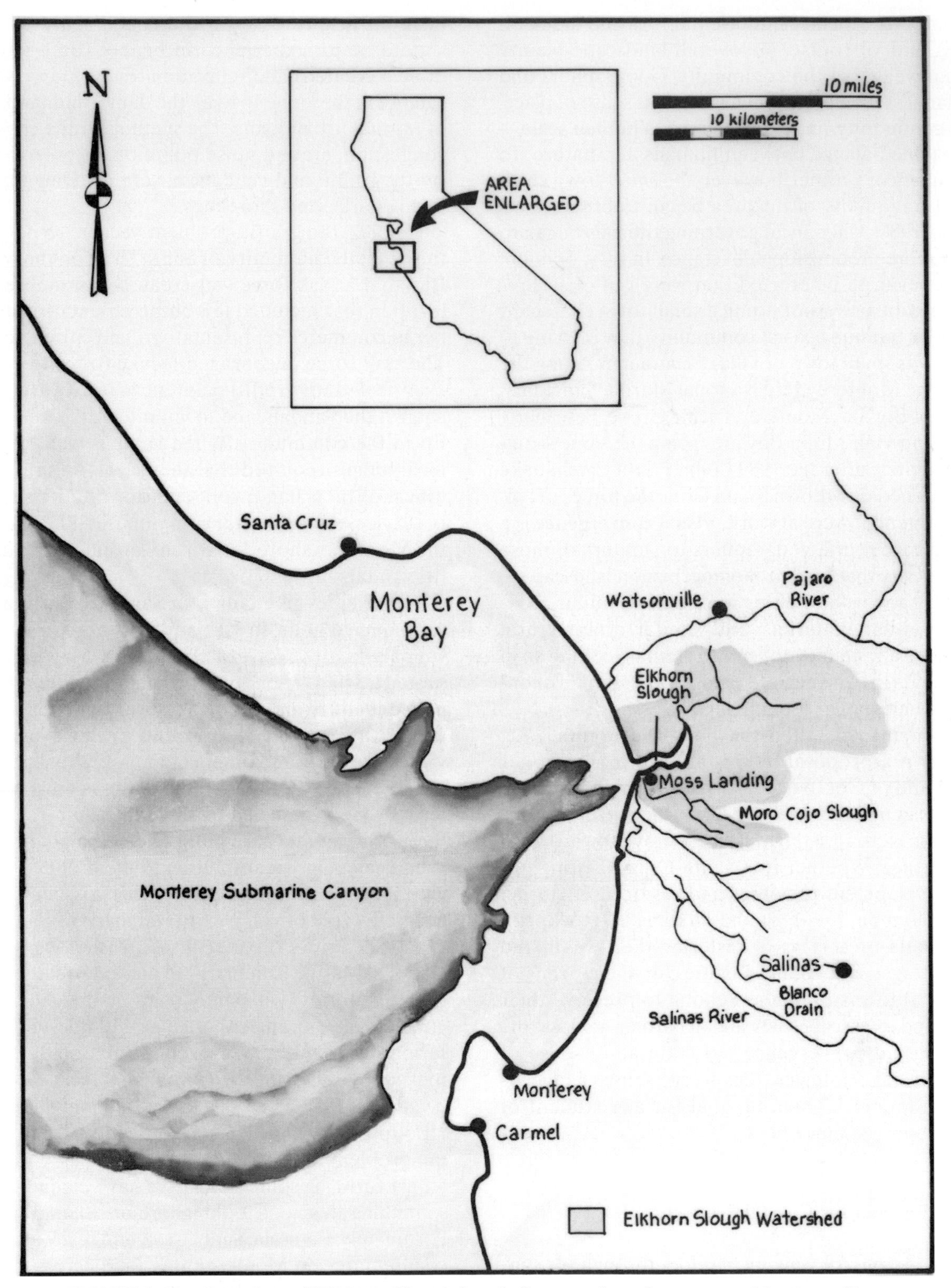

FIGURE 8-14 Location of Elkhorn Slough in relation to submarine canyon, Monterey Bay, California. (*Source:* Map by Susan Giles)

FIGURE 8-15 Aerial photograph of Elkhorn Slough, California, with Monterey Bay in background. Note sediment plume coming out of mouth of Elkhorn Slough. (*Source:* Courtesy of Kenton Parker, Elkhorn Slough Reserve)

case study of some of the other major problems facing open space managers.

CASE STUDY: ELKHORN SLOUGH

Elkhorn Slough is one of the most important wetlands in California. It is located on the middle of the curve of Monterey Bay, and, consequently, is ecologically linked to the Monterey Bay National Marine Sanctuary. (See Figure 8-14, previous page). This slough is the largest coastal wetland between San Francisco Bay to the north and Morro Bay to the south.

Aerial photographs of the wetland area reveal a mix of land types and human uses. (See Figure 8-15). Paramount, of course, is the main channel which meanders for over 11.2 km (7 mi). It is flanked by rolling hills with native coast live oak and exotic eucalyptus groves as well as agricultural fields and old dairy barns; the channel's tributaries (tidal creeks) are surrounded by mud flats and salt marshes with boardwalks for walking and bird observation. In 1998, several marsh restoration projects were in progress. Industrial plants, a harbor, a boatyard, research facilities, and a variety of small businesses are at the mouth of the estuary.

Elkhorn Slough is important enough to be included in the *National Estuarine Research Reserve System (NERRS).* (See Figure 8-16). This system of estuarine sanctuaries was established by the United States Congress as part of the Coastal Zone Management Act of 1972. Seven years later, in 1979, 364 ha (900 acres) of Elkhorn Slough received designation as an Estuarine Sanctuary (since renamed "National Estuarine Research Reserve"). Designated a California "State Ecological Reserve" as well, the reserve has grown to 567 ha (1,400 acres). In February 1998, through the efforts of conservationists with the financial help from the David and Lucile Packard Foundation, a pristine, 172 ha (425acre) wildland in the hills above Elkhorn Slough was purchased. According to the Elkhorn Slough Foundation, the Big Sur Land Trust, and other

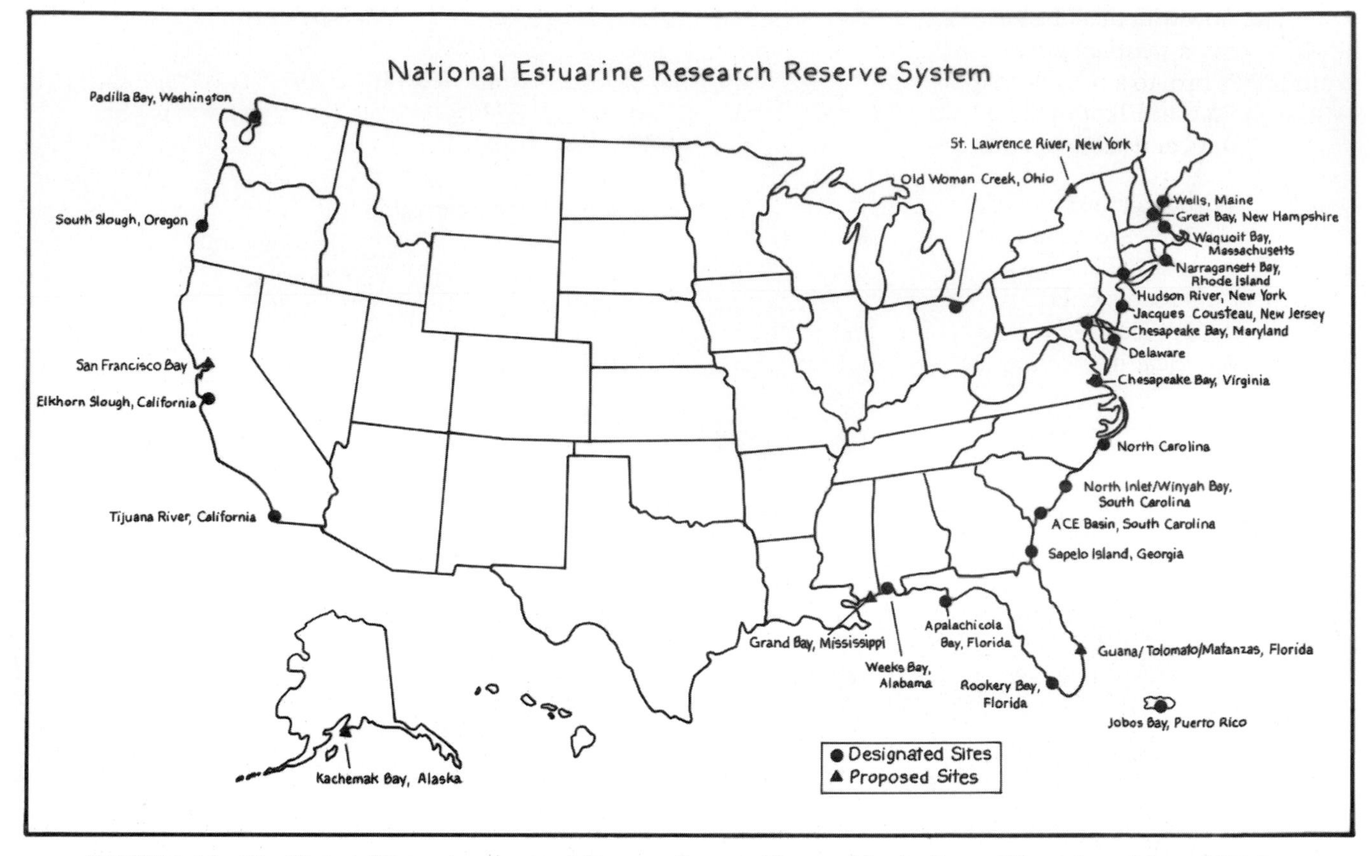

FIGURE 8-16 The National Estuarine Research Reserve System. (*Source:* Map by Susan Giles, adapted from *1996 Program Overview for the National Marine Sanctuaries: National Estuarine Research Reserves.* Silver Springs, Maryland: National Oceanic and Atmospheric Administration, p. 4)

principal players involved, it is hoped that this site will eventually be annexed to the Elkhorn Slough Reserve, further adding to the reserve's total acreage.

As it does for the National Marine Sanctuary Program, the National Oceanic and Atmospheric Administration (NOAA) has the responsibility for designating estuarine reserves and administering the system. NOAA's goal is to establish, manage, and protect these estuarine reserves from pollution and the pressures of coastal development. Long-term environmental monitoring and scientific research as well as environmental education (e.g., school programs, guided tours, citizen workshops) are integral functions of the reserves. Although additional sites are under consideration to expand the system, Elkhorn Slough is one of only twenty-two reserves across the nation and Puerto Rico. As of February 1996, the NERRS collectively managed 178,068 ha (440,000 acres) in 18 states and Puerto Rico for environmental protection, research, and education. At the time of this writing six reserves were in the process of designation.

The management and enhancement of any open space like Elkhorn Slough cannot be understood without first having an in-depth knowledge of its geological and human history. For the purpose of this chapter, however, a brief mention of the natural and cultural resources of the Elkhorn Slough must suffice. Much of the information for this case study is drawn from two excellent works: ABA Consultants (1989) and Silberstein and Campbell (1989).

Natural Habitats

What is the difference between a wetland, an estuary, and a slough? The general public uses all three terms interchangeably when referring to Elkhorn Slough. Technically, however, Elkhorn Slough is a **wetland** that is a "full-time **slough**" and a "part-time **estuary**" (Silberstein and Campbell 1989). It is a wetland because it fits this broad term that includes many types of water/land environments. It is only a part-time estuary because the slough does not have a year round freshwater supply. By contrast, San Francisco Bay is a "full-time estuary" because it has the continual freshwater flow from the San Joaquin and Sacramento rivers. Because it only receives freshwater flow with the infusion of winter rains, Elkhorn Slough is only a *seasonal estuary.* With those definitions in mind, let's turn to a brief description of the four major natural habitats at Elkhorn Slough—the uplands, waterways, mud flats, and salt marshes.

Uplands. Although the *uplands* are technically not part of the slough, they are nevertheless included since the two are firmly interlinked—water runs off the hills, bringing nutrients and sediments to the slough. The waterways of the slough provide food for many upland animals and moderate the local microclimate. Elkhorn Slough has several upland habitats, including rolling grassy hills with coast live oaks *(Quercus agrifolia),* dense stands of tough chaparral plants (e.g., manzanita [*Arctostaphylos* spp.], coyote bush [*Baccharis pilularis*], and California lilac [*Ceanothus* spp.]), and even *sand dunes* that parallel the shoreline. Also within the uplands of Elkhorn Slough are tall stands of eucalyptus trees—brought from Australia to California as potential lumber trees in the mid-1800s. (See Figure 8-17).

The upland zone provides resting, escape, and nesting cover for various estuarine birds, such as kestrels (*Falco sparverius*), red-tailed hawks (*Buteo jamaicensis*),

(a)

(b)

FIGURE 8-17 Upland Habitats of Elkhorn Slough, California. (a) Oak-grassland hills; (b) Chaparral plants;

(c)

FIGURE 8-17 (Continued) (c) Sand dunes; (d) Eucalyptus groves. (*Source:* Photos by author)

(d)

and California quail (*Lophortyx californicus*). A number of roosting birds utilize the eucalyptus groves and Monterey Pine trees, such as great blue herons (*Ardea herodias*), cormorants (*Phalacrocorax* spp.) and great egrets (*Casmerodius albus*). Hummingbirds, bees, and other nectar-feeders also use these non-native tree species. It is here in the slough's upland areas that one can find such mammals as coyote (*Canis latrans*), red fox (*Vulpes fulva*), and deer (*Odocoileius hemionus*).

Waterways. The second major natural habitat in Elkhorn Slough is its *waterways*—the main channel and tidal creeks. (See Figure 8-18). At its mouth, the main channel is relatively deep—approximately 7.6 m (25 ft) deep, and 213 m (700 ft) across. There is water at all times in the channel, even when the tide is at its lowest. Tidal action provides oxygen-rich water and nutrients for the myriad of invertebrate animals that ultimately become food for fish, birds, and mammals, as well as a nourishing fishery resource for humans.

The upper and lower slough waters and associated plants and animals are quite different. Waters in the upper slough are saltier during summer months and fresher during the winter season than lower slough waters. Consequently, this region attracts "backwater" flora and fauna. Here, the great blue heron (*Ardea herodias*) and the snowy egret (*Egretta thula*) stalk fish along the shallow water at the channel's edge. By contrast, the lower slough waters—toward the mouth of the slough—are more heavily influenced by the sea, and attract "oceanic species." On the main channel near the shore, for example, predominant bird species include brown pelicans (*Pelecanus occidentalis*), cormorants (*Phalacrocorax* spp.), and loons

FIGURE 8-18 Waterways of Elkhorn Slough, California. (a) Waterways habitat and pickleweed; (b) Pelican in flight above waterways; (c) Amtrak passenger train traveling along one of the many dikes across the estuarine waterways. (*Source:* Photos by author)

(*Gavia* spp.). The two most conspicuous mammals swimming in the harbor mouth are sea otters (*Enhydra lutris*) and California sea lions (*Zalophus californianus*).

Overall, the waterways of Elkhorn Slough act as a nursery for a variety of fish found in Monterey Bay. Typical fishes within the waterways include the leopard shark (*Triakis semifasciata*), bat ray (*Myliobatis californica*), and several schools of midwater swimmers, including northern anchovies (*Engraulis mordax*) and Pacific sardines (*Sardinops sagax*).

Mud flats. The *mud flats* constitute the third major natural habitat at Elkhorn Slough. (See Figure 8-19). At low tide, both sides of the main channel are flanked by **mud flats,** formed by the layering of fine sediments washed down from the uplands. Some of this mud has been deposited over thousands of years, and can be as much as 6 m (20 ft) deep (Silberstein and Campbell 1989). From a distance, the mud flats look barren of life. A close inspection, however, will reveal an underworld habitat full of living organisms, including ghost shrimp (*Callianassa californiensis*), mud crabs (*Hemigrapsus oregonensis*), fat innkeeper worms (*Urechis caupo*), and gaper clams (*Tresus nuttallii*). When the tide is out, flocks of shorebirds can be seen feeding on the exposed mud flats. Predominate bird species that feed on the flats include avocets (*Recurvirostra americana*), marbled godwits (*Limosa fedoa*) and black-bellied plovers (*Pluvialis squatarola*). Harbor seals (*Phoca vitulina*) also regularly haul out on the low-tide mud flats.

Salt marshes. The fourth and final major natural habitat at Elkhorn Slough are the *salt marshes.* (See Figure 8-20). As the deposition of fine silt continues from the main channel and nearby creeks, some of the mud flats build in elevation and evolve into a **salt marsh**—a former mud flat that has risen until it is inundated by saltwater only at high tide. When the level of the "young" marsh has been raised sufficiently, it is colonized by **halophytes**—plants that can withstand periodic inundation of saltwater. At Elkhorn Slough, the dominant halophyte on this higher ground is the succulent pickleweed (*Salicornia virginica*). It thrives uncontested in the lower half of the slough. On the upper half of the slough, there is a transition area between marsh and uplands where the pickleweed must also compete with other halophytes, such as saltbush (*Atriplex* spp.), salt grass (*Distichlis spicata*), and alkali heath (*Frankenia grandifolia*). For some unknown reason, cordgrass (*Spartina* spp.), a prominent plant in East Coast marshes and in San Francisco Bay, is not found in Elkhorn Slough (Silberstein and Campbell 1989).

Cultural History

Humans have had a long and complex history with Elkhorn Slough. This history can be divided into three periods of human occupancy—the Coastanoan Period, the Reclamation Period, and the Harbor Period to Present.

Coastanoan period. During the Coastanoan Period (beginning around 6,000 years ago), the Coastanoan (coastal dwellers) group of Indians (also called Ohlone, pronounced "oh-*lone*-nee") fed heavily on the local marine populations at Elkhorn Slough, but they appear to have made no lasting long-term impact on the number or distribution of marine and coastal species within the slough

FIGURE 8-19 Birds feeding in mudflats, Elkhorn Slough, California. A major landmark in the area is the twin towers of Duke Energy (formerly the Pacific Gas & Electric power plant.) (*Source:* Photo by author)

FIGURE 8-20 Salt marsh of Elkhorn Slough, California. (*Source:* Photo by author)

itself (Dondero 1984). Geographer Burton Gordon (1987), however, has noted that the Coastanoans *did* impact the landscape around the slough, primarily by using fire to maintain open grasslands where they hunted deer, elk, and other large game. Today, the only readily observable cultural remnants from this period are the numerous Indian middens (refuse heaps) around the slough, and a nearby Indian cemetery. The middens are filled with everything from the shells of clams and other marine invertebrates to the bones of mammals and birds.

When the Spanish began settling missions in the 1700s, the traditional Indian communities and their way of living gently with the land began to break down. The Spanish introduced cattle that overgrazed the oak-covered hills and carried foreign seeds on their hoofs to the soil. From that point on, the landscape of Elkhorn Slough began to change dramatically.

Reclamation period. Whereas conservation, preservation, and restoration may be the ideologies and buzzwords of coastal resource managers today, "reclamation" was the key approach to managing the land back in the 1800s when American settlers moved into coastal central California. Today, the term **reclamation** has many meanings, but then, it meant "taking-back" from nature a piece of perceived wasteland (e.g., coastal wetland) and improving upon it for *human use* (e.g., the drainage of a wetland for agriculture). Chinese immigrants were the first settlers to do just that in the 1880s and 1890s (Lydon 1985). They diked, ditched, and drained many of the wetlands around the slough for agriculture, especially for sugar beet production. Americans expanded the reclamation process, much of it for cattle grazing, extracting salt, or for duck hunting (Dickert and Tuttle 1980).

As wetlands were converted to crop or grazing lands, wildlife habitats were lost. According to ABA Consultants (1989), 90 percent of the coastal wetlands in the Monterey Bay area were lost during the Reclamation Period. The mid-1800s were also a period when loggers harvested trees from the hillsides, whalers hunted migrating whales off Moss Landing, and Charles Moss hoped to start his own town by building shipping facilities at the mouth of Elkhorn Slough.

This was also the period when the mouth of the Salinas River was purposefully *diverted* to protect wetlands that had been reclaimed for agriculture. In the mid 1870s, just prior to the invasion of Chinese immigrants and American "reclaimers," the Salinas River merged with Elkhorn Slough near its mouth before they both drained into Monterey Bay. Around 1908, the area's farmers made a new mouth for the Salinas River 8 km (5 mi) south of Moss Landing. (See Figure 8-21).

This action, of course, caused severe changes in the slough's hydrology and wildlife habitats. Without the Salinas River bringing freshwater to Elkhorn Slough, the slough became a *tidal embayment*—an area saltier and less estuarine in character than before. Furthermore, with less freshwater in the area, and increasing pumping of groundwater, the level of the groundwater table in and around Moss Landing fell dramatically. This naturally led to increased saltwater intrusion problems. As the composition of the wetland vegetation changed, it is estimated that millions of waterfowl and shorebirds

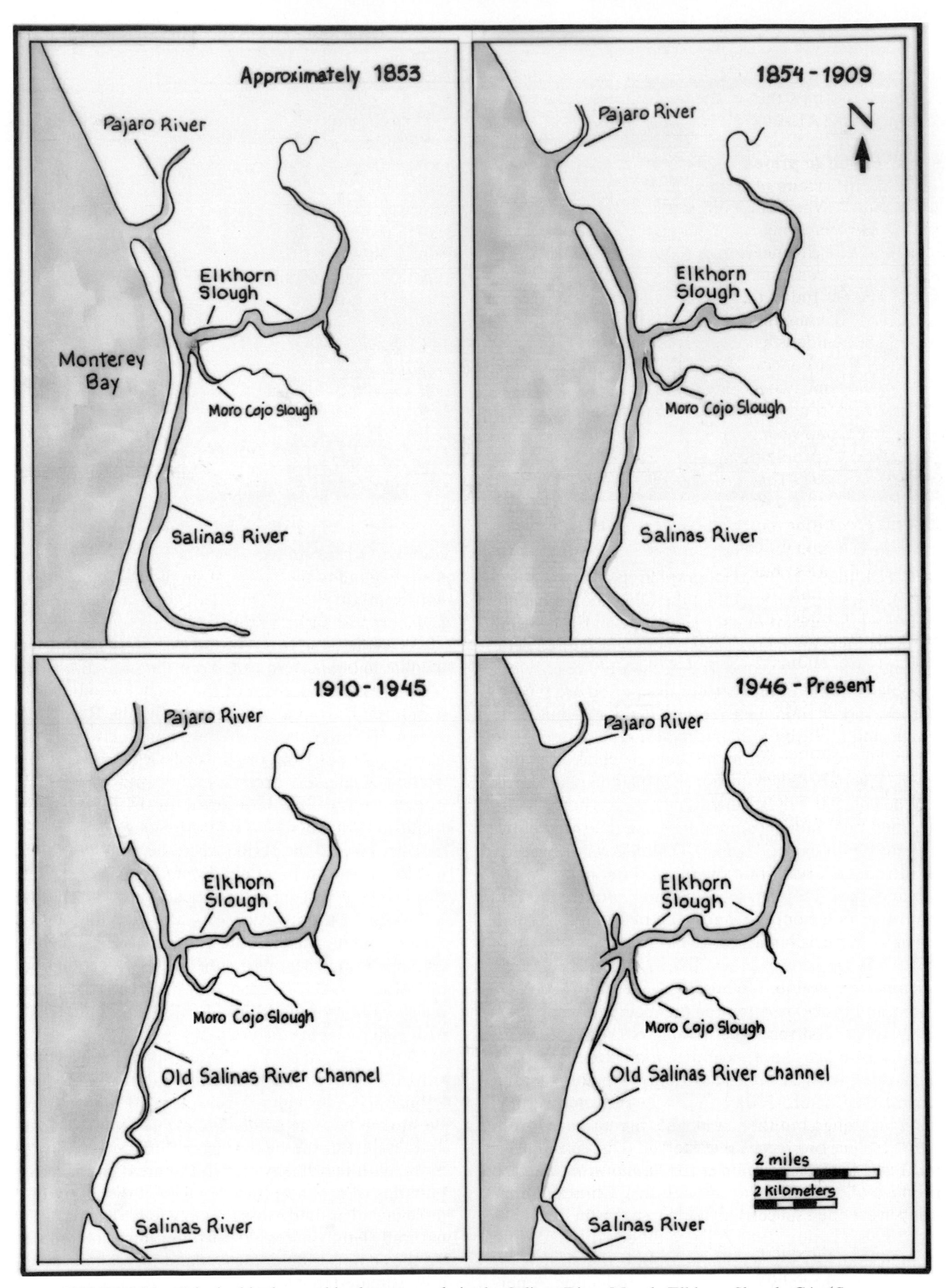

FIGURE 8-21 Principal hydrographic changes made in the Salinas River Mouth, Elkhorn Slough, CA. (*Source:* Map by Susan Giles, adapted from Schwartz 1983, p. 69, and Gordan 1987, p. 235.)

disappeared, as well as the salmon, steelhead, and introduced striped bass that normally used the slough and local wetlands (ABA Consultants 1989).

Harbor period to present. The Harbor Period begins with the construction of Moss Landing Harbor between 1946 and 1947. Whereas Elkhorn Slough originally flowed north before emptying into the bay, the Army Corps of Engineers cut a channel straight across the coastal dunes to the bay, then constructed berths for boats on both sides of the channel. Today, the harbor has slips for over 600 commercial fishing and pleasure boats. Naturally, this caused a tremendous hydrologic and environmental impact. Prior to harbor construction, Elkhorn slough was a shallow, quiet embayment. After the harbor entrance was constructed, strong tidal currents began to impact the slough and it is now considered to be the most important environmental problem facing Elkhorn Slough (ABA Consultants 1989). (This "tidal scouring" problem will be discussed in further detail below).

It was also during the 1940s that heavy industry began to come to the slough. Kaiser Industries (now called National Refractories) built a factory at Moss Landing to extract magnesium from sea water; the plant uses the magnesium to make the heat-resistant bricks required for high-temperature ovens. In the 1950s, Moss Landing's highly visible landmark—the twin 152 km (500 ft) tall stacks of the Pacific Gas and Electric Company were built. In the mid-1960s, research facilities began to arrive, such as the California State University's Moss Landing Marine Laboratories (which were destroyed by the 1989 Loma Prieta Earthquake, and are now being rebuilt in a nearby location) and the Monterey Bay Aquarium Research Institute (MBARI) facilities.

Environmental Concerns and Management Strategies

The history of Elkhorn Slough has been one of land manipulation, especially the "3Ds"—Diking, Dredging, and Diverting. As alluded to above, this land manipulation has brought about profound environmental changes to slough hydrology, to the percentage of wetland habitat for wildlife, and of course, to the number and types of wildlife species. ABA Consultants (1989) identified four major environmental issues that are being addressed in the Elkhorn Slough area: (a) erosion and sedimentation, (b) water quality, (c) wetland enhancement, and (d) public access. Of these four, the two biggest problems in Elkhorn Slough are clearly erosion and sedimentation, and water quality. We will now briefly look at these two problems and how local coastal open space managers are trying to cope.

Erosion and sedimentation. *Erosion problems* come in various forms at Elkhorn Slough. *Erosion of wetland habitats* due to tidal scouring is perhaps the biggest management problem. **Tidal scouring** can be defined as the washing away of sediments by strong tidal currents. When Moss Landing Harbor was built in 1946, the land-water interface at the site was geomorphically changed. A negative consequence of that change has been increased tidal scouring. The twice daily strong tidal currents have transformed the slough mouth that was once 1.2 m (4 ft) deep and 6 m (20 ft) wide to its present depth of 9.1 m (30 ft) and width of 91 m (300 ft). With a wider and deeper slough mouth, the tidal scouring is reaching further up the main channel, eroding banks and marshlands. (See Figure 8-22).

Elkhorn Slough managers must also contend with *erosion of agricultural lands* (e.g., the strawberry fields that surround the slough), *erosion within large gullies* (e.g., the steep vegetated slopes above the west slough, and "*erosion blight*" (e.g., the visual ugliness of erosion scars and unvegetated fans). (See Figure 8-23).

Sedimentation problems, of course, are often the direct result of erosion. At Elkhorn Slough, major sediment deposition problems have occurred in the form of *deposition in waterbodies* (e.g., Carneros Creek, as well as the small freshwater ponds above the slough), *deposition in marshes* (e.g., the pickleweed marsh along the west slough), and *deposition along public and private roads* (e.g., Elkhorn Road). Dealing with these erosion and sedimentation problems is not an easy task.

In 1998, local officials were still evaluating the various management strategies recommended by ABA consultants, the U.S. Natural Resources Conservation Service, and a more recent tidal hydraulics erosion study by Philip Williams & Associates (1992). One major strategy has been recommended to help reduce the tidal scouring problem: the construction of a submerged rock sill. The strategy is to construct a *rock sill* (a type of erosion control structure) underneath the Highway 1 bridge near the slough's mouth. This, according to Philip Williams & Associates, would replicate the pre-harbor entrance channel conditions when the channel was more narrow and shallow. However, such a project would be extremely expensive—$3,650,000 (1992 dollars), plus the cost for feasibility studies, detailed modeling, and environmental review.

In addition to the high price tag that the harbor district could not afford, the "rock sill strategy" raised serious questions about its impact on slough hydraulics and ecology. Although the intent was to replicate pre-harbor entrance channel conditions, there were no guarantees that this would succeed. Basic questions were left unanswered, such as how would the sill alter the migration patterns of the fish and mammals that normally traverse up and down the main channel? According to many authorities involved, the construction of a sill was too expensive, and too environmentally radical an idea for the area concerned.

(a)

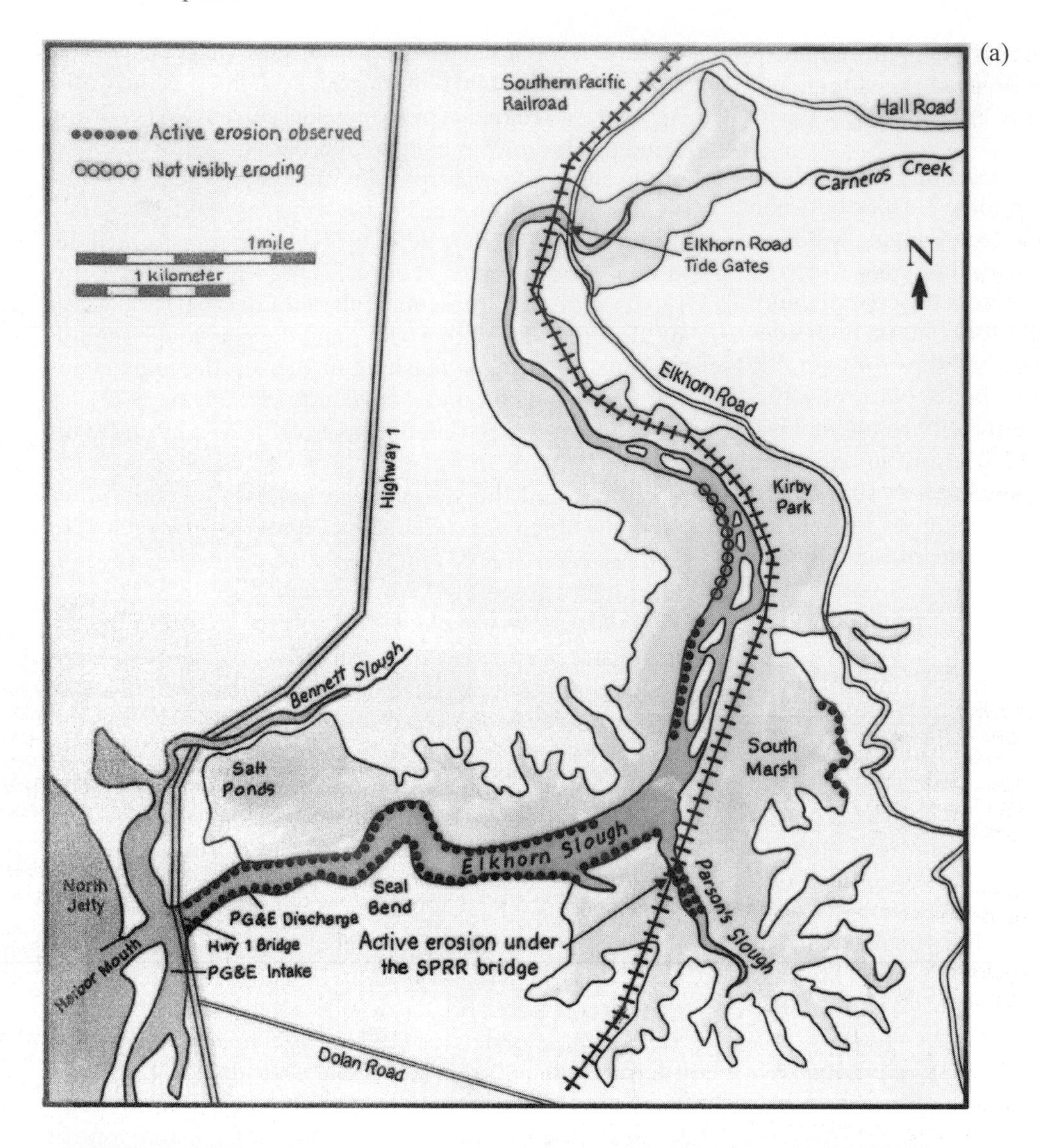

(b)

FIGURE 8-22 (a) Map: Areas of active erosion, positions of Pacific Railroad (formerly Southern Pacific Railroad), and locations in and around Elkhorn Slough, California. (*Source:* Map by Susan Giles, adapted from ABA Consultants 1989, p. 15); (b) Active bank erosion by tidal scouring in Elkhorn Slough. (*Source:* Photo by author)

(a)

(b)

FIGURE 8-23 Elkhorn Slough must deal with the problems of soil erosion and pesticide runoff from adjacent strawberry fields, Elkhorn Slough, California.; (a) Strawberry fields can be seen on slopes above the waterway; mud flats are visible in foreground. (b) Field laborers on way to weed strawberry fields; Elkhorn Slough waterway can be seen in the background. (*Source:* Photos by author)

The remaining erosion and sedimentation problems are controlled by managing runoff from surrounding agricultural fields. According to ABA Consultants (1989), the goal should be to reduce erosion to an acceptable level of 5 tons/acre/year. Techniques for retarding soil erosion on steep slopes are well known, such as *contour planting* and the use of surface and *underground drainage pipes* to collect and transport excess water to catchment basins. The problem, however, is that Elkhorn Slough managers are not in control of the surrounding farmlands; the farms are operated by independent landowners or leasees. Consequently, the above soil conservation techniques are all done voluntarily.

Various strategies were considered to convince local farmers to participate, including *financial incentives* (e.g., an improved federal cost-sharing program), "*arm-twisting*" (e.g., better enforcement of existing soil erosion ordinances), *ordinance revision* (e.g., revising the Erosion Control Ordinance with an emphasis on wetland protection), *changes in land use* (e.g., converting land use designation from "agriculture" to "low-density rural housing"), and *environmental education* (i.e., educating local farmers about both the economics and environmental consequences of soil loss). Naturally, each of these strategies had a long list of advantages and disadvantages. In 1994, only a few of the farmers participated in soil conservation techniques because of the high capital costs, limited public financing, and hard economic times. By 1998, however, there were signs that real progress was being made with local farmers, particularly through an outreach program in environmental education instigated by researchers working in conjunction with the staffs at Elkhorn Slough Reserve and U. S. Natural Resources Conservation Service. For further information, see Mountjoy 1993.

Water quality. Elkhorn Slough has both surface and groundwater problems. In terms of surface waters, the biggest problems are the high levels of persistent insecticides (e.g., *DDT, dieldrin, and toxaphene*), and the herbicide *dacthal.* The residues of the above mentioned pesticides are old, since those compounds have not been used for some time. Additional restrictions on their use would be of little or no benefit in reducing their presence in the slough. According to Elkhorn Slough scientists, the long-term solution is to keep the soil on the hillsides and out of the waterways.

Surprisingly, most of these pesticides are not coming from the strawberry fields that surround the slough (i.e., the immediate slough watershed), but rather from the Salinas River which carries pesticide-laden runoff from the major agricultural center of Monterey County. A sand bar quite often forms at the new mouth of the Salinas River, thereby forcing the current to flow into the Old Salinas River channel up into Moss Landing Harbor and eventually into Elkhorn Slough (Refer back to Fig. 8-21). For the short term, ABA Consultants (1989) has recommended an "alternative drainage plan" to reduce the transport of Salinas River pesticides into the slough. In other words, the area's residents are once again considering a new "plumbing design" for Elkhorn Slough.

The long-term answer to future types of pesticide contamination is to either reduce pesticide use, substitute more environmentally-friendly pesticides, or find alternatives to pesticides. Fortunately, research is already underway. Before the development of synthetic pesticides, farmers used such *cultural controls* as crop rotation, strip cropping, field fallowing (using strips of uncultivated land), and debris removal (removing potential breeding places for insects) to curtail insect infestation. All of the above cultural controls modify the crop ecosystem to hinder pest species. Some farmers are now taking a second look at those familiar practices. Other alternatives are *biological control* (the reintroduction or development of new natural predators, parasites, and pathogens to combat specific insects), *attractants* (the use of sex attracting chemicals, light, or sound to lure pests into toxic traps or to upset their mating process), *sterilization* (the use of radiation or chemicals to sterilize the male of the insect pest species), *hormones* (the use of insect hormones to prevent pest insects from reaching maturity or reproducing), and *resistant crop varieties* (the breeding of plants that are resistant to diseases, insects, and fungi). Each technique, of course, has its own set of advantages and disadvantages in terms of pest control, economic cost, and environmental impact.

Integrated pest management (IPM) is a control program that combines cultural, biological, and chemical methods in a specified sequence. It emphasizes the control of pests rather than their total elimination, the use of careful field monitoring, and the use of different pesticides at different times to avoid the emergence of genetic resistance. Dozens of successful control programs illustrate that IPM can control pests without the use of conventional insecticides, while maintaining high yields at reduced costs.

Interfacing with Sanctuary Management

A number of agencies and organizaions oversee the management of Elkhorn Slough; including (1) NOAA, which also administers the adjacent Monterey Bay National Marine Sanctuary; (2) the California Department of Fish & Game (CDFG); (3) The Elkhorn Slough Foundation (ESF), (4) The Reserve Advisory Committee (RAC); (5) The Elkhorn Slough Interpretetive Guide Association (ESIGA); (6) The Nature Conservancy; and (7) The Big Sur Land Trust. Since Elkhorn Slough Reserve is part of NOAA's National Estuarine Research Reserve System, it already interacts with NOAA officials in the management of the slough. However, when NOAA designated a Monterey Bay National Marine Sanctuary at the mouth of Elkhorn Slough, it became necessary for "estuarine managers" to interface on a regular basis with "marine sanctuary managers."

For example, NOAA officials administrating Monterey Bay National Marine Sanctuary developed a Water Quality Protection Program (WQPP), an interagency committee, to design a comprehensive program to enhance and protect the sanctuary's physical, biological, and chemical resources. The basis of the program is an ecosystem-based approach that addresses a number of issues that relate to sanctuary water quality. Urban runoff, agricultural activities, boating and marinas, point sources, and water management are issues of concern. The WQPP uses a knowledge-based, concensus process which brings together resource managers, scientists, businesses, and public groups to identify problems and develop effective solutions.

Further Information on Elkhorn Slough

Since Elkhorn Slough is a part of the NERRS, it has been (and continues to be) heavily researched. Hundreds of scientific and government reports, dozens of books, and several master's theses have been produced about this area. The Elkhorn Slough Foundation (1700 Elkhorn Road, Watsonville, CA 95076) maintains a computerized update of the "Elkhorn Slough Bibliography." This extensive document lists both published scientific and management papers and unpublished reports. Copies of these documents are available upon request for a small fee.

GENERAL POLICY RECOMMENDATIONS

As has been shown, preserving open space requires tackling a wide range of problems, including disappearing prime agricultural lands, loss of wildlife habitat, declining recreational resources, loss of village or small town character, and even public nuisances such as graffiti and vandalism. What follows is a set of general recommendations that will hopefully guide coastal communities as they strive to maintain (or restore) the quality of their open space.

Agricultural Lands

1. *Farmland preservation*
   - Americans want farmland protected. Although farmlands are being converted to nonagricultural uses *nationwide* at an alarming rate, a 1980 Harris Poll for the U.S. Department of Agriculture revealed that many Americans want prime farmlands protected. They rated the loss of good farmland as a "very serious" problem, and by a seven-to-one margin they were willing to accept federal action to protect farmlands from erosion. Of course, the landowners would probably feel quite differently about the prospect of federal intervention. Coastal farmlands are protected to a degree under various state coastal acts, but coastal farmland acreage continues to decline.
   - *Work toward eliminating loopholes and inequities in existing techniques for preserving farmland, such as preferential assessment, transfer of development rights, relief from inheritance and estate taxes, agricultural districting, the delineating of urban limit lines, and community land trusts.* These are innovative methods that have met with some success across the country, especially in Wisconsin, Oregon, California, Iowa, and Michigan. Yet they are too often marked by loopholes, inequities, and a general failure to preserve prime agricultural lands against the onslaught of urban encroachment. Further refinement of these existing techniques (as well as the development of new innovative techniques) is needed.
   - *Experiment with innovative ways to make farmlands more economically viable.* For example, some dairy farmers in Marin County, California, convert manure (a waste product of their dairy operation) into three resources: (a) an odorless solid they can use for cattle bedding material or as fertilizer, (b) an odorless liquid they can use as fertilizer and potential feed supplement, and (c) methane gas which can be used to generate electricity. The cow manure and barn washings are "digested" with the help of anaerobic bacteria in a methane digester bag. (See Figure 8-24).

     The dairy owners have shown that with a methane digester and a power generator, they can produce all the electricity they need for their dairy and household requirements. They can also sell surplus power to the local utility company, thereby earning a marginal profit. Not only does recycling manure (a "waste" product) help keep the farming operation solvent, it also helps keep those nutrients from washing down into coastal wetlands.

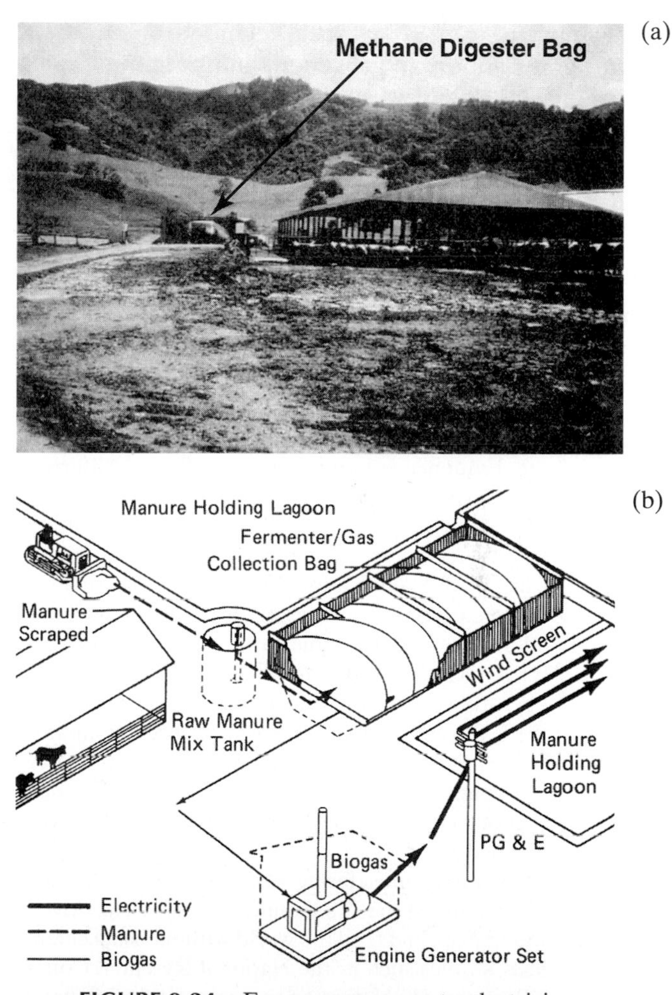

FIGURE 8-24 From cow manure to electricity. (a) A methane digester bag next to a milking barn on a Central California dairy farm. (Photo by author) (b) A diagram of Marindale Dairy's methane digester system. (*Source:* From *Northern California Sun,* vol. 10, no. 5 (September–October 1984), p. 13)

Coastal farmers in Marin County, California have used other strategies as well to keep their farm operations and agricultural lands intact. For an excellent summary of what these "farmers on the edge" did (and did not) do, see Hart (1991).

2. *Erosion and sedimentation control*
   - *Work closely with local soil and water conservation districts and the U.S. Natural Resources Conservation Service to design and implement comprehensive management plans for erosion control.* Several techniques are available for controlling soil erosion. "Engineering techniques" make use of tools and mechanical equipment. "Biological techniques" call for the manipulation of natural vegetation or domesticated crops. Combining the two techniques in a well-orchestrated management plan produces the best results.

3. *Pesticide control*
   - *Advocate the use of integrated pest management (IPM) on farmlands.* Agricultural pesticides are a serious threat to our coastal areas (Pait, De Souza, and Farrow 1992). Although no one has found a way to prevent pesticides from washing down into the soil and into our coastal waterways, integrated pest management promises to minimize the pesticide problem. Since 1950, several integrated programs have reduced the use of insecticides by 50 to 75 percent, while increasing crop yields and reducing overall costs.
   - *Encourage the movement towards alternative farming.* Evidence is mounting that suggests "alternative" farming, which incorporates certain features of IPM, may prove more efficient than farming techniques that use chemical pesticides and fertilizers. In 1989, the National Research Council—one of the most prestigious scientific bodies in the United States—reported that farmers who practice alternatives to high-input industrial agriculture are operating successfully in all the climatic regions of the U.S., that their yields per hectare are comparable to industrial agriculture, and that their negative environmental impacts are significantly less than that of conventional agriculture (National Research Council 1989).

     Although most farmers continue to embrace the old belief that the U.S. cannot increase yields, lower food prices, and feed the world without using chemicals, studies such as the National Research Council's report are beginning to turn the tide. Big firms ("the establishment") are even beginning to "get high" on organic farming. For example, Dole Foods Company, a subsidiary of Sunkist Growers and Castle & Cooke Incorporated have started to grow organic produce. The shift in farming techniques has clearly begun. It just needs your support. Request and buy organic when possible.

Wildlife Habitat

1. *Habitat acquisition and preservation*
   - *Acquire and preserve as many coastal wildlife habitats and seashores as possible.* If the encroachment of urbanization upon wildlife habitat is a major cause of loss of open space for humans (not to mention the major cause of wildlife extinction in America), then slowing the rate of habitat elimination is of prime importance in open space (and wildlife) preservation. Ownership, of course, may be either public or private.

     In terms of *public* ownership and administration, the U.S. National Park Service and the U.S. Fish and Wildlife Service generally receive the highest marks from environmental groups. In addition to managing our *national parks* (e.g., Yosemite National Park) and *national recreation areas* (e.g., Golden Gate National Recreation Area), the National Park Service also owns and manages *national seashores* and *national lakeshores.* These latter categories of protection were created as a response to the American love for shorelines, coasts, lakes, and sea.

     In 1992, the National Park Service administered 11 national seashores and four national lakeshores. Placing areas in "seashore parks" helps, but does not guarantee protection. For example, Cape Hatteras National Seashore is threatened by off-road vehicles that damage fragile ecosystems. Padre Island, the Gulf Islands, and Fort Jefferson National Seashores are experiencing visual pollution and threats of oil spills from offshore oil drilling. Natural processes, such as the distribution of beach sand, are being disturbed by coastal development. At Cape Lookout National Seashore in North Carolina, for example, the beach erosion is so severe that the park's historic lighthouse is being undermined.

     The U.S. Fish and Wildlife Service administers the *National Wildlife Refuge System*—a collection of lands and waters selected for their value to America's wildlife populations. Today, it includes over 450 refuges covering 36 million ha (90 million acres), many of which are on our nation's shores. In addition to protecting land for wildlife, these refuges provide open space and recreation for the general public (e.g., hiking, sightseeing, birdwatching, biking, and sometimes boating and swimming).

     In addition to the federal system, state, county, and city governments administer their own pieces of open space along America's coastlines.

     In terms of private ownership and administration of open space/wildlife habitats, the Nature

Conservancy, the Trust for Public Land, and the Izaak Walton League are very active in protecting lands for the public good. The Nature Conservancy is considered by many to have the best track record for land acquisition and consequent protection of ecosystem diversity. It is a national, nonprofit, membership-sponsored conservation organization. Over 1.5 million acres (660 preserve in 49 states, the Virgin Islands, Canada, and the Caribbean) have been preserved through the efforts of the Conservancy and its members since the organization was formed in 1950.

2. *Habitat improvement or restoration*
   - *Promote "habitat protection" over "habitat mitigation."* Several states have established mitigation measures to counter coastal wetland destruction. For example, a developer who wants to build on one unit of coastal wetlands in California must turn over four units of other wetlands to the state government. If they do not own other wetlands for exchange purposes, the developers must purchase other lands, convert them into wetlands, then release these newly created or restore wetlands to the state.

   The problem is that many artificial (human created) wetlands turn out to be *inferior* to the natural ones, especially in terms of biomass production and species richness (Race and Christie 1982). Consequently, until humans can create wetlands as rich and diverse as natural ones, it would behoove us to choose habitat protection programs over the promises of habitat mitigation. Reverting many former reclaimed farmlands back to wetlands is now a conservation strategy occurring in the Netherlands. When the Dutch plan is completed, about 10 percent of the country's farmlands will be restored to wetlands. (See Figure 8-25).

   For two recent studies about wetland protection or mitigation programs along the shores of the Monterey Bay National Marine Sanctuary, see Dyste 1995; and Zenk 1996.

(a)

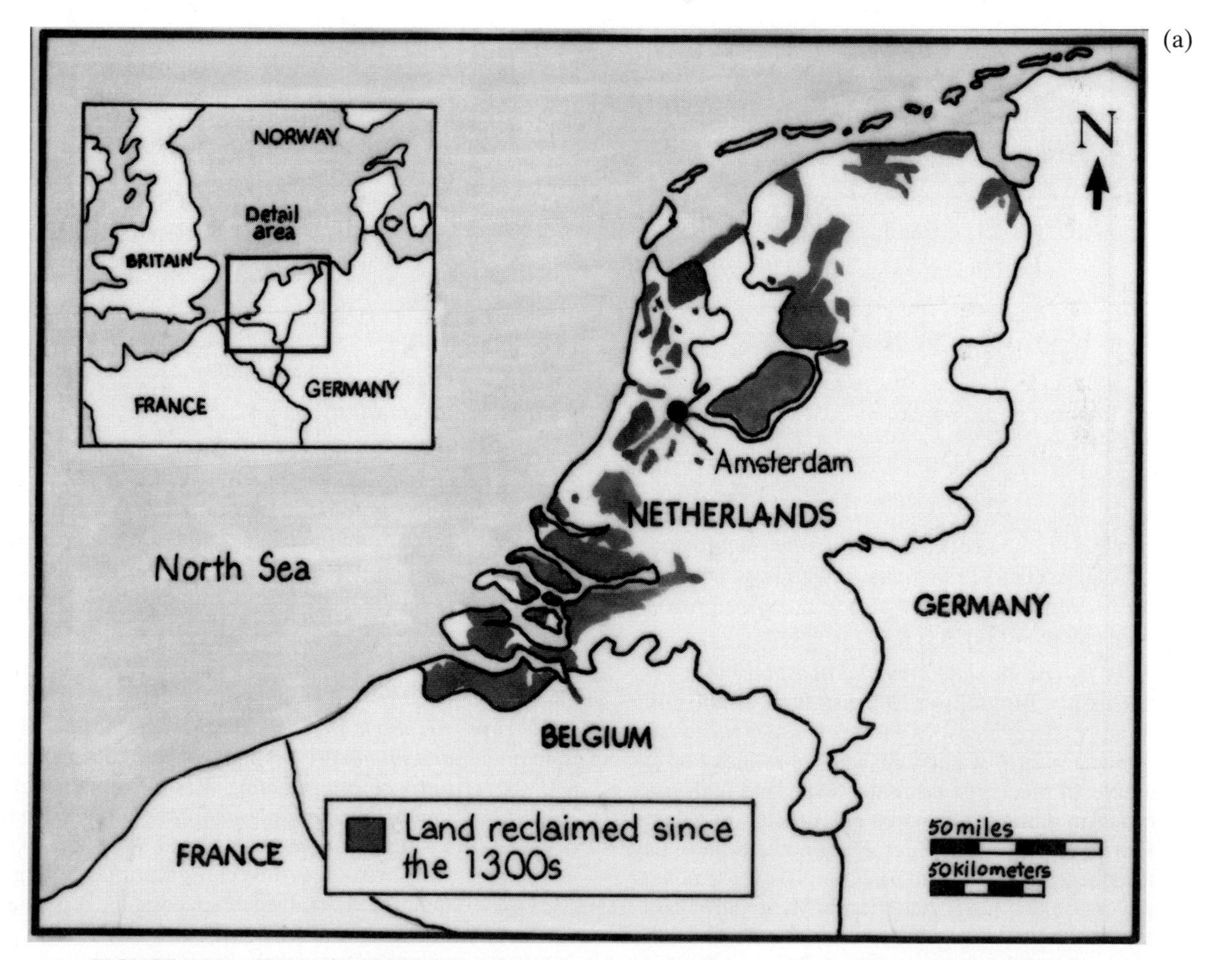

FIGURE 8-25 Reverting former reclaimed farmlands back to wetlands, Netherlands. (a) Land reclaimed since 1300s. (Continued)

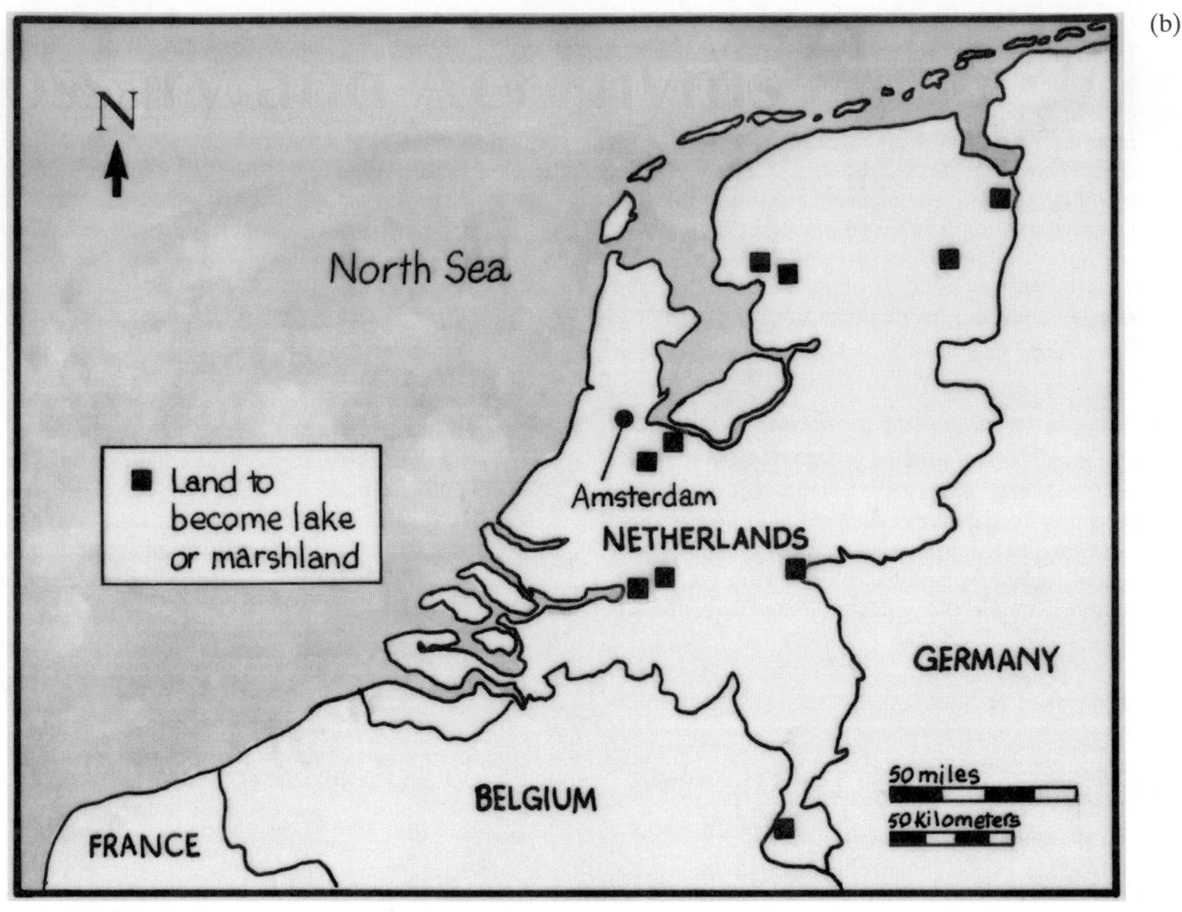

FIGURE 8-25 (Continued) (b) When the Dutch plan is complete, about 10 percent of the country's farmland will have been returned to nature. (*Source:* Maps by Susan Giles, adapted from *The New York Times,* Sunday Punch section, March 21, 1993)

Recreational Resources & Access

1. *Access*
   - *Legislate access.* Legislation is one major way of gaining public access to waterfronts. California's Coastal Act of 1976 was passed not only to help protect the environment but also to meet the growing public demand for access to the shore.
   - *Link federal or state financial assistance to access easements.* In addition to legislation, establishing "tradeoff agreements" is another strategy to increase public access to waterfronts and shorelines. For example, in the 1970s rising sea-levels and high risks of storm damage prompted city officials of Miami, Florida, to apply for federal assistance to help pay for a beach nourishment program. To overcome taxpayers objections, a condition of the federal assistance was that the public must have access to the restored beach (i.e., no public access = no financial assistance [Carter 1988]). By this strategy, the citizens of Miami gained another public beach.

2. *Water quality*
   - *Establish national bathing beach standards.* The country has no mandatory water quality standards for swimming in coastal waters. Water quality for recreational uses is regulated by the states, subject to federal oversight. In other words, criteria are only recommended (not mandated) by EPA for marine and freshwater bodies.

3. *Visitor impact*
   - *Respect human values as well as the integrity of flora and fauna.* Preserving a piece of open space for the purpose of protecting wildlife habitat is extremely important, if not paramount. However, ignoring the wishes of humans can trigger a public backlash that will threaten future conservation funding. Consequently, coastal open space managers must try to balance *biocentric* (plant and animal-centered) management with *anthropocentric* (human-centered) management. Of course, this is not always easy to do. For example, is allowing "dune-buggies" and other

forms of off-road vehicles in coastal dune "sacrifice areas" (zones designated strictly for use of off-road vehicles) a balanced approach to managing visitor impact? Many open space managers would argue "Yes!" Others would argue "Absolutely not!"

- *Manage According to the Carrying Capacity of the Area.* This is also no easy task, since **carrying capacity**—the amount of use that an area can tolerate without suffering an unacceptable impact—is an elusive term. The concept specifies no exact number, but rather a range of activity for which an area can be used. Moreover, carrying capacity varies from time to time. For example, during a drought, when the biological communities are experiencing physical stress, the carrying capacity of an area declines and human use should be curtailed.
- *Encourage "ecotourism," not "high impact" tourism.* If tourism is being promoted for a coastal area, then encourage what is increasingly being called "ecotourism." **Ecotourism** is nature-based tourism that stresses conservation of natural resources, while at the same time supporting sustainable economic progress. The emphasis is on protecting wildlife and ecosystems, on maintaining rural aesthetic character, on providing economic alternatives to resource extraction activities, and on gaining income for local communities.

Sense of Place

- *Identify and preserve those aspects of your coastal village, town, or city that give it a sense of seaside community.* Seaside communities all have one thing in common—that edge where land meets sea. At that edge, there are the physical components, such as the sandy beaches and dunes, the marshy lagoons and estuaries, the rugged cliffs and headlands, the salty air, sea winds and fog, and the barking harbor seals and noisy seagulls. There are also the cultural and historical components, such as the rickety wood-planked wharves, the thriving fishing, powerboat, and sailboat marinas, the old hotels and nineteenth century white church steeples, and so on. From a geographer's or environmental scientist's point of view, eliminating elements of a community that give sense of place *reduces quality of life*—it moves a community towards *"no-place USA."*
- *Resist (or at least seriously question) those elements that have a tendency to break down sense of place.* Carmel, California remains a quaint seaside community because the majority of its citizens fight to keep it that way. Although traffic is bumper to bumper during the summer tourist season, the citizens of Carmel refuse to blacktop one foot of land unnecessarily or to install parking meters. Except in the business district, few sidewalks or street lights grace Carmel. It would be simpler to destroy a building than a tree. Indeed, it takes City Council approval to touch so much as a limb. An ordinance dating from 1916—the year Carmel was founded—prohibits the cutting, mutilation or removal of trees or shrubbery on city property. Today, Carmel has a General Plan, Local Coastal Plan, and Architectural Plan that helps set the tone for town development—giving historic preservation and maintaining sense of place a high priority.

Simply put, question those who want to "widen and pave" for the convenience of the automobile culture. Question those that want to tear down historical buildings to construct shiny new ones that they maintain will be "more functional." Carmel is one of the loveliest (and wealthiest) seaside communities in the world because it has *resisted change,* it has hung onto its quaintness, it has preserved its sense of place. At the turn of the century, real-estate agents were selling lots in Carmel for $20. In 1992, "fixer-uppers" (if you could find one) started at $350,000. And, "choice" (high-quality) houses sold for over $4 million. In sum, preserving sense of place = environmental protection = cultural preservation = high financial return.

For additional examples of coastal communities and their efforts to maintain sense of place, see Petrillo and Grenell editors. (1985). For general strategies on maintaining sense of place, see Garnham (1985).

Respect for Property

- *Use a combination of techniques to address what might be called anti-social behavior.* There is no single technique or "quick-fix solution" to reducing such anti-social behavior as graffiti and vandalism, littering, trespassing, drunken and noisy behavior, and so on. Yet, these problems must be addressed because they clearly take away from the open space experience. Resolving these problems will require a multi-pronged approach, including *education* (e.g., environmental education in elementary and secondary schools); *psychological strategies* (e.g., strategically placed rubbish bins); *regular cleanup patrols* (e.g., quick cleanup of "tagging," "territorial markings," and other forms of graffiti [experience has shown that a clean environment encourages leaving it clean; a graffiti/litter filled environment encourages more of the same]); *legislation* (e.g., anti-graffiti ordinances); *strong enforcement* (e.g., judges and prosecutors that consider graffiti-painting a serious form of vandalism); *social analysis* (e.g., social workers and police officers working with gang members), *incentives* (e.g., bounties placed on graffiti vandals); *volunteerism* (e.g., merchants who use their own time and money to pick up litter or repaint graffiti-painted surfaces), and *further research* (e.g., there appears to be relatively little research on techniques to counter public nuisance behavior, especially graffiti vandalism).

CONCLUSION

The increasing pressure of our growing population threatens the diversity of our coastal environment. Coastal agricultural land, wildlife habitat, and recreational areas are being displaced by the sterile urban environment typical of suburban sprawl. In order to preserve coastal habitat, economic productivity and self-sufficiency of coastal areas, and quality living space, a mutually beneficial balance must be achieved between competing ecological and human defined interests.

As with most social issues, the solutions to our coastal open space management problems lie with people and their ability to communicate and work together toward a common goal. Only with a proactive, cooperative effort between the principal players will regional, long-range planning efforts ever succeed. Such an approach will be needed to improve resource conservation, maintain elements of "human scale," and assure good stewardship of the coastal and marine environment. These concepts (and more) will help move us *toward integrated coastal and marine sanctuary management,* our concluding chapter.

REFERENCES

ABA Consultants. 1989. *Elkhorn Slough Wetland Management Plan.* Capitola, California: ABA Consultants.

Ballantine, Kent. 1994. Criteria and Standards Division, Water Quality, EPA, personal communication, July 28.

Bandy, P. J., and R. D. Tabler. 1974. "Forest and Wildlife Management: Conflict and Coordination." In *Wildlife and Forest Management in the Pacific Northwest,* edited by H.C. Black. Corvallis, Oregon: School of Forestry, Oregon State University, pp. 21–26.

Berry, Wendell. 1978. *The Unsettling of America: Culture and Agriculture.* New York: Avon Books.

Buttery, Robert F., and Paul W. Shields. 1975. "Range Management Practices and Bird Habitat Values." In *Proc. Symp. Mgt. Forest and Range Habitats for Nongame Birds: Forest Service General Technical Report WO-1,* by U.S. Department of Agriculture. Washington, D.C.: U.S. Department of Agriculture, pp. 183–189.

Carter, R. W. G. 1988. *Coastal Environments.* New York: Academic Press.

Culliton, Thomas J., Maureen A. Warren, Timothy R. Goodspeed, Davida G. Remer, Carol M. Blackwell, and John J. McDonough III. 1990. *50 Years of Population Change Along the Nation's Coasts, 1960–2010.* Rockville, Maryland: National Oceanic and Atmospheric Administration.

Dahl, T. E., and C. E. Johnson. 1991. *Wetlands Status and Trends in the Conterminous United States Mid-1970s to mid-1980s.* Washington, D.C.: Fish and Wildlife Service, U.S. Department of Interior.

Dickert, T. G., and A. Tuttle. 1980. *Elkhorn Slough Watershed: Linking the Cumulative Impacts of Watershed Development to Coastal Wetlands.* Berkeley, California: Institute of Urban and Regional Development, University of California, Berkeley.

Dondero, S. 1984. *Preliminary Report on Archaeological Testing, CA-Mnt-229, Elkhorn Slough, California.* Sacramento, California: Office of Environmental Analysis, California Department of Transportation.

Dyste, Rosemary. 1995. *Wetland Buffers in the Monterey Bay Region: A Field Study of Function and Effectiveness.* M.S. Thesis. San José State University.

Garnham, Henry L. 1985. *Maintaining the Spirit of Place.* Mesa, Arizona: PDA Publishers Corporation.

Goldberg, Edward D. 1993. "Competitors for Coastal Ocean Space." *Oceanus.* Vol. 36, No. 1: pp. 12–18.

Gordon, Burton L. 1987. *Monterey Bay Area: Natural History and Cultural Imprints.* 2d ed. Pacific Grove, California: The Boxwood Press.

Gosselink, J. G., and R. H. Baumann. 1980. "Wetland Inventories: Wetland Loss Along the United States Coast," *Zeitschrifte für Geomorphologie. N.E. Supplement.* Vol. 34, pp. 173–187.

Gosselink, J. B., and E. Maltby. 1990. "Wetland losses and gains," in *Wetlands: A Threatened Landscape,* M. Williams, ed., Oxford, England: Basil Blackwell Ltd., pp. 296–322.

Gwyn, Thomas W. 1992. "The Colors of Green." *California Coast and Ocean.* Winter/Spring, Vol. 8, No. 1: pp. 6–12.

Hammitt, William E., and David N. Cole. 1987. *Wildland Recreation: Ecology and Management.* New York: Wiley.

Hart, John. 1991. *Farming on the Edge.* Berkeley, California: University of California Press.

Hyman, Rick. 1993. "Saving Carmelita Cottages." *California Coast and Ocean.* Winter/Spring, Vol. 9, No. 1–2: pp. 35–39.

Klee, Gary A. 1991. *Conservation of Natural Resources.* Englewood Cliffs, New Jersey: Prentice Hall.

Lydon, Sandy. 1985. *Chinese Gold: The Chinese in the Monterey Bay Region.* Capitola, California: Capitola Book Company.

Mitsch, William J., and James G. Gosselink. 1993. *Wetlands.* 2d ed. New York: Van Nostrand Reinhold.

Mohney, David and Keller Easterling, eds. 1991. *Seaside: Making a Town in America.* New York: Princeton Architectural Press, Inc.

Monroe, Michael W., Judy Kelly, and Nina Lisowski. 1992. *State of the Estuary: A Report on Conditions and Problems in the San Francisco Bay/Sacramento-San Joaquin Delta Estuary.* Oakland, California: Association of Bay Area Governments.

Monterey County Farm Bureau. 1988. *Food for Thought.* Salinas, California: Monterey County Farm Bureau.

Mountjoy, Daniel C. 1993. *Farming Practices Survey and Outreach Recommendations for the Elkhorn Slough Water Quality Management Plan: Final Report.* Monterey, California: Association of Monterey Bay Area Governments.

National Oceanic and Atmospheric Administration. 1994. *Workshop Summary: Preliminary Identification of Issues and Strategies.* National Ocean Service, October, Washington D.C.: National Oceanic and Atmospheric Administration.

National Research Council. 1989. *Alternative Agriculture.* Washington, D.C.: National Academy Press.

Orange County Environmental Management Agency. 1994. *Revised Draft Environmental Impact Report for the Bolsa Chica Project.* County Project Number 551, State Clearinghouse Number 93-071064. Santa Ana, California: Orange County Environmental Management Agency.

Pait, Anthony S., Alice E. De Souza, and Daniel R. G. Farrow. 1992. *Agricultural Pesticides in Coastal Areas: A National Summary.* Rockville, Maryland: National Oceanic and Atmospheric Administration.

Petrillo, Joseph E., and Peter Grenell, eds., (1985). *The Urban Edge: Where the City Meets the Sea.* Los Altos, California: The California State Coastal Conservancy & William Kaufmann, Inc.

Philip Williams & Associates. 1992. *Elkhorn Slough Tidal Hydraulics Erosion Study.* San Francisco: Philip Williams & Associates, Ltd.

Race, M. S. and Christie, D. R. 1982. "Coastal Zone Development: Mitigation, Marsh Creation, and Decision Making." *Environmental Management.* Vol. 6, No. 4: pp. 317–328.

Schwartz, David Lee. 1983. *Geologic History of Elkhorn Slough, Monterey County, California.* M.S. thesis, San José State University.

Silberstein, Mark, and Eileen Campbell. 1989. *Elkhorn Slough.* Monterey, California: Monterey Bay Aquarium.

Thompson, Edward, Jr., Ann Sorensen, John Harlan, and Richard Greene. 1994. *Farming on the Edge: A New Look at the Importance and Vulnerability of Agriculture Near American Cities.* Washington, D. C.: American Farmland Trust.

U.S. Department of Commerce. 1988. *Coastal Management: Solutions to Our Nation's Coastal Problems.* Technical Assistance Bulletin No. 101. Washington, D.C.: National Oceanic and Atmospheric Administration.

Williams, Michael., ed. 1990. *Wetlands: A Threatened Landscape.* Oxford, UK: Basil Blackwell, Ltd.

Zenk, Teri U. 1996. *An Evaluation of Selected California Coastal Commission Wetland Projects.* M.S. Thesis, San José State University.

FURTHER READING

Altieri, Miguel A., et al. 1987. *Agroecology: The Scientific Basis of Sustainable Agriculture.* Boulder, Colorado: Westview Press.

Bildstein, Keith L., G. Thomas Bancroft, Patrick J. Dugan, David H. Gordon, R. Michael Erwin, Erica Nol, Laura X. Payne, and Stanley E. Senner. 1991. "Approaches to the Conservation of Coastal Wetlands in the Western Hemisphere." *Wilson Bulletin.* Vol. 103, No. 2: pp. 218–254.

Boicourt, William C. 1993. "Estuaries: Where the River Meets the Sea." *Oceanus.* Vol. 36, No. 2: pp. 29–37.

Bowers, Keith. 1993. "What is Wetlands Mitigation?" *Land Development.* Winter; Vol. 5, No. 3: pp. 28–33.

Breen, Ann and Dick Rigby. 1994. *Waterfronts: Cities Reclaim Their Edge.* New York: McGraw-Hill, Inc.

Brooke, Jan, and Catriona Paterson. 1993. "Estuary Management: The British Experience." In *Coastal Zone '93: Proceedings of the Eighth Symposium on Coastal and Ocean Management,* edited by Orville T. Magoon, W. Stanley Wilson, Hugh Converse, and L. Thomas Tobin. New York: American Society of Civil Engineers, pp. 56–67.

Cohen, Andrew Neal. 1991. *An Introduction to the Ecology of the San Francisco Estuary.* 2d ed. San Francisco: San Francisco Estuary Project.

Culliton, Thomas J., John J. McDonough III, Davida G. Remer, and David M. Lott. 1992. *Building Along America's Coast: 20 Years of Building Permits, 1970–1989.* Rockville, Maryland: National Oceanic and Atmospheric Administration.

Field, Donald W., Anthony J. Reyer, Paul V. Genovese, and Beth D. Shearer. 1991. *Coastal Wetlands of the United States: An Accounting of a Valuable National Resource.* Rockville, Maryland: National Oceanic and Atmospheric Administration.

French, Peter W. 1997. *Coastal and Estuarine Management.* New York: Routledge.

Goering, Peter, Helena Norberg-Hodge, and John Page. 1993. *From the Ground Up: Rethinking Industrial Agriculture.* London: Zed Books.

Goldman-Carter, Jan. 1989. *A Citizens' Guide to Protecting Wetlands.* Washington, D.C.: The National Wildlife Federation.

Hairston, Ann J. 1992. *Wetlands: An Approach to Improving Decision Making in Wetland Restoration and Creation.* Covelo, California: Island Press.

Hammer, Donald A. 1991. *Creating Freshwater Wetlands.* Boca Raton, Florida: Lewis Publishers.

Hendee, John C., George H. Stankey, and Robert C. Lucas. 1990. *Wilderness Management.* Golden, Colorado: North American Press.

Imperial, Mark T., Donald Robadue Jr., and Timothy M. Hennessey. 1992. "An Evolutionary Perspective on the Development and Assessment of the National Estuary Program." *Coastal Management.* Vol. 20: pp. 311–341.

Kier, William M. 1994. *Fisheries, Wetlands and Jobs: The Value of Wetlands to America's Fisheries.* (A report issued in cooperation with the Pacific Coast Federation of Fishermen's Associations, the Atlantic States Marine Fisheries Commission, the Southeastern Fisheries Association, the East Coast Fisheries Federation, and the Ocean Trust). Oakland, California: Campaign to Save California Wetlands.

Kusler, John A., and Mary E. Kentula, eds. 1990. *Wetland Creation and Restoration: The Status of the Science.* Covelo, California: Island Press.

Lyon, John G. and Jack McCarthy, eds. 1995. *Wetland and Environmental Applications of GIS.* Boca Raton, Florida: Lewis Publishers.

Marble, Anne D., ed. 1992. *A Guide to Wetland Functional Design.* Boca Raton, Florida: Lewis Publishers.

McCreary, Scott, Robert Twiss, Bonita Warren, Carolyn White, Susan Huse, Kenneth Gardels, and Dominic Roques. 1992. "Land Use Change and Impacts on the San Francisco Estuary: A Regional Assessment with National Policy Implications." *Coastal Management.* Vol. 20: pp. 219–253.

National Oceanic and Atmospheric Administration. 1991. *National Estuarine Research Reserve System: Site Catalog.* Washington, D.C.: National Oceanic and Atmospheric Administration.

National Oceanic and Atmospheric Administration. 1995. *Watershed Restoration: A Guide for Citizen Involvement in California.* Silver Spring, Maryland: Coastal Ocean Office, National Oceanic and Atmospheric Administration.

National Research Council. 1992. *Restoration of Aquatic Ecosystems: Science, Technology, and Public Policy.* Washington, D.C.: National Academy Press.

Office of Technology Assessment. 1992. *Science and Technology Issues in Coastal Ecotourism.* Washington D.C.: Government Printing Office.

Payne, Neil F. 1992. *Techniques for Wildlife Habitat Management of Wetlands.* New York, NY: McGraw-Hill.

Poincelot, Raymond P. 1986. *Toward a More Sustainable Agriculture.* Westport, Conn.: AVI Publishing Company, Inc.

San Francisco Estuary Project. 1993. *Comprehensive Conservation and Management Plan.* San Francisco: San Francisco Estuary Project.

Strategic Assessment Branch. 1990. *Estuaries of the United States: Vital Statistics of a National Resource Base.* Rockville, Maryland: National Oceanic and Atmospheric Administration.

Young, Terry F., Chelsea H. Congdon. 1994. *Plowing New Ground: Using Economic Incentives to Control Water Pollution from Agriculture.* Oakland, CA: Environmental Defense Fund.

9 Toward Integrated Coastal And Marine Sanctuary Management

Coasts are boundaries between land and water where the geologic nature of the land is unstable and often fragile, and where the environment is constantly changing under the dynamic power of climate, coastal processes, sediment budget, relative sea level, and increasing human activity. (See Figure 9-1).

The vibrant beauty of shorelines is attracting more and more homeowners, businesses, and recreationists to coastal regions. Although the population explosion in the coastal zone is a worldwide phenomenon, it is especially acute in the United States, where a variety of government subsidies over the past 50 years have enabled widespread and often unwise development to occur. As has been illustrated in this book, it has led to major concerns in the areas of coastal hazards, coastal pollution, and loss of open space. These coastal problems are further exacerbated by the practices of ocean dumping and offshore oil development.

This book has attempted to illustrate not only that intelligent stewardship of our coastal and marine resources will require a balancing of human needs and expectations with environmental realities, but that it will also require a more integrated system of management based on the

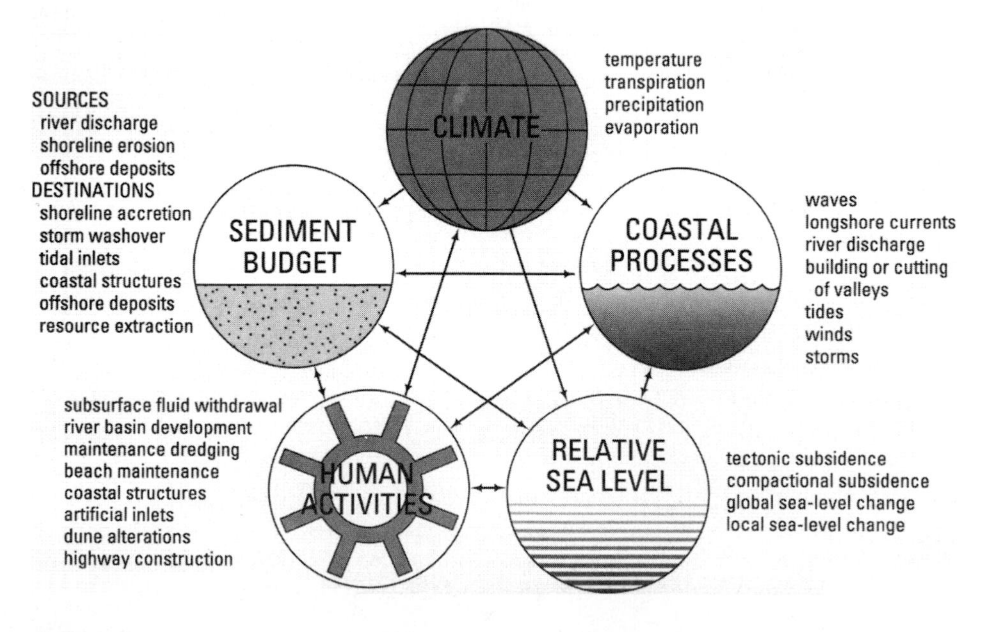

FIGURE 9-1 Factors affecting coastal environments: A summary. (*Source: Coasts in Crisis*, 1990 by Williams, Jefress, Dodd and Gohn. U. S. Geological Survey. Used with permission.)

concepts of traditional environmentalism (i.e., traditional [indigenous] systems of resource management [Refer back to Chapter 1, Appendix 1]). This last chapter will address the level of integration of the U.S. CZM program, summarize the need to incorporate traditional environmental concepts into the planning and management process, then end with some overall recommendations regarding coastal and marine sanctuary management.

HOW INTEGRATED IS THE U.S. CZM PROGRAM?

The Four Dimensions of Integrated Coastal Management (ICM)

As first described in Chapter 2, coastal researchers have identified four kinds of integration associated with the concept of Integrated Coastal Management (ICM): (a) Intergovernmental, (b) Land-water interface, (c) Intersectoral, and (d) Interdisciplinary. The question now is just how much is the U.S. CZM program in compliance with these four kinds of integration? Much of what follows is based on the findings of Finnell (1985), Knecht and Archer (1993), Kenchington and Crawford (1993), and personal observations.

The intergovernmental dimension. If you recall, this dimension of ICM refers to the integration of various levels of government into coastal resource management. In the United States, this involves the national, state, regional, and local level. The federal CZM legislation clearly requires that "intergovernmental coordination" occur: The "federal consistency" provision within the statutes requires that federal agencies act in concert with state CZM programs; local governments are required to act consistently with the approved state CZM program. Of course, government entities do not always comply with their stated goals. This is where public participation and the ever-vigilant environmental group (e.g., Save our Shores, Center for Marine Conservation, Friends of Chesapeake Bay) can bring legal action to ensure that intergovernmental coordination does occur.

The land-water interface dimension. Integrated Coastal Management also requires understanding how coastal waters and their uses affect the shoreline (e.g., impact of oil and gas development; mineral development; ocean dumping; ocean incineration; fisheries activities), as well as how uses and disturbances of the shoreline impact coastal waters (e.g., nonpoint pollution within coastal watersheds; elimination of wetlands; disturbance of riparian corridors). *Both directions* of the traverse between land and water must be studied and understood, and as argued in this book, this must include the "land-marine sanctuary" and "marine sanctuary-land" phenomena. There would be no field of study, planning, or management strategy known as "coastal resource management" if it were not for the land-water interface dimension. Whereas urban planners may study only the land, and marine scientists and oceanographers limit their studies to the sea, *the* distinguishing characteristic of coastal resource managers is their interest in the land-water interface.

The 1972 CZM Act acknowledges that integrating the land-water interface is essential to effective coastal management, though few states have done much management of the coastal ocean. As illustrated in this book, this is now changing, since more marine sanctuaries are being created, and their managers are becoming increasingly active in seeking proper management of the land adjacent to their sanctuaries.

The intersectoral dimension. Integrated Coastal Management also requires that various sectoral activities (e.g., urban development, mineral extraction, fishing, marine sanctuary management) be integrated into discussions and policy decisions in order to resolve multiple use conflicts. For example, the dumping of dredge spoils in offshore waters can potentially affect coastal fisheries. Consequently, both sectors—dredging as part of harbor management, and coastal fisheries—must be integrated into the coastal management program. The CZM program in the U.S. *directly* addresses such sectoral activities as coastal development, land use, water use, public access, citing of governmental energy facilities, and waterfront restoration. Through the federal consistency provisions, it also *indirectly* addresses such sectoral activities as ocean dumping of wastes, management of fisheries, offshore oil and gas development, and offshore hard minerals activities. However, identifying sectors seems to have done more for helping intergovernmental coordination than improving intersectoral management. According to Knecht and Archer (1993), the U.S. coastal program needs a more effective intersectoral management mechanism.

The interdisciplinary dimension. This last major dimension of Integrated Coastal Management recognizes that the coastal environment is more than a compilation of natural resources (biotic and abiotic) that would traditionally be studied by natural and physical scientists (e.g., biologists, marine scientists, oceanographers, geologists, etc.). The coastal zone also includes social, cultural, economic, political, legal, and ethical issues that must be addressed. Consequently, integrated coastal management must be holistic and interdisciplinary, including the perspectives most often studied by economists, political scientists, sociologists, geographers, anthropologists, and other members of the social sciences.

To date, there has been uneven "in-put " of information from the various disciplines to the U.S. CZM program, with the sciences being dominant. This has been the case primarily because CZM programs have placed an emphasis on natural resource allocation or environmental protection. Coastal management programs are now beginning to move beyond merely allocating or protecting a natural resource (e.g., wetland), and are broadening out to incorporate cultural resources (e.g., historic preservation) as well. As the breadth of coastal management programs broadens, so will the need for the those specially trained to be *generalists (e.g., environmental scientists, environmental planners, geographers)*. These individuals are trained to think holistically—to evaluate and integrate the natural *and* social component of any landscape.

Barriers to Integrated Coastal Management

Fragmentation. With the exception of the Coastal Zone Management Act, America's coastlines are for the most part regulated and managed on a single purpose, use-by-use basis. As was introduced in Chapter 2, *Coastal Management: The Principal Players,* and repeatedly illustrated in the following chapters, there are separate pieces of legislation that are often managed by different government agencies with regard to marine protected areas (Chapter 3), coastal hazards (Chapter 4), coastal pollution (Chapter 5), ocean dumping (Chapter 6), offshore oil development (Chapter 7), and open space preservation and management (Chapter 8). This administrative fragmentation makes integrated coastal management a difficult concept to achieve, and consequently, may never be fully achievable.

Powerful constituencies. A second barrier to ICM is the development of powerful constituencies. As institutions evolve (whether they be a government agency, a university department, or a private business), they tend to want to expand "their turf"—their jurisdiction in their particular area of expertise. After a while, they begin to lose sight of their stated mission (e.g., fisheries, offshore oil development, wildlife protection), and begin to concentrate (perhaps unconsciously) more on expanding their domain, the size of their annual budget, and the total number of employees. Once that happens, it is difficult to get these constituencies to participate in a broader administrative structure. Many future masters thesis and Ph.D. dissertation topics should concentrate on the following important questions: *How can single purpose constituents (e.g., offshore oil development, fisheries, endangered species, marine transportation, water quality, housing development, etc.), be convinced that a broader, better integrated approach to coastal and marine sanctuary management is not only good for the environment, but will also serve their long-term interests as well? Once convinced, what are some innovative strategies for achieving the goal of integrated coastal and marine sanctuary management?*

THE NEED TO APPLY TRADITIONAL ENVIRONMENTAL CONCEPTS

Fortunately, both domestic and international political forces are pushing the U.S. toward a more integrated strategy of coastal resource management. As these programs evolve, it would help if the principal coastal and marine sanctuary players keep in mind those traditional environmental concepts first introduced in Chapter 1. What follows is a summary of how these important environmental concepts can be applied to the five major coastal issues addressed in this book: coastal hazards, coastal pollution, ocean dumping, offshore oil development, and open space preservation and management. Although most of the traditional environmental concepts noted in Appendix 1 of Chapter 1 are applicable to each coastal issue, for the sake of brevity, only five concepts per coastal issue will be identified in the following summary tables.

Natural Hazards

The increasing media attention being given to the dangers of El Niño, coastal hurricanes, and the growing scientific concern over sea level rise, are heightening public awareness of the issue of natural hazards. This may force coastal states and local communities to pursue a more integrated response strategy based on traditional environmental concepts. (See Table 9-1).

Striving for the traditional concept of *conflict resolution* (to resolve anticipated coastal problems), for example, Maine established wetland buffer zones as a protective measure against anticipated sea level rise (Klarin and Hershman 1990). Attempting to have better *harmony with nature*, some coastal states are seeking a more balanced use of engineering v. softer and more ecologically acceptable approaches to preventing coastal erosion. For example, North Carolina banned the construction of new seawalls and other shore-hardening structures (the engineering approach), in favor of establishing one of the toughest erosion-based setbacks in the country (the softer approach [North Carolina Division of Coastal Management 1988]).

Coastal Pollution

Another positive sign that we are moving closer to Integrated Coastal Management is in the area of coastal pollution. (See Table 9-2). Most scientists agree that

TABLE 9-1
Coastal Hazards

Traditional Environmental Concepts	*Application to Coastal Hazards*
Recycling and Reuse	Use of high quality dredged sand for beach nourishment (e.g., Santa Cruz Small Craft Harbor uses sand dredged from the harbor entrance to nourish Twin Lakes Beach, California).
Conflict resolution	Methods to resolve anticipated coastal problems (e.g., Maine established wetland buffer zones to anticipate migration in response to sea level rise [Klarin and Hershman 1990]).
Environmental education	Public education programs as part of a national policy on coastal erosion.
Resource conservation	Personal lifestyle changes to reduce global warming and possible sea level rise.
Harmony with nature	Balanced use of engineering v. softer and more ecologically acceptable approaches to preventing coastal erosion (e.g., North Carolina has banned the construction of new seawalls and other shore-hardening structures; instead, the state has one of the toughest erosion-based setbacks—a distance of 30 times the average annual rate of erosion, measured from the first stable line of vegetation [N.C. Division of Coastal Management 1988]; Nags Head, North Carolina changed its zoning regulations to reduce the extent of high- density development, thereby reducing the number of lives exposed to coastal hazards [Beatley, Brower, and Schwab 1994]).

TABLE 9-2
Coastal Pollution

Traditional Environmental Concepts	*Application to Coastal Pollution*
Cooperation	The cooperation of agencies to control nonpoint source pollution (Monterey Bay National Marine Sanctuary's Memorandum of Agreement signed by eight federal, state, and regional agencies).
Habitat Protection	Maintaining buffer zones (e.g., natural forested lands along a shore's edge) to help filter pollutants and take up excess nutrients.
Ecological stability	Use of Best Management Practices (BMPs) in agriculture (e.g., contour plowing), urban development (e.g., stormwater collection ponds), and construction (e.g., phased land clearance).
Holistic thinking	Such regional planning approaches as The National Estuary Program (Imperial et al 1993), The Great Lakes Program (Caldwell 1985), and various watershed protection approaches to solving non-point source pollution (U.S. EPA. 1991).
Recycling and reuse	Coastal communities that reduce, recycle, or reuse sewage waste rather than transport it elsewhere—the "out-of-site, out of mind" management strategy (e.g., San Francisco's use of reclaimed water to irrigate golf courses; reuse or recycling of dredge spoils).

nonpoint source pollution (polluted runoff) impairs more waterbodies than any other single pollution source in the country. Even Congress acknowledged this fact by enacting the Coastal Nonpoint Program, otherwise known as Section 6217 of the Coastal Zone Act Reauthorization Amendments of 1990 (CZARA). The Act calls for the traditional concepts of *holistic thinking* and *cooperation.* Specifically, it requires that coastal states develop effective and enforceable runoff control programs (U.S. E.P.A. 1991). This necessitates that a holistic watershed approach be used. A watershed approach, in turn, requires greater *cooperation* between agencies, since water quality issues and land use issues are generally handled by different agencies. Other examples of regional planning approaches (which require holistic thinking and interagency cooperation) include The National Estuary Program (Imperial et al 1993) and The Great Lakes Program (Caldwell 1985).

Ocean Dumping

There are also some positive signs that coastal communities are increasingly applying traditional environmental concepts related to ocean dumping. (See Table 9-3). For example, the concept of *harmony with nature* is applied when a society only dumps materials into the ocean that are compatible with the marine environment, such as outdated ships that have been stripped down to remove fuel and anything else that might prove harmful to the environment; the ship's "carcass" can serve to make a good reef.

In June 1995, for example, German society questioned, and eventually halted, the proposed dumping (sinking) of the *Brent Spar*—an obsolete 300,000-barrel storage platform that was previously used to warehouse oil drilled from the North Sea until it could be shipped via oil tanker. When European environmental officials and

TABLE 9-3
Ocean Dumping

Traditional Environmental Concepts	*Application to Ocean Dumping*
Holistic thinking	The development of a regional strategy to deal with dredge spoils (e.g., Oakland, California, Marshland restoration project).
Quality over Quantity	Coastal dwellers that attempt to reduce the flow of matter and energy through their lives.
Renewable resource use	Coastal communities that stress "better use" options for their coastal waters (e.g., Hawaii's pioneering research into the beneficial uses of cold seawater, their ocean Thermal Energy Conversion project).
Wastelessness	Coastal communities that shift from waste management to waste prevention (e.g., Denmark's ban of certain plastic items).
Harmony with nature	The ocean dumping of materials that are only compatible with the marine environment (e.g., outdated ships that serve to make good reefs); the maintenance of the existing U.S. ban on all ocean dumping of sewage sludge, radioactive waste.

Greenpeace learned that its owners, Royal Dutch Shell and Exxon, proposed to pull the giant offshore oil rig to a watery grave 241km (150 mi) off the coast of Scotland, the news ignited a frenzied environmental debate across much of Europe, especially in Germany. At the urging of Greenpeace, German motorists boycotted Shell gasoline stations across the country for the duration of the debate, leading to a 10 percent drop in sales. The debate even turned violent, with a Shell station in Hamburg being firebombed and one in Frankfurt being riddled with gun shot.

The "Brent Spar Debate" centered around two questions: (1) would the sunken oil platform be compatible with the marine environment, and more importantly (2) would this act set a precedence for the future dumping of similar obsolete rigs: Shell officials argued that pulling the rig to land and dismantling it (their normal practice with outdated rigs) was risky in this case. They maintained that the platform had major structural defects in two of its storage tanks, and the prospect of pushing the 140 m (460-ft) tall steel structure on its side for dismantling proposed a very high risk (i.e., it could break in several pieces and be a hazard to ships within the region). On the other hand, the European environmental community and Greenpeace argued that the rig contained 100 tons of sand, oil sludge, heavy metals, and some low-grade radioactive waste that could damage the food chain for fish in the area, leading to a reduction in the fishing stocks. Furthermore, they argued that disposal of such a rig by sinking was a dangerous precedent because over the next two decades more than 400 rigs could become obsolete as oil in the North Sea runs out. Due to the astronomical costs of disposal of the rigs on land, the European environmental community argued the companies were merely beginning a practice of taking the cheapest way out.

The positive aspect of this story is not necessarily the stopping of the sinking of the Brent Spar, because the debate continues as to whether the sunken oil rig would have actually been harmful to the marine environment. What seems positive, however, is that societies around the world are increasingly concerned about the environmental consequences of dumping into the coastal and marine environment, and thus, are seeking a more harmonious relationship with nature.

Offshore Oil Development

One can also identify some positive trends related to the concern over offshore oil development, especially as it relates to such traditional environmental concepts as energy efficiency, wastelessness, habitat protection, renewable resource use, and human scale. (See Table 9-4). For example, there are a growing number of coastal (and inland) communities that are working to promote, or protect, the notion of *human scale* which conserves energy and lessens the need for additional offshore oil development.

At a 1995 Brookings Institution conference on "Alternatives to Sprawl," participants were electrified by a report from the Bank of America endorsing the formerly elitist view that sprawl in California has created enormous environmental, economic, and social costs—costs which until now had been hidden, ignored, or quietly borne by society. One of the biggest environmental costs, of course, being the enormous consumption of fossil fuels. Similar conferences and discussions have inspired a growing corps of visionaries that are looking at the "village" as a model for future development. A village can be defined as a cluster of houses around a central place that is the focus of civic life. Since villages generally do not extend more than a quarter-mile from the center to the edge, emphasis is on walking and bicycling, not reliance on automobiles and fossil fuels. "Village living" is most definitely living at the human scale.

TABLE 9-4
Offshore Oil Development

Traditional Environmental Concepts	*Application to Offshore Oil Development*
Energy efficiency	Coastal communities that design streets and shopping areas that encourage alternatives to the automobile—walking, biking, light rail.
Wastelessness	The use of Urban Growth Boundaries (UGBs) to promote more efficient, less land-consumptive development patterns (e.g., Canon Beach, Oregon); developing cleaner ways to drill for oil (e.g., the drilling platform Glomar Highland IV off Dauphin Island, Alabama).
Habitat Protection	Coastal communities that support the extention of federal jurisdiction boundary to create a larger buffer zone against offshore oil development (e.g., H.R. bill #2583, California Ocean Protection Act of 1993).
Renewable resource use	Coastal communities that want to reverse the nation's dependence on fossil fuels by stressing a diversity of energy sources—a mixture of traditional nonrenewable energy (e.g., oil, natural gas, coal) and renewable energy (e.g., solar, wind, geothermal, tidal).
Human Scale	Coastal communities that promote the notion of human scale—return of the walkable "village" (e. g., Seaside, Florida).

TABLE 9-5
Open Space Preservation And Management

Traditional Environmental Concepts	*Application to Open Space Management*
Sense of Community (Place)	Protecting community character through urban and architectural design review standards and processes (e.g., Seaside, Florida; Carmel, California; Canon Beach, Oregon; Nantucket, Massachusetts; Hilton Head, South Carolina).
Diversity	Use of land acquisition as a strategy to preserve coastal diversity (e.g., Florida's Conservation and Recreation Lands Program; Land acquisition initiatives have also occurred in Little Compton, R.I., Martha's Vineyard, Mass., and Hilton Head, S.C.). Designing coastal communities for a rich cultural life is also important.
Holistic thinking	Expanding the geographical and ecological scope of coastal management to include unique offshore marine habitats (e.g., Oregon's Ocean Management Plan [Hout 1990]; new town plans based on holistic thinking—Haymount town on Virginia's Rappahannock River; Dewees Island, South Carolina).
Recycling and Reuse	Port communities that revitalize their waterfronts and rejuvenate their downtowns (e.g., Baltimore's Harborplace).
Carrying capacity	Comprehensive planning based on the capacity of a coastal area to sustain new growth (e.g., Sanibel Island, Florida).

One of the first applications of this "neotraditionalism" or "new urbanism" came when Miami-based architects Andres Duany and his wife and partner, Elizabeth Plater-Zyberk, designed Seaside—a small coastal resort town on the Florida panhandle. Since then, other neotraditional developments have been developed, such as Harbor Town on an island in the Mississippi, five minutes from downtown Memphis; Kentlands in the suburbs of Maryland; and Laguna West on the outskirts of Sacramento, California. Another test of this idea may be the Disney Company's first planned community in Celebration, Florida—a 2,024 ha (5,000 acre) piece of land near Disney World. Whereas Disney officials describe their 1996 planned development as "traditional little-town America," some critics maintain that this project is not really neotraditional development, but rather an attempt to lure citizens into living in a "theme park." Regardless of the outcome of this Disney project, there seems to be a growing number of people interested in exploring urban alternatives that can help promote *energy efficiency*.

Open Space Preservation and Management

There appears to be an increasing number of traditional environmental concepts being applied to the field of open space preservation and management. (See Table 9-5). *Holistic thinking*, for example, is being applied to two leading edge community developments—Haymount, Virginia; and Dewees Island, South Carolina. Haymount is a new town planned for a site near Virginia's Rappahannock River, 80 km (50 mi) south of Washington D.C. in Caroline County, Virginia. It is being planned with the belief that sustainable development is a "Trilogy" that integrates ecology, sociology, and economics—all for the purpose of blending nature with the human spirit. The plan calls for 4000 homes, as well as office and commercial space. However, the developed area of the town will be compact, so as to preserve two-thirds of the 688-ha (1700-acre) site, which is a perpetual forest and wildlife habitat. Although the number one intention of this development project is "to sell real estate," the traditional concepts of *holistic thinking*, *integration*, and *sustainability* have been incorporated to a certain degree—at least more than would be the case if this development project occurred 25 or more years ago.

Habitat protection and *environmental restoration* are also underlying principles of the development of Dewees Island—a private, boat access, oceanfront island retreat located 19 km (12 mi) northeast of Charleston, SC. The barrier island is 488 ha (1,206 acres), with covenants prohibiting the construction of more than 150 homes. With two full-time environmentalists stationed at the site, habitat protection is taken seriously. More importantly, agreements with the state assure that 65 percent of the island is protected from any form of development, and 142 ha (350 acres) on the northeastern end of the island is designated as a wildlife refuge. There will be no paved roads, no cars, no private docks, no golf courses, and no link to the mainland except by ferry from the nearby Isle of Palms. Landscaping plants must be indigenous to the sea islands, and cisterns will be located at each house to collect rain water so as to help reduce use of the island's wells. Although the S.C. Coastal Conservation League and other local environmental groups would have preferred to see the entire island protected as a nature preserve, they generally admit that if development must occur, the Dewees Island Project represents holistic thinking in its planning and maximum sensitivity to the environment.

FINAL THOUGHTS

The world we have created today as a result of our thinking thus far has problems which cannot be solved by thinking the way we thought when we created them.

—Albert Einstein

The purpose of this book is to introduce university students and interested laypersons to the major environmental issues facing the coastal zone. It is also designed to be an informative reference book for individuals working at natural resource management agencies that oversee coastal areas or adjacent marine sanctuaries. It is hoped that all will contemplate certain questions when they next visit the shore: To what degree are the coastal communities discussing, or planning for, possible sea level rise? What strategies are being used to deal with cliff erosion? Are humans doing anything to cause the beach to narrow? Where is all this marine debris coming from, and how can it be prevented? Are these coastal communities increasing in population, and how are they dealing with the issue of public access? Is there a coastal protected area such as a national marine sanctuary nearby, and are the sanctuary managers attempting to integrate their conservation strategies with the concerns of land-based businesses and resource management agencies? When contemplating these scenes, it is important to remember that these individuals are witnessing firsthand the direct results of the conflicting uses of the coastal zone.

Several general recommendations emerge from the more topical discussions in this book:

- *Important to bridge the land/ocean (especially marine sanctuary) interface.* Bridging the land/ocean interface is not easy, and will require an outreach program to help assemble the technical expertise and opinions of government agencies, the private sector, and the general public. This outreach program could be partially modeled after the Integrated Coastal Management approach used by NOAA to develop the Florida Keys National Marine Sanctuary Management Plan and currently being used to develop a water quality management plan for the Monterey Bay National Marine Sanctuary.
- *Important to argue the economic benefits of the coast.* Coastal advocates rarely argue the economic benefits of protecting a coastal area. Rather, they rely on the coast's inarguable splendor as reason enough to protect it. This is a mistake, since the activities and resources within the coastal zone contribute to the state and national economies. When advocating for coastal protection, be sure to include data on coastal resource-dependent industries and the revenues they generate. This would include such information on commercial and sport fishing, aquaculture operations, marine supply stores, boat yards, surf shops, and so on. The economic argument should most certainly include revenues generated by tourism—an increasingly important contributor to local and national economies. In the "Age of Deficits," where governmental budgets continue to tighten, it is critical that environmentalists and natural resource managers strengthen their arguments with the economic benefits of coastal protection.
- *Important to bridge gap between the academic community and mission agencies.* The scientific community can no longer work in isolation from those mission agencies

(e.g., coastal commisions) that are responsible for management and policy. The old notion that "good science and policy don't mix" must give way to a new perspective that incorporates complex scientific findings with appropriate policy decisions and management strategies. Partnerships based on the traditional environmental concept of cooperation should help break down the implicit differences between these two "cultures."

- *Important to bridge the gap between the sciences and social sciences.* Tension also traditionally exists between academic schools within universities. In particular, there are often "academic turf wars" between the sciences (e.g., physics, chemistry, biology, geology, meteorology) and the social sciences (e.g., geography, history, economics, political science,) when it comes down to who should "rule over" the environmental curriculum on the campus. Some universities, such as Duke University, have increased cooperation by creating separate multidisciplinary/ interdisciplinary colleges or schools of the environment within the university. In such an administrative unit, all interested academicians participate in concert for a common goal. Likewise, coastal studies require that there be greater and more effective communication between natural/ physical scientists, social scientists, and engineers.
- *Important to increase public awareness and citizen empowerment, and broaden the coastal constituency.* Individual citizens must begin to take responsibility for restoring, not just degrading, such national coastal treasures as Chesapeake Bay, Maryland, or Monterey Bay, California. Residents along Chesapeake Bay, for example, are beginning to fight back against coastal destruction, and the Chesapeake Bay Program is seen by many as an international model of intergovernmental cooperation and public support to reverse years of coastal destruction (Wheeler 1994). Furthermore, it illustrates the power of cooperation and citizen empowerment. According to Shelley (1993), citizen action groups play a critical role in defining, setting priorities for, and resolving the environmental (including coastal) issues in our daily lives.

  Coastal communities are beginning to discuss a variety of ways to increase public awareness, citizen empowerment, and broadening their coastal constituency. In California, for example, coastal protection advocates are working on such strategies as public media campaigns, shore watch networks, coastal resource education for traditional disenfranchised school children, marine and coastal career development curricula, and state legislator orientation programs (Notthoff and Fischer 1994). One such program is *Los Marineros* marine education program cosponsored by the Channel Islands National Marine Sanctuary and the Santa Barbara Museum of Natural History. According to Jennifer, a Los Marineros student, *"It gives the children more knowledge about the ocean, helps them become aware of the damage being done to our earth and our oceans, and, what really counts to kids, it's fun!* (Cushman 1993, p. 41).

In the final analysis, whether our coastal areas and marine sanctuaries remain environmentally healthy and thus inviting 20 years from now depends on our ability to adapt those traditional environmental concepts to 21st century coastal living. (See Figure 9-2). As identified in this book, some excellent programs are already underway. Since the priorities for coastal ecosystem science have already been identified (National Research Council 1994), it is actually more a matter of national will whether or not this country is ready to intensify its efforts at coastal protection.

FIGURE 9-2 Will this shoreline and marine sanctuary off Capitola, California look this inviting 20 years from now? (*Source:* Photo by author)

REFERENCES

Beatley, Timothy, David J. Brower, Ann K. Schwab. 1994. *An Introduction in Coastal Zone Management.* Washington, D.C.: Island Press.

Caldwell, Lynton K., ed. 1985. *Perspectives on Ecosystem Management for the Great Lakes.* New York: State University of New York Press.

Cushman, Sheila. 1993. "Los Marineros: An Investment in the Future of Our Oceans—and Our Children." *Oceanus.* Vol. 36, No. 3, pp. 41–48.

Environmental Protection Agency (EPA). 1991. *The Watershed Protection Approach: An Overview.* Washington, DC: Environmental Protection Agency.

Finnell, Gilbert L., Jr. 1985. "Intergovernmental Relationships in Coastal Land Management." *Natural Resources Journal.* Vol. 25, No.1, pp. 31–60.

Hout, Eldon. 1990. "Ocean Policy Development in the State of Oregon." *Coastal Management.* Vol 18, No. 3, pp. 255–266.

Imperial, Mark T., Tim Hennessey, and Donald Robadue, Jr. 1993. "The Evolution of Adaptive Management for Estuarine Ecosystems: The National Estuary Program and Its Precursors." *Ocean and Coastal Management.* Vol. 20, No. 2, pp. 147–180.

Kenchington, Richard, and David Crawford. 1993. "On the Meaning of Integration in Coastal Zone Management." *Ocean & Coastal Management.* Vol. 21, No. 1–3, pp. 109–127.

Klarin, Paul, and Marc Hershman. 1990. "Response of Coastal Zone Management Programs to Sea Level Rise in the United States." *Coastal Management.* Vol. 18, No. 2, pp. 143–165.

Knecht, Robert W., and Jack Archer. 1993. "Integration" in the U.S. Coastal Zone Management Program. *Ocean & Coastal Management.* Vol. 21, No. 1–3, pp. 183–199.

National Research Council. 1994. *Priorities for Coastal Ecosystem Science.* Washington, D.C.: National Academy Press.

North Carolina Division of Coastal Management. 1988. *A Guide to Protecting Coastal Resources through the CAMA Permit Program.* Raleigh, North Carolina: North Carolina Division of Coastal Management.

Notthoff, Ann, and Michael Fischer. 1994. "Coastal Agenda 2000: Protecting and Managing California's Coast into the 21st Century." Prepared for The David and Lucille Packard Foundation, February 1994. Unpublished manuscript.

Shelley, Peter. 1993. "The Role of Citizen Groups in Environmental Issues." *Oceanus.* Vol. 36, No. 1, pp. 77–84.

Williams, S. Jeffress, Kurt Dodd, and Kathleen Krafft Gohn. 1990. *Coasts in Crisis.* Washington, D. C.: U. S. Geological Survey, p.26.

Wheeler, T. 1994. "Saving the Bay." *People and the Planet.* Vol. 3, No. 1, pp. 24–25.

FURTHER READING

Kildow, Judith. 1997. "The Roots and Context of the Coastal Zone Movement." *Coastal Management.* Vol. 25, No. 3, July–Sept., pp. 231-263.

Knecht, Robert W., Biliana Cicin-Sain, and Gregory W. Fisk. 1997. "Perceptions of the Performance of the State Coastal Zone Management Programs in the United States. II. Regional and State Comparisons." *Coastal Management.* Vol. 25, No. 3, July–Sept., pp. 325–343.

The Resources Agency. 1997. *California's Ocean Resources; An Agenda for the Future.* Sacramento, California: The Resources Agency of California.

Sorensen, Jens. 1997. "National and International Efforts at Integrated Coastal Management: Definitions, Achievements, and Lessons." *Coastal Management.* Vol. 25, No. 1, pp. 3–41.

Suman, Daniel. 1997. "The Florida Keys National Marine Sanctuary: A Case Study of an Innovative Federal-State Partnership in Marine Resource Management." *Coastal Management.* Vol. 25, No. 3, pp. 293–324.

Coastal Directory

FEDERAL

Designated National Marine Sanctuaries

Channel Islands NMS
113 Harbor Way
Santa Barbara, CA 93109
(805) 966-7107

Cordell Bank NMS/
Gulf of the Farallones NMS
Fort Mason, Building #201
San Francisco, CA 94123
(415) 561-6622

Fagatele Bay NMS
P.O. Box 4318
Pago Pago, American Samoa 96799
011-684-633-7354

Florida Keys NMS (Administration)
P.O. Box 500368
Marathon, FL 33050
(305) 743-2437

Florida Keys NMS/Upper Region (Key Largo)
P.O. Box 1083
Key Largo, FL 33037
(305) 451-1644

Florida Keys NMS/Lower Region (Looe Key)
216 Ann Street
Key West, FL 33040
(305) 292-0311

Flower Garden Banks NMS
216 W. 26th St., Suite 104
Bryan, Texas 77803
(409) 779-2705

Gray's Reef NMS
10 Ocean Science Circle
Savannah, GA 31411
(912) 598-2345

Gulf of the Farallones NMS
Fort Mason, Building 201
San Francisco, CA 94123
(415) 561-6622

Hawaiian Islands Humpback Whale NMS
726 South Kihei Rd.
Kihei, HI 96753
(808) 879-2818

Monitor NMS
c/o The Mariners' Museum
100 Museum Drive
Newport News, VA 23606-3759
(757) 596-2222

Monterey Bay NMS
299 Foam St., Suite D
Monterey, CA 93940
(408) 647-4201

Stellwagen Bank NMS
14 Union Street
Plymouth, MA 02360
(508) 747-1691

Proposed National Marine Sanctuaries

Northwest Straits NMS
NOAA/SRD (Seattle)
7600 Sand Point Way, NE
Bin C 15700
Seattle, WA 98115
(206) 526-4293

Thunder Bay NMS
2205 Commonwealth Blvd.
Ann Arbor, MI 48105
(734) 741-2270

Selected Federal Agencies

Army Corps of Engineers (U.S.)
Flood Plain Management Services and Coastal Resources Branch
20 Massachusetts Ave., N.W.
Washington, DC 20314-1000
(202) 272-0169

Capital Information Switchboard
(202) 224-3121
(For federal information operator)

Coast Guard (U.S.)
(Check phone book for local offices)

Consolidated Farm Service Agency (CFSA) (formerly Agricultural Stabilization and Conservation Service)
Conservation and Environmental Programs Division/ STOP 0513
1400 Independence Ave., S.W.
Washington, DC 20250
(202) 720-6221

Council on Environmental Quality
360 Old Executive Office Bldg., NW.
Washington, DC 20501
(202) 456-6224

Department of Agriculture
14th St. and Independence Ave., SW
Washington, DC 20250
(202) 720-8732

Department of Energy
1000 Independence Ave., SW
Washington, DC, 20585
(202) 586-494 (Public Affairs)
www: http://www.doe.gov

Department of the Interior
Interior Bldg.
1849 C St., NW
Washington, DC 20240
(202) 208-3100

Department of State
2201 C St., NW
Washington, DC 20520
(202) 647-4000 (General Inquiries)
(703) 235-2948 (U.S. Man and the Biosphere Program [MAB])
(202) 647-2396 (Office of Oceans)
(202) 647-2335 (Office of Marine Conservation)

Environmental Protection Agency (U.S.) (EPA)
401 M Street, SW
Washington DC, 20460
(202) 260-2090 (General Inquiries)
(202) 260-7166 (Wetlands, Oceans, and Watersheds)
www: http://www.epa.gov
(Check telephone book, library, or internet for regional offices)

Federal Emergency Management Agency (FEMA)
(Check phone book/library/internet for regional office)

Fish and Wildlife Service (U.S.) (USFWS)
Department of the Interior
Branch of Coastal and Wetland Resources
4401 North Fairfax Drive
Arlington, Virginia 22203
(703) 358-2201

Fish and Wildlife Service (U.S.) (USFWS)
Department of the Interior
Washington, DC 20240
(202) 208-3736
(Check telephone book, library, or internet for regional offices)

Forest Service (U.S.) (USFS)
P.O. Box 96090
Washington, DC 20090-6090
(202) 205-0957
Check your phone book, library, or internet for local offices.

Geological Survey (U.S.) (USGS)
U.S. National Center
Reston, VA 22092
(703) 648-4000
(Excellent source for topographic maps and aerial photographs, as well as information on geology and geologic hazards. Check telephone book, library, or internet for regional offices).

National Estuarine Research Reserves, NOAA
Department of Commerce
1305 East-West Highway, 12th Floor
Silver Springs, Maryland 20910
(301) 713-3132

National Park Service
Interior Bldg.
P.O. Box 37127

Washington, DC 20013-7127
(202) 208-4747
(Check your phone book, library, or internet for regional offices)

Natural Resource Conservation Service (formerly Soil Conservation Service)
U.S. Department of Agriculture
14th and Independence Ave., SW
P.O. Box 2890
Washington, DC 20013
(202) 720-3210
(Check your phone book/library/internet for regional offices)

Minerals Management Service (MMS)
U.S. Department of the Interior
1849 C Street, N.W.
Washington, DC 20240
(202) 208-3985
(Check phone book, library, or internet for regional offices)

National Oceanic and Atmospheric Administration (NOAA)
Herbert C. Hoover Bldg., Rm. 5128
14th and Constitution Ave., NW
Washington, DC 20230
(202) 482-3384
(For Headquarters)

NOAA's National Marine Fisheries Service
Silver Spring Metro Center 3
1335 East-West Hwy
Silver Spring, MD 20910
(301) 713-2239
(For information involving living marine resources)

NOAA's National Ocean Service (NOS)
1305 East-West Hwy, Rm 13609
Silver Spring, MD 20910
(301) 713-3074
(Includes Directors of Marine Sanctuaries and Reserves; Ocean and Coastal Resource Management; Ocean Resources Conservation and Assessment)

NOAA's Coastal Ocean Program Office
Silver Spring Metro Center 3, Rm 15140
1315 East-West Hwy.,
Silver Spring, MD 20910-3223
(301) 713-3338
(Coordinates and integrates NOAA's Coastal Ocean Program)

National Seashores
[California, Florida, Georgia, Maryland, Massachusetts, Mississippi, New York, North Carolina, and Texas have National Seashores. Check phone book, library, or Internet for appropriate state]

STATE

Association of State Floodplain Managers
P.O. Box 7921
Madison, WI 53707
(608) 266-1926

Association of State Wetlands Managers
P.O. Box 2463
Berne, NY 12034
(518) 872-1804

Check your phone book, library, or internet for similar type state agencies:
Coastal Commission (e.g., California Coastal Commission)
Department of Boat and Waterways
Department of Conservation
Department of Fish and Game
Department of Forestry and Fire Protection
Department of Parks and Recreation
Department of Transportation
Department of Water Resources
Division of Mines and Geology
Energy Commission
State Lands Commission
State Water Resources Control Board
State Senate Office of Research
Equalization Board of Hazardous Substances
Hazardous Waste Materials
State Resources Agency
Office of Oil Spill Prevention and Response
Sea Grant Marine Advisory Program
State Coastal Conservancy
State Agency Coordinator Oil and Hazardous Materials
Wildlife Conservation Board
Water Quality Control Board

REGIONAL

Check you phone book, library, or internet for names of similar type commissions, boards, etc.:
Association of [e.g., Monterey Bay Area] Governments (AMBAG)

Bay Conservation and Development Commission (BCDC) (e.g., San Francisco Bay Conservation and Development Commission)
Regional Studies Program (e.g., Central Coast Regional Studies Program)
Regional Water Quality Control Board

COUNTY

Check your phone book, library, or internet for similar sounding county offices:
Agricultural commissioner
Health Department (Environmental Health/Toxic Materials)
Planning and Building Inspection Department
Transportation Agency
Water Resources Agency
Air Pollution Control District
Parks and Open Spaces and Cultural Services
Planning Department
Public Works
Recycling Programs
Septic Tank Inspection and Permits

CITY/LOCAL

Check your phone book, library, or internet for similar sounding city/local offices or buildings:
Historical Museum (e.g., Capitola Historical Museum)
Planning Department
Recreation/Community Activities
Harbor Master
Parks Maintenance and Forestry
Planning Department
Public Works Department
Risk Management
Water Department
Zoning Regulations and Enforcement

PRIVATE AGENCIES/INSTITUTIONS

Scripps Institution of Oceanography
University of California, San Diego
Public Affairs Office, Dept. 0233
La Jolla, CA 92093
(619) 534-3624

Woods Hole Oceanographic Institution
193 Oyster Pond Road
Woods Hole, MA 02543-1525
(508) 289-2398

Check your phone book, library, or Internet for similar sounding agencies/institutions:
Land Trusts (e.g., Big Sur Land Trust)
Marine Reserves (e.g., Bodega Marine Reserve)
Marine Science Centers (e.g, Catalina Marine Science Center)
Marine Stations (e.g., Hopkins Marine Station)
Marine Laboratories (e.g., Moss Landing Marine Laboratories)

COASTAL/MARINE EDUCATION & RESEARCH

Centre for Coastal Management
Southern Cross University
P.O. Box 157
Lismore NSW 2480
Australia

Centre for Tropical Coastal Management Studies
Department of Marine Sciences & Coastal Management
University of Newcastle upon Tyne
Newcastle upon Tyne NE1 7RU
United Kingdom

Center for Marine Studies and Sea Grant Program
14 Coburn Hall
University of Maine
Orono, ME 04469
(207) 581-1435

Center for the Study of Marine Policy
Graduate College of Marine Studies
University of Delaware
Newark, Delaware 19716
(302) 831-8086

Coastal Ecology Institute
203 Old Coastal Studies Bldg.
Louisiana State University
Baton Rouge, Louisiana 70803-2606
(225) 388-6515

Coastal Education and Research Foundation, Inc. (CERF)
4310 NE 25th Ave.
Fort Lauderdale, FL 33308
finkl@acc.fau.edu
(Produces *Journal of Coastal Research*)

Coastal and Oceanographic Engineering
University of Florida
P.O. Box 116590
Gainesville, FL 32611-6590
(352) 392-1436

Coastal Research Center
Stockton State College
Pomona, NJ 08240
(609) 652-4245

Coastal Studies Institute
306 Old Geology Bldg.
Louisiana State University
Baton Rouge, LA 70803
(504) 388-2395

California Sea Grant College
University of California
9500 Gilman Dr.
La Jolla, CA 92093-0232
(619) 534-4440

College of Marine Studies
University of Delaware
Newark, Delaware 19716
(302) 831-2841

Department of City and Regional Planning
University of North Carolina at Chapel Hill
CB#3140, New East Building
Chapel Hill, North Carolina 27599-3140
(919) 962-3983

Department of Oceanography and Coastal Sciences
Louisiana State University
153 Howe-Russell Bldg.
Baton Rouge, Louisiana 70803
(225) 388-6308

Director, Coastal Graduate Research
Department of Environmental Studies
San José State University
San José, California 95192-0115
(408) 924-5455

Florida Sea Grant College Program
Bldg. 803, Rm 4
P.O. Box 110400
University of Florida, Gainesville, FL 32611-0400
(904) 392-5870

Earth & Ocean Science
Duke University
Durham, N.C. 27708
(919) 684-5847

Hopkins Marine Station
100 Ocean View Blvd.
Pacific Grove, CA 93950
(831) 655-6200

Institute of Marine and Coastal Science
Rutgers University
71 Dudley Road
New Brunswick, NJ 08901
(732) 932-6555

Institute for Coastal and Marine Resources
Department of Sociology and Anthropology
East Carolina University
Greenville, NC 27858-4353
(252) 328-6766

Wrigley Institute for Environmental Studies
University of Southern California, AF-232
Los Angeles, CA 90089-0371
(213) 740-6780

Institute of Marine Sciences (Formerly Center for Coastal and Marine Studies)
University of California
Santa Cruz, CA 95064
(831) 459-2464

Office of Marine Programs
University of Rhode Island
Narragansett Bay Campus
South Ferry Road
Narragansett, RI 02882
(401) 874-6211

Marine Education Center
Humacao University
HUC Station,
Humacao, PR 00791
(787) 850-9360

Marine Laboratory
Florida State University
Rt. 1
Box 219A
Sopchoppy, FL 32358
(904) 697-4095

Marine Science Center
Oregon State University
2030 S. Marine Science Drive
Newport, Oregon 97365
(541) 867-0100

Marine Sciences Research Center
State University of New York
Stony Brook, NY 11794
(516) 632-8700

Massachusetts Sea Grant College Program
Woods Hole Oceanographic Institution
193 Oyster Pond Rd.
Woods Hole, MA 02543-1525
(508) 289-2398

Monterey Bay Aquarium
886 Cannery Row
Monterey, CA 93940
(831) 648-4888

Monterey Bay Aquarium Research Institute (MBARI)
P.O. Box 628
Moss Landing, CA 95039-0628
(831) 775-1700

Moss Landing Marine Labs
P.O. Box 450
Moss Landing, CA 95039-0450
(831) 633-3304

Naval Postgraduate School
Department of Oceanography
Monterey, CA 93943
(831) 656-2552

New Hampshire Sea Grant Program
Kingman Farm
University of New Hampshire
Durham, NH 03824
(603) 749-1565

New York Sea Grant Program
121 Discovery Hall
State University of New York at Stony Brook
Stony Brook, NY 11794-5001
(516) 632-6905

North Carolina Sea Grant Program
Box 8605 105-1911 Bldg.
North Carolina State University
Raleigh, NC 27695-8605
(919) 515-2454

Oregon Sea Grant College Program
A500
Oregon State University
Corvallis, OR 97331-2131
(541) 737-2714

School of Marine Affairs, HF-05
University of Washington
3707 Brooklyn Ave., NE
Seattle, Washington 98105
(206) 543-7004

Texas Sea Grant College Program
Texas A&M University
1716 Briarcrest, S-702
Bryan, TX 77802
(409) 845-3854

Virginia Sea Grant Program
Virginia Graduate Marine Science Consortium
170 Rugby Rd., Madison House
University of Virginia,
Charlottesville, VA 22903
(804) 924-5965

For additional sources of training in coastal/ocean management, contact Ms C. Casullo, United Nations University, UNESCO, Room M132 - 1, rue Miollis, 75015, Paris, France. Ms. Casullo oversees a database on this type of training.

ENVIRONMENTAL ORGANIZATIONS/ NONPROFIT GROUPS

There are hundreds of environmental organizations/ nonprofit groups. Many deal with coastal/marine issues (e.g., Sierra Club, Friends of the Earth, The Environmental Defense Fund, National Wildlife Federation, etc.). For a comprehensive listing of environmental organizations, see the National Wildlife Federation's annual *Conservation Directory*. The list below only includes those organizations that concentrate on coastal/marine issues.

Alliance for Chesapeake Bay
660 York Rd., Suite 100
Baltimore, MD 21212
(410) 377-6270

American Cetacean Society
P.O. Box 1391
San Pedro, CA 90733-0391
(310) 548-6279

American Oceans Campaign
725 Arizona Ave., Suite 102
Santa Monica, CA 90401
(310) 576-6162

Baykeeper, San Francisco
Bldg. A, Fort Mason
San Francisco, CA 94123
(415) 561-2299

Center for Marine Conservation
1725 DeSales St. NW, Suite 600

Washington, DC 20036
(202) 429-5609

Cetacean Society International
P.O. Box 953
Georgetown, CT 06829
(203) 544-8617

Chesapeake Bay Foundation, Inc.
162 Prince George St.,
Annapolis, MD 21401
(410) 268-8816

Clean Ocean Action
P.O. Box 505
Sandy Hook, NJ 07732
(732) 872-0111

Coastal Conservancy [California]
1330 Broadway, Suite 1100
Oakland, CA 94612
(510) 286-1015

Coast Alliance
215 Pennsylvania Ave., SE
Washington, DC 20003
(202) 546-9554

Coastal Conservation Association, Inc.
4801 Woodway, Suite 220 West
Houston, TX 77056
(713) 626-4234

Coastal Society (The)
P.O. Box 25408
Alexandria, VA 22313-5408
(703) 768-1599

Coastal Advocates
263 N. Santa Cruz Avenue, Suite 237A
Los Gatos, CA 95030
(408) 395-9116

Coral Reef Alliance
64 Shattuck Sq., Suite 200
Berkeley, CA 94704
(510) 848-0110

Cousteau Society (The)
870 Greenbrier Circle, Suite 402
Chesapeake, VA 23320
(757) 523-9335

Friends of the Sea Otter
2150 Garden Rd., B-4
Monterey, CA 93940
(408) 373-2747

Greenpeace USA, Inc.
1436 U Street, NW
Washington, DC 20009
(202) 462-1177

Maine Coast Heritage Trust
169 Park Row
Brunswick, ME 04011
(207) 729-7366

Marine Environmental Research Institute (MERI)
772 W. End Ave.
New York, NY 10025
(212) 864-6285

Marine Mammal Center (The)
Marine Headlands
Golden Gate National Recreation Area (GGNRA)
Sausalito, CA 94965
(415) 289-7325

National Coalition for Marine Conservation
3 W. Market St
Leesburg, VA 20176
(703) 777-0037

National Marine Educators Association
P.O. Box 51215
Pacific Grove, CA 93950
(produces *Current*)
(408) 648-4841

National Watershed Coalition
9150 W. Jewell Ave., Suite 102
Lakewood, CO 80232-6469
(303) 988-1810

National Waterways Conference, Inc.
1130 17th St., NW
Washington, DC 20036
(202) 296-4415

North American Wetlands Conservation Council
4401 North Fairfax Dr.
Arlington, VA 22203
(703) 358-1784

Ocean Voice International
P.O. Box 37026
3332 McCarthy Rd.,
Ottawa, Ontario, Canada
K1V 0W0
(613) 990-8819

Puget Soundkeeper Alliance
1415 W. Drayus
Seattle, WA 98119
(206) 286-1309

San Francisco Estuary Project
c/o RWQCB
1515 Clay St., Suite 1400
Oakland, CA 94612
(Publishes *Estuary*)
(510) 622-2300

Save the Dunes Council
444 Barker Rd.
Michigan City, IN. 46360
(219) 879-2937

Save the Harbors/Save the Bay
25 West St., 4th Fl
Boston, MA 02111
(617) 451-2860

Save San Francisco Bay Association
1736 Franklin St., 4th Floor
Oakland, CA 94612
(510) 452-9261

Save Our Shores
O'Neil Building, Suite 5A
2222 E. Cliff Drive
Santa Cruz, CA , 95062
(831) 462-5660

Sea Shepherd Conservation Society
P.O. Box 628
Venice, CA 90294
(310) 301-7325

Surfrider Foundation
P.O. Box 2704 #86
Huntington Beach, CA 92647
(949) 631-6273

JOURNALS/MAGAZINES/NEWSLETERS

There are also hundreds of journals/magazines/newsletters that have articles on coastal/marine issues. The selected list below only identifies those that concentrate on coastal/marine issues.

Bulletin of the Coastal Society
Bulletin of Marine Science
Calypso Log
Campnet
Coastal Erosion Bulletin
Coastline: European Union for Coastal Conservation
Coastlines
Coastal Management
Coastal Management in Tropical Asia
Coastal Zone Management
Coastal News
Current: The Journal of Marine Education
EMECS Newsletter [Environmental Management of Enclosed Coastal Seas]
Estuaries
Estuarine Chemistry
Estuarine and Coastal Marine Sciences
Estuarine Coastal and Shelf Science
Estuary: Your Bay-Delta News Clearinghouse
Faro. (In Spanish). A Newsletter for Coastal Zone Managers
ICCOPS Newsletter (International Centre for Coastal and Ocean Policy Studies)
Intercoast Network Newsletter
Journal of Coastal Conservation
Journal of Coastal Research
Limnology and Oceanography
Littoral
Marine Biology
Marine Connection (The)
Marine Conservation News
Marine Fisheries Review
Marine Fish Management
Marine Mammal News
Marine Policy Reports
Marine Resource Economics
Maritime Humanities Newsletter
Maritime Policy and Management
Moss Landing Marine Laboratories Technical Publications
Ocean and Arctic Engineering
Ocean and Coastal Management
Ocean Development & International Law
Oceans Magazine
Oceanus
Progress in Oceanography
Reefkeeper Network
Sanctuary Currents
Shore and Beach
Wetlands

INTERNET RESOURCES FOR COASTAL PROFESSIONALS

There is a growing number of internet websites. The following resource list is intended only to offer a starting point for exploration in the electronic media.

American Shore and Beach Preservation Association
http://www2.ncsu.edu/ncsu/CIL/ncsu_kenan/shore_beach/

Coastlines
http://www.epa.gov/nep/coastlines/coastlines.html.

Department of Agriculture (U.S.)
http://www.usda.gov

Department of Energy (U.S.)
http://www.doe.gov

NOAA's Office of Ocean and Coastal Resource Management
http://www.nos.noaa.gov/ocrm/

National Sea Grant College
http://www.mdsg.umd.edu/NSGO/index.html

OTHER DIRECTORIES

Federal agencies (e.g., U.S. Fish and Wildlife Service), legislative offices (e.g., Assemblymember or Senate), regional governments (e.g., Association of Monterey Bay Area Governments), marine sanctuary offices (e.g., Monterey Bay National Marine Sanctuary), coastal commissions (e.g., California Coastal Commission), and education centers (e.g., Coastal Education and Research Foundation [CERF]) often put out their own resource directories. Try contacting them for their latest editions.

Glossary and Acronyms

AEC (Ch. 2). Areas of Environmental Concern.

Algal bloom (Ch. 5). A periodic explosion of algae populations on the surface of stagnant streams, ponds, or lakes. The process can shade out other aquatic plants as well as use up the water's oxygen supply as the plants decompose. Fish kills may result. Blooms are often caused by pollution from excessive nutrient input, such as nitrates and phosphates.

ATBA (Ch. 7). Area To Be Avoided.

ATOC (Ch 5). Acoustic Thermometry of Ocean Climate.

Atoll (Ch. 1). A ringlike coral reef island surrounding a shallow lagoon. The circular coral reef developed on the flanks of a subsiding volcano. Compare *barrier reef* and *fringing reef.*

Backshore zone (Ch. 1). Back or upper zone of the beach (shore), which extends from cliffs or sand dunes down to high water mark (HWM). This zone is normally beyond the reach of ordinary waves or tides; generally only wetted by storm tides. The backshore zone may have sand dunes, berms (ridges of wave-heaped sand and/or gravel), or marshes. Compare *nearshore* and *offshore.*

Backwash (Ch. 1). Water receding seaward after waves have broken on a beach. The backwash sweeps sediment seaward.

Barrier beaches (Ch. 1). A long, narrow beach rising slightly above the high tide level, generally parallel to the shore, and separated from the mainland by a lagoon, bay, or river mouth. This term is only appropriate for beaches of barrier islands and barrier spits. Compare *pocket beaches* and *mainland beaches*.

Barrier islands (Ch. 1). Elongate islands of sand (offshore bars) parallel to the shore, separated from the mainland by a bay or lagoon. They often appear as part of barrier island chains—a series of *barrier islands, barrier spits,* and *barrier beaches* extending along a coast a considerable distance.

Barrier reef (Ch. 1). A long, narrow coral reef generally parallel to a coast and separated from it at some distance by a considerably wide and deep lagoon or bay. Breaks or openings in the reef enable movement of shipping. Compare *fringing reef* and *atoll.*

Barrier spit (Ch. 1). A long tongue-like depositional landform (sand or gravel) extending from a headland and partially crossing a bay.

Bay barrier (Ch. 1). A barrier spit of sand or gravel that eventually extends all the way across a bay, cutting it off from the sea and forming a lagoon; produced by littoral drift and wave action. Sometimes called a *baymouth bar.*

BCDC (Ch. 2). San Francisco Bay Conservation and Development Commission.

Beach (Ch. 1). The strip of unstable unconsolidated material (e.g., sand, gravel) bordering the sea that lies between the backshore and low water marks. This gently sloping shore of unconsolidated material is constantly washed by waves, wind, and tidal currents.

Beach drift (Ch. 1). Sand, gravel, shells, and other unconsolidated materials that are moved *on the beach* by the littoral (longshore) current in the effective direction of the waves.

Beach nourishment (Ch. 4). Replenishing a beach with sand that was lost to erosion. This process may be done by such methods as piping in clean dredge spoils from nearby harbor entrances, trucking in sand from distant sand dunes or sand quarries, or constructing groins to capture littoral drift.

Benthic (Ch. 3). Pertaining to at or near the bottom of oceans, lakes, or rivers, and the organisms that inhabit this bottom zone. Opposite of *pelagic.*

Benthos (Ch. 7). Sedentary plant and animal organisms (e.g., seaweed or coral) that live in the *benthic* (bottom-dwelling) zone of oceans, lakes, or rivers.

Bioaccumulation (Ch. 5). The absorption and concentration of toxic chemicals (e.g., heavy metals or chlorinated hydrocarbons) in living organisms.

Biodiversity (Ch. 3). The variety of flora and fauna species living within a given ecosystem, habitat, or region. A principle of ecology: the greater the biodiversity of a region, generally the greater its overall stability.

Biogenic (Ch. 1). Originating from organic living organism activity, such as the growth of plants (*phytogenetic coasts,* such as mangrove or reed coasts) or animals (*zoogenetic coasts,* such as coral coasts).

Biological oxygen demand (B.O.D.) (Ch. 6). A measure of the dissolved oxygen utilized by aerobic microorganisms in water rich with organic matter. In otherwords, a measure of the amount of oxygen required for respiration and organic breakdown in a water body.

Bioremediation (Ch. 7). The use of microorganisms for the decontamination of ocean water, groundwater, or soil. For example, the use of special bacteria strains to help degrade (clean up) oil spills that occur in bays.

Bioturbation (Ch. 6). The alteration of the ocean bottom sediment by benthic organisms. Benthic organisms cause changes by moving on the surface sediment, and from burrowing into deeper layers.

Blowout (Ch. 7, in terms of oil). An uncontrolled flow of gas, oil, or other fluids from a well platform into the surrounding air and water. (Ch. 1, in terms of dunes). Wind hollows or basins within sand dunes.

BPD (Ch. 7). Barrels Per Day.

Breaker (Ch. 1). A sea-surface wave that has become so steep that it becomes unstable and finally collapses, forming *breakers* that "break" (plunge, spill, or surge) on the shore, over a reef.

Breaker zone (Ch. 1). The area within the *nearshore zone* where *breakers* occur. See *breaker.*

Breakwater (Ch. 4). An artificial offshore structure aligned parallel to shore, usually to protect a harbor, anchorage, or shore area from waves.

Bulkhead (Ch. 4). A steel, wood, or concrete wall erected parallel to and near the high water mark for the purpose of protecting adjacent uplands from waves and current action. It can also serve as a retaining wall.

Carrying capacity (Chs. 1, 8). The maximum population size (e.g., number of cattle) that a given habitat (e.g., range of a given acreage) will support indefinitely under normal environmental conditions (e.g., without drought).

CBRS (Ch. 2). Coastal Barrier Island Resources System.

CCA (Ch. 2). California Coastal Act.

CCC (Ch. 2). California Coastal Commission.

CCMP (Ch. 2). California Coastal Management Program.

CEQA (Ch. 2). California Environmental Quality Act.

Chumming (Ch. 3). The traditional practice of throwing animal waste products (e.g., blood and intestines) into the ocean as bait. It is a controversial practice to attract sharks for "thrill diving."

CINMS (Ch. 3). Channel Islands National Marine Sanctuary.

Cluster development (Ch. 2). A subdivision strategy for conserving open space. By allowing building lots to be smaller, and grouping them together, the remaining land (open space) can be used for recreational or conservation purposes. Usually, the overall housing density remains the same as with a conventional subdivision plan.

Coast (Ch. 1). The area from the *shoreline* (the line of intersection of a water body with land) extending inland to the limit of tidal or sea-spray influence. It also includes the *nearshore zone* (the waters from the shoreline to the outer limit of the *continental shelf).* Synonymous with the term *coastal zone*. See *shoreline*, *nearshore zone*, and *continental shelf.*

Coastal area (Chs. 1, 8). A term used to distinguish the coastal zone according to political or cultural elements, such as states or counties.

Coastal barrier (Ch. 1). A general term used for natural geomorphic features that form offshore and help protect the mainland from the brunt of storms. Types of coastal barriers include *barrier spits, bay barriers,* and *barrier islands.*

Coastal bluff (Ch. 1). A coastal bluff (also called *sea cliff*) is a steep coastal slope created by waves, eroding rock and sediment at its base.

Coastal hazards (Ch. 4). Storms, flooding, hurricanes, tsunamis, earthquakes, and other natural phenomena that cause damage to natural resources, property, and human populations in coastal regions.

Coastal inlet (Ch. 1). A narrow channel or passage-way that leads inland. Although the term *coastal inlet* connotes flow in one direction, most act in terms of bidirectional flow between the land and sea. Three major types of coastal inlets are *lagoons, deltas,* and *estuaries.*

Coastal [zone] management (Ch. 2). Term used in the United States in its 1972 Coastal Zone Management Act. Implied management of all uses in the coastal zone. However, U.S. coastal management programs (CZMs) initially emphasized management of dryside (land), not wetside (ocean), uses in the coastal zone. Integration among uses and policies was not emphasized in the 1972 Act.

Coastal pollution (Ch. 5). Various forms of contaminants in the coastal area, such as *water pollution* (e.g., sediment, toxic wastes) and *marine debris* (e.g., medical waste and garbage).

Coastal zone (Ch. 1). The interface between land and sea, where each has an influence on the other. A controversial term in that there are many interpretations. Whereas some planners may define the *coastal zone* according to complete natural (biogeographical) systems, politicians often use political boundaries. For example, the Resource Agency of California refers to: "the area extending inland approximately 1/2 mile (1 km) from the mean low tide and seaward to the outermost limit of the state boundaries." Coastal scientists generally think of the coastal zone as having the following characteristics: (1) variable width; (2) extending seaward to the edge of the continental shelf; (3) no set landward demarcation; and (4) a fragile zone where humans can easily interrupt or destroy natural ecosystems and natural processes (e.g., chemical, biological, physical).

Coastline (Ch. 1). See *Shoreline.*

CoBRA (Ch. 2). The Coastal Barrier Resources Act.

COE (Ch. 2). U.S. Army Corps of Engineers.

Composting (Ch. 6). A controlled process of degrading organic residues or a mixture of organic residues and soil by microorganisms. By letting organic wastes decompose in the presence of air, a nutrient-rich humus, or compost, is the resulting product.

Consistency (Ch. 2). Consistency provision of the Coastal Zone Management Act of 1972 (Section 1456) allowing states to review and control federal activities which may adversely impact the state's coastal zone, resulting in a violation of that state's federally approved coastal management program. In otherwords, the term *consistency* refers to the right of coastal states to determine if proposed federal actions are consistent with their own approved CZM programs.

Continental shelf (Ch. 2). The portion of the sea floor adjacent to a continent or an island, covered by shallow water that is 200 m (656 ft) or less in depth; gently sloping (1 degree or less); generally excellent fishing grounds.

Cryogenic (Ch. 1). Originating from processes involving frozen liquids. For example, a coastline carved out by moving ice sheets. Compare to *Biogenic.*

CSCC (Ch. 2). California State Coastal Conservancy.

Cultural Eutrophication (Ch. 5). Acceleration by humans of the natural process of enrichment (aging) of water bodies. Agricultural, urban, and industrial discharge are the primary human-made sources that overnourish aquatic ecosystems. Compare *Eutrophication.*

CWA (Ch. 2). Clean Water Act.

CZMA (Ch. 2). Coastal Zone Management Act. The 1972 federal act designed to "protect, preserve, develop and, where possible, restore, or enhance the resources of the nation's coastal zone."

Deep ocean disposal (Ch. 6). Refers to the controversial waste management scheme to use non-coastal waters (ocean waters 914 m [3,000 ft] deep or more) to dispose of sludge, industrial chemicals, and/or radioactive wastes.

Deltas (Ch. 1). A fan-shaped depositional deposit where a river enters a lake or an ocean.

Derrick (Ch. 7). The triangular-shaped component of an offshore oil rig that is above the ocean surface.

Dredging (Ch. 6). The removal of underwater bottom materials (e.g., sand, silt, rock, coral) by suction, scooping, or other dredging technique. Many harbor inlets need to be periodically dredged in order to keep them open.

Drill mud (Ch. 7). A special mixture of water or refined oil, clay, and chemical additives that are used to cool the drill bit of an offshore oil well operation.

Drillship (Ch. 7). A self-contained, self-propelled, ship equipped with a derrick for drilling wells in deep water.

DWT (Ch. 7). Dead Weight Tons.

Ebb tide (Ch. 1). A falling or lowering tide during the daily tidal cycle. Compare *flood tide.*

Ecotone (Ch. 1). A transition zone where two adjacent ecosystems merge together, rather than change abruptly. The zone has characteristics of both kinds of neighboring vegetation, yet it possesses characteristics of its own.

Ecotourism (Ch. 8). An "eco-friendly" (low-impact) form of tourism that stresses environmental and cultural conservation.

EPA (Ch. 2). U.S. Environmental Protection Agency.

ERV (Ch. 7). Emergency Offshore Rescue Vessel.

ESA (Ch. 2). Endangered Species Act.

Estuary (Ch. 8). A region in which fresh water from a river or stream mixes with salt water from the sea; this semi-enclosed, funnel-shaped tidal mouth of a river valley is an important nursery ground for many marine animals..

Eutrophication (natural eutrophication) (Ch. 5). The natural aging process by which all lakes "die" (i.e., evolve into a marsh and eventually disappear). This natural process is often accelerated by human activities. Compare *cultural eutrophication.*

FBNMS (Ch. 3). Fagetele Bay National Marine Sanctuary.

FCMP (Ch. 2). Florida Coastal Management Program.

FDCA (Ch. 7). Florida Dept. of Community Affairs.

FDEP (Ch. 7). Florida Dept. of Environmental Protection.

FDPA (Ch. 2). The Flood Disaster Protection Act.

FEMA (Ch. 2). Federal Emergency Management Agency.

FIRWD (Ch. 6). Farallon Islands Radioactive Waste Dump.

Fjord (Ch. 1). A coastal valley that was carved out by a glacier then flooded by the sea. These glacial troughs are characterized by partial submergence, steep parallel walls, truncated spurs, and hanging valleys.

Flood tide (Ch. 1). The period when the tide is rising during the daily tidal cycle. Compare *ebb tide.*

Flume (Ch. 5). A long, narrow, artificial channel or chute to direct a stream of water. Flumes can be designed to control water levels in an artificial lagoon, to help fish migrate upstream and back to sea, or to convey logs from a logging site to a road for transport.

Foredune (primary dune) (Ch. 4). A dune or line of dunes nearest the sea. The first, frontal, or primary dune is periodically eroded and then built up again. Coastal scientists and conservationists generally agree that foredunes should be off- limits to development.

FORT (Ch. 7). Fishermen's Oil Response Team.

FOSC (Ch. 7) Federal On-Scene Coordinator.

Fringing reef (Ch. 1). A coral reef built directly against the shore of an island or continent. Its rough, table-like surface is exposed at low tide. There may be a shallow channel or lagoon between the reef and the coast, although strictly there is no body of water between the reef and the land upon which it is attached. Compare *barrier reef* and *atoll.*

FWPCA (Ch. 2). Federal Water Pollution Control Act.

FWS (Ch. 2). U.S. Fish and Wildlife Service.

GFNMS (Ch. 6). Gulf of the Farallones NMS.

GLP (Ch. 2). Great Lakes Program.

GLNPO (Ch. 2). Great Lakes National Program Office.

Green marketing (Ch. 6). The buying and promotion of commercial products that are "gentle" or beneficial to the environment. For example, boat owners that wash their boats in harbor slips with biodegradable "boat soap" are practicing green marketing, in that they are using a commercial product that is benign to the environment.

Groin (Ch. 4). An elongate wall constructed of rock, concrete, or wood built perpendicular to the shore into the surf zone. Its purpose is to reduce localized beach erosion or build up a beach by intercepting littoral (longshore) drift.

Gunnite (Ch. 4). A shoreline protective strategy. First, a framework or other type of structure is built around the eroding sea cliff. Concrete in slurry form is then sprayed on to the framework, to which it adheres and hardens. Like most shoreline protective devices, gunnite is only a temporary solution and is most often aesthetically displeasing.

Habitat (Ch. 8). The natural environment of a particular plant or animal. It includes all the essentials of life for that organism, such as food, water, and cover.

Halophytes (Ch. 8). Plants that can grow in soils that have a high salt content, such as those found in coastal areas.

HIHWNMS (Ch. 3). Hawaiian Islands Humpback Whale National Marine Sanctuary.

Hinterland (Ch. 1). A coastal hinterland refers to the upland areas behind the coast (e.g., watersheds, water course, hills, plains) that "feed" materials (e.g., fluvial sediment; toxic chemicals; human debris) to the coast.

Human scale (Ch. 8). A human settlement that has a good balance between humans and nature. Most environmentalists would argue that villages and towns are more at a *human scale* than giant "out of scale," (out of control?) places like New York City, Los Angeles, or Chicago, where humans and their artificats dominate the landscape.

Hydromulching (Ch. 4). The application of a layer of mulch to the ground by spraying a slurry of organic materials such as straw, leaves, plant residues, or sawdust. This technique of mulch application is often used in coastal dune restoration projects.

ICM (Ch. 2). Integrated Coastal Management. A continuous and dynamic process designed to manage coastal and marine areas and resources whereby participation by all affected economic sectors, government agencies, and non-government organizations is involved. It is an effort to overcome the fragmentation inherent in coastal resource management. The concept of ICM is more of a "goal to move toward" than something that may ever be achieved.

IMO (Ch. 2). International Maritime Organization.

Incineration (Ch. 6). A management strategy to dispose of municipal solid waste. Combustible solid, liquid, or gaseous wastes are burned and changed into noncombustible gases.

Integrated solid waste system (Ch. 6). The *integrated* (combined) waste management strategy that uses conventional systems (e.g., landfills and incineration) along with nonconventional systems (e.g., precycling and resource recovery).

IPM (Chs. 5, 8). Integrated Pest Management. A method of pest management that uses a combination of biological and cultural controls with a minimum use of chemicals.

Jacket (Ch. 7). The "legs" (supporting structure) of an offshore oil platform. The jacket consists of large-diameter pipe welded together to form a multilegged stool-like structure. The jacket is secured to the ocean floor with pilings (heavy beams of steel driven into the earth).

Jackup rig (Ch. 7). A floatable barge which supports an offshore oil drilling rig. Once the barge is at the desired location, extendable legs from the barge are lowered and fastened to the ocean floor.

Jetty (Ch. 4). An elongated structure built of rock, cement, or steel projecting into a body of water. Jetties are built in pairs usually at the mouth of a harbor or river intersecting the coast. The purpose is to stabilize a channel for navigation.

KLNMS (Ch. 3). Key Largo National Marine Sanctuary.

Lagoon (Ch. 1). A shallow stretch of seawater between the mainland and a narrow, elongate strip of land, such as a reef, barrier island, sandbar, or spit; especially the water body between an offshore coral reef and an island. Lagoons may have one or more of the following characteristics: extends roughly parallel to the coast; restricted circulation, and consequently may be stagnant; limited fresh water input; and high salinity.

LAMP (Ch. 2). Lakewide Management Plans.

LCP (Ch. 2). Local Coastal Program.

LDC (Ch. 6). London Dumping Convention.

Lease Sale (Ch. 7). The process by which the Minerals Management Service (MMS) offers specific Outer Continental Shelf (OCS) tracts to oil companies for lease.

Liquification (Ch. 4). The process by which loosely compacted soils become "liquid" due to the high frequency shaking caused by an earthquake. The subsequent fluid motion of the ground results in severe damage to structures built in this area.

Littoral drift (Chs. 1, 4). The movement of sand and other material by littoral (longshore) currents along the shore in a direction parallel to the beach.

LKNMS (Ch. 3). Looe Key National Marine Sanctuary.

LOS (See **UNLOSC**).

Mainland beaches (Ch. 1). The type of beach that stretches unbroken for several miles along the edges of a mainland or major land mass. Compare *pocket beaches* and *barrier beaches.*

Mangrove (Ch. 1) *Mangles,* or thick tangles of woody shrub and tree roots of various taxonomic groups. Mangroves are most noted for their spectacular display of prop (above ground) roots. Mangroves are limited geographically to the tropical (frost free) zone of 25 degrees N and 25 degrees S latitude.

Marine and coastal protected area (Ch. 3). Areas of water (e.g., bays, lagoons, estuaries) or coastal land (e.g., headlands, dunes, beaches) that are specially designated to protect biological diversity, increase environmental awareness, provide sites for recreation, as well as study areas for research and monitoring.

Marine debris (Ch. 5). One type of coastal pollution. This term refers to all the types of wastes (floatable debris or litter) that accumulate on our beaches and shores, such as plastic and glass bottles, cigarette butts, soft-drink cans, tires, plastic bags, etc. Much of this *marine debris* comes from urban runoff (via creeks and streams within the adjacent watershed) and constant use by humans of the local beaches. Compare *water pollution.*

Marine reserve (Ch. 3). In general, *marine reserves* are designed to be exclusive (i.e., public access and use are restricted). These areas are *replenishment zones* for fish, marine mammals, or other organisms, and, consequently, can be thought of as a nonexploitation sanctuary. Compare *marine sanctuary.*

Marine sanctuary (Ch. 3). An area protected under the federal Marine Protection Research and Sanctuaries Act of 1972. Although considered "protected" to some degree, these areas remain *exploitation zones* in that humans can have access and use its natural resources (e.g., commercial and recreational fishing, kelp harvesting, jet skiing, kayaking, sailing, etc.). Compare *marine reserve.*

Marine terrace (Ch. 1). An elevated (uplifted sea floor), seaward-sloping platform, that was cut and eroded by waves. Multiple terraces commonly exist at different elevations . Also called *coastal terrace* or *wave-cut bench.*

MARPOL (Chs. 2, 6). International Convention for the Prevention of Pollution from Ships.

MBARI (Ch. 3). Monterey Bay Aquarium Research Institute.

MBNMS (Ch. 3). Monterey Bay National Marine Sanctuary.

MMRP (Ch. 5). Marine Mammal Research Program.

MMPA (Ch. 2). Marine Mammal Protection Act.

MPA (Ch. 3). Marine and Coastal Protected Area.

MPPRCA (Ch. 6). Marine Plastics Pollution Research & Control Act.

MPRSA (Chs. 2, 3). Marine Protection, Research and Sanctuaries Act.

Motorized personal water craft (Ch. 8). This term usually refers to jet skiis—the highly controversial "motorcycles on water" that are super fast, highly noisy, and especially dangerous to other boaters, not to mention marine wildlife.

MSCFD (Ch. 7). Million Standard Cubic Feet Per Day.

Mud flats (Ch. 8). A relatively flat foreshore composed of fine silt, usually exposed at low tide but covered at high tide, occurring in sheltered estuaries or *behind coastal barriers,* such as *sand spits.* Some plants, such as *mangroves* within tropical waters, help trap particles of mud and bind it with their roots.

Nearshore Zone (Ch. 1). One of the subdivisions of the coast. It lies between the *backshore zone* and the *offshore zone.* Unlike the *backshore* the nearshore is mostly underwater.

NEPA (Ch. 2). National Environmental Policy Act.

Neritic (Ch. 6). A zone of relatively shallow water above the *continental shelf* that includes the *intertidal zone,* extends from the high tide mark to the edge of the continental shelf, and contains extensive neritic deposits derived from the remains of shellfish, sea urchins, and coral.

NFIA (Ch. 2). National Flood Insurance Act.

NFIP (Ch. 2). National Flood Insurance Program.

NMFS (Ch. 2). National Marine Fisheries Service.

NMSP (Ch. 3). National Marine Sanctuaries Program.

NOAA (Ch. 2). National Oceanic and Atmospheric Administration.

Nonpoint Source (pollution) (Ch. 5). Pollution arising from a broad, ill-defined source rather than a discrete point. Nonpoint sources of pollution include runoff from grazing lands, cultivated fields, and manure disposal areas. Leaking sewerage systems and saltwater intrusion are also nonpoint sources of pollution. Compare *point source.*

NPS (Ch. 2). National Park Service.

Ocean Dumping (Ch. 6). The disposal of waste materials (e.g., garbage, sewer sludge, industrial waste) by barge or ship to a particular site in the ocean.

OCNMS (Ch. 3). Olympic Coast National Marine Sanctuary.

OCS (Ch. 7). Outer Continental Shelf. The seaward part of the continental shelf beyond the state-controlled three-mile limit.

ODBA (Ch. 6). Ocean Dumping Ban Act.

OECD (Ch. 2). Organization of Economic Cooperation and Development.

Offshore zone (Ch. 1). That geographic area that lies seaward of the *nearshore zone.* Since it is "offshore," it is not considered part of the littoral (coastal) area. Compare *nearshore zone.*

Open Space (Ch. 8). Any land on which buildings or other forms of development do not dominate the landscape. Wilderness areas, parks, and farmlands are major types of open space.

OS&T (Ch. 7). Offshore Storage and Treatment Vessel.

OTEC (Ch. 6). Ocean Thermal Energy Conversion.

Platform (Ch. 7). An offshore oil drilling structure. It consists of a *jacket* (underwater frame or legs attached to the sea floor) and attached *deck* (visable area above water where drilling activities occur).

Pocket beaches (Ch. 1). A small narrow beach formed in a "pocket"—an enclosed or sheltered place along a coast such as between two points or headlands; often at the mouth of a coastal stream. Compare *mainland beaches* and *barrier beaches.*

Point Source (pollution) (Ch. 5). Pollution arising from a discernible, discrete origin, usually stationary, such as the discharge from an industrial site, a sewer pipe, a contaminated well, or a drainage ditch. Compare *nonpoint source.*

POTWs (Ch. 6). Publicly Owned Treatment Works.

Precycling (Ch. 6). One form of waste minimization (i.e., source reduction). Consciously purchasing merchandise that has less waste to discard or recycle (e.g., a product with minimal packaging). Compare *recycling.*

Primary treatment (Ch. 5). The first stage in wastewater treatment, wherein most floating debris and solids are mechanically removed by screening and sedimentation; often includes chlorination. This process removes only about 30 percent of BOD from domestic sewage, and less than half of the metals and toxic organic substances. Compare *secondary treatment* and *tertiary treatment.*

Processing (Ch. 7). In terms of oil and gas processing, this term refers to the preparation of the product (e.g., dewatering) that occurs before oil and gas can be sent to refineries.

RAP (Ch. 2). Remedial Action Plans.

Reclamation (Ch. 8). A term that has changed meaning over time. In the 1800s, reclamation referred to *"reclaiming, or taking back"* a piece of land from nature that was considered a wasteland (e.g., wetland), and transforming it for human use (e.g., draining and filling a marsh for agriculture). Today, the term may also mean *restoring* a piece of land that has been disturbed by a natural disaster such as a fire or flood.

Recycling (Ch. 6). One form of waste minimization. A conservation technique that goes beyond mere reuse of a product (e.g., returnable bottles). Recycling involves the recovery, reprocessing, and refabrication of components from society's industrial waste products (e.g., aluminum obtained from aluminum cans). Compare *precycling.*

Replenishment zone (Ch. 3). A *non-exploitation sanctuary* to enhance replenishment of fishery stocks. Compare *marine sanctuary.*

Resource recovery (Ch. 6). One form of waste minimization. It involves the recovery of useful materials (e.g., metals, paper, glass) or energy from solid wastes before their ultimate disposal. There are two major types of resource recovery: recycling and composting. Compare *precycling.*

Restoration (environmental restoration) (Ch. 4). Repair or rehabilitation of an ecological habitat (e.g., wetland) to restore its structure and ecological processes. Wetlands replication is a major subfield of restoration ecology.

Revetment (Ch. 4). A structure built to protect a shoreline bluff or embankment from erosive wave action. The structure is usually constructed of stones that are laid with a sloping face. Generally better designed than *riprap,* since it employs the concept of layering and rock gradation. Compare *riprap.*

Riprap (Ch. 4). Boulders (1–5 tons) or rubble stacked along the shore to protect a cliff, bluff, dune, house or other structure from wave attack. Less sophisticated than a revetment. Compare *revetment.*

Rocky headlands (Ch. 2). A high, rocky projection of the land into the sea.

Rolling setback (Ch. 4). One type of land use control device to help protect houses and other structures from coastal hazards. This regulation requires that the seaward-most develop line be slowly moved landward away from the hazardous zone (e.g., eroding cliff). In other words, new buildings would have to be built further back from a hazardous area than previously built houses. Compare *stringline setback.*

ROW (Ch. 7). Right of Way.

SALM (Ch. 7). Single Anchor Leg Mooring.

Salt marsh (Ch. 8). A type of marshland that is rooted in saline soils, alternatively inundated and exposed by tides, and occurs on the shores of bays. These low, wet, muddy areas are usually covered with a mat of grassy halophytic plants.

Saltwater Intrusion (Ch. 5). The phenomenon occurring when a body of salt water, because of its greater density, invades a body of fresh water. Although it can occur in either surface-water or groundwater bodies, the term is most often associated with underground displacement of freshwater by saltwater. This encroachment of saltwater in underground situations may be from natural causes, or from human induced changes, such as over pumping of groundwater for drinking or agricultural purposes.

SAMP (Ch. 2). Special Area Management Planning.

Sand dunes (Ch. 1). Any mound, ridge, or hill of loose sand, heaped up by the wind. Dunes are generally classified as *primary* or *foredune* (most seaward), or *secondary* (landward, behind the primary) dunes. Most coastal scientists agree that houses or other structures should not be built on primary dunes, due to their unstable nature.

Scrubber (Ch. 6). An air pollution control device to reduce flue gas emissions.

SEA (Ch. 5). Strategic Environmental Assessments.

Seamount (Ch. 3). An isolated peak rising from the floor of an ocean. Most are extinct volcanoes that have been worn down and submerged below sea level.

Seawall (Ch. 4). A concrete wall built parallel to the coastline to prevent erosion and other damage by wave action; normally more massive and better engineered than other types of shoreline protective devices.

Secondary treatment (Ch. 5). The second stage of wastewater treatment, wherein the wasteflow is subjected to (1) bacteria, which consumes the organic wastes and (2) chlorination, which disinfects (kills) disease-carrying bacteria and some viruses. Secondary treatment may remove up to 90 percent of BOD. Compare *primary treatment* and *tertiary treatment.*

Semisubmersible (Ch. 7). One type of offshore oil drilling rig. It has a barge-like hull with buoyant pontoons. This type of rig is designed to operate in ocean depths of 91 to 457 m (300 to 1,500 ft).

Sense of Place (Ch. 8). A term used by cultural geographers, urban planners, and others that refers to a village, town, or geographical region's special character, personality, or distinctiveness. These are generally qualities that local planners want to preserve.

Setbacks (Ch. 2). A legally prescribed boundary seaward of which no new building or other structure is allowed to be built. A setback (buffer or exclusion zones) is a coastal resource management technique to keep developers from building too close to hazardous or sensitive features, such as an eroding seacliff.

Shoaling (Ch. 4). The process by which a water body becomes gradually shallow. For example, harbor entrances often become "shoaled" by littoral drift. To keep the harbor from becoming too shallow for navigation purposes, the shoaled material (sand and silt) needs to be periodically dredged. See *dredging.*

Shoreline (Ch. 1) A *shoreline* (or *coastline*) is the line of intersection of the land with the sea.

SIOSC (Ch. 7). The State (California) Interagency Oil Spill Committee.

Slough (Ch. 8). A marsh, swamp, pond, or bog which is part of an inlet or backwater. Usually consists of mudflats, tidal inlets, and marsh grasses.

Sludge (Chs. 5, 6). The solid or semisolid by-product of wastewater that remains after municipal, commercial, or industrial wastewater treatment. Usually associated with sedimentation at sewage treatment plants.

Snow fence (Ch. 4). A fence erected for the purpose of breaking the force of wind moving snow or sand. Snowfences are used on coastal sand dunes where resource managers attempt to keep sand from moving onto adjacent paths or roadways.

Soft soil (Ch. 4). Recent coastal deposits or fills, considered high risk areas for *liquification* resulting from earthquake-induced ground shaking. See *liquification.*

Source reduction (waste minimization) (Ch. 6). An umbrella term that includes reducing the volume of material goods, eliminating unnecessary packaging, and decreasing the amount of toxic substances in products. Includes such techniques as *precycling* and *resource recovery.*

Storm flooding (Ch. 1). Heightened *flooding* due to hurricanes and coastal storms.

Storm surge (Chs. 1, 4). Heightened *wave action* due to hurricanes and coastal storms (i.e., rise above normal water or tidal level on the open coast).

Storm tide (Ch. 4). *Storm surge* supplemented by tidal condition. For example, if high tides are normally at 1.8 m (6 ft), a 5.2 m (17 ft) storm surge creates a 7 m (23 ft) storm tide. See *storm surge.*

Stringline set-back (Ch. 4). A land use strategy to regulate distance structures should be from hazardous areas on the coast. According to this system, future houses and other structures cannot be built any closer to a hazardous area (e.g, eroding coastal cliff) than existing structures. Compare *rolling set-back.*

Subduction (Ch. 6). The process in which, when two of Earth's crustal plates converge, one "dives" beneath the other crustal plate.

Subsidence (Ch. 4). Sinking of the ground surface due to the removal of large quantities of underground water or oil.

Surf zone (Ch. 1). The area between the *breaker zone* and the *swash zone.* It is here where broken waves travel toward the shore. Compare *breaker zone* and *swash zone.*

Swash (Ch. 1). The wave uprush on a sloped beach within the *swash zone.* Compare *backwash.*

Swash zone (Ch. 1). The sloping part of the beach that is alternately covered with the uprush of waves (*swash*) and immediately uncovered (*backwash*).

Tertiary Treatment (Ch. 5). Advanced cleansing of waste water beyond the primary (mechanical) and the secondary (biological) treatments. Nutrients (e.g., phosphorus and nitrogen) and a high percentage of the suspended solids are removed from the water. Compare *primary treatment* and *secondary treatment.*

Tetrapod (Ch. 4). Multi-ton concrete blocks with four appendages that are used for coastal protection. These interlocking blocks are often used instead of large boulders in the construction of *groins* and *jetties.*

Tidal range (Ch. 1). The average difference in water level between high and low tide. See *Tides.*

Tidal scouring (Ch. 8). The washing away of sediments by strong tidal currents. For example, the banks of tidal inlets can be "scoured" if tidal flushing is heightened by some form of human disturbance (e.g., the building of a harbor at the mouth of an estuary).

Tidal Wave (Ch. 4). A heightened wave associated with the astronomical tidal forces of the sun and moon. The term tidal wave is *not* another word for tsunami, as is often believed. Compare *tsunami.*

Tides (Ch. 1). The daily rising and falling of sea level and bodies of water connected to them (e.g., estuaries, lagoons, bays) caused chiefly by the gravitational pull of the sun and moon, as well as the rotation of the earth.

Tributyltin (TBT) (Ch. 5). An organic tin-containing compound that is an ingredient of anti-fouling paints used on the hulls (bottoms) of ships and small boats.

Tsunami (Chs. 1, 4) Seismic sea waves. Japanese term for a large wave, potentially catastrophic, caused by seismic disturbances like submarine fault movements, underwater landslides, or volcanic eruptions. Commonly misnamed a *tidal wave.* Compare *tidal wave.*

UNCTAD (Ch. 2). United Nations Conference on Trade and Development.

UNEP (Ch. 2). United Nations Environmental Program.

UNLOSC, or LOSC (Ch. 2). United Nations Law of Sea Convention.

Upwelling (Ch. 3). The movement of cold, nutrient-rich bottom water to the surface as a consequence surface divergence (induced by the Coriolos force) or by offshore winds. It results in abnormally low surface-water temperatures that are rich in nutrients, thus creating a favorable environment for plant and animal growth.

Urban Growth Boundary (Ch. 2). A land use management tool used by urban planners. Specifically, it is an imaginary line surrounding an urban area beyond which new development is not allowed.

Visual convergence (Ch. 8). A term used by cultural geographers that refers to those forces in society that cause landscapes in the United States (and the world) to look alike (e.g., the building fronts of U.S. owned hamburger chains that can be found in California, Japan, Taiwan, and elsewhere). Compare the opposite concept, *sense of place.*

Waste minimization (Source reduction) (Ch. 6). An umbrella term that includes *precycling* (i.e., consciously purchasing merchandise that has a minimal adverse effect on the environment) and *resource recovery* (e.g., *recycling, composting*).

Waste-to-energy incinerators (Ch. 6). An incinerator that burns garbage to produce heat and steam, which can then be used to produce electricity.

Water pollution (Ch. 5). One of two major forms of coastal pollution, the other being *marine debris.* The addition of harmful (e.g., oil or industrial wastes) or objectionable (e.g., sediment) materials to water, causing a degradation of water quality. Water pollution can be caused by *natural* (e.g., heavy rains washing sediment into an estuary) or *human* forces (e.g., industrial outfalls). Compare *marine debris.*

Watershed (Drainage or catchment basin) (Ch. 2). The *gathering ground* of a single river system. The crest of a ridge often serves to divide one river system and its watershed from another.

Watershed management (Ch. 2). One form of water resource management. The comprehensive management of a particular watershed—its use, regulation, and treatment of water and land resources. Other forms of water resource management are *floodplain management* (the management of floodplains) and *groundwater management* (the management of groundwater resources).

Waves (Ch. 1). An undulating ridge of water moving across the surface of a body of water, such as an ocean, sea, or lake. Generally caused by wind action.

Wetland (Ch. 8). A transitional vegetated lowland between terrestrial and aquatic systems. Wetlands occur in many coastal areas and estuaries worldwide. Characteristics of wetlands include: highly productive ecosystem; flooded at a sufficient frequency to support vegetation adapted for life in saturated soils; have a water table near or above the surface (shallow covering of water); hydric soils; and prevalence of hydrophytic plants. Examples of wetlands are sloughs, swamps, and marshes.

WHO (Ch. 2). World Health Organization.

Wildlife (Ch. 8). A term that has both a narrow and broad definition. The narrow definition is that wildlife is all undomesticated (free-ranging) vertebrate animals, except fish. The broad definition is that wildlife is all undomesticated species, including plants in wild ecosystems.

WQPP (Ch. 5). Water Quality Protection Program.

Zoning (Ch. 2). A regulatory process used by urban planners to divide a given geographical area into zones (subareas), each of which is designated for a particular use or uses (e.g., agricultural, recreation, moderate and low income housing.

Index